Geophysical Monograph Series

Including

IUGG Volumes
Maurice Ewing Volumes
Mineral Physics Volumes

Geophysical Monograph 136

The Central Atlantic Magmatic Province
Insights from Fragments of Pangea

W. Hames
J. G. McHone
P. Renne
C. Ruppel
Editors

American Geophysical Union
Washington, DC

Published under the aegis of the AGU Books Board

Library of Congress Cataloging-in-Publication Data
The Central Atlantic magmatic province : insights from fragments of Pangea / W. Hames
... [et al.], editors.
p. cm. -- (Geophysical monograph ; 136)
Includes bibliographical references.
ISBN 0-87590-995-7
1. Rifts (Geology)--North Atlantic Ocean. 2. Continental margins--North Atlantic
Ocean. 3. Magmatism--North Atlantic Ocean. 4. Geology, Stratigraphic--Triassic. 5.
Geology, Stratigraphic--Jurassic. I. Hames, W. (Willis) 1960- II. Series

QE606.5.N724 C46 2003
551.8'72--dc21   2002038310

ISSN 0065-8448
ISBN 0-87590-995-7

CONTENTS

CONTENTS

PREFACE

A singular event in Earth's history occurred roughly 200 million years ago, as rifting of the largest and most recent supercontinent was joined by basaltic volcanism that formed the most extensive large igneous province (LIP) known. A profound and widespread mass extinction of terrestrial and marine genera occurred at about the same time, suggesting a causal link between the biological transitions of the Triassic-Jurassic boundary and massive volcanism. A series of stratigraphic, geochronologic, petrologic, tectonic, and geophysical studies have led to the identification of the dispersed remnants of this Central Atlantic Magmatic Province (CAMP) on the rifted margins of four continents. Current discoveries are generally interpreted to indicate that CAMP magmatism occurred in a relative and absolute interval of geologic time that was brief, and point to mechanisms of origin and global environmental effects. Because many of these discoveries have occurred within the past several years, in this monograph we summarize new observations and provide an up-to-date review of the province.

Certainly several previous works support the current monograph, and contain matter that is closely related and relevant to the topic. However, while these volumes provide important syntheses that emphasize stratigraphic and petrologic data of the North American portion of the CAMP, more recent research in several areas has promoted recognition and definition of the CAMP in a circum-Atlantic context, including: (1) precise absolute geochronology of CAMP igneous rocks; (2) high-resolution stratigraphic studies that also may combine detailed paleontology with magneto-stratigraphy and cyclo-stratigraphy; (3) petrologic and paleomagnetic studies that characterize and relate distal portions of the CAMP; (4) geochemical and paleontologic studies that suggest global climate change at the Triassic-Jurassic boundary; and (5) recognition of the important role for LIPs in crustal development and as a record of modes for mass and energy transfer throughout Earth's history.

In this light, we conceived the present monograph to provide insights unique to a circum-Atlantic perspective of the CAMP and its possible effects on Earth's climate and biological systems from an international group of authors.

Some authors present syntheses of earlier work, and discussions of models proposed for the CAMP, in combination with new data and in a format that is more detailed than previously available. Stratigraphic relationships that define the relative timing of CAMP magmatism, and correlations to important biostratigraphic units are a major theme for some papers. Other papers draw inferences and correlations to basaltic sub-provinces that may be members of the CAMP, such as the buried continental flood basalts and the seaward dipping reflector sequences (SDRs) of eastern North America. Detailed reviews and analysis of radioisotopic data for CAMP igneous rocks are an important component of some papers in this volume, with conclusions that bear on the timing and duration of CAMP magmatic activity. Although consistent themes with respect to timing and duration of magmatism emerge from these papers, variances in opinion are also evident, with some authors preferring to view CAMP basalts as a temporal benchmark, whereas others interpret the available age data to represent a measurable range of duration for CAMP magmatism.

The basaltic remnants of the CAMP have been identified in North and South America, West Africa and Europe, over regions with a reconstructed total area of about 10 million square kilometers. Thus, reviews that report and synthesize observations for each of these regions constitute critical aspects of this volume. Papers here also provide new data and tectonic interpretations for the CAMP in South America, where CAMP basalts have been identified roughly 2000 km within the continent's interior. The deeply eroded plexus of basaltic dikes exposed in eastern North America, with their related sills and flows of Newark Supergroup basins, are a source for much of the petrologic and geochemical data of the CAMP. New data and syntheses of previous studies of the North American CAMP presented in this volume suggest extensive involvement of the subcontinental lithosphere, and passive mechanisms for generation of CAMP magmas, as compared to models that invoke deep mantle plumes. CAMP basaltic rocks in West Africa and Portugal constitute an eastern boundary of the province. New data for this eastern part of the CAMP are presented within the context of a regional synthesis.

Identification of CAMP basalts in the Brittany region of France also emphasizes the northern extent of the province, as reported with new data.

The results of other studies examine differing characteristics of the CAMP that point to underlying causes and probable effects of the province. The anisotropy of magnetic susceptibility in basaltic dikes and flows has proven very useful in analysis of magma transport directions and the location of magma sources, as reported in an analysis of dike and flow patterns from the Fundy basin. It is clear that the volatile emissions of the CAMP province were sufficient to induce major, global environmental changes, as also reviewed and considered here. The role of CAMP magmatism as a driving force behind global Triassic-Jurassic environmental changes and mass extinctions is discussed and carefully considered, with comparisons of terrestrial and marine paleontologic records to the evolution of the CAMP and other possible causative mechanisms.

The present volume provides definite advances in our understanding of the CAMP that will be useful to scholars and researchers with interests in the magmatism of this province, events associated with the Triassic-Jurassic boundary, the nature of mass extinctions, and large igneous provinces in general. Although the availability of literature for the CAMP has surged in the past few years, as this volume attests, we are still in a reconnaissance stage for understanding the CAMP with new discoveries emerging on multiple research fronts. Thus, papers in this volume also direct the reader to areas for further research with a rich potential for understanding one of Earth's most significant magmatic episodes.

This volume arose from discussions following a theme session in the Fall 1999 meeting of the American Geophysical Union (The Earliest Magmatism of the Circum-Atlantic Large Igneous Province) and a workshop in the summer of 2000 (International Workshop for a Climatic, Biotic, and Tectonic, Pole-to-Pole Coring Transect of Triassic-Jurassic Pangea). We gratefully acknowledge the inspiration and support from many colleagues who participated in those meetings. We are particularly grateful to the many reviewers of papers in this volume (as indicated in the individual acknowledgements), who provided an exceptionally high level of professional peer review. We also acknowledge the editorial assistance and oversight of the American Geophysical Union book department staff.

Willis E. Hames
Auburn University

J. Gregory McHone
University of Connecticut and Wesleyan University

Paul R. Renne
Berkeley Geochronology Center and
University of California, Berkeley

Carolyn Ruppel
Georgia Institute of Technology

Introduction

W. Hames

Department of Geology and Geography, Auburn University, Auburn, Alabama

J. G. McHone

University of Connecticut, Department of Geology and Geophysics, Storrs, Connecticut
Wesleyan University, Department of Earth and Environmental Sciences, Middletown, Connecticut

P. Renne

Berkeley Geochronology Center, Berkeley, California
University of California, Department of Earth and Planetary Science, Berkeley, California

C. Ruppel

School of Earth and Atmospheric Sciences, Georgia Institute of Technology, Atlanta, Georgia

OVERVIEW OF RESEARCH ON THE CENTRAL ATLANTIC MAGMATIC PROVINCE

Large igneous provinces represent catastrophic out-pourings of immense quantities of basaltic lava, the origin of processes that are hidden deep within Earth's interior and obscured by the movements of tectonic plates. Scientists from many different fields are combining their efforts to understand the underlying mechanisms by which large igneous provinces' form, the distribution and timing of such provinces, and the effects and consequences of the associated volcanism. This book presents a collection of thirteen papers that delve into such fundamental questions for one of Earth's most calamitous events – the profound and extensive magmatism that accompanied the breakup of Pangea.

Dramatic relics of the magmatism that accompanied breakup of Pangea are preserved on every continent of the circum-Atlantic region, in exposures of igneous intrusions that once provided magma conduits, as partially eroded

The Central Atlantic Magmatic Province:
Insights from Fragments of Pangea
Geophysical Monograph 136
Copyright 2003 by the American Geophysical Union
10.1029/136GM01

and buried flood basalts that once covered large areas of the land, and in the sedimentary strata that host the volcanic remnants. Scientists collect forensic evidence in rock exposures from coastal Brittany to Brazilian rainforests, in strata buried deeply beneath Texas and on plateaus in sub-Saharan Mali in the effort to describe this province and determine its origin. Indeed, over the past several years we have learned through research that much of the magmatism in this province was very sudden, although it may have spread diachronously. The absolute age and stratigraphic juxtaposition of this province seems coincident with the transition from the Triassic to the Jurassic period. Many scientists are now studying this magmatic province in an effort to understand how its effects may relate to the global climate changes and profound mass extinction at the end of the Triassic (when perhaps 80% of Earth's terrestrial and marine faunas perished). And yet such a catastrophic distress for some may provide opportunity to others, as the Jurassic rise of dinosaurs to dominate Earth may have begun with global climate changes wrought by this large igneous province. Thus, the research on this individual magmatic province is part of a broader and contentious scientific trial to determine whether massive volcanic events, meteorite impact events, or combinations of both are responsible for the major mass extinction events that have modulated the course of life on Earth for the past 300 million years.

Recognition of the Central Atlantic Magmatic Province

Although geologists first recognized the extent and significance of Triassic-Jurassic basalts around the Atlantic region in the late nineteenth century (Russell, 1880), it is only within the past few years that the timing and extent of magmatism was sufficiently constrained to define what we now refer to as the Central Atlantic Magmatic Province [CAMP; Marzoli et al., 1999]. The known distribution of CAMP dikes, sills and flows is represented in Figure 1 (adapted from McHone, 2000). Rifting of Pangea and deposition of clastic sediments in intermontane basins began by about 230 million years ago, reactivating more ancient crustal boundaries and leading to a configuration of incipient continental fragments as shown in Figure 1 by roughly 200 million years ago. CAMP basalts and associated intrusions developed near the center of this evolving rift setting and close to an equatorial latitude [Olsen, 1997]. Much of the perimeter for the CAMP coincides with orogenic belts and deep boundaries in the continental lithosphere, such as the western limit of the CAMP and the Blue Ridge to Piedmont province boundary of the southeastern Appalachians.

Characteristics of the CAMP

The CAMP is dominated by tholeiitic basalts, with an extent and major geochemical relationships that continue to be defined by new discoveries. CAMP dikes have been identified in northwestern France [Jordan et al., this volume], that are similar in age and overall chemistry to CAMP basalts that are 2000 km within continental South America [De Min et al., this volume], although regional heterogeneities in basalt composition exist and seem related to differences in regional lithospheric composition. CAMP basalts in eastern Brazil are also more extensive than previously recognized, where paleomagnetic data also suggest local basaltic magmatism significantly before and after a main stage at 200-190 Ma [Ernesto et al., this volume]. The CAMP volcanics in Morocco and Portugal comprise a significant proportion of subaerial lavas and pyroclastics interbeded with shallow-water clastic sequences, and the geochemistry of dikes and flows in the region is also compatible with a lithospheric mantle source [Youbi et al., this volume]. The broadly uniform composition of many CAMP basalts, the extent of CAMP magmatism, the relationship to thickened lithosphere and crustal boundaries of Pangea, and geochemical trends are collectively most compatible with "passive" models involving upper mantle geodynamics [De Min et al., this volume; Jourdan et al., this volume; Puffer, this volume; Salters et al., this volume].

Relationships of the CAMP to the Seaward Dipping Reflector Sequences of the Central Atlantic Basin

A seam that runs near the central axis of the CAMP, along the plate boundary that would become the Atlantic basin, is defined by the seaward-dipping reflector sequences (SDRS, Figure 1). The SDRS are considered from seismic studies to represent thick, wedge-shaped accumulations of basalt flows [Holbrook and Kelemen, 1993; Oh et al., 1995] that are deeply buried within the continental margin. Extrusion of basalts in the SDRS marks an end to CAMP magmatism in the continental realm and the birth of the central Atlantic basin. However, the timing of formation for SDRS in the central Atlantic, and whether they should be considered as part of the CAMP, is unclear. For comparison, the SDRS of the North Atlantic Igneous Province seem to have formed several years after the continent-based magmatism [Saunders et al., 1999]. Considering the CAMP, Benson [this volume] reviews stratigraphic interpretations of the oldest drift-stage rocks of the North American margin and of early Atlantic sea-floor spreading rates suggestive of much younger ages for the SDRS and earliest Atlantic sea floor. In contrast, Schlische et al. [this volume] present a model for diachronous rift-drift CAMP magmatism, in which the SDRS form in an earlier magmatic pulse with the CAMP to the south and after the CAMP through a separate magmatic pulse to the north. Olsen et al. [this volume] review and test earlier evidence for the age of the CAMP and drilled units correlated with the SDRS, and suggest they could be of the same age. Both Benson and Olsen et al. [this volume] present discussions to highlight the need for drilling of basalts correlated with the SDRS in order to evaluate the alternate models for timing of the SDRS in the Central Atlantic and their relationship to CAMP magmatism.

The CAMP as an Agent of Planetary Catastrophe

Through recorded human history we have come to understand that volcanic events can cause catastrophes, but that the effects are generally local. The violent and explosive eruption of Mount Vesuvius was a catastrophe for the citizens of Pompeii as recorded by Pliny the Younger (and experienced by Pliny the Elder!), although this event was inconsequential for the climate and life of Earth beyond a small region of western Italy. It is difficult, therefore, to imagine volcanic events, even the massive outpourings of basalt in large igneous provinces, as constituting "planetary catastrophes" as posed by Courtillot [1999]. If we consider the scale of large igneous provinces (Figure 2), however, the magnitude of their potential effects is more readily understood.

The Columbia River Flood Basalt Province (Figure 2) is approximately one tenth the size of the other large ig-

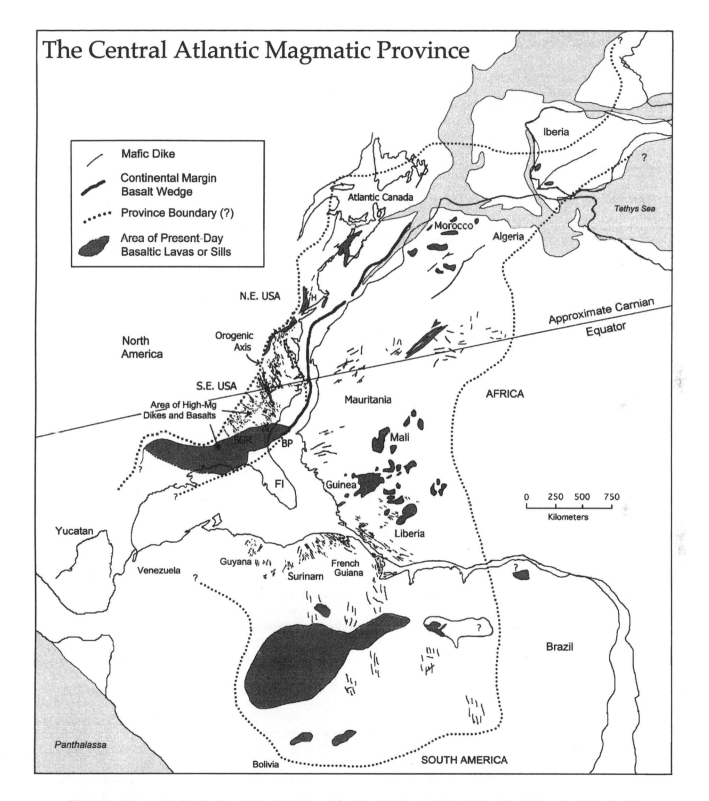

Figure 1. Geographic distribution of the CAMP (modified from McHone, 2000). SGR = South Georgia rift basin; BP = Blake Plateau.

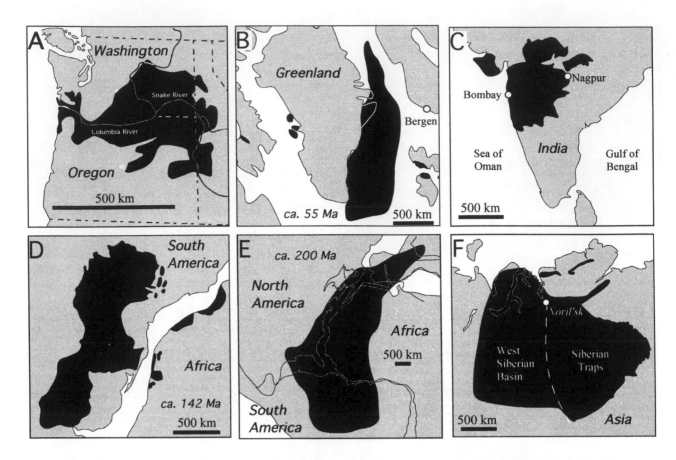

Figure 2. Comparison of the distribution of the CAMP to other Phanerozoic Large Igneous Provinces. Note the 500 km scale bar in each of the sketches. The black shading in each simplified map represents an entire extent for the province, as defined by authors cited for networks of dikes and other plutonic rocks, flows, and seaward-dipping reflector sequneces, and does not imply continuous surface coverage. Note also that some of the provinces are shown in tectonic reconstructions with ages as indicated. A: Columbia River Flood basalt Province [from Hooper, 1999], B: North Atlantic Igneous Province [from Saunders et al., 1999], C: Deccan Traps [from Courtillot, 1999], D: Parana-Etendeka Province [from Peate, 1999]; E: The CAMP [from McHone, 2000, see figure 1], F: the Siberian Traps [from Reichow et al., 2002].

neous provinces shown, yet it is relatively young, well preserved and exposed, and thus its characteristics are relatively easy to discern. A single basalt flow (the Pomona flow) in this province was channeled for hundreds of kilometers from its source in Idaho down the Snake River and Columbia River canyons and into the Pacific ocean [Wells, 1989; Hooper, 1999]. (Note that flow directions and sources can also be reconstructed for the more ancient CAMP, as reported by Ernst, this volume.) The Columbia River Flood Basalt Province is built of more than 300 such flows, with a total volume of basalt of about 175,000 km3 [Tolan et al., 1989]. In contrast, the Siberian and Deccan Traps (Figure 2) are estimated to contain basalt on the order of millions of cubic kilometers. The total volume of the CAMP is difficult to evaluate, as the province is eroded, fragmented and dispersed, because of the ambiguities in relationship to the SDRS, and because much of its extent is defined by deeply

eroded dikes that may never have reached the surface. However, estimates by McHone (this volume) suggest a total volume on the order of two million cubic kilometers. Thus, the effects of the CAMP magmatism would presumably be comparable to the effects of the Deccan and Siberian Traps.

Basaltic volcanism of large igneous provinces releases copious amounts of sulfur, fluorine, and chlorine that would have significant effects on the global atmosphere and ecosystems [Self et al., 1999]. The largest historic eruption of basaltic lavas occurred near Mount Laki in southeastern Iceland, when about 12 km^3 of basaltic lavas erupted from fissures within a few months of 1783-1784 [Thordarson and Self, 1997]. Although this eruption is minute in comparison to single flows of large igneous provinces (that may reach 2000 to 4000 km^3 in the Columbia River Flood Basalt Province, for example) the gases emitted from the Laki eruption (some tens of mil-

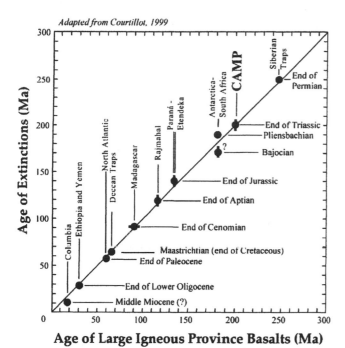

Figure 3. Correlation of the timing for eruption of various large igneous provinces to the timing of mass extinction events over the past 250 million years, Adapted from Courtillot, 1999.

lions of metric tons) were sufficient to decimate local farmland and livestock and lead to the most severe famine the island has ever experienced, and caused regional climate change and a particularly harsh winter of 1783-1784 in Europe. Total volatile emissions of the CAMP, in contrast, are estimated by McHone [this volume] to be on the order of a trillion metric tons, a figure comparable to estimates for emissions from the Deccan and Siberian Traps.

The timing of eruption for most Phanerozoic large igneous provinces can also be related to times of climate change and ecological distress, particularly through the record of mass extinctions [Figure 3; Courtillot, 1999]. Considering the last 300 million years, the largest of the mass extinctions are the Permian-Triassic event, the Triassic-Jurassic event, and the Cretaceous-Tertiary event [e.g., Sepkowski, 1996]. The Permian-Triassic event, in which roughly 90% of Earth's species perished, was synchronous with basaltic eruptions of the Siberian Traps and the Cretaceous-Tertiary event, which saw the demise of the dinosaurs and rise of mammalian faunas, was synchronous with both the culmination of Deccan volcanism and the Chicxulub meteorite impact event. In the case of the CAMP, radioisotopic dating studies indicate a timing for a dominant stage of volcanism very close to 200 Ma [Baksi, this volume; Palfy, this volume] and stratigraphic studies suggest a duration for CAMP magmatism (as preserved in the Newark Supergroup basins) of less than

600,000 years [Olsen et al., this volume]. The Triassic-Jurassic mass extinction event constituted a loss of approximately 80% of marine species and a simultaneous crisis of terrestrial species [see discussions in Palfy, this volume, and Olsen et al., this volume]. Palfy concludes (this volume) that the timing of CAMP magmatism, patterns of extinction at the Triassic-Jurassic boundary, isotopic changes observed at the boundary, and lack of compelling evidence for a bolide impact (major iridium anomaly, shocked minerals or tektites) are most compatible with the hypothesis that CAMP magmatism forced climate changes that resulted in the Triassic-Jurassic mass extinctions.

REFERENCES

Baksi, A. K., Evaluation of radiometric ages for the Central Atlantic Magmatic Province; timing, duration and possible migration of magmatic centers, in *The Central Atlantic Magmatic Province*, edited by W. E. Hames, J. G. McHone, P. R. Renne and C. Ruppel, American Geophysical Union, Washington, D.C., 2002 (this volume).

Benson, R. N., Age estimates of the seaward-dipping volcanic wedge, earliest oceanic crust, and earliest drift-stage sediments along the North American Atlantic continental margin, in *The Central Atlantic Magmatic Province*, edited by W. E. Hames, J. G. McHone, P. R. Renne and C. Ruppel, American Geophysical Union, Washington, D.C., 2002 (this volume).

Courtillot, V., Evolutionary Catastrophes: The science of mass extinction, second edition, 173 p., University Press, Cambridge, 1999.

de Min, A., E. M. Piccirillo, A. Marzoli, G. Bellieni, P. R. Renne, M. Ernesto, L. S. Marques, The Central Atlantic Magmatic Province in Brazil ... Geodynamic Implications, in *The Central Atlantic Magmatic Province*, edited by W. E. Hames, J. G. McHone, P. R. Renne and C. Ruppel, American Geophysical Union, Washington, D.C., 2002 (this volume).

Ernesto, M., G. Bellieni, E.M. Piccirillo, L.S. Marques, A. de Min, I.G. Pacca, G. Martins, J.W.P. Macedo, Paleomagnetic and geochemical constraints on the timing and duration of the CAMP activity in northeastern Brazil, in *The Central Atlantic Magmatic Province*, edited by W. E. Hames, J. G. McHone, P. R. Renne and C. Ruppel, American Geophysical Union, Washington, D.C., 2002 (this volume).

Ernst, R.E., J. Z. de Boer, P. Ludwig, T. Gapotchenko, Emplacement of the North Mountain basalts of the 200 Ma CAMP event: evidence from the magnetic fabric, in *The Central Atlantic Magmatic Province*, edited by W. E. Hames, J. G. McHone, P. R. Renne and C. Ruppel, American Geophysical Union, Washington, D.C., 2002 (this volume).

Holbrook, W. S. and P. B. Kelemen. 1993. Large igneous province on the U.S. Atlantic margin and implications for magmatism during continental breakup. Nature 364:433-436.

Hooper, P.R., The Columbia River flood basalt province: Current status, in *Large Igneous Provinces*, edited by J.J. Mahoney and M.F. Coffin, p. 1-28, American Geophysical Union, Washington, D.C., 1999.

Jourdan, F., A. Marzoli, H. Betrand, M. Cosca, D. Fontignie, The northernmost CAMP: 40Ar/39Ar age, petrology and Sr-Nd-Pb isotope geochemistry of the Kerforne dike, Brittany, France, in *The Central Atlantic Magmatic Province*, edited by W. E. Hames, J. G. McHone, P. R. Renne and C. Ruppel, American Geophysical Union, Washington, D.C., 2002 (this volume).

Marzoli, Andrea, P. R. Renne, E. M. Piccirillo, M. Ernesto, G. Bellieni, and A. De Min, Extensive 200-Million-Year-Old Continental Flood Basalts of the Central Atlantic Magmatic Province, Science, v. 284, pp. 616-618, 1999.

McHone, J. G., 2000, Non-plume magmatism and tectonics during the opening of the central Atlantic Ocean: Tectonophysics, v. 316, pp. 287-296.

McHone, J.G., Volatile emissions from Central Atlantic Magmatic Province basalts: Mass Assumptions and environmental consequences, in *The Central Atlantic Magmatic Province*, edited by W. E. Hames, J. G. McHone, P. R. Renne and C. Ruppel, American Geophysical Union, Washington, D.C., 2002 (this volume).

Oh, Jinyong, J. A. Austin Jr., J. D. Phillips, M. F. Coffin, and P.L. Stoffa, Seaward-dipping reflectors offshore the southeastern United States: Seismic evidence for extensive volcanism accompanying sequential formation of the Carolina trough and Blake Plateau basin, Geology, v. 23, pp. 9-12, 1995.

Olsen, P. E., Stratigraphic record of the early Mesozoic breakup of Pangea in the Laurasia-Gondwana rift system. Ann. Rev. Earth Planet. Sci., v. 25, pp. 337-401, 1997.

Olsen, P.E., D. V. Kent, M. Et-Touhami, J. Puffer, Cyclo, Magneto-, and Bio-stratigraphic Constraints on the Duration of the CAMP Event and its Relationship to the Triassic-Jurassic Boundary, in *The Central Atlantic Magmatic Province*, edited by W. E. Hames, J. G. McHone, P. R. Renne and C. Ruppel, American Geophysical Union, Washington, D.C., 2002 (this volume).

Palfy, J., Volcanism of the Central Atlantic Magmatic Province as a potential driving force in the end-triassic mass extinction, in *The Central Atlantic Magmatic Province*, edited by W. E. Hames, J. G. McHone, P. R. Renne and C. Ruppel, American Geophysical Union, Washington, D.C., 2002 (this volume).

Peate, D.W., The Parana-Etendeka province, in *Large Igneous Provinces*, edited by J.J. Mahoney and M.F. Coffin, p. 217-246, American Geophysical Union, Washington, D.C., 1999.

Puffer, J. H., A reactivated back-arc source for CAMP magma, in *The Central Atlantic Magmatic Province*, edited by W. E. Hames, J. G. McHone, P. R. Renne and C. Ruppel, American Geophysical Union, Washington, D.C., 2002 (this volume).

Reichow, M.K., A.D. Saunders, R.V. White, M.S. Pringle, A.I. Al'Mukhamedov, A.I. Medvedev, N.P. Kirda, 40Ar/39Ar dates from the west Siberian basin: Siberian Flood Basalt Province doubled, Science, v. 296, pp. 1846-1849, 2002.

Russel, I.C., 1880, On the former extent of the Triassic formation of the Atlantic states: American Naturalist, v. 14, p. 703-712.

Salters, V.J.M., P.C. Ragland, W.E. Hames, C. Ruppel, K. Milla, Temporal chemical variations within lowermost Jurassic tholeiitic magmas of the Central Atlantic Magmatic Province, in *The Central Atlantic Magmatic Province*, edited by W. E. Hames, J. G. McHone, P. R. Renne and C. Ruppel, American Geophysical Union, Washington, D.C., 2002 (this volume).

Saunders, A.D., J. G. Fitton, A. C. Kerr, M. J. Norry, and R. W. Kent, The North Atlantic Igneous Province, in *Large Igneous Provinces*, edited by J.J. Mahoney and M.F. Coffin, p. 45-93, American Geophysical Union, Washington, D.C., 1999.

Schlische, R.W., M. O. Withjack, P. E. Olsen, Relative timing of CAMP, rifting, continental breakup, and basin inversion: tectonic significance, in *The Central Atlantic Magmatic Province*, edited by W. E. Hames, J. G. McHone, P. R. Renne and C. Ruppel, American Geophysical Union, Washington, D.C., 2002 (this volume).

Self, S., T. Thordarson, L. Keszthelyi, Emplacement of continental flood basalt flows, in *Large Igneous Provinces*, edited by J.J. Mahoney and M.F. Coffin, p. 381-410, American Geophysical Union, Washington, D.C., 1999.

Sepkoski, J. J., Patterns of Phanerozoic extinction: a perspective from global data bases, in Golbal Events and Event Stratigraphy in the Phanerozoic, edited by O.H. Wallister, pp. 35-51, Springer, Berlin, 1996.

Thordarson, T. L., S. Self, N. Oskarsson, T. Hulsebosch, Sulfur, chlorine and fluorine degassing and atmospheric loading by the 1783-1784 Laki (Skaftar Fires) eruption in Iceland, Bull. Volcanol., 58, 205-225, 1996.

Tolan, T. L., S. P. Reidel, M. H. Beeson, J. L. Anderson, K. R. Fecth, and D. A. Swanson, Revisions to estimates of the aerial extent and volume of the Columbia River Basalt Group, in *Volcanism and Tectonism in the Columbia River Flood-Basalt Province*, Spec. Pap. 239, edited by S.P. Reidel and P.R. Hooper, pp. 1-20, Geological Society of America, Boulder, CO, 1989.

Wells, R. E., R. W. Simpson, R. D. Bentely, M. H. Beeson, M. T. Mangan, and T. L. Wright, Correlation of Miocene flows of the Columbia River basalt group from the central Columbia River Plateau to the Coast of Oregon and Washington, in *Volcanism and Tectonism in the Columbia River Flood-Basalt Province*, Spec. Pap. 239, edited by S.P. Reidel and P.R. Hooper, pp. 113-130, Geological Society of America, Boulder, CO, 1989.

Youbi, N., L. T. Martins, J. M. Munhá, H. Ibouh, J. Madeira, E. A. Chayeb, A. El Boukhari, The Late Triassic-Early Jurassic Volcanism of Morocco and Portugal in the Framework of the Central Atlantic Magmatic Province: an Overview, in *The Central Atlantic Magmatic Province*, edited by W. E. Hames, J. G. McHone, P. R. Renne and C. Ruppel, American Geophysical Union, Washington, D.C., 2002 (this volume).

Cyclo-, Magneto-, and Bio-Stratigraphic Constraints on the Duration of the CAMP Event and its Relationship to the Triassic-Jurassic Boundary

Paul E. Olsen[1], Dennis V. Kent[2], Mohammed Et-Touhami[3], John Puffer[4]

Early Mesozoic tholeiitic flood basalts of the Central Atlantic Magmatic Province (CAMP) are interbedded throughout much of their extent with cyclical lacustrine strata, allowing Milankovitch calibration of the duration of the extrusive episode. This cyclostratigraphy extends from the Newark basin of the northeastern US, where it was first worked out, to Nova Scotia and Morocco and constrains the outcropping extrusive event to less than 600 ky in duration, beginning roughly 20 ky after the Triassic-Jurassic boundary, and to within one pollen and spore zone and one vertebrate biochron. Based principally on the well-known Newark astronomically calibrated magnetic polarity time scale with new additions from the Hartford basin, the rather large scatter in recent radiometric dates from across CAMP (>10 m.y.), centering on about ~200 m.y., is not likely to be real. Rather, the existing paleomagnetic data from both intrusive and extrusive rocks suggest emplacement of nearly all the CAMP within less than 3 m.y. of nearly entirely normal polarity. The very few examples of reversed magnetizations suggest that some CAMP activity probably occurred just prior to the Triassic-Jurassic boundary. Published paleomagnetic and $^{40}Ar/^{39}Ar$ data from the Clubhouse Crossroads Basalt are reviewed and with new paleomagnetic data suggest that alteration and possible core misorientation could be responsible for the apparent differences with the CAMP. The Clubhouse Crossroads Basalt at the base of the Coastal Plain of South Carolina and Georgia provides a link to the volumetrically massive volcanic wedge of seaward dipping reflectors present in the subsurface off the southeastern US that may be part of the same igneous event, suggesting that the CAMP marks the formation of the oldest Atlantic oceanic crust.

1. INTRODUCTION

The Early Mesozoic age Central Atlantic Magmatic Province (CAMP) of eastern North America, southwestern Europe, West Africa, and South America (Figure 1) is possibly the largest known example of a continental flood

[1]Lamont-Doherty Earth Observatory of Columbia University, Rt. 9W, Palisades, NY 10964, USA

[2]Department of Geological Sciences, Rutgers University, Piscataway, NJ 08854-8066, USA, and Lamont-Doherty Earth Observatory of Columbia University, Rt. 9W, Palisades, NY 10964, USA

[3]LGVBS, Département des Sciences de la Terre, Université Mohamed Premier, 60, 000 Oujda, Morocco, and Lamont-Doherty Earth Observatory of Columbia University, Rt. 9W, Palisades, NY 10964, USA

[4]Department of Geological Sciences, Rutgers University, 195 University Avenue, Boyden 407, Newark, N.J. 07102

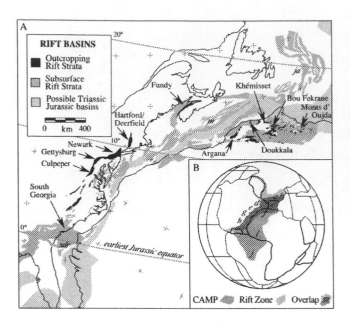

Figure 1. Distribution of rift basins in eastern North America and Morocco and the distribution of the CAMP. A, Eastern North American rift basins. Abbreviations are: ja, Jeane d'Arc basin of the Grand Banks area; m, Mohican basin on the Scotian Shelf; sdr,seaward dipping reflectors an the southeastern United States continental margin. B, Pangea in the earliest Jurassic showing the distribution of rifts, the Central Atlantic Magmatic Province(CAMP) and areas discussed in text. Abbreviations are: a, pyroclastics and ?sills of the Aquitaine basin of southwestern France; b, alkali basalt pyroclastics and flows in Provence, c, the Ecrins-Pelvoux ofthe external massif of the Alps, France; d, flows in Iberia, including the Pyrennes; e, flows in the basins on the Grand Banks region, Canada; f, Fundy basin area of the Maritime provinces, Canada; g, flows in Morocco, Tunisia, and Algeria; h, flows in the major rifts in the eastern United States; i, flows in the South Georgia rift and the offshore seaward dipping reflectors; j, ultrabasic layered plutons and dikes of Liberia, Mali, and Senegal; k, flows of Brazil.

basalt. Indeed, according to Marzoli et al. (1999), this Central Atlantic Magmatic Province (CAMP) may be the largest large igneous province (LIP) of all, extending over 7×10^6 km^2 prior to the formation of the Atlantic and covering significant continental portions of at least four present-day tectonic plates. While there has been tremendous progress in constraining the duration of this huge volcanic event (e.g. Deckart et al., 1997; Marzoli et al., 1999; Hames et al., 2000), existing radiometric dating techniques have not yet been able to produce reproducible (intra-laboratory) and internally consistent dates within the extremely narrow precision limits needed to understand the possible relationship of the CAMP to the Triassic-Jurassic mass extinctions, climate change, and formation of the earliest Atlantic oceanic crust (e.g. Turrin, 2000). Indeed, the present stated analytical uncertainties are one to two orders of magnitude

better than their demonstrable accuracy as judged by independent geological criteria.

In this paper we elaborate on the cyclostratigraphic constraints on the duration of the CAMP lavas, described by Olsen et al. (1996b) using new cyclostratigraphic, biostratigraphic and igneous geochemical data from Nova Scotia and Morocco. We provide new paleomagnetic limits on the duration of the intrusive and extrusive parts of the CAMP using the Newark basin astronomically calibrated geomagnetic polarity timescale (Kent and Olsen, 2000a; Olsen and Kent, 1999), updated using new data from the Hartford basin (Kent and Olsen, 1999b). We provide new geochemical data from Moroccan basalts and new paleomagnetic data from the Clubhouse Crossroads Basalt, and we place all of the constraints on the duration of the CAMP basalts within the framework of the Triassic-Jurassic mass-extinction.

2.   CYCLOSTRATIGRAPHIC CONSTRAINTS

Based primarily on the results of the Newark Basin Coring Project and the cores recovered by the Army Corps of Engineers, the cyclostratigraphy of sedimentary strata of the extrusive zone of the Newark basin was described and interpreted as being controlled by Milankovitch climate cycles (Olsen et al., 1996b) (Figure 2). Because it is known from continuous core, the Newark basin sections provides a basis of comparison for the other rift basin sections, and the cyclostratigraphy and its interpretation is reviewed below (Figure 3) (Kent et al., 1995; Olsen et al., 1996b; Olsen and Kent, 1999; Kent and Olsen, 2000a). We have recently also produced a preliminary cyclostratigraphy and paleomagnetic reversal stratigraphy for the lower half of the post-extrusive Portland Formation of the Hartford basin (Figure 3) that places an upper bound on the duration of the normal polarity zone encompassing the CAMP lavas (Kent et al., 1999b).

2.1 Newark Basin

A permeating hierarchy of sedimentary cycles that reflect changes in lake depth characterizes the predominately lacustrine portions of the Newark basin section. This cyclicity is apparently controlled by Milankovitch climate cycles in precipitation and evaporation (Figure 2). While the classic section for this cyclicity is the Lockatong Formation (Van Houten, 1962, 1964, 1980), it has also proved prevalent in the underlying upper part of the Stockton Formation (Raven Rock Member: Olsen and Kent, 1999; Kent and Olsen, 2000a), and the overlying Passaic Formation and within and above the extrusive zone itself (Olsen, et al., 1996a, 1999). Indeed similar cyclicity (with variations) typify the sediments within and around the extrusive zone of all of the Central Atlantic Margin Basins (CAM) basins (Olsen, 1997; Olsen and Kent, 2000).

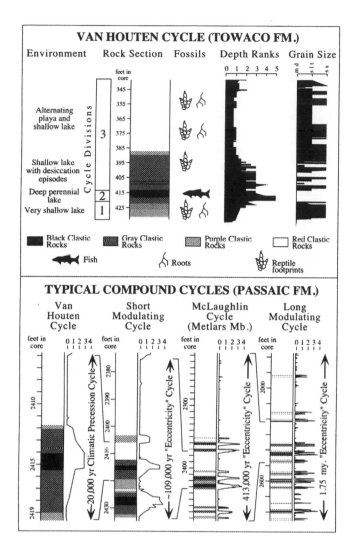

Figure 2. Van Houten and modulating (compound) cycles. Modified from Olsen et al. (1996). Depth Ranks is an ordination of sedimentary facies along a very shallow (0) to very deep (5) relative lake depth gradient.

The most obvious outcrop-scale sedimentary cycle in the Newark basin section is the Van Houten lake level cycle, named by Olsen (1986) after its discoverer (Van Houten, 1962, 1964, 1969) (Figure 2). Each Van Houten cycle consists of a transgressive division (1), a high-stand division (2), and a regressive and low-stand division (3), produced by the rise and fall of lakes under control of the ~20 ky cycle of climatic precession. Vertical sequences of Van Houten cycles show a hierarchical variation of expression making up three larger scale lithological cycles attributed to modulation by a hierarchy of the three main cycles of eccentricity: the short modulating lithological cycle (~100 ky eccentricity cycle), the McLaughlin lithological cycle (404 ky eccentricity cycle), and the long modulating cycle (~2 m.y. eccentricity cycle) (Figure 2). These modulating cycles are expressed as changes in the degree of development or absence of various desiccation features. As interpreted by Olsen and Kent (1996, 1999), high amplitude climatic precession cycles, occurring at times of maximum eccentricity, tend to have produced Van Houten cycles with the least desiccation features, a division 2 with very well developed black, laminated shales preserving whole fish, and overall a more gray and black color, even in division 3. In contrast, low amplitude climatic precession cycles, occurring during times of low eccentricity, have pervasive desiccation features, a poorly developed division 2, itself with desiccation features, and an overall red color, sometimes with evaporite pseudomorphs. The best developed Van Houten cycles, with the best developed black shales, thus occur during the overlap of maxima between the three eccentricity cycles and their lithological counterparts. It is consistent modulation of the expression of the Van Houten cycles that is the fingerprint of the control by Milankovitch climate cycles.

The physical stratigraphy of the lacustrine strata of the Newark basin extrusive zone has been described by Olsen et al. (1996b) (Figure 4). The uppermost Passaic Formation and overlying units are characterized by Van Houten cycles that are two to six times the thickness of those in the older strata, but still show the same kind of hierarchy of modulating cycles as in the rest of the basin lacustrine section. These modulating cycles, as well as the basic Van Houten cycles are what permit the cyclostratigraphic calibration of the duration of the extrusive zone as well as demonstrate the synchronicity in different basins of the flow sequences themselves.

Olsen (1984) and Olsen and Kent (1996) developed a quasi-quantitative classification of the sedimentary facies within Van Houten cycles called depth ranks that permits a single-variable description of the inferred water depth, and hence numerical analysis (e.g. Olsen and Kent, 1996, 1999). A microlaminated, usually black, shale with no signs of desiccation and common preservation of fossil fish is interpreted as the deepest water facies (rank 5). In contrast, a usually red massive mudstone with abundant superimposed desiccation cracks is interpreted as the driest facies (rank 0). The depth rank curve for the uppermost Passaic, extrusive zone strata, and overlying strata is shown in Figure 4.

Taken together with the uppermost Passaic Formation, the depth rank curve of the sedimentary formations of the extrusive zone and Boonton marks out a clear hierarchical pattern consisting of about three McLaughlin cycles (Figure 4). This pattern of cyclicity is clearly a continuation of the overall pattern seen in the Passaic and Lockatong formations, albeit with an increase in accumulation rate. The pattern of short modulating and Van Houten cycles is clearest in cores in the Towaco and Boonton formations. Van Houten cycles and short modulating cycles are muted

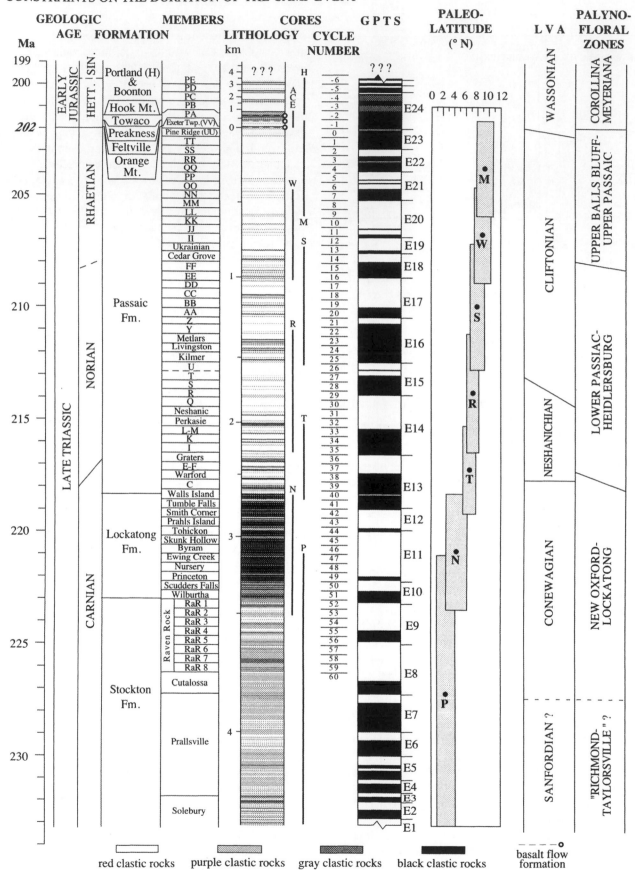

red clastic rocks   purple clastic rocks   gray clastic rocks   black clastic rocks   basalt flow formation

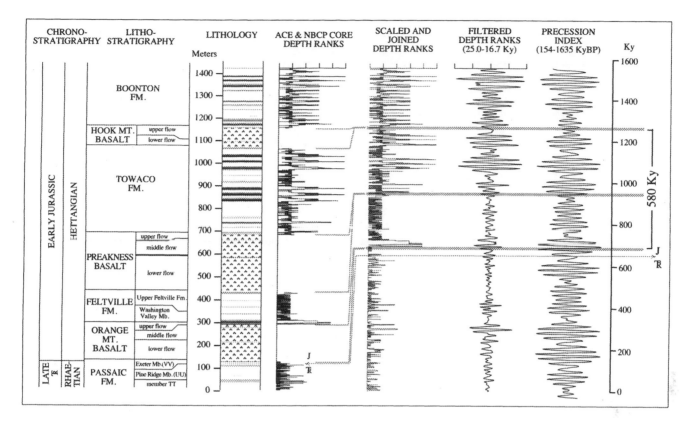

Figure 4. Cyclostratigraphic calibration of the Triassic-Jurassic boundary and succeeding extrusive zone flows and interbedded and overlying sedimentary strata (adapted from Olsen et al., 1996b). Depth ranks are a numerical classification of sedimentary facies sequences in order of increasing interpreted relative water depth (see Olsen and Kent, 1996). Comparison of the depth rank curves with an arbitrary segment of a precession index curve indicates that it is not necessary to assume any significant time is represented by the lava flow formations themselves and that the entire flow sequence as probably deposited during an interval of less than 600 ky. Note also that the Triassic-Jurassic boundary (correlated to the Jacksonwald syncline by Magneto- and lithostratigraphy) lies about 20 ky below the Orange Mountain basalt. The depth rank record from the strata above the Preakness Basalt is based on the ACE cores, while that from the Passaic and Feltville Formations are based on the Martinsville no. 1 core of the NBCP.

in the Feltville Formation, except at its base where there are two very well developed limestone-bearing Van Houten cycles. Correlative sections of the uppermost Passaic Formation with much better expressed cyclicity provide supplementary data, especially critical for the Triassic-Jurassic boundary (Figure 5).

Because of uncertainties in the chaotic behavior of the planets, the recession of the moon, and dynamics of the Earth's interior, it is not yet possible to construct target curves of insolation for direct astronomical tuning of the Newark basin depth rank series (Laskar, 1999). However it is possible to compare the depth rank curve with arbitrary

Figure 3. Time scale for the Late Triassic and Early Jurassic based on geomagnetic polarity time scale (GPTS) and astronomical calibration from the Newark Basin Coring Project (Kent and Olsen, 1999a; Olsen and Kent, 1999), the ACE cores (Fedosh and Smoot, 1988; Olsen et al., 1996), and preliminary results from the Hartford basin (Kent and Olsen, 1999b). Biostratigraphic data from Huber et al. (1996), Lucas and Huber (2002), Cornet (1977), and Cornet and Olsen (1985). For the GPTS, black is normal polarity, white is reversed polarity, and gray represents intervals for which there is incomplete sampling (Hartford basin section only). Abbreviations are: ACE, ACE (Army Corps of Engineers) cores; H, Hartford basin section; HETT., Hettangian; L.V.A., Land Mammal Ages; M, Martinsville (NBCP) core; N, Nursery (NBCP) core; P, Princeton (NBCP) cores; R, Rutgers (NBCP) cores; S, Somerset (NBCP) cores; SIN., Sinemurian; T, Titusville (NBCP) core; W, Weston Canal (NBCP) cores. Cycle number refers to the 404 ky cycle of eccentricity with lines placed at the calculated minima.

segments of appropriate length of insolation curves for the last 10 m.y. (Laskar, 1990). To do this we first assume that the Van Houten cycle does indeed represent the ~20 ky climatic precession cycle. Second we use Fourier analysis to quantitatively determine the thickness period of the Van Houten cycles for each sedimentary formation. The depth rank curves in thickness are then scaled to time independently for each formation assuming a ~20 ky duration for the average Van Houten cycles. Then using the largest envelope of the curves, they are matched to an insolation curve, constrained by phase relationships of the two curves.

It is apparent that the stacked and scaled depth rank sections resemble three successive 404 ky cycles. The highest amplitude parts of these cycles occur in the lower Feltville and uppermost Passaic formations, the middle to upper Towaco Formation, and finally the upper Boonton Formation, as represented in the cores. Detailed comparison of the depth rank curves to different insolation curves suggests that it not necessary to assume that the basalt flows in the exposed Pangean basins represent any significant time, although they could represent as much as 100 ky in total, depending on the insolation curve chosen (Olsen et al., 1996a). Olsen et al. (1996a) thought that about 40 ky was represented above the Triassic-Jurassic boundary, but examination of new outcrops in the Jacksonwald syncline of the southwestern Newark basin suggests that value should be closer to 20 ky, as only one Van Houten cycle is represented (Olsen et al., 2002a) (Figures 4, 5).

As noted by Olsen et al. (1996a), for really major gaps to be represented by the basalt flows, the cyclostratigraphy of the various formations would either have to be offset from each other by multiples of the 404 ky cycles to maintain the phase relationships with the insolation curves, or the fundamental cyclicity itself would have to be misidentified, neither of which seem likely. No other central Atlantic margin rift basin is known in as much stratigraphical detail as the Newark basin; however specific parts of other basins are known in enough detail to compare with parts of the Newark basin depth rank curve for the extrusive zone.

2.2 *Other Exposed Rift Basins in the US*

Sedimentary strata immediately below, and interbedded with, basalts in other basins in the United States, show a pattern of cyclicity very similar to that in the Newark basin. Generally Van Houten cycles are minimally several meters thick (10 - >100 m thick) and black laminated shales are commonly present in the better expressed cycles. The sedimentary strata within and above the extrusive zone is known in the most detail in the Hartford basin. From what is available in outcrop, and sparse core, the Hartford basin section appears cyclostratigraphically extremely similar to the Newark basin. Intervals of core are available for the

lower Portland Formation (Pienkowski and Steinen, 1995) and it is clear that the stratigraphy of the lowermost Portland is nearly identical to the lower Boonton Formation. However the similarity to the Newark basin stratigraphy is most clearly seen in the middle to upper East Berlin Formation, which is virtually identical to the middle and upper Towaco Formation (Figure 5). Long sections through the underlying Shuttle Meadow Formation are unknown, but it is clear that two limestone-bearing Van Houten Cycles are present at the base of the formation, very similar to the two at the base of the Feltville Formation. In addition, there is at least one gray sequence below the oldest basalt (Heilman, 1987; Olsen et al., 2002c), but otherwise the pre-basalt sequence in the same area consists primarily of red and brown sandstone and conglomerite (Figure 5). Overall the depth rank curve of the available outcrops and cores is extremely close to that of the Newark basin, and hence the basalt flow formations appear to have erupted in synchrony. Although the section is far from completely known, there is very little evidence suggesting that that the basalt formations themselves represent significant time (e.g. greater than a few tens of thousands of years). The cyclostratigraphic correlations are in complete agreement with the geochemistry of the three distinctive basalt formations in each basin (Puffer and Philpotts, 1989) and the distinctive paleomagnetic excursion in the lower Preakness and Holyoke basalts (Prevot and McWilliams, 1989).

The Deerfield basin has a stratigraphy that is closely comparable to that of the Hartford basins, except that only one basalt flow formation is present (the Deerfield Basalt). The only part of the Deerfield section that is known in detail is the 180 m section continuously exposed at Turners Falls in the lower Turners Falls Formation in contact with the underlying Deerfield Basalt. The cyclostratigraphy and depth rank section of this outcrop is very similar to the middle East Berlin and the middle and lower Towaco formations, and based on cyclostratigraphy, this section represents about 300 to 345 ky. Thus, the Deerfield Basalt correlates extremely closely with the Holyoke Basalt of the Hartford basin and the Preakness Basalt of the Newark basin, which is in agreement with available geochemical and paleomagnetic data (Puffer and Philpotts, 1988; Prevot and McWilliams, 1989). No other basalt units are present in the Deerfield basin. The Fall River beds below the Deerfield basalt resemble the upper Shuttle Meadow Formation, but there is no sign of the lower two limestone-bearing Van Houten cycles typical of that formation. Based on their absence and the very abrupt change in facies between the underlying Sugarloaf Formation and an apparent angular relationship with the Fall River beds of Olsen et al. (1992), Smoot (in Olsen, 1997) and Hubert and Dutcher (1999) suggest that there is a minor unconformity. In addition, there is virtually no exposure for hundreds of meters above the beds exposed at Turners Falls and there is no basalt present where one would be expected on the basis of cyclostratigraphy. Cornet (1977) argued that there was evidence of an unconformity within the Turners Falls Formation equivalent to the Hampden Basalt of the Hartford basin. However, as discussed by Olsen et al. (1992), this evidence is based on pollen, from beds not demonstrably in superposition, and from levels well above where the equivalent of the Hampden Basalt should be.

The Pomperaug basins is situated between the Newark and Hartford basins. The strata between the "East Hill basalt" and "Orenaug basalt" has two limestone-bearing Van Houten cycles at its base, again very similar to the Feltville Formation of the Newark basin and the Shuttle Meadow Formation of the Hartford basin (LeTourneau and Huber, 2001). Hence, a correlation between the "East Hill basalt" and the Orange Mountain Basalt and the Talcott Basalt is suggested.

Nothing in detail is known about the cyclostratigraphy of the one basalt flow formation (Aspers Basalt) in the Gettysburg basin because of poor outcrop.

Poor outcrop and relative lack of study also limits what is known about the Culpeper basin extrusive zone cyclostratigraphy (Figure 5). However, two limestone-bearing Van Houten cycles are present above the lowest basalt formation (Mt. Zion Church Basalt) in the basal Midland Formation, suggesting correlation of the latter to the Orange Mountain, Rattlesnake Hill and Talcott basalts. The Hickory Grove Basalt overlies the Midland Formation, and is itself overlain by the Turkey Run Formation, which has one black shale-bearing Van Houten cycle near its base, but is otherwise very poorly known. The Waterfall Formation above the succeeding Sander Basalt is also very poorly known stratigraphically, but is known to have very thick Van Houten cycles, several with a division 2 that is thick and microlaminated (Lindholm, 1979). Based on this poorly-known stratigraphy, it is not possible to tell exactly where the Hickory Grove and Sander basalts fall within the Newark basin stratigraphy, with one plausible correlation having the Van Houten cycle at the base of the Turkey Run Formation as equivalent to the one in the middle Shuttle Meadow Formation, or having that Van Houten cycle correlating with the weakly-developed cycles in the lower Towaco or Turners Falls formations. In the first case, the Sander Basalt would be the equivalent of the Preakness Basalt, and the Hickory Grove Basalt would belong in the middle Feltville or Shuttle Meadow formations, the correlation favored by Olsen et al. (1996a). In the second case, the Hickory Grove Basalt would correlate with the Preakness Basalt and the Sanders would fall somewhere below the middle Towaco and East Berlin formations. In either case, there is a difference in the basalt stratigraphy between the Culpeper basin and the Newark, Hartford, and Deerfield basins. This differ-

ence also shows up in the basalt chemostratigraphy and magnetics (see below).

2.3 Fundy and Moroccan Basins

The outcropping Fundy (Nova Scotia and New Brunswick, Canada) and Moroccan basins have cyclical sequences surrounding and interbedded with basaltic lavas that were deposited under much more arid conditions and with much lower accumulation rates, than in the more southern basins (Smoot and Olsen, 1988; Kent and Olsen, 2000b). There is only one basalt formation in the Fundy basin. The North Mountain Basalt lies between the predominately red Blomidon and McCoy Brook formations. There is a thin red, gray, and black cyclical sequence containing the palynologically identified Triassic-Jurassic boundary termed the "Partridge Island member" (Fowell and Traverse, 1995; Olsen et al., 2000, 2002a) below the basalt and two limestone-bearing Van Houten cycles overlying the basalt called the Scots Bay Member. Based on the position of the Triassic-Jurassic boundary, Olsen et al. (2002a, 2002c) hypothesized that the uppermost Blomidon Formation cyclical sequence correlates with the Van Houten cycles in the Exeter Township Member of the Passaic formation as shown in Figures 5 and 5. The two obvious Van Houten cycles of the Scots Bay Formation should correlate with the lower Feltville Formation and its equivalents (Figure 5). Thus, based on cyclostratigraphy, the North Mountain Basalt should correlate with the Orange Mountain Basalt and its correlatives in the Newark and more southern outcropping basins.

It is important to note the considerable lateral variations seen in the basal Van Houten cycles in the Feltville, Shuttle Meadow, and McCoy Brook formations of the Newark, Hartford, and Fundy basins. In these three basins, the lowest cycle has at its base a red and green, predominately clastic sequence where the accumulation rate is the highest (e.g., Birney De Wet and Hubert, 1989; Olsen et al., 1989; Olsen et al. 1996a;). However, laterally, in areas of lower accumulation rate, division 2 of the lowest cycle rests directly on the lava, and in some areas the entire lower cycle can be missing. In addition, there are substantial changes in facies laterally in division 2, passing from microlaminated black and gray calcareous shale to massive gray or pink limestones. These lateral variations are important when comparisons are made to isolated sections in the Moroccan basins (see below).

Moroccan Triassic-Jurassic basins south of the Rif (northernmost Morocco, opposite Gibraltar) share many of their lithologic features in outcrop with the Fundy basin, suggesting a preliminary cyclostratigraphic correlation (Figure 5). Sequences remarkably similar to the Partridge Island member of the Fundy basin occur below the lowest basalt in at least the Argana, Khemisset, Berrechid (east of and Khemisset), Bou Fekrane basins and in the and Monts

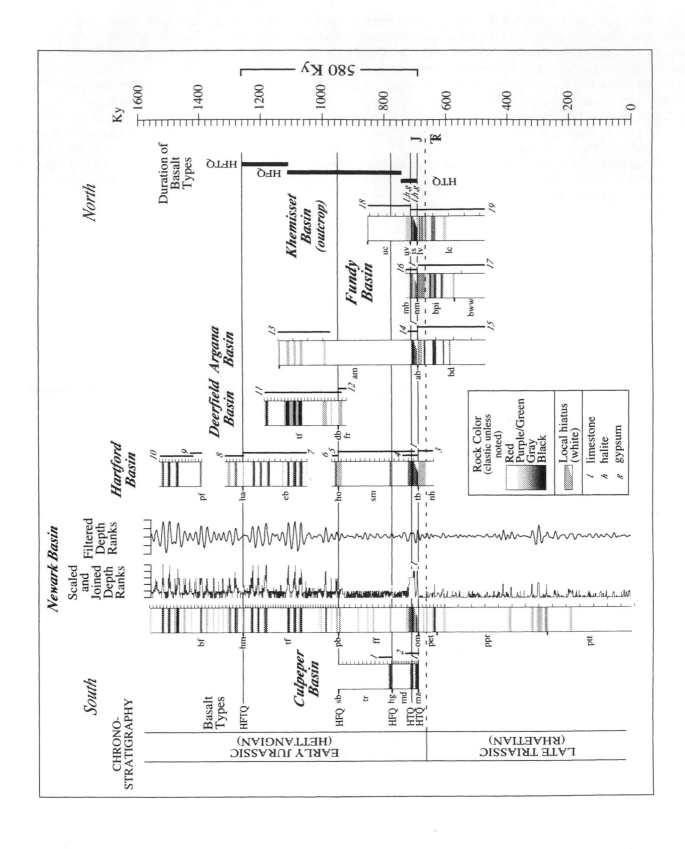

d'Oujda and Beni Snassen (north of Monts d'Oujda) areas. The palynologically-identified Triassic-Jurassic boundary occurs in the same cyclostratigraphic position in at least the Argana basin (Olsen et al, 2000; 2002a, 2002c) relative to the oldest basalt. The stratigraphy and lateral variations (still incompletely known) in the basal carbonate rich sequence above the Argana Basalt and the carbonate-rich sequences between the upper and lower basalt sequences in the Khemisset and Bou Fekrane basins, and in the Beni Snassen and Monts d'Oujda area, are consistent with the limestone-bearing Van Houten cycles in the Newark, Hartford, and especially Fundy basins. Based on cyclostratigraphy, and especially these carbonate-rich sequences, the Argana Basalt, and the older basalts in the Khemisset and Bou Fekrane basins, and the lower basalt in the Monts d'Oujda area should correlate with the Orange Mountain Basalt. However, the basalts directly overlying the carbonate-rich cycles in the Khemisset and Bou Fekrane basins and in the Monts d'Oujda area, would seem to correlate with a position above the limestone-bearing Van Houten cycles in the Feltville Formation, and below the rest of the formation (Figure 5).

New geochemical data is available for several key Moroccan basalts, for which we present cyclostratigraphic data (Figure 6, Table 1). These data are in agreement with the cyclostratigraphic data in suggesting a previously unrecognized basalt flow episode of initial Pangean composition

(HTQ of Puffer, 1992; types 1 and 2 of Bertrand et al., 1982) that occurs prior to the eruption of the HFQ type, but after the other known HTQ flows of more southern basins, in at least the Khemisset and Bou Fekrane basins. There is as yet no evidence of basalts of HFQ or LTQ composition in Morocco.

Much of this tentative correlation with Morocco disagrees with published scant biostratigraphy. According to the reviews by Oujidi et al. (2000a. 2000b) and Oujidi and Et-Touhami (2000) the lower Khemisset and Bou Fekrane basalts are of Carnian or Ladinian age, and the lower basalt in the Monts d'Oujda area is of Ladinian age, while the upper basalts in the these basins are Carnian and Norian age (see also Oujidi et al., 2000c). This variation in age and the nominal differences between the Argana and Fundy basins is surprising given the close lithological homotaxality of the sequences and their relative proximity to one another. It seems clear that the relative ages of these units will require additional biostratigraphic, magnetostratigraphic, and geochronologic data to work out, although for this paper we clearly favor the cyclostratigraphic arguments.

Of considerable interest is that fact that in the deep parts of the Moroccan basins, notably in the Doukkala, Khemisset, and Berrechid basins, the strata resembling the Partridge Island member of the Fundy basin apparently pass laterally into bedded halite-dominated evaporites (Et-Touhami, 2000; Oujidi, et al., 2000b). In the Doukkala basin, foraminifera

Figure 5. Cyclostratigraphic calibration of the 1.6 m.y. around the Triassic-Jurassic boundary and CAMP extrusive zone in eastern North America and Morocco. Basin sections are arranged in paleogeographical position showing the distribution of basalt geochemical types, measured sections showing basic lithologies and cyclostratigraphies of individual basin sections. Only the parts of the basin sections for which there is measured cyclostratigraphic data are shown. Basalts are shown as lines with essentially no duration. A, Overall cyclostratigraphic calibration of CAMP basalts.

Formation names are as follows: mz, Mount Zion Church Basalt; md, Midland Formation; hg, Hickory Grove Basalt; tr, Turkey Run Formation; sb. Sander Basalt; ptt, Passaic Formation (member TT); ppg, Passaic Formation (Pine Grove Member); pet, Passaic Formation (Exeter Township Member); om, Orange Mountain Basalt; ff, Feltville Formation; pb, Preakness Basalt; tf, Towaco Formation; hm, Hook Mountain Basalt; bf, Boonton Formation; nh, New Haven Formation; tb, Talcott Basalt; sm, Shuttle Meadow Formation; ho, Holyoke Basalt; eb, East Berlin Formation; ha, Hampden Basalt; pf, Portland Formation; fr, Falls River beds (Sugarloaf Formation); db, Deerfield basalt; tf, Turners Falls Formation; bd, Bigoudine Formation; ab, Argana Basalt; am, Ameskroud Formation; bww, Blomidon Formation ("White Water member"); bpi, Blomidon Formation ("Partridge Island member"); nm, North Mountain Basalt; mb, McCoy Brook Formation; lc, lower clay formation; lv, lower basalt formation; is, interbedded sedimentary unit; uv, upper basalt formation; uc, upper clay formation. Sections are: 1, Turkey Run, Casanova, VA; 2, section formerly exposed at Licking Run Dam, Midland, VA; 3, Cinque Quarry, North Haven, CT; 4, Parmalee Brook at Stagecoach Rd., Durham, CT; 5, US Rt. 1, Branford, CT; 6, small creek south of Bluff Head, Guilford, CT; 7, intersection of State routes 9 and 15, East Berlin, CT; 8, river, stream, and railroad track outcrops adjacent to the Footprint Preserve, Holyoke, MA; 9; railroad cuts and Connecticut River bluffs, near Cedar Knob, northern Holyoke, MA; 10, Stony Brook, East Granby, CT; 11, Rt. 2, Greenfield, MA; 12, Connecticut River, Turners Falls and Gill, MA; 13, Bigoudine, Morocco (Hofmann et al., 2000); 14, Bigoudine, Morocco (same section as 12, but not described by Hofmann et al. (2000); 15, Argana, Morocco (from Olsen et al., 2002c); 16, coves on southeast side of Scots Bay, Nova Scotia (summarized from Birney De Wet and Hubert, 1989 and Olsen et al., 1989); 17, Partridge Island (from Olsen et al., 2002c); 18, well sections described by Tourani et al (1999); 19, section at Nif Gour, Morocco (from Olsen et al., 2002c).

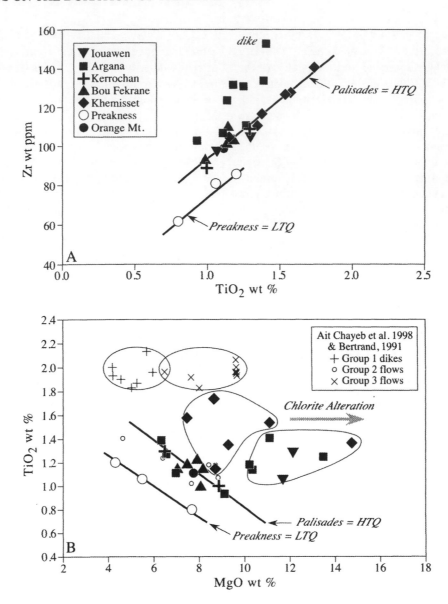

Figure 6. Geochemical data from basalts from Morocco compared to the Palisades/Orange Mountain Basalt (HTQ) trend and the Preakness Basalt (LTQ) trend. Note that, apart from the effects of alteration, all of the new samples lie on the HTQ trend and none lie on the LTQ trend. This includes samples of both lower and upper flow sequences in the Kerrouchen, Bou Fekrane, Khemisset, and Iouawen basins.

indicate that these strata contain the Triassic-Jurassic boundary (Slimane and El Mostaine, 1997), agreeing with the palynological assessment of cyclical red, gray and black strata in the Argana basin. Overlying the basalts in the subsurface of the Berrechid, and Khemisset basins are also halite beds, although their correlation both with the surface and elsewhere is less certain. It is worth noting that the specific pattern of bundling of laterally extensive mud-halite cycles in the salt beds above the basalts in the Berrechide and Khemisset described by Et-Touhami (2000) bear a strong resemblance to the cyclostratigraphy of the Towaco

Formation and its correlatives in eastern North America. More stratigraphic, especially magnetostratigraphic, work is needed to assess the age-significance of this tantalizing similarity, however.

2.4 Iberia and France: a Marine Connection

A little known direct connection between CAMP extrusives interbedded with continental strata and probable CAMP lavas extruded into marine Triassic-Jurassic strata is present in six areas in southern Europe. These tholeiitic to alkali basalt flows and pyroclastics occur in southern Iberia,

Table 1. Geochemical data from Moroccan Basalts.

Basin[1]	Kem.	Kem.	Kem.	Kem.	Kem.	Kem.	Bou.	Bou.	Bou.	Bou.	Weo.	Weo.	Ker.	Ker
Unit[2]	lbst	ubst	ubst	lbst	lbst	ubst	lbst	ubst	lbst	lbst	lbst	ubst	ubst	lbst
Sample#	21-01	21-02	21-03	21-16	21-18	21-20	22-07	22-08	22-11	22-19	22-20	22-22	23-01	23-02
Comment[3]	ves.									alt.	alt.			
Associated with[4]														

Major Elements (%)

SiO2	52.4	5 1.5 1	52.8	52.03	53.1	5 1.68	52.4	5 1.09	5 1.53	52	50.46	46.2	49.23	53.3
TiO2	1.58	1.38	1.15	1.54	1.74	1.35	1.14	0.99	1.14	1.18	1.3	1.07	1.0	1.29
Al2O3	13.94	10.04	13.18	11.89	13.76	13.43	14.31	14.99	14.26	14.38	11.07	13.58	13.99	14.82
Fe2O3	9.85	9.95	10.88	10.07	9.63	10.25	9.9	9.52	10.88	10.88	11.17	9.25	10.46	9.93
MnO	0.16	0.16	0.18	0.17	0.16	0.17	0.17	0.16	0.18	0.18	0.18	0.16	0.17	0.16
TiO2	1.58	1.38	1.15	1.54	1.74	1.35	1.14	0.99	1.14	1.18	1.3	1.07	1.0	1.29
MgO	7.46	14.75	8.72	11.12	8.65	9.26	8.17	8.08	7.06	7.51	12.12	11.67	8.86	6.46
CaO	9.63	6.8	10.33	8.32	9.18	10.1	10.85	11.25	10.76	10.5	8.8	11.88	11.16	10.59
Na2O	2.37	1.33	2.03	1.93	2.32	1.98	2.08	2.23	2.19	2.07	1.44	2.01	2.04	2.23
K2O	0.72	0.2	0.38	0.46	0.62	0.38	0.4	0 53	0.4	0.48	0.18	0.51	0.3	0.49
P2O5	0.21	0.16	0.14	0.18	0.23	0.16	0.14	0.15	0.14	0.16	0.15	0.15	0.15	0.18
LOI	2.21	3.76	0.89	1.87	0.97	1.87	0.87	0.95	0.75	1.6	2.71	3.22	2.2	1.71
totals	102.11	101.42	101.83	101.12	102.1	101.98	101.57	100.93	100.43	102.12	100.88	100.77	100.56	102.45

Trace Elements (ppm)

Rb	18		17	26	19	16	21	15		14	14		14	18
Ba	170		185	190	275	125	115	170	120	183	150	320	310	
Sr	237	207	182	256	261	193	174	180	175	176	191	207	178	187
Zr	128	117	105	127	141	111	110	93	102	103	106	98	89	109
Ni	86	108	97	104	88	92	108	92	96	94	99	99	86	79
Cr	247	268	264	303	228	238	357	212	261	229	255	225	217	237

Basin	Arg.	Arg.	Arg.	Arg.	Arg.	Arg.	Arg.	Arg.	OMB	Preak.	Preak.	Preak.
Unit	flow 7	flow 6	flow 5	flow 4	flow 3	flow 2	flow 1	dike[5]	flow 1	flow 1	flow 2	flow 3
sample#	27-04	27-05	27-06	27-07	27-08	27-9	27-10	27-11				
Comment					ves.	ves.	alt.					
Associated with				zeo.	horn.	horn.						

Major Elements (%)

SiO2	53.85	50.9	52.55	55.79	50.56	46.8	49.36	48.8	52.27	53	53	5 1
TiO2	1.27	0.93	1.11	1.39	1.18	1.25	1.14	1.41	1.11	1.06	1.2	0.8
Al2O3	14.13	13.98	14.75	13.65	13.37	12.26	12.46	12.3	14.14	14.1	13.8	15.2
Fe2O3	10.84	10.28	10.97	8.65	11.16	11.71	11.17	11.93	11.19	13.5	14	11.5
MnO	0.18	0.17	0.18	0.15	0.18	0.19	0.18	0.2	0.2	0.2	0.19	
TiO2	1.27	0.93	1.11	1.39	1.18	1.25	1.14	1.41	1.11	1.06	1.2	0.8
MgO	6.54	9.11	6.98	6.32	10.2	13.45	10.33	11.09	7.75	5.5	4.3	7.7
CaO	9.32	8.94	9.76	8.71	7.81	5.99	6.09	6.75	10.72	10	9.4	10
Na2O	2.76	2.7	2.65	3.21	3.94	3.21	4.16	3.84	2.43	3	2.8	3
K2O	1.08	0.94	0.94	1.58	0.73	0.83	0.69	0.5	0.43	0.8	0.73	0.18
P2O5	0.23	0.2	0.22	0.2	0.17	0.22	0.23	0.18	0.13	0.15	0.14	0.09
LOI	0.81	2.22	0.88	0.75	0.86	2.51	2.18	1.37				
totals	102.28	101.3	102.1	101.79	101.34	99.67	99.13	99.78				

Trace Elements (ppm)

Rb	18	17	17	21	11	17	16	8	15	18	25	8
Ba	158	174	165	275	240	155	160	80			179	
Sr	185	190	186	238	230	249	207	184	192	140	146	142
Zr	111	103	107	134	132	131	124	153	99	81	86	62
Ni	79	92	8 1	96	89	1 09	93	110	99	30	25	7 1
Cr	201	224	203	248	271	270	249	321	309	30	16	213

[1] Abbreviations: Kem, Khemisset; Bou, Bou Fekrane; Weo, Weone; Ker, Kerrouchen; Arg., Argana; OMB, Newark basin; Orange Mountain Basalt; Preak, Newark basin, Preakness Basalt.
[2] Abbreviations: lbst, lower basalt sequence; ubst, upper basalt sequence.
[3] Abbreviations: ves., vesicular; alt., altered.
[4] Abbreviations: zeo., zeolites; horn., ?hornfels.
[5] Dike is very thin (<30 cm) and wholly within the Argana basalt.

the Pyrenees of France and Spain, the Aquitaine basin of southwestern France, the Provence area of France, and in the French western Alps.

In southern Iberia, Palain (1977) and Puffer (1993, 1994) describe tholeiitic flows interbedded with red mudstones. In southern Portugal in particular, the St. Bartolomé de Messines section (Palain, 1977) is remarkably similar to that seen in Morocco, especially the Khemessit basin, with two basalt flow sequences separated by a thin limestone sequence that reportedly contains marine Early Jurassic invertebrates (Sopeña et al., 1988).

Flows (and more numerous intrusive sills) occur in the Spanish and French Pyrennes (eastern Corbières) (Azambre et al., 1981; Sopeña et al., 1988), some of which are apparently alkali basalt flows (Azambre and Fabriès, 1989). Most are strongly deformed and altered, but some are fairly fresh. These igneous rocks are probably related to units present in the subsurface in the adjacent Aquitaine basin of France. There, sheets of tholeiitic "ophites" are interbedded within "Keuper" red mudstones and evaporites of apparent Norian age. These are sometimes cited as flows (e.g., Orti Cabo, 1983) but according to Stévaux and Winnock (1974) the under- and overlying strata are metamorphosed near the contact, strongly suggesting they are sills, similar to those in the Pyrenees. However, in the same basin, there are extrusive volcanics that make up a regional marker bed (the tuff of Dubar, 1925) interbedded within Rhaeto-Liassic marine carbonates (Carcans Formation) (Barthe and Stévaux, 1971). The "tuff" is an excellent candidate to be part of the extrusive suite of the CAMP, although detailed study is lacking.

Alkali basalt flows and pyroclastics outcrop in Provence, France, north of Toulon (Lacroix, 1893; Azambre and Fabriès, 1989) and in the Ecrins-Pelvoux (Dauphiné) external massif of the Alps in France (Dumont, 1998; Bainchi et al., 1999). The published biostratigraphic ages of the strata surrounding the alkali basalt flows and pyroclastics in the Ecrins-Pelvoux and Provence areas is Rhaeto-Liassic (Moret and Manquat, 1948; Barthe and Stévaux, 1971; Dumont, 1998), and there are K-Ar dates of 201 and 197 Ma for these, respectively (Baubron, 1974; Baubron in Buffet, 1981). In particular, alkali flows and tuffs in the Dauphiné area are underlain by Triassic dolomites and overlain by fossiliferous Hettangian neritic limestone. According to Moret and Manquat (1948), the most basal of the overlying beds have some invertebrates known from the Rhaetian, but Hettangian forms dominate, and typical Rhaetian forms (e.g., *Avicula contorta*) are absent. These assemblages clearly require additional study; it is unclear whether they are equivalent to the "pre-planorbis" beds, the traditional basal Hettangian (marked by *P. planorbis*), or the Rhaetian, although we favor the two former possibilities.

2.5 Offshore Nova Scotia, Grand Banks, and Morocco

CAMP basalt flows have been encountered offshore in wells on the Scotian Shelf and Grand Banks (Pe-Piper et al., 1992; Holser et al., 1988) and offshore Morocco (Oujidi and Et-Touhami, 2000). They are interbedded with evaporites, and thus are closely comparable to basalt occurrences in the deeper parts of the Moroccan basins. However, no details relevant to cyclostratigraphy are available.

2.6 Brazil

Lava flows evidently part of the CAMP occur in eastern and southwestern Brazil (Marzoli et al., 1999). The basalts examined by Marzoli et al. (1999) do not seem to occur in Triassic-Jurassic sedimentary basin sequences, although at least one is close to such a sequence (Botucatu Formation) that may be of Triassic or Early Jurassic age (Leonardi, 1994). Thus, no cyclostratigraphic constraints on these deposits exist.

3. MAGNETOSTRATIGRAPHIC CONSTRAINTS

Early paleomagnetic studies of the Newark Supergroup basins concentrated mainly on the strongly magnetized basaltic lavas and associated igneous intrusions and found that the rocks carried a predominantly normal polarity magnetization (Dubois et al., 1957; Opdyke, 1961; Irving and Banks, 1961; Beck, 1965, 1972; Larochelle and Wanless, 1966; De Boer, 1967, 1968; Carmichael and Palmer, 1968). These results contributed strongly to the notion of a long normal polarity "superchron" during the Late Triassic and/or Early Jurassic (e.g., Graham interval of McElhinny and Burek, 1971; Newark interval of Pechersky and Khramov, 1973; TRN interval of Irving and Pulliah, 1976), a concept which persists in some recent data compilations (Haq et al., 1988; Johnson et al., 1995; Algeo, 1996). Indeed, with only very few exceptions (e.g., Smith, 1987; Phillips, 1983), virtually all paleomagnetic determinations deemed reliable on hundreds of sampling sites in basaltic lavas and intrusions associated with the Newark Supergroup throughout eastern North America yield normal polarity (e.g., Smith and Noltimier, 1979; Hodych and Hayatsu, 1988; Prevot and McWilliams, 1989).

As summarized by Marzoli et al. (1999), normal polarity is a virtual signature of CAMP tholeiites from the entire circum-Atlantic realm, including South America, West Africa, and southern Europe, as well as eastern North America. As such, CAMP stands alone among flood basalts as having virtually no polarity reversals (i.e., Columbia River, Ethiopian, Greenland, Deccan, Parana, Karoo, Siberian, and Emeishan are each documented to have 2 or more polarity intervals; see references in Courtillot et al., 1999). This may be due to some combination of low reversal frequency (e.g., Johnson et al., 1995), remagnetization (e.g., Witte and Kent, 1990), very short duration of the igneous activity (e.g., Olsen et al., 1996b), or inadequate sampling resolution.

The constant normal polarity of the igneous rocks contrasts with the mixed polarities associated with the thick sequence of sedimentary rocks underlying, overlying, and intruded by the Newark igneous rocks (Figure 3), although the sediments interbedded with the basalts are uniformly of normal polarity. As described above, the sediments, usually cyclical and representing about 30 m.y.. of the Late Triassic, clearly record polarity reversals at an average rate of about 2 per m.y.. (McIntosh et al., 1985; Witte et al., 1991; Kent et al., 1995), similar to the reversal rate in marine sediments of Late Triassic age (Gallet et al., 1992). According to the astronomically tuned GPTS for the Late Triassic based on the Newark data (Kent and Olsen, 1999a), the commencement of consistent normal polarity precedes the basaltic lavas by at most ~1 m.y.. (Chron 23n). The very short reverse interval (Chron E23r) immediately underlying the basalts and the Triassic/Jurassic boundary (in Chron E24n) are considerable closer to the basalts (>20 ky) (Olsen et al., 1996a) (Figures 3, 5). There is limited exposure of the post-extrusive Boonton Formation in the Newark basin which is of normal polarity where sampled (McIntosh et al., 1985; Witte and Kent, 1990). However, we have found what are probably the oldest Hettangian to Sinemurian reversed intervals in the Hartford basin Portland Formation, the older part of which correlates with the Boonton Formation, and there are frequent reversals by the end of the Hettangian and into the Sinemurian in marine sediments of Early Jurassic age from the Paris basin (Yang et al., 1996) (Figures 3, 8). As is the case for the Newark basin the strata interbedded with the flows are of normal polarity (Witte and Kent, 1991). Therefore, the bracketing occurrences of frequent reversals in the Late Triassic and Early Jurassic combined with the estimated length of the Hettangian (<4 m.y.. according to the recent time scale of Gradstein et al. [1995]) limit the total duration of any predominantly normal polarity interval in the latest Triassic to earliest Jurassic to less than about 4 million years (Figure 7).

The lavas and intrusions associated with the Newark Supergroup had often been assumed to have been emplaced over a prolonged period, such as from 170 Ma to 204 Ma in the widely used Irving and Irving (1982) paleomagnetic pole compilation, or as two distinct igneous events and corresponding paleopoles at around 175 Ma and 190 Ma (Smith and Noltimier, 1979). However, the Newark Supergroup igneous rocks are now thought more likely to represent a small number of events over a short emplacement episode in the earliest Jurassic (~201 Ma; Sutter, 1988; Dunning and Hodych, 1990). Nonetheless, there is still

considerable scatter in the radiometric dates as shown by Marzoli et al. (1999) in which their summary of dates spans some 37 million years (Figure 7), with most dates close to 200 Ma.

Some assessment of the geological significance of this scatter in ages can be made by a comparison of the polarity record and the distributions of the actual dates. Using the probability function for all CAMP rocks presented by Marzoli et al. (1999) as a sampling function for smoothing the paleomagnetic record, it is clear that expected frequency of normal polarity for the entire Late Triassic and early part of the Jurassic should be about 0.5, or 50% (Figure 7). It is not until this sampling interval is compressed to 5 m.y. or less that we reach the observed >95% normal polarity for this probability distribution. Less "peaked" functions would have to be even narrower in their duration. Based on this probabilistic argument, it is likely that the scatter in ages reflects some kind of distortion of the actual crystallization ages by as yet poorly documented geological processes (e.g. Sutter, 1988; Seidemann, 1988, 1991), and thus the cited analytical precession estimates are one to two orders of magnitude greater than the accuracy of the dates. Similarly, it is most likely that most CAMP rocks were emplaced or extruded during chron E 23 or E24, based on this analysis (Figure 7), over an interval of not more than 3 m.y..

3. BIOSTRATIGRAPHIC CONSTRAINTS

The most distinctive and perhaps most useful biostratigraphic constraints on the duration of the CAMP basalts is the Triassic-Jurassic boundary. The palynologically identified boundary is marked by the last appearance of many pollen and spore taxa, most importantly the distinctive pollen taxon *Patinasporites densus*, which is quite common until its last appearance (Cornet, 1977; Cornet and Olsen, 1985; Fowell et al., 1994; van Veen et al., 1995; Fowell and Traverse, 1995). The succeeding survivor assemblage contains few if any first appearances initially, and is characterized by the preponderance of the pollen taxon *Corollina meyeriana*, which makes its first appearance in eastern North America about 20 m.y. earlier in the Carnian (Cornet, 1977; Cornet and Olsen, 1985).

Cornet (1977) recognized the Passaic-Heidlersburg palynoflora as the uppermost of the biostratigraphic divisions of Triassic age in eastern North America, with its top being the Triassic-Jurassic boundary (Cornet and Olsen, 1985). All of the CAMP lavas in the outcropping basins of eastern North America fall into the succeeding zone of Cornet (1977), the Corollina meyeriana Zone (Cornet and Olsen, 1985). The next zone, the Corollina torosus Zone (Cornet, 1977; Cornet and Olsen, 1985), does not begin until well into the post-basalt formations.

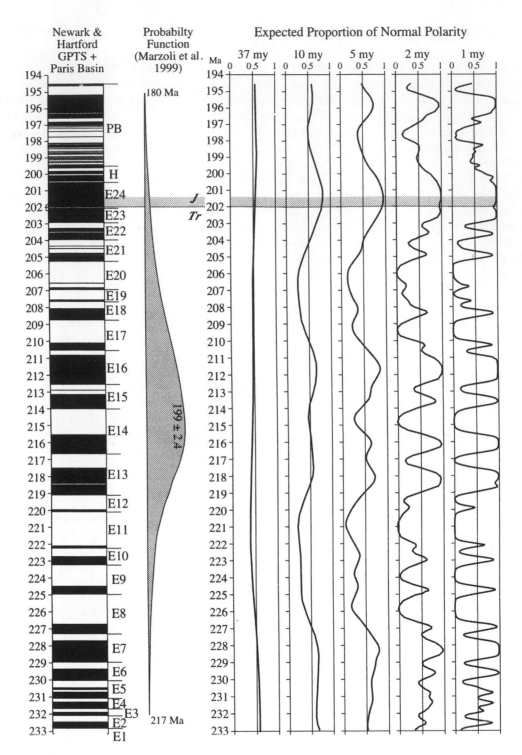

Figure 7. Paleomagnetic constraints on the duration of the CAMP event showing the proportion of the CAMP that would be expected to be of normal polarity if the timing of the emplacement followed the shape of the probability function of Marzoli et al. (1999) with total durations of 37, 10, 5, 2, and 1 million years. Note that the predicted proportion of normal polarity matches the observed only if the duration of the event is less than 5 m.y.. In Marzoli et al. (1999) this probability function had a total duration of 37 million years (180-217 Ma.) with the maximum probability at 199±2.4 Ma. The probability curves were calculated basically as a moving average with a window shaped as the Marzoli et al. probability function with a lag of 100 ky and with a value of 1 for normal and 0 for reversed polarity in the polarity time scale. The polarity time scale is based on Figure 3 spliced onto the polarity record from the Paris Basin (PB) Early Jurassic of Yang et al. (1996) based on the overlapping Hettangian age portions of the Newark (E1-E24) and Hartford (H) polarity records: white is reversed, black is normal, and gray (Hartford basin only) is unsampled (counted here as normal), ages for the Sinemurian based in part on Palfy (2000).

A major biotic turnover is also seen in vertebrate taxa, both osseous and ichnologic, at or very near the Triassic-Jurassic boundary (Colbert, 1958; Olsen et al., 1989, 1990; Olsen et al., 2002). The last appearance of the procolophonid parareptile *Hypsognathus fenneri* occurs in the upper Passaic Formation about two McLaughlin cycles below the Triassic-Jurassic boundary, along with the first appearance of the crocodylomorph cf. *Protosuchus* sp. (Olsen et al., 2002b). *Hypsognathus fenneri* is the index fossil of the Cliftonian Land Vertebrate Age (LVA) (Huber et al. 1993). Immediately below the palynologically defined Triassic-Jurassic boundary are the last appearances of the suchian ichnogenera *Brachychirotherium* and *Apatopus* (Szajna and Hartline, 2002; Olsen et al., 2002b), which occur though nearly the entire Late Triassic in eastern North America.

Directly above the palynologically dated boundary occurs the first appearance of the dinosaurian ichnospecies *Eubrontes giganteus*, which is the index fossil for the next LVA, the Wassonian (Lucas and Huber, 2002; Olsen et al., 2002b). All of the CAMP lavas in eastern North America occur within the Wassonian LVA. Within the Wassonian, there is very little taxonomic turnover, with the exception of the appearance of the ornithischian ichnotaxon *Anomoepus*, in the Feltville Formation and its equivalents (Olsen and Rainforth, 2002), and the prosauropod ichnogenus *Otozoum* in the East Berlin Formation and its equivalents. The latter taxon so far occurs only in the Hartford basin and north (Rainforth, 2000, 2002).

Thus, all CAMP extrusives in outcropping basins in eastern North America occur within one palynological biozone (Corollina meyeriana Zone) and one Land Vertebrate Age (Wassonian). The Corollina meyeriana Zone is apparently entirely Hettangian in age (Cornet, 1977; Cornet and Olsen, 1985). The Hettangian is widely regarded as the shortest stage of the Jurassic (<4 m.y.; Gradstein et al., 1995; Palfy et al., 2000). While the fact that the CAMP extrusives are limited to a single LVA (Wassonian) is suggestive of a limited time span, the top of the Wassonian is unconstrained and its duration not independently well calibrated. Nonetheless, it is clear that the available biostratigraphic data suggested a short duration for the North American outcropping CAMP extrusives, probably less than 2 m.y..

5. COMPARISON WITH RADIOMETRIC DATES

Many radiometric dates, from a variety of systems and vintages (K-Ar, $^{40}Ar/^{39}Ar$, U-Pb, Rb-Sr), have been published for igneous rocks included or possibly included within the CAMP. These will not be reviewed here because they have been the subject of recent reviews (Ragland et al., 1992; Marzoli et al., 1999; Hames et al., 2000; Baksi, this volume). However, it is worth noting that the present consensus (e.g. Marzoli et al., 1999) is that the duration of at least most of the CAMP is very brief (<2 m.y.) centered on about 200 Ma. Indeed, for example, Hames et al. (2000) describe dates from the oldest and youngest basalt flows in the Newark basin (Orange Mountain and Hook Mountain basalts, respectively) that are analytically indistinguishable from one another averaging about 200 Ma. However, there are still inconsistencies among dating systems, different laboratories and standards, and even within minerals within the same rock (e.g. Turrin, 2000) that must be reconciled before a real understanding of the geological significance of these dates can be fully realized and approach the precision of astronomical calibration.

6. RELATIONSHIP OF CAMP TO TRIASSIC-JURASSIC BOUNDARY

Where evidence is available, all CAMP basalt flows lie above the biostratigraphically identified Triassic-Jurassic boundary, beginning within 20 ky after it based on astronomical calibration (Olsen et al., 1996b, 1999, 2002c). Thus, there is no evidence from superposition that directly allows for CAMP to be the cause of the mass extinctions (Olsen, 1999; Olsen et al., 2002b). Nonetheless, the very close proximity in time is suggestive, and because the outcropping flows are restricted to the northernmost part of the CAMP, it is very possible that a more southerly initiation of this igneous episode occurred prior to the Triassic-Jurassic boundary (Olsen, 1999; Hames et al., 2000).

There is in fact indirect evidence that at least some CAMP igneous activity might predate the boundary. Given that we do know where at least many of the CAMP basalts fall in the paleomagnetic record, the most likely interval of reversed polarity captured by the CAMP intrusions is the remarkably short E23r (Figures 3, 5, 8). The best known example of a CAMP unit with reversed polarity is one of the north-south trending dikes in North Carolina reported by Smith (1987; site 22). The north-south trending system includes dikes with typical CAMP ages of ~200 Ma (e.g., Ragland et al., 1992), but specific independent confirmation of the CAMP origin of the dike with reversed magnetizations is needed. However, if we accept this dike as CAMP, it most likely indicates some CAMP igneous activity just prior to the Triassic-Jurassic boundary (i.e., during E23r). However, whether the dike was emplaced before or after the boundary, the reversed polarity of this dike definitely indicates that there was some CAMP activity at a different time than all of the studied flows.

Other than the one dike of reported reversed polarity in the Carolina Piedmont (Smith, 1987), the most interesting reported intrusions with reversed polarity are several dikes in Liberia and the Freetown layered intrusion of Sierra Leone.

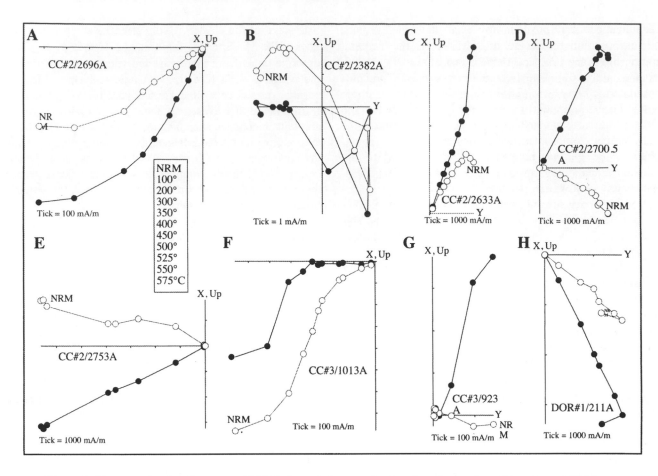

Figure 8. Vector end-point diagrams of NRM thermal demagnetization data from selected samples of Clubhouse Crossroads Basalts from USGS test holes CC#2, CC#3, and DOR#1. Open/closed symbols are plotted on vertical/horizontal planes where vertical is assumed to be the drill-core axis and the horizontal axes are arbitrary because the drill-core was unoriented with respect to geographic north. List of thermal demagnetization steps applies to all samples. Results of principal component analysis are shown in Table 2.

Dalrymple et al. (1975) reported on K-Ar and $^{40}Ar/^{39}Ar$ dates and paleomagnetic data from diabase dikes in Liberia. About half of the dikes intruded into Precambrian basement show reversed polarity, while those intruded into younger sedimentary rocks are completely of normal polarity. Pole positions from the two areas of intrusion are indistinguishable and compatible with other west African CAMP poles. Existing K-Ar dates are broadly compatible with CAMP (excluding those thought to have excess argon by Dalrymple et al., 1975). If the reversed dikes prove to be part of the CAMP they also indicate some magmatism distinctly older or younger that than the known extrusives of the CAMP.

The Freetown pluton is one of several layered intrusions in west-central Africa. It has a reported Rb-Sr age of 193±3 Ma (Beckinsale et al., 1977). Hargraves et al. (1999) report that out of 13 sites they sampled, nine are of reversed and 4 are of normal polarity, which is definitely anomalous com-

pared to CAMP. However, according to Hargraves et al. (1999) the magnetization post-dates deformation of the complex intrusion and thus must post-date the intrusion by some amount of time, in agreement with a pole position that would be anomalous for the latest Triassic or earliest Jurassic. Therefore the presence of a significant proportion of reversed polarity has no obvious bearing on the duration of the CAMP event. There are other layered intrusions in the region (Sierra Leone, Liberia, Mali, Guinea, Senegal). The chemistry of these intrusions is compatible with CAMP, and in general newer $^{40}Ar/^{39}Ar$ ages of the layered intrusions in the region are appropriate for the CAMP (Diallo, et al., 1992; Deckart et al., 1997). In addition, one of the Guinea intrusions has yielded a U-Pb age of 201.5 +/-0.5 Ma (from duplicated fractions of concordant zircon; G. R. Dunning pres. comm., 2000). Thus, these west African layered intrusions are probably part of the CAMP, although their magnetizations may post-date the event, and thus provide little information on the duration of the CAMP or their relationship to the Triassic-Jurassic boundary.

Thus, there is some evidence that some of the CAMP may precede the Triassic-Jurassic boundary, and therefore a causal relationship could be possible (Olsen, 1999; Hames et al., 2000) between the two events. However, the best candidate for massive volcanism prior to the Triassic-Jurassic boundary in the CAMP lies the still largely unconstrained volcanic wedge of seaward dipping reflectors (Holbrook and Kelemen, 1993) that may be volumetrically as large as the rest of the CAMP (Olsen, 1999).

7. CLUBHOUSE CROSSROADS BASALT, SEAWARD DIPPING REFLECTORS, AND CAMP

Large areas of the Coastal Plain of southeastern Georgia and South Carolina are underlain by very shallow dipping tholeiitic flows, locally referred to as the Clubhouse Crossroads Basalt, described as interbedded quartz normative and olivine normative basalts (Gottfried et al., 1983; McBride et al, 1989), with most of the observed chemical variation being ascribed to extensive alteration (Gottfried, et al., 1983). These basalts were cored in the early 1980's in three holes by the USGS (CC#1, CC#2 and CC#3 at ~33°N 80°W) near Charleston, South Carolina, and have provided both $^{40}Ar/^{39}Ar$ dates (Lanphere, 1983) and paleomagnetic data (Phillips, 1983). The Clubhouse Crossroad basalts are critical because they have been traced in seismic profiles to the seaward dipping reflectors (SDRs) interpreted as massive edifices of basalt formed during the initial formation of oceanic crust, and are the only direct constraints on the age of the SDRs (McBride et al., 1989; Austin et al., 1990; Holbrook and Kelemen, 1993; Holbrook et al., 1994; Oh et al., 1995; Talwani et al., 1995). The age cited for these basalt flows is generally 184 Ma based on dates by Lanphere (1983), similar to the predicted age of the oldest Atlantic oceanic crust as extrapolated from magnetic anomalies (Klitgord and Shouten, 1986). However, we believe that the radiometric dating and magnetic data that seem to show that the Clubhouse Crossroads Basalt is not part of CAMP are ambiguous, and hence we suggest that the basalts are indeed part the CAMP (as originally thought by Lanphere, 1983), although extensive recoring and redating are required to confirm this. If the Clubhouse Crossroads basalts are part of the CAMP, so may be the seaward dipping reflectors and the age of the transition to oceanic crust may be closer to 200 Ma, not 184 Ma..

Lanphere (1983) dated three samples of Clubhouse Crossroads Basalt, obtaining whole rock $^{40}Ar/^{39}Ar$ incremental release spectra, isochron ages, and total fusion ages.

The incremental heating ages have saddle-shaped spectra, with "plateau" ages of 187±1.3, 161±3.1, and 187±3, isochron ages of 192±15.4, 167±2, and 184±3.3, and total fusion ages of 296±2.9, 172±4.5, and 182±2.8 respectively. Lanphere argued that the most reliable date was the incremental age of 184±3.3 Ma because of the relatively concordant isochron and total fusion ages of the same sample. Interestingly, Lanphere, also argued that, "this age is in good agreement with reliable ages of tectonically related lower Mesozoic diabase intrusions in eastern North America and Liberia.". The latter ages, of course are now thought to be close to ~200 Ma, but the dates on the Clubhouse Crossroads Basalt did agree with the best whole rock ages of similar and older vintage (e.g. Dallmeyer, 1975; Dalrymple et al., 1975; Sutter and Smith, 1979; Seidemann, et al., 1984; Seidemann, 1989). Significantly, Sutter (1988) was able to show that feldspar separates from many eastern North American tholeiites give young ages (175 to 178 Ma) caused by young, possibly hydrothermal alteration, compared to amphibole and biotite separates from the same rocks (200 to 202 Ma), suggesting that many whole rock dates, such as those from the Clubhouse Crossroads, may be mixing ages of these two end members (see also Turrin, 2000). Gottfried et al. (1983) indeed described such alteration in these basalts. Whole rock determinations may never be able to accurately date these fine-grained, altered basalts, but additional work on dating mineral separates of single crystals is clearly needed to better constrain accurate radiometric dates of the Clubhouse Crossroads Basalt and understand the origin of the uncertainties.

Phillips (1983) provided NRM and stepwise thermal demagnetization data that suggested the Clubhouse Crossroads Basalt was of mixed magnetic polarity and characterized by relatively high mean (absolute) paleomagnetic inclinations (35°±3.2°) indicating a paleolatitude of 19.5°N. Magnetic polarity results from the NBCP predicts that if it is part of the CAMP, the Clubhouse Crossroads Basalt should be of nearly uniformly normal polarity and should have been extruded close to the equator and hence characterized by much lower paleomagnetic inclinations (e.g. 2-8°N). Superficially, Phillips' results from the Clubhouse Crossroads Basalt would strongly suggest they are not part of the CAMP regardless of their apparent radiometric age. However, Phillips now considers these results as unreliable because of possible core misorientation (pers. comm. in Olsen et al., 1996b). Thus, it is possible that these basalts are of entirely or nearly entirely normal magnetization, and the high paleomagnetic inclinations could be due to unresolved Middle Jurassic or later magnetic overprints.

In order to test this hypothesis we resampled the Clubhouse Crossroads Basalt (USGS CC#2 and CC#3 cores) and

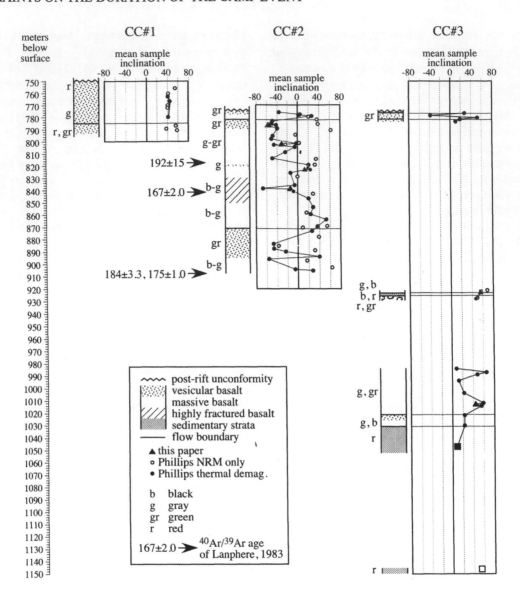

Figure 9. Paleomagnetic data from the Clubhouse Crossroads Basalt from Phillips (1983) and new data from this work plotted on the lithological section of the Clubhouse Crossroads cores from Gottfried et al. (1983).

a basalt encountered in the St. George, South Carolina area (USGS-St. George No. 1 core- DOR-211), and subjected the samples to progressive AF and thermal demagnetization (Figures 8, 9). The results of our analysis on 5 samples from CC#2 and 2 samples from CC#3 are completely compatible with those of Phillips (1983) including samples with apparent reversed polarity (Figure 9). There is little evidence of a systematic overprint in the thermal demagnetization data (Figure 8). Using the vector isolated between 300° and 500°C, 3 of the samples had stable negative inclinations and 3 had stable positive inclinations. One sample (CC#2/2382A) was judged to be magnetically unstable. The absolute mean inclination of the 6 stable thermally demagnetized sample magnetizations is 28.1°±6.3° (standard error)

which is indistinguishable from the 35° reported by Phillips and consistent with a Middle Jurassic pole (mean sample paleolatitude of 14.9°N) and the radiometric ages of Lanphere (1983).

However, several lines of evidence point towards the possibility of a combination of alteration and core misorientation leading to the apparent differences between CAMP and the Clubhouse Crossroads Basalt. First, with one exception, all of the polarity transitions occur within what Gottfried et al. (1983) identified as single flows (i.e. single cooling units) (Figure 9). For example, based on the thermally demagnetized samples of both Phillips (1983) and our own data, three polarity zones (2 reversed and 1 normal) occur in flow 2 of core CC#2. One polarity transition in particular, at 840 m, occurs in an interval in which Gottfried et al. (1983) specifically says there is, "no indication of any flow boundary". If these two polarity transitions really occur within one flow it implies that cooling of the flow took several thousand years, comparable to the 7.9±4.5 ky mean duration of Newark polarity transition (Kent and Olsen, 1999a) and estimates of 4 to 10 ky for the Brunhes/Matuyama transition (Clement and Kent, 1984), which seems very unlikely because it would require implausibly slow cooling rates for the individual flows. It seems even more unlikely that such events would be captured by two other flows in the same sequence, one of which is 4.8 m thick (CC#3: flow 2). It should be noted that although Gottfried et al. (1983) recognized only 12 flows based on physical and chemical criteria among all three coring sites, Phillips (1983) argued that there are 26 flows based on the paleomagnetic polarity data, that we (and he) believe are unreliable.

Second, there is little correspondence in polarity between what should be the same flows in adjacent cores near the top of each of cored sequence (Figure 9). Because only small portions of the upper part of the flow sequence were cored in CC#1 and CC#3, physical correlation between the flows is speculative. However, the upper parts of all three cores are similar in being highly vesicular and of similar chemistries (Gottfried et al. 1983). In specific, CC#1 shows no signs of the reversed polarity zones seen in the upper parts of the basalt in CC#2 and CC#3. Both the first and second points are most easily explained by some of the core segments having been oriented upside down while being boxed - an extraordinarily easy mistake for which we have personal experience. Regrettably, there is no obvious overprint that can be consistently isolated by thermal demagnetization in these cores, preventing a check on their orientation, unlike the Newark basin sedimentary cores that could be oriented post hoc (e. g. Witte and Kent, 1991; Kent et al., 1995).

Third, there appears to be a correlation between the higher (absolute) inclinations and the degree of alteration. In our six samples, the lightest colored (most altered) samples have the steep inclinations (Table 2). These correspond to intervals that Gottfried et al. (1983) argued were highly altered. The least vesicular and darkest (i.e. freshest looking) samples have the lowest inclinations. (Table 2). Although, difficult to test with Phillips' (1983) data, it appears that the higher inclinations again appear to be most often associated with vesicular and altered zones. In addition, in the most altered core sampled (CC#1: Gottfried et al., 1983), there is essentially no change in inclination from the top to the bottom of the core (~45 m), including the uppermost samples which are red, vesicular, and soft. It is very difficult to understand how these inclinations can be primary, yet their average of 45.6° is similar to many of the higher inclination

measured throughout the cores. If just the four freshest appearing of our samples are used (DOR#1/211A; CC#22753A; CC#2/2700.5A; CC#2/2696A), the mean (absolute) inclination is 16.3° corresponding to a paleolatitude of 8.3°N. This compares to an expected inclination for the Clubhouse Crossroads Basalt of 14.8° (paleolatitude of 7.5°N) for the time during which the Orange Mountain Basalt cooled. Unfortunately, lack of significant time averaging in flows and secular variation makes it very difficult to directly access the meaning of the variation in inclination without additional detailed paleomagnetic, petrographic, and geochemical work (e. g. Witte and Kent, 1990).

We note also that the single red bed result that was subjected to thermal demagnetization by Phillips (1983) has an inclination of 9°, corresponding to a paleolatitude of 4.6°. This compares to the expected inclination of 4.5° (2.3°N paleolatitude) for the Clubhouse Crossroads Basalt based on the Late Triassic-Early Jurassic sedimentary strata in the Newark basin Martinsville core (Kent et al., 1995). We plan to examine these underlying red beds for tests of the origin of the magnetization of the overlying basalts themselves. Thus, we feel that while our new data provide only a slight amount of support for the Clubhouse Crossroads Basalt being part of CAMP, the vagaries and ambiguities in all of the present data allow for that possibility. Moreover, we also argue that the only way to reliably assess the age of the Clubhouse Crossroads Basalt is by additional and much more complete coring and study of these critical basalts themselves.

Oh et al. (1995) argue that the J reflector of Schilt et al. (1983) and Dillon et al. (1983), which is supposedly the signature of the Clubhouse Crossroads Basalt (McBride et al., 1989), overlies the seaward dipping reflectors off the southeastern US. At the present, the Clubhouse Crossroads Basalt is the best constraint on its age. With the radiometric dates and its apparent reversed polarity intervals in question, it is entirely plausible that the seaward dipping reflectors are part of the CAMP event as argued by Withjack et al (1998). Like the rest of the CAMP event, the duration of the emplacement of the seaward dipping reflectors may have been very brief, occurring largely during chron E24n. Indeed Talwani et al. (1995) have argued that the East Coast Magnetic anomaly is a consequence of the volcanic wedge of seaward dipping reflectors being of entirely (or nearly entirely) normal polarity, completely in agreement with its hypothesized origin as part of the CAMP, but not in agreement with the apparent mixed polarity of the Clubhouse Crossroads Basalt to which they are supposed to be tied.

8. REMAINING CHALLENGES

Five unresolved issues loom large in the cyclo-, magneto- and bio-stratigraphic constraints on the duration of the

Table 2. Summary of thermal demagnetization analysis of NRM of selected samples from Clubhouse Crossroads Basalt in USGS drill cores CC#2, CC#3, and DOR#1 near Charleston, South Carolina. Simple arithmetic mean (absolute) inclination of 'c' component for 7 acceptable samples is 28.1° with a standard error of 6.3°.

Sample	Cmp[1]	N[2]	Range	V[3]	°MAD[4]	Dec	Inc	Color
CC#2/2382A	a	4	NRM - 300	F	17.8*	173.0	54.9	light green &
CC#2/2382A	c	4	300 - 450	A	3.1	179.2	-56.6	vesicular
CC#2/2633A	a	4	NRM- 300	F	5.4	277.5	1.9	greenish-brown &
CC#2/2633A	c	5	300 - 500	A	1.6	289.5	-30.1	vesicular
CC#2/2696A	a	4	NRM - 300	F	8.8	155.2	16.4	medium gray
CC#2/2696A	c	5	300 - 500	A	1.7	118.2	18.8	
CC#2/2700.5A	a	4	NRM - 300	F	21.8*	34.S	39.1	dark gray
CC#2/2700.5A	c	5	300 - 500	A	1.7	297.5	13.2	
CC#2/2753A	a	4	NRM - 300	F	3.5	152.7	-16.1	dark gray
CC#2/2753A	c	5	300 - 500	A	4.7	154.4	-14.2	
CC#3/923A	a	4	NRM - 300	F	3.8	287.2	4.4	dark greenish-
CC#3/923A	c	5	300 - 500	A	30.6*	135.1	55.9	gray
CC#3/1013A	a	4	NRM - 300	F	12.6	124.2	32.5	greenish-gray
CC#3/1013A	c	5	300 - 500	A	7.1	175.9	44.9	
DOR#1/211A	a	4	NRM - 300	F	21.3*	78.0	14.6	black
DOR#1/211A	c	5	300 - 500	A	1.8	63.1	19.1	

[1] Cmp is magnetization component described by principal component analysis (Kirschvink, 19801. Low unblocking temperature 'a' components are likely to be overprints; high unblocking temperature 'c' components are best estimate of characteristic magnetization for each sample.

[2] N is number of demagnetization data analyzed over thermal demagnetization interval in °C.

[3] V is whether best-fit vector is free (F) or anchored (A) to the origin.

[4] °MAD is maximum angular departure (*excluded from overall mean because of high uncertainty); Dec is declination (arbitrary azimuth) and Inc is inclination (assuming drill core axis is vertical and lava flows are horizontal) of best-fit component vector. *MAD values greater than 15° indicate poorly resolved component.

CAMP event: 1) demonstration, in detail, of how all of the more northern extrusives (Nova Scotia, Scotian Margin, Grand Banks, Morocco, Iberia, and France) fit into the cyclo-, magneto-, and bio-stratigraphy of the Triassic-Jurassic transition interval; 2) completion of a paleomagnetic census of all of the flows; 3) investigation of the stratigraphic constraints on CAMP basalts in South America; 4) description of the relationship between the Clubhouse Crossroads Basalt, the cyclo- and magnetostratigraphy stratigraphy of locally underlying rift strata and the Triassic-Jurassic boundary (if preserved there), and the magnetization of the basalts themselves; and 5) determination of the magnetostratigraphy and age of the seaward dipping reflectors. It is apparent that, while surface work can address issues 1 and 2 in outcropping units, most of these questions will require continuous coring to fully resolve, especially the last two.

9. CONCLUSIONS

Based on cyclo-, magneto- and biostratigraphic calibration of the outcropping basalt flows of the CAMP, the duration of igneous activity lasted less than 600 ky. Based on the known distribution of normal and reversed polarity zones through the Late Triassic and Early Jurassic, the apparent ages of all CAMP rocks, spanning >10 m.y., is not likely to be real, and is likely to be an order of magnitude less. However, the existence of at least some CAMP intrusions of reversed polarity suggest that some igneous activity may have begun prior to the Triassic-Jurassic boundary, and could be involved in the mass-extinctions. The existing data suggesting that the Clubhouse Crossroads Basalt of South Carolina and Georgia is of middle Jurassic age and was not part of the CAMP should probably be discounted pending much additional work. Thus, while the known extent of

CAMP rocks is very large in area $(7 \times 10^6 \text{ km}^2)$, its true magnitude may be much greater, because the major volcanic wedge of seaward-dipping reflectors lying off the southeastern US, which is tied to the Clubhouse Crossroads Basalt, may in fact be part of the CAMP as well.

Acknowledgments. Work on this project by Olsen, Et-Touhami, and Kent was funded by grants from the US National Science Foundation to Olsen and Kent (EAR-98-14475 and EAR-98-04851), Kent and Olsen (EAR-98-04851, EAR-00-00922), and a grant from the Lamont Climate Center to Olsen. Work in Morocco was aided by logistical support from ONAREP, for which we are very grateful. Et-Touhami was supported during work on this project by a fellowship from the Fulbright Foreign Student Program (MACECE). We thank G. S. Gohn for access to the Clubhouse Crossroads Basalt housed at the USGS in Reston, VA. The paper was reviewed by Emma C. Rainforth, which greatly improved the paper.

REFERENCES

Ait Chayeb, E. H., Youbi, N., El Boukhari, A., Bouabdelli, M., and Amrhar, M., Le volcanisme permien et mésozoic inférieur du bassin d'Argana IHaut Atlas occidental, Maroc): un magmatisme intraplaque associé à l'overture de l'Atlantique Central. *Journal of African Earth Sciences*, v. 26, n°. 4, p, 499-519, 1998.

Algeo, T. J., Geomagnetic polarity bias patterns through the Phanerozoic, *Journal of Geophysical Research*, v. 101, p. 2785-2814, 1996.

Azambre, B. and Fabriès, J., Mesozoic evolution of the upper mantle beneath the eastern Pyrenees; evidence from xenoliths in Triassic and Cretaceous alkaline volcanoes of the eastern Corbières (France), *Tectonophysics*, v. 170(3-4), p. 213-230, 1989.

Azambre, B., Rossy, M., and Elloy, R., Les dolerites triasiques (ophites) des Pyrenees; donnees nouvelles fournies par les sondages petroliers en Aquitaine, *Bulletin de la Societe Geologique de France*, v. 23(3), p. 263-269, 1981.

Austin, J. A., Jr., Stoffa, P. L., Phillips, J. D., Oh, J., Sawyer, D. S, Purdy, G. M., Reiter, E., and Makris, J., Crustal structure of the Southeast Georgia Embayment-Carolina Trough; preliminary results of a composite seismic image of a continental suture(?) and a volcanic passive margin, *Geology*, v. 18, p. 1023-1027, 1990.

Bainchi, G. W., Martinotti, G., Oberhaensli, R. E., Tethys phases of rifting in the realm of the Western Alps; some observations deduced by field and geochemical data. in 4th workshop on Alpine geological studies, edited by .B. Szekely, I. Dunkl, J. Kuhlemann, and W. Frisch, *Tuebinger Geowissenschaftliche Arbeiten, Reihe A, Geologie, Palaeontologie, Stratigraphie*, v. 52, pp. 127, 1999.

Barthe, A. and Stevaux, J., Le bassin à évaporites du Lias inférieur de l'Aquitaine. *Bull. Centre Rech. Pau-SNPA*, v. 5(2), p. 363-369, 1971.

Baubron, J. C., Sur l'âge triassique du "volcan de Rougiers" (Var).

Méthode potassium-argon, Comptes Rendu Acad Sci., Paris, v. 279D, p. 1159-1162, 1974.

Beck, M. E., Paleomagnetic and geological implications of magnetic properties of the Triassic diabase of southeastern Pennsylvania, *Journal of Geophysical Research*, v. 70, p. 2845-2856, 1965.

Beck, M. E., Paleomagnetism of Upper Triassic diabase from Pennsylvania: further results, *Journal of Geophysical Research*, v. 77, p. 5673-5687, 1972.

Beckinsale, R. D., Bowles, J. F. W., Pankhurst, R. J., and Wells, M. K., Rubidium-strontium age studies and geochemistry of acid veins in the Freetown Complex, Sierra Leone, *Mineralogical Magazine and Journal of the Mineralogical Society*, v. 41, no. 320, p. 501-511, 1977.

Bertrand, H., The Mesozoic tholeiitic province of Northwest Africa; a volcanotectonic record of the early opening of Central Atlantic, In *Magmatism in extensional structural settings; the Phanerozoic African Plate*, edited by A. B. Kampunzu, and R. T. Lubala, Springer-Verlag. Berlin, p. 147-188, 1991.

Bertrand, H., Dostal, J., and Dupuy, C., Geochemistry of early Mesozoic tholeiites from Morocco, *Earth and Planetary Science Letters*, v. 58, 2, p. 225-239, 1982

Birney De Wet, C. C., Hubert, J. F., The Scots Bay Formation, Nova Scotia, Canada, a Jurassic carbonate lake with silica-rich hydrothermal springs, *Sedimentology*, v. 36(5), p. 857-873, 1989.

Buffet, G., Notice feuille St. Bonnet, Carte géologique de la France à 1/50.000, Coordination M. Gidon, 1981.

Carmichael, C. M. and Palmer, H. C., Paleomagnetism of the Late Triassic, North Mountain basalt of Nova Scotia, *Journal of Geophysical Research*, v. 73, p. 2811-2822, 1968.

Clement, B. M. and Kent, D. V., A detailed record of the lower Jaramillo polarity transition from a Southern Hemisphere, deep-sea sediment core, *Journal of Geophysical Research*, B, v. 89, no. 2, p. 1049-1058, 1984.

Colbert, E. H., Tetrapod extinctions at the end of the Triassic period, *Proceedings of the National Academy of Sciences of the United States of America*, v. 44. p. 973-977, 1958.

Cornet, B., *The Palynostratigraphy and Age of the Newark Supergroup*, Ph.D. thesis, Pennsylvania State University, University Park, 1977.

Cornet, B. and Olsen, P. E., A summary of the biostratigraphy of the Newark Supergroup of eastern North America, with comments on early Mesozoic provinciality, In *Symposio Sobre Flores del Triasico Tardio st Fitografia y Paleoecologia, Memoria. Proc. II) Latin-American Congress on Paleontology (i984)*, edited by R. Weber, Instituto de Geologia Universidad Nacional Autonoma de Mexico, Mexico City, p. 67-81, 1985.

Courtillot, V., Jaupart, C., Manighetti, I., Tapponnier, P., and Besse, J., , On causal links between flood basalts and continental breakup, *Earth and Planetary Science Letters*, v. 166, p. 177-195, 1999.

Dallmeyer, R. D., The Palisades sill; a Jurassic intrusion? Evidence from ^{40}Ar/^{39}Ar incremental release ages, *Geology*, v. 3, p. 243-245, 1975.

Dalrymple, G. B., Gromme, C. S, and White, R. W., Potassium-argon age and paleomagnetism of diabase dikes in Liberia; initiation of central Atlantic rifting, *Geological Society of America Bulletin*, v. 86; 3, Pages 399-411, 1975.

De Boer, J., Paleomagnetic-tectonic study of Mesozoic dike swarms in the Appalachians: Journal of Geophysical Research, v. 72, p. 2237-2250, 1967.

De Boer, J., Paleomagnetic differentiation and correlation of the Late Triassic volcanic rocks in the central Appalachians (with special reference to the Connecticut Valley): Geological Society of America Bulletin, v. 79, p. 609-626, 1968.

Deckart, K., Feraud, G., and Bertrand, H., Age of Jurassic continental tholeiites of French Guyana, Surinam and Guinea; implications for the initial opening of the central Atlantic Ocean, *Earth and Planetary Science Letters*, v. 150, p. 205-220, 1997.

Diallo, D., Bertrand, H., Azambre, B., Gregoire, M., and Caseiro, J., Le complexe basique-ultrabasique du Kakoulima (Guinee-Conakry); une intrusion tholeiitique tholeiitique stratifiee liee au rifting de l'Atlantique central, *Comptes Rendus de l'Academie des Sciences, Serie 2, Mecanique, Physique, Chimie, Sciences de l'Univers, Sciences de la Terre*, v. 314, p. 937-943, 1992.

Dillon, W. P., Klitgord, K. D, and Paull, C. K., Mesozoic development and structure of the continental margin off South Carolina, *U.S. Geological Survey Professional Paper 1313*, p. N1-N16, 1983.

Dubar, J., Le Lias de Pyrénées françaises. Mèm. Soc. Gèol. Nord., v. 9, p. 38-66. 1925.

DuBois, P. M., Irving, E., Opdyke, N. D., Runcorn, S. K., and Banks, M. R., The geomagnetic field in Upper Triassic times in the United States, *Nature*, v. 180, p. 1186-1187, 1957.

Dumont, T., Sea-level changes and early rifting of a European Tethyan margin in the western Alps and southeastern France, in *Mesozoic and Cenozoic sequence stratigraphy of European basins*, edited by P.-C. de Gracianshy, J. Hardenbol, T. Jacquin, and P. R. Vail, SEMP Special Publication No. 60, pp. 623-640, 1998.

Dunning, G. R., and Hodych, J. P., U/Pb zircon and baddeleyite ages for the Palisades and Gettysburg sills of the northeastern United States: Implications for the age of the Triassic/Jurassic boundary, *Geology*, v. 18, p. 795-798, 1990.

Et-Touhami, M., Lithostratigraphy and depositional environments of Lower mesozoic evaporites and associated red beds, Khemisset Basin, northwestern Morocco, in *Epicontinental Triassic, Volume 2*, edited by G. Bachmann and I, Lerche, Zentralblatt fur Geologie und Palaontologie, VIII, p. 1217-11241. 2000.

Fedosh, M. S., and Smoot, J. P., A cored stratigraphic section through the northern Newark basin, New Jersey, *U.S. Geological Survey Bulletin 1776*, p. 19-24, 1988.

Fowell, S. J., Cornet, B., and Olsen, P. E., Geologically rapid Late Triassic extinctions: Palynological evidence from the Newark Supergroup, in *Pangaea: Paleoclimate, Tectonics and Sedimentation During Accretion, Zenith and Break-up of a Supercontinent*, edited by G. D. Klein, Geological Society of America Special Paper 288, p. 197-206, 1994.

Fowell, S. J. and Traverse, A., Late Triassic palynology of the Fundy basin, Nova Scotia and New Brunswick, *Review of Palaeobotany and Palynology*, v. 86, p. 211-233, 1995.

Gallet, Y., Besse, J., Krystyn, L., Marcoux, J., and Theveniaut, H., Magnetostratigraphy of the Late Triassic Bolucektasi Tepe section (southwestern Turkey): Implications for changes in mag-

netic reversal frequency, *Physics of the Earth and Planetary Interiors*, v. 73, p. 85-108, 1992.

Gottfried, D., Annell, C. S., and G. R. Byerly, Geochemistry and tectonic significance of subsurface basalts from Charleston, South Carolina: Clubhouse Crossroads test holes #2 and #3, *US Geol. Survey Professional Paper 1313*, p. A1-A19, 1983.

Gradstein, F. M., Agterberg, F. P., Ogg, J. G., Hardenbol, J., van Veen, P. V., Thierry, J., and Huang, Z., A Triassic, Jurassic and Cretaceous time scale, *SEPM Special Publication no. 54*, p. 95-126, 1995.

Hames, W. E., Renne, P. R., and Ruppel, C., New evidence for geologically instantaneous emplacement of earliest Jurassic Central Atlantic magmatic province basalts on the North American margin, *Geology*, v. 28, p. 859–862, 2000.

Haq, B. U., Hardenbol, J., and Vail, P. R., Mesozoic and Cenozoic chronostratigraphy and cycles of sea-level change, *Society of Economic Paleontologists and Mineralogists Special Publication*, v. 42, p. 71-108, 1988.

Hargraves, R. B., Briden, J. C., and Daniels, B. A., Paleomagnetism and magnetic fabric in the Freetown Complex, Sierra Leone, *Geophysical Journal International*, v. 136, p. 705-713, 1999.

Heilman, J. J., That catastrophic day in the Early Jurassic, *Connecticut Journal of Science Education*, v. 25, p. 8-25, 1987.

Hofman, A., Tourani, A., and Gaupp, R., 2000, Cyclicity of Triassic to Lower Jurassic continental red beds of the Argana Valley, Morocco: implications for paleoclimate and basin evolution. Palaeogeography, Palaeoclimatology, Palaeoecology, v. 161, p. 229-266, 2000.

Hodych, J. P., and Hayatsu, A., Paleomagnetism and K-Ar isochron dates of Early Jurassic basaltic flows and dikes of Atlantic Canada, *Canadian Journal of Earth Sciences*, v. 25, p. 1972-1989, 1988.

Holbrook, W. S. and Kelemen, P. B., Large igneous province on the US Atlantic margin and implications for magmatism during continental breakup, *Nature*, v. 364, p. 433-437, 1993.

Holbrook, W. S., Reiter, E. C., Purdy, G. M., Sawyer, D., Stoffa, P. L., Austin, J. A., Jr., Oh, J., Makris, J., Deep Structure of the U.S. Atlantic continental margin, offshore South Carolina, from coincident ocean bottom and multichannel seismic data, *Journal of Geophysical Research*, v. 99(B5), p. 9155-9178, 1994.

Holser, W. T., Clement, G. P., Jansa, L. F., Wade, J. A., Evaporite deposits of the North Atlantic Rift. in *Triassic-Jurassic rifting; continental breakup and the origin of the Atlantic Ocean and passive margins, volume B*, edited by W. Manspeizer, Developments in Geotectonics 22(A-B), p. 525-556, 1988.

Huber, P., Lucas, S. G., Hunt, A. P., Vertebrate biochronology of the Newark Supergroup Triassic, eastern North America, in *The Nonmarine Triassic*, edited by S. G. Lucas and M. Morales, Bulletin of the New Mexico Museum of Natural History and Science 3, p. 179-186, 1993.

Hubert, J. F. and Dutcher, J. A., Sedimentation, volcanism, stratigraphy, and tectonism at the Triassic-Jurassic boundary in the Deerfield basin, Massachusetts, *Northeastern Geology and Environmental Sciences*, v. 21, p. 188-201, 1999.

Irving, E. and Banks, M. R., Paleomagnetic results from the Upper Triassic lavas of Massachusetts, *Journal of Geophysical Research*, v. 66, p. 1935-1939, 1961.

Irving, E. and Irving, G. A., Apparent polar wander paths Carboniferous through Cenozoic and the assembly of Gondwana, *Geophysical Surveys*, v. 5, p. 141-188, 1982.

Irving, E. and Pullaiah, G., Reversals of the geomagnetic field, magnetostratigraphy, and relative magnitude of paleosecular variation in the Phanerozoic, *Earth Science Reviews*, v. 12, p. 35-64, 1976.

Johnson, H. P., Van Patten, D., Tivey, M. A., and Sager, W. W., Geomagnetic polarity reversal rate for the Phanerozoic, *Geophysical Research Letters*, v. 22, p. 231-234, 1995.

Kent, D. V. and Olsen, P. E., Magnetostratigraphy and paleopoles from the Late Triassic Dan River-Danville basin: interbasin correlation of continental sediments and a test of the tectonic coherence of Newark rift basins in eastern North America, *Geological Society of America Bulletin*, v. 109(3), p. 366-377, 1997.

Kent, D. V. and Olsen, P. E., Astronomically tuned geomagnetic polarity time scale for the Late Triassic, *Journal of Geophysical Research*, v. 104, p. 12,831-12,841, 1999a.

Kent, D. V. and Olsen, P. E., Search for the Triassic/Jurassic long normal and the J1 cusp: *Eos, Transactions, American Geophysical Union, Supplement*, v. 80(46), p. F306, 1999b.

Kent, D. V. and Olsen, P. E., Magnetic polarity stratigraphy and paleolatitude of the Triassic--Jurassic Blomidon Formation in the Fundy basin (Canada): implications for early Mesozoic tropical climate gradients. *Earth And Planetary Science Letters*, v. 179, no. 2. p. 311-324, 2000a.

Kent, D. V. and P. E. Olsen, Implications of a new astronomical time scale for the Late Triassic, In *Epicontinental Triassic, Volume 3*, edited by G. Bachmann and I. Lerche, Zentralblatt fur Geologie und Palaontologie, VIII, p. 1463-1474, 2000b.

Kent, D. V., Olsen, P. E., and Witte, W. K., Late Triassic-earliest Jurassic geomagnetic polarity sequence and paleolatitudes from drill cores in the Newark rift basin, eastern North America, *Journal of Geophysical Research*, v. 100, p. 14,965-14,998, 1995.

Klitgord, K. D. and H. Schouten, *Plate kinematics of the central Atlantic*, 351-378 pp., Geological Society of America, Boulder, CO, 1986.

Lacroix, A., Les enclaves des roches volcaniques, *Ann. Acad. Macon*, v. 10, 710 pp., 1893.

Lanphere, M. A., $^{40}Ar/^{39}Ar$ ages of basalt from Clubhouse Crossroads test hole near Charleston, South Carolina, *US Geological Survey Professional Paper 1313*, p. B1-B8, 1983.

Larochelle, A. and Wanless, R. K., The paleomagnetism of a Triassic diabase dike in Nova Scotia, *Journal of Geophysical Research*, v. 71, p. 4949-4953, 1966.

Laskar, J., The chaotic motion of the solar system: a numerical estimate of the size of the chaotic zones. *Icarus*, v. 88, p. 266-291, 1990.

Laskar, J., The limits of Earth orbital calculations for geological time-scale use. *Philosophical Transactions of the Royal Society of London (series A)*, v. 357, no. 1757, p. 1735-1759, 1999.

Leonardi, G, *Annotated atlas of South America tetrapod footprints (Devonian to Holocene)*, Brasilia, Companhia de Pesquisa de Recursos Minerals, 248 p., 1994.

LeTourneau, P. M. and Huber, P., Early Jurassic eolian dune field, Pomperaug rift basin, Connecticut: implications for Pangean paleoclimate and paleogeography (in review) *Sedimentary Geology*, 2001.

Lindholm, R. C., Geologic history and stratigraphy of the Triassic-Jurassic Culpeper Basin, Virginia. Geological Society of America Bulletin v. 90(11), p. I 995-I 997, II 1702-II 1736, 1979.

Lucas, S. G. and Huber, P., Vertebrate Biostratigraphy and Biochronology of the nonmarine Late Triassic, in *The Great Rift Valleys of Pangea in Eastern North America, vol .2: Sedimentology, Stratigraphy, and Paleontology*, edited by P. M. LeTourneau and P. E. Olsen, Columbia University Press, New York (in press), 2002.

Marzoli, A., Renne, P. R, Piccirillo, E. M., Ernesto, M., Bellieni, G., and De-Min, A., Extensive 200-million-year-old continental flood basalts of the Central Atlantic Magmatic Province, *Science*, v. 284, p. 616-618, 1999.

McBride, J. H., Nelson, K. D., Brown, L. D., Evidence and implications of an extensive early Mesozoic rift basin and basalt/ diabase sequence beneath the southeast coastal plain, *Geological Society of America Bulletin*, v. 101, p. 512-520, 1989.

McElhinny, M. W. and Burek, P. J., Mesozoic paleomagnetic stratigraphy, *Nature*, v. 232, p. 98-102, 1971.

McIntosh, W. C., Hargraves, R. B., and West, C. L., Paleomagnetism and oxide mineralogy of upper Triassic to lower Jurassic red beds and basalts in the Newark Basin, *Geological Society of America Bulletin*, v. 96, p. 463-480, 1985.

Moret, L. and Manquat, G., Sur un gisement fossilifère remarquable du Lias inférier du Grand Serre, près de Grenoble: Comptes Rendus Somaires Société géologiques France, v. 38, p. 316-317, 1948.

Oh, J., Austin, J. A., Jr., Phillips, J. D., Coffin, M. F., and Stoffa, P. L., Seaward-dipping reflectors offshore the Southeastern United States; seismic evidence for extensive volcanism accompanying sequential formation of the Carolina Trough and Blake Plateau basin, *Geology*, v. 23p. 9-12, 1995.

Olsen, P. E., Periodicity of lake-level cycles in the Late Triassic Lockatong Formation of the Newark Basin (Newark Supergroup, New Jersey and Pennsylvania). In *Milankovitch and Climate, NATO Symposium, Pt. 1*, edited by A. Berger, J. Imbrie, J. Hays, G. Kukla, and B. Saltzman, D. Reidel Publishing Co., Dordrecht, p. 129-146, 1984.

Olsen, P. E., A 40-million-year lake record of early Mesozoic climatic forcing. *Science*, v. 234, p. 842-848, 1986.

Olsen, P. E., Stratigraphic record of the early Mesozoic breakup of Pangea in the Laurasia-Gondwana rift system. *Annual Reviews of Earth and Planetary Science*, v. 25, p. 337-401, 1997.

Olsen, P. E., Giant Lava Flows, Mass Extinctions, and Mantle Plumes. *Science* v. 284, no. 5414, p. 604 - 605, 1999.

Olsen, P. E., Cyclostratigraphic controls on the duration and correlation of the Triassic-Jurassic mass extinction and associated basalts, *Geological Society of America, Abstracts with Programs*, v. 32(1), p. A-63, 2000.

Olsen, P. E, Et-Touhami, M.; Kent, D. V., Fowell, S. J., Schlische, R. W., Withjack, M. O., LeTourneau, P. M., and Smoot, J. P., Stratigraphic, paleoclimatic, and tectonic implications of a revised stratigraphy of the Fundy rift basin, Maritimes, Canada and a comparison with the Argana Rift Basin, Morocco (Triassic-Jurassic), *Canadian Journal of Earth Science* (in prep.), 2002a.

Olsen, P. E., Fowell, S. J., and Cornet, B., The Triassic-Jurassic boundary in continental rocks of eastern North America: a progress report: in Global Catastrophes in Earth History; an Interdisciplinary Conference on Impacts, Volcanism, and Mass Mortality, edited by V. L. Sharpton and P. D. Ward, *Geological Society of America Special Paper 247*, p. 585-593, 1990.

Olsen, P. E. and Kent, D. V., Milankovitch climate forcing in the tropics of Pangea during the Late Triassic. *Palaeogeography, Palaeoclimatology, and Palaeoecology*, v. 122, p. 1-26, 1996.

Olsen, P. E. and Kent, D. V., Long-period Milankovitch cycles from the Late Triassic and Early Jurassic of eastern North America and their implications for the calibration of the early Mesozoic time scale and the long-term behavior of the planets, *Philosophical Transactions of the Royal Society of London (series A)*, v. 357, p. 1761-1787, 1999.

Olsen, P. E. and Kent D. V., High resolution early Mesozoic Pangean climatic transect in lacustrine environments, in *Epicontinental Triassic, Volume 3*, edited by G. Bachmann and I, Lerche, Zentralblatt fur Geologie und Palaontologie, v. 1998, n°. 11-12,, p. 1475-1496, 2000.

Olsen, P. E., Kent, D V., Cornet, B., Witte, W. K., and Schlische, R. W., High-resolution stratigraphy of the Newark rift basin (Early Mesozoic, Eastern North America). *Geological Society of America*, v. 108, p. 40-77, 1996a.

Olsen, P. E. and Kent, D. V., Fowell, S. J., Schlische, R. W., Withjack, M. O., and LeTourneau, P. M., Implications of a comparison of the stratigraphy and depositional environments of the Argana (Morocco) and Fundy (Nova Scotia, Canada) Permian-Jurassic basins, in *Le Permien et le Trias du Maroc, Actes de la Premièr Réunion su Groupe Marocain du Permien et du Trias*, edited by M. Oujidi and M. Et-Touhami, Hilal Impression, Oujda, p. 165-183, 2000.

Olsen, P. E., Kent, D. V., H.-D. Sues,, Koeberl, C., Huber, H., Montanari, A., Rainforth, E. C., Fowell, S. J., Szajna, M. J., and Hartline, B. W., Ascent of Dinosaurs Linked to an Iridium Anomaly at the Triassic-Jurassic Boundary, *Science*, v. 296, p. 1305-1307, 2002b.

Olsen, P. E., Koeberl, C., Huber, H., Montanari, A., Fowell, S. J., EtTouhami, M., and Kent, D. V., The continental Triassic-Jurassic boundary in central Pangea: recent progress and discussion of an Ir anomaly, *Geological Society of America Special Paper* (in press), 2002c.

Olsen, P. E., McDonald, N. G., Huber, P., and Cornet, B., Stratigraphy and Paleoecology of the Deerfield rift basin (Triassic-Jurassic, Newark Supergroup), Massachusetts, in *Guidebook for Field Trips in the Connecticut Valley Region of Massachusetts and Adjacent States (vol. 2): New England Intercollegiate Geological Conference 84th Annual Meeting, Contribution no. 66*, edited by P. Robinson and J. B. Brady, Department of Geology and Geography, University of Massachusetts, Amherst, Massachusetts, p. 488-535, 1992.

Olsen, P. E. and Rainforth, E., The Early Jurassic ornithischian dinosaurian ichnite *Anomoepus*. in *The Great Rift Valleys of Pangea in Eastern North America, vol .2: Sedimentology, Stratigraphy, and Paleontology*, edited by P. M. LeTourneau and P. E. Olsen, Columbia University Press (in press), 2002.

Olsen P. E, Schlische R. W., and Fedosh M. S., 580 ky duration of the Early Jurassic flood basalt event in eastern North America estimated using Milankovitch cyclostratigraphy, in *The Continental Jurassic*, edited by M. Morales, Museum of Northern Arizona Bulletin 60, p. 11-22, 1996b.

Olsen, P. E., Schlische, R. W., and Gore, P. J. W. (eds.), *Field Guide to the Tectonics, stratigraphy, sedimentology, and paleontology of the Newark Supergroup, eastern North America*, International Geological Congress, Guidebook for Field Trips T351, 1989.

Olsen, P. E., Shubin, N. H. and Anders, P. E., New Early Jurassic tetrapod assemblages constrain Triassic-Jurassic tetrapod extinction event, *Science*, v. 237, p. 1025-1029, 1987.

Olsen, P. E., Sues, H.-D., and Kent, D. V., Constraining the Timing and Magnitude of the Triassic-Jurassic Mass Extinction in Continental Ecosystems. *Eos, Transactions, American Geophysical Union, Supplement*, v. 80(46), p. F50, 1999.

Opdyke, N. D., The paleomagnetism of the New Jersey Triassic: A field study of the inclination error in red sediments, *Journal of Geophysical Research*, v. 66, p. 1941-1949, 1961.

Orti Cabo, F., Evaporites of the Western Mediterranean, in *Triassic of the Mediterranean Region, Volume I, Geological framework, concepts and exploration considerations*, edited by G. F. Stewart and J. W. Shelton, and others, ERICO, London, pp. I199-I244, 1983.

Oujidi, M., Courel, L., Benaouiss, N., El Mostaine, M., El Youssi, M., Et Touhami, M., Ouarhache, D., Sabaoui, A. Tourani, A., Triassic series of Morocco: stratigraphy, palaeogeography and structuring of the southwestern peri-Tethyan platform. An overview, in *Peri-Tethys Memoir 5: New Data on Peri-Tethyian Sedimentary Basins*, edited by S. Crasquin-Soleau and E. Barrier, Mém. mus. natn. Hist. nat., v. 182, p. 23-38, 2000a.

Oujidi, M., Courel, L., Benaouiss, N., El Mostaine, M., El Youssi, M., Et Touhami, M., Ouarhache, D., Sabaoui, A., and Tourani, A., Moroccan palaeogeographic maps during Early Mesozoic times, in *Le Permien et le Trias du Maroc*, edited by M. Oujidi and M. Et-Touhami, Actes de la Première Réunion du Groupe Marocain du Permien et du Trias, Hilal Impression, Oujda, p. 15-24, 2000b.

Oujidi, M. and Et-Touhami, M., Stratigraphy of Perman and Triassic systems in Morocco: an overview, in *Le Permien et le Trias du Maroc*, edited by M. Oujidi and M. Et-Touhami, Actes de la Première Réunion du Groupe Marocain du Permien et du Trias, Hilal Impression, Oujda, p. 1-13, 2000.

Oujidi, M., Et-Touhami, M., and Azzouz, O., Contexte géodynamique stratigraphie et sédimentologie des formations triassiques des Monts d'Oujda (Moroc oriental): Livret guide de l'excursion, in *Le Permien et le Trias du Maroc*, edited by M. Oujidi and M. Et-Touhami, Actes de la Première Réunion du Groupe Marocain du Permien et du Trias, Hilal Impression, Oujda, p. 145-164, 2000c.

Palain, C., Age et paléogéographique de la base du Mésozoiques (Série des Gres de Silves) de l'Agarve-Portugal Meridional. Cuad. Geol. Ibérica, v. 4, p. 259-268, 1977.

Palfy, J., Smith, P. L., Mortensen, J. K., A U-Pb and $^{40}Ar/^{39}Ar$ time scale for the Jurassic, *Canadian Journal of Earth Sciences*, v. 37, p. 923-944, 2000.

Pechersky, D. M. and Khramov, A. N., Mesozoic palaeomagnetism scale of the USSR, *Nature*, v. 244, p. 499-501, 1973.

Pe-Piper, G., Jansa, L. F., and Lambert, R. St-J., Early Mesozoic magmatism of the Eastern Canadian margin; petrogenetic and tectonic significance, in *Eastern North American Mesozoic magmatism*, edited by J. H. Puffer and P. C. Ragland, Geological Society of America Special Paper, v. 268, p. 13-36, 1992.

Pienkowski, A. and Steinen, R., P., *Geological Society of America, Abstracts with Programs*, v. 27(1), p. 74, 1995.

Phillips, J. D., Paleomagnetic investigations of the Clubhouse Crossroads Basalt, *US Geol. Survey Professional Paper 1313*, p. C1-C18, 1983.

Prevot, M. and McWilliams, M., Paleomagnetic correlation of the Newark Supergroup volcanics, *Geology*, v. 17, p. 1007-1010, 1989.

Puffer, J. H., Eastern North American flood basalts in the context of the incipient breakup of Pangea. in *Eastern North American Mesozoic magmatism*, edited by J. H. Puffer and P. C. Ragland, Geological Society of America Special Paper 268, p. 95-118, 1992.

Puffer, J. H., Early Jurassic basalts of the Algarve basin, Portugal: *Carboniferous to Jurassic Pangea, Program and Abstracts, Canadian Society of Petroleum Geologists, Annual Convention, Calgary, Canada*, Canadian Society of Petroleum Geologists, Calgary, p. 254, 1993.

Puffer, J. H., 1994, Initial and secondary Pangaean basalts, in *Pangaea: Global Environments and Resources*, edited by A. F. Embry, B. Beauchamp, and D. J. Glass, Canadian Society of Petroleum Geology Memoir 17, p. 85-95, 1994.

Puffer, J. H., and Philpotts, A. R., Eastern North American quartz tholeiites: Geochemistry and petrology, in *Triassic-Jurassic rifting; continental breakup and the origin of the Atlantic Ocean and passive margins, volume B*, edited by W. Manspeizer, Developments in Geotectonics 22(A-B), p. 579-605, 1988.

Ragland, P. C., Cummins, L. E., and Arthur, J. D., Compositional patterns for early Mesozoic diabases from South Carolina to central Virginia, *Geological Society of America Special Paper 268*, p. 309-331, 1992.

Rainforth, E. C., The Early Jurassic ichnogenus *Otozoum*, *Geological Society of America, Abstracts with Programs*, v. 32, no. 1, p. 67, 2000.

Rainforth, E. C., Revision and reevaluation of the Early Jurassic prosauropod ichnogenus Otozoum, *Palaeontology* (in press), 2002.

Ratcliffe, N. M., Reinterpretation of the relationship of the western extension of the Palisades sill to the lava flows at Ladentown, New York, based on new core data, *US Geological Survey Bulletin*, v. 1776, p. 113-135, 1988.

Schilt, F. S., Brown, L. D., Oliver, J. E., and Kaufman, S., Subsurface structure near Charleston, South Carolina; results of CO-CORP reflection profiling in the Atlantic Coastal Plain, *US Geological Survey Professional Paper 1313*, p. H1-H19, 1983.

Seidemann, D. E., Masterson, W. D., Dowling, M. P., and Turekian, K. K., K-Ar dates and $^{40}Ar/^{39}Ar$ age spectra for Mesozoic basalt flows of the Hartford Basin, Connecticut, and the Newark Basin, New Jersey, *Geological Society of America Bulletin*, v. 95, p. 594-598, 1984.

Seidemann, D. E., The hydrothermal addition of excess ^{40}Ar to the lava flows from the Early Juraassic in the Hartford Basin (northeastern U.S.A.): Implications for the time scale, *Chemical Geology (Isotope Geoscience Section)*, v. 72, p. 37-45, 1988.

Seidemann, D. E., Age of the Triassic / Jurassic boundary; a view from the Hartford Basin, *American Journal of Science*, v. 289, p. 553-562, 1989.

Seidemann, D. E., Comment and reply on "U/Pb zircon and baddeleyite ages for the Palisades and Gettysburg sills of the northeastern United States: Implications for the age of the Triassic/Jurassic boundary", *Geology*, v. 19, p. 766-767, 1991.

Slimane, A. and El Mostaine, M., Observations biostratigraphiques au niveau des formations rouges del al sèquence synrift dans les bassins de Doukkala et Essaouira. in *1ém Réunion du Groupe marocain du Permien et du Trias, Oujda, Maroc.*, Faculté des Sciences, Université Mohammed I, Oujda, p. 54, 1997.

Smith, T. E. and Noltimier, H. C., Paleomagnetism of the Newark Trend igneous rocks of the north central Appalachians and the opening of the central Atlantic Ocean, *American Journal of Science*, v. 279, p. 778-807, 1979.

Smith, W. A., Paleomagnetic results from a crosscutting system of northwest and north-south trending diabase dikes in the North Carolina Piedmont, *Tectonophysics*, v. 136, p. 137-150, 1987.

Smoot, J. P.; Olsen, Paul, E., Massive mudstones in basin analysis and paleoclimatic interpretation of the Newark Supergroup. In *Triassic-Jurassic rifting; continental breakup and the origin of the Atlantic Ocean and passive margins, volume B*, edited by W. Manspeizer, Developments in Geotectonics 22(A-B), p. 249-274, 1988.

Sopeña, A., López, J., Arche, A., Pérez-Arlucea, Ramos, A., Virgilli, C. and Hernando, S., Permian and Triassic rift basins of the Iberian Peninsula. In *Triassic-Jurassic rifting; continental breakup and the origin of the Atlantic Ocean and passive margins, volume B*, edited by W. Manspeizer, Developments in Geotectonics 22(A-B), p. 757-786, 1988.

Stévaux, J. and Winnock, E., Les basins du Trias et du Lias inférieur d,Aquitaine et leurs épisodes évaporitiques, Bull. Soc. Géol. France, v. 7a. Sér., 16, p. 679-695, 1974.

Sutter, J. F., Innovative approaches to the dating of igneous events in the early Mesozoic basins of the Eastern United States, *US Geological Survey Bulletin 1776*, p. 194-200, 1988.

Sutter, J. F., and Smith, T. E., $^{40}Ar/^{39}Ar$ Ages of Diabase Intrusions from Newark Trend Basins in Connecticut and Maryland: Initiation of Central Atlantic Rifting, *American Journal of Science*, v. 279, p. 808-831, 1979.

Szajna, M. J. and Hartline, B. W., A New vertebrate footprint locality from the Late Triassic Passaic Formation near Birdsboro, Pennsylvania, in *The Great Rift Valleys of Pangea in Eastern North America, vol .2: Sedimentology, Stratigraphy, and Paleontology*, edited by P. M. LeTourneau, P.M. and P. E. Olsen, Columbia University Press, New York (in press), 2002.

Talwani, M., Sheridan, R. E., Holbrook, W. S., and Glover, L., III, The Edge Experiment and the U.S. East Coast Magnetic Anomaly, in *Rifted Ocean-Continent Boundary*, edited by E. Banda, E., M. Torne, and M. Talwani, Kluwer Academic Publishers, Norwell, p. 155-181, 1995.

Turrin, B. D., $^{40}Ar/^{39}Ar$ mineral ages and potassium and argon systematics from the Palisade sill, New York, *EOS, Transactions, American Geophysical Union*, v. 81, no. 48, p. F1326, 2000.

Van Houten, F. B, Cyclic sedimentation and the origin of analcime-rich Upper Triassic Lockatong Formation. west central New Jersey and adjacent Pennsylvania. *American Journal of Science*, v. 260, p. 561-576., 1962

Van Houten, F. B., Cyclic lacustrine sedimentation, Upper Triassic

Lockatong Formation, central New Jersey and adjacent Pennsylvania. In *Symposium on Cyclic Sedimentation*, edited by D. F. Meriam, Kansas Geological Survey Bulletin, v. 169, p. 497-531, 1964.

Van Houten, F. B., Late Triassic Newark Group, north-central New Jersey and adjacent Pennsylvania and New York. In *Geology of Selected Areas in New Jersey and Eastern Pennsylvania and Guidebook of Excursions*, edited by S, Subitzki, Geol. Soc. Am., Field Trip 4, Atlantic City, NJ. Rutgers University Press, New Brunswick, p. 314-347, 1969.

Van Houten, F. B. Late Triassic part of Newark Supergroup, Delaware River section, west central New Jersey. In *Field Studies of New Jersey Geology and Guide to Field Trips , 52nd Annual Meeting, Rutgers University, Newark, NJ.* edited by W. Manspeizer, New York. State Geological Association, New York, p. 264-269, 1980.

van Veen, P. M., Fowell, S. J., and Olsen, P. E., Time calibration of Triassic/Jurassic microfloral turnover, eastern North America; discussion and reply, *Tectonophysics*, v. 245, p. 93-99, 1995.

Withjack, M. O., Schlische, R. W., and Olsen, P. E., Diachronous rifting, drifting, and inversion on the passive margin of Eastern North America: An analog for other passive margins, *American Association of Petroleum Geologists Bulletin*, v. 82(5A), p. 817-835, 1998.

Witte, W. K. and Kent, D.V., The paleomagnetism of red beds and basalts of the Hettangian Extrusive Zone, Newark Basin, New Jersey, *Journal of Geophysical Research*, v. 95, p. 17,533-17,545, 1990.

Witte, W. K., and Kent, D. V., Tectonic implications of a remagnetization event in the Newark Basin, *Journal of Geophysical Research*, v. 96, p. 19,569-19,582, 1991.

Witte, W. K., Kent, D. V., and Olsen, P. E., Magnetostratigraphy and paleomagnetic poles from Late Triassic-earliest Jurassic strata of the Newark Basin, *Geological Society of America Bulletin*, v. 103, p. 1648-1662, 1991.

Yang, Z., Moreau, M.-G., Bucher, H., Dommergues, J.-L., and Trouiller, A., Hettangian and Sinemurian magnetostratigraphy from Paris Basin, *Journal of Geophysical Research*, v. 101, p. 8025-8042, 1996.

Relative Timing of CAMP, Rifting, Continental Breakup, and Basin Inversion: Tectonic Significance

Roy W. Schlische and Martha Oliver Withjack

Department of Geological Sciences, Rutgers University, Piscataway, New Jersey

Paul E. Olsen

Department of Earth and Environmental Sciences and Lamont-Doherty Earth Observatory of Columbia University, Palisades, New York

Short-duration CAMP magmatic activity at ~200 Ma in eastern North America provides a temporal benchmark for assessing the relative timing of rifting, drifting, and basin inversion. In the southeastern United States, rifting ceased and shortening/inversion began before CAMP magmatism. In the northeastern United States and southeastern Canada, rifting continued during and after CAMP magmatism. Rifting ceased in the northeastern United States and southeastern Canada by the early Middle Jurassic, after CAMP magmatic activity. Shortening/inversion occurred in southeastern Canada before or during the Early Cretaceous. The available geological, geophysical, and geochronological data favor a diachronous rift-drift transition (seafloor spreading began earlier in the south) rather than the traditional synchronous rift-drift transition along the entire central North Atlantic margin. In this scenario, there are two magmatic pulses. The first includes CAMP and the formation of seaward-dipping reflectors (SDR's) near the continent-ocean boundary during the rift-drift transition along the southern margin segment. The second, younger magmatic pulse is associated with the formation of SDR's during the rift-drift transition along the northern margin segment. We believe that the widespread magmatism and shortening/inversion in eastern North America are related to active asthenospheric upwelling that culminated during the rift-drift transition. Inversion is a common feature along many volcanic passive margins and is associated with a change in the strain state from extension at a high angle to the margin during rifting to shortening at a high angle to the margin during drifting. The presence of dikes oriented at a high angle to the trend of the margin (e.g., the dike swarms in the southeastern United States, southeastern Greenland, offshore northwest Europe, and South America) may reflect this change in strain state associated with inversion.

The Central Atlantic Magmatic Province:
Insights from Fragments of Pangea
Geophysical Monograph 136
Copyright 2003 by the American Geophysical Union
10.1029/136GM03

1. INTRODUCTION

The Mesozoic rift system and passive margin of eastern North America (Figure 1) are arguably the best exposed and best studied of the rift systems and passive margins that formed during the progressive (diachronous) breakup

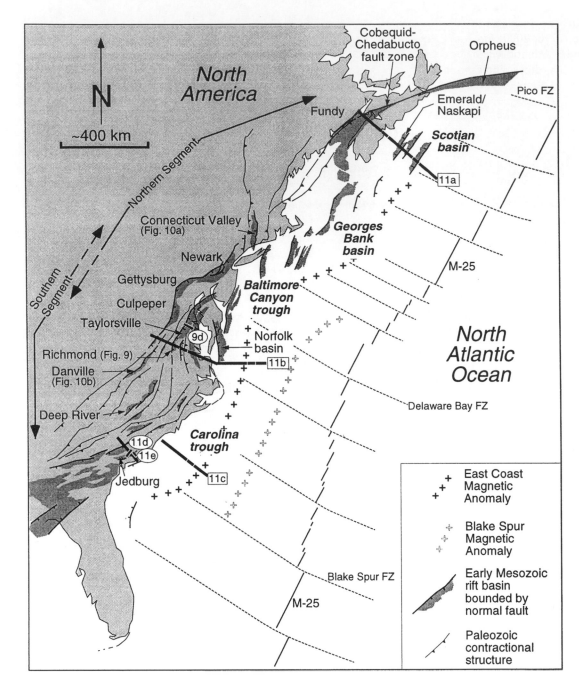

Figure 1. Major Paleozoic contractional structures, early Mesozoic rift basins of eastern North America, and key tectonic features of the eastern North Atlantic Ocean. Thick dashed lines and rectangles with notation show location of transects in Figure 11; solid lines and ellipses with notation show location of sections in Figures 9 and 11. Modified from *Withjack et al.* [1998].

of Pangea [e.g., *Dietz and Holden*, 1970]. The rift system and passive margin provide a natural laboratory for studying the extensional structures and rift basins associated with continental rifting, breakup, and initiation of seafloor spreading [e.g., *Manspeizer*, 1988; *Sheridan and Grow*, 1988; *Schlische*, 1993; *Withjack et al.*, 1998; *Schlische*, 2002]. The early Mesozoic is also a critical period in Earth history, as the Triassic-Jurassic boundary marks the second largest mass extinction [e.g., *Sepkowski*, 1997] and is associated with one of the largest (at least in terms of area) large-igneous provinces (LIP's) in the world [e.g., *McHone*, 1996, 2000; *Marzolli et al.*, 1999; *Olsen*, 1999]. The Central Atlantic Magmatic Province (CAMP) includes flood basalts, dikes, and associated intrusive sheets generally dated at about 200 Ma [e.g., *Olsen et al.*, 1996b; *Olsen*, 1999]. The eastern North American passive margin is also host to volcanic/volcaniclastic wedges (seaward-dipping reflectors, SDR's) [e.g., *Sheridan et al.*, 1993; *Holbrook and Kelemen*, 1993; *Oh et al.*, 1995; *Talwani et al.*, 1995] that make it a volcanic passive margin, although the exact temporal relationships between CAMP and the SDR's are not yet clear.

In recent years, a growing body of evidence indicates that the tectonic history of the eastern North American rift system and passive margin is not as simple as previously thought. *Withjack et al.* [1995, 1998] drew attention to a phase of contractional deformation and basin inversion that occurred during and/or shortly after the rift-drift transition. Although postrift contractional deformation has long been recognized in the Mesozoic rift system [e.g., *Shaler and Woodworth*, 1899; *Sanders*, 1963; *deBoer and Clifton*, 1988; *Wise*, 1993], the structures documented by *Withjack et al.* [1995, 1998] are much more widespread and represent much more shortening than previously reported. More importantly, *Withjack et al.* [1998] proposed that the initiation of postrift contractional deformation and basin inversion was diachronous along the central North American margin. Rifting ended and postrift deformation began earlier in the south than in the north. *Withjack et al.* [1998] hypothesized that the postrift contractional deformation is associated with processes occurring during the rift-drift transition. This has two important implications: 1. If the initiation of postrift contractional deformation and basin inversion is diachronous along the margin, then the initiation of seafloor spreading is also diachronous. 2. The causes of inversion may be temporally and causally related to the processes contributing to the formation of SDR's at the rift-drift transition.

This paper explores the early Mesozoic tectonic history of eastern North America for the region between Georgia and Nova Scotia. We first review basic concepts regarding the tectonic history of eastern North America and define key terminology. We then describe critical geological and geophysical data and observations from eastern North America regarding CAMP, the initiation and cessation of rifting, the initiation of postrift deformation and basin inversion, and the initiation of seafloor spreading; this section updates and extends the findings of *Withjack et al.* [1998]. We then discuss two possible interpretations of these geological and geophysical data, the possible relationship between inversion and a Middle Jurassic hydro-

thermal event, and uncertainties regarding the shortening direction. We tentatively apply our model to other passive margins, and discuss the causes of inversion in passive-margin settings.

2. BASIC CONCEPTS AND DEFINITIONS

The eastern North American rift system (Figure 1), which formed as a result of the continental extension preceding the separation of North America and Africa, consists of a series of exposed and buried rift basins that extends from the southeastern United States to the region of the Grand Banks, Canada [e.g., *Olsen*, 1997]. The exclusively continental strata and lava flows that fill the exposed basins are collectively known as the Newark Supergroup (Figure 2) [*Froelich and Olsen*, 1984]. We divide the rift system into a northern segment and a southern segment. The diffuse boundary between the two segments separates rift basins containing Early Jurassic-age strata (northern segment) from those lacking Early Jurassic-age strata (southern segment) (Figure 2). Offshore, the Blake Spur magnetic anomaly dies out to the north and the East Coast magnetic anomaly shifts to the east in the boundary region (Figure 1).

The basins of the eastern North American rift system are mostly half graben bounded by predominantly normal-slip border faults, although some basins or subbasins are bounded by oblique-slip faults [e.g., *Manspeizer*, 1988; *Olsen and Schlische*, 1990; *Schlische*, 1993, 2002] (Figure 3). The rocks within and surrounding the rift basins may be subdivided into prerift, synrift, and postrift units. Prerift rocks were present prior to the start of basin subsidence, and include Precambrian basement and mildly to strongly deformed Paleozoic sedimentary and metasedimentary units. These rocks were deformed in the Appalachian orogenies preceding the assembly of Pangea, and many of the border faults of the rift basins are reactivated Paleozoic structures [e.g., *Lindholm*, 1978; *Swanson*, 1986; *Ratcliffe and Burton*, 1985; *Ratcliffe et al.*, 1986]. Synrift units are rocks that accumulated while the basins were actively subsiding. These units generally dip and thicken toward the border fault and display a progressive decrease in dip upsection, thereby defining a wedge-shaped unit. Coarse-grained facies may be present in synrift strata immediately adjacent to the border-fault system. Postrift units accumulated after the end of fault-controlled subsidence during thermal subsidence following breakup.

Although controversy surrounds the precise definition of basin inversion [*Williams et al.*, 1989], it is generally agreed that a reversal in the deformational style produces inversion structures. Specifically, an extensional phase

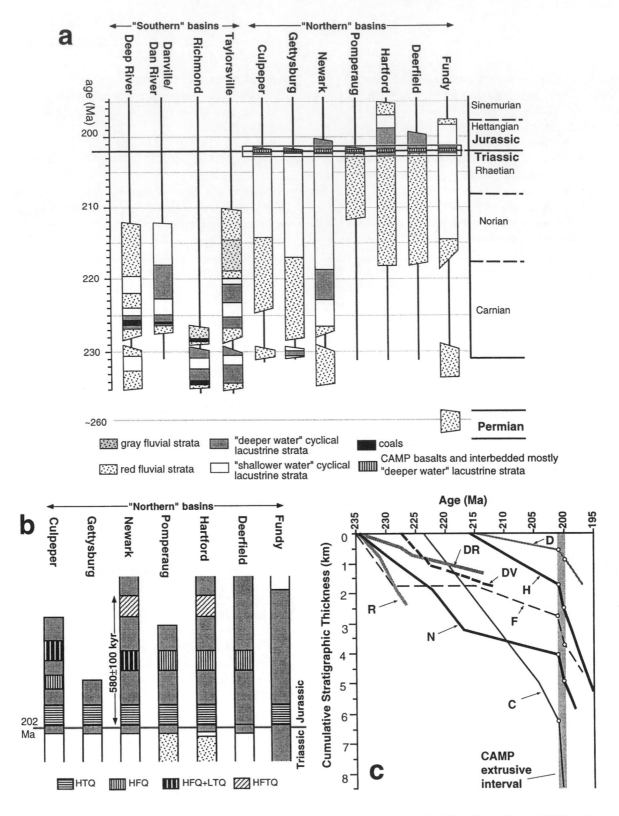

Figure 2. **a)** Stratigraphic architecture of eastern North American rift basins. Modified from *Olsen* [1997] and *Schlische* [2002]. **b)** Detail of earliest Jurassic extrusive interval (CAMP) showing correlation of basalt types by geochemistry. HTQ is high-titanium quartz-normative; HFQ is high-iron quartz-normative; LTQ is low-titanium quartz-normative; and HFTQ is high-iron and titanium quartz-normative. The duration of extrusive interval in the Newark basin, 580±100kyr, is based on cyclostratigraphy [*Olsen et al.*, 1996b]. Modified from *Olsen* [1997]. **c)** Cumulative stratigraphic thickness versus geologic age for various exposed rift basins in eastern North America. The Culpeper (C), Deerfield (D), Fundy (F), Hartford (H), and Newark (N) basins all show pronounced increases in stratal thickness in earliest Jurassic time (CAMP extrusive interval). Other abbreviations are DR, Deep River; DV, Danville; and R, Richmond basins. Modified from *Schlische and Anders* [1996].

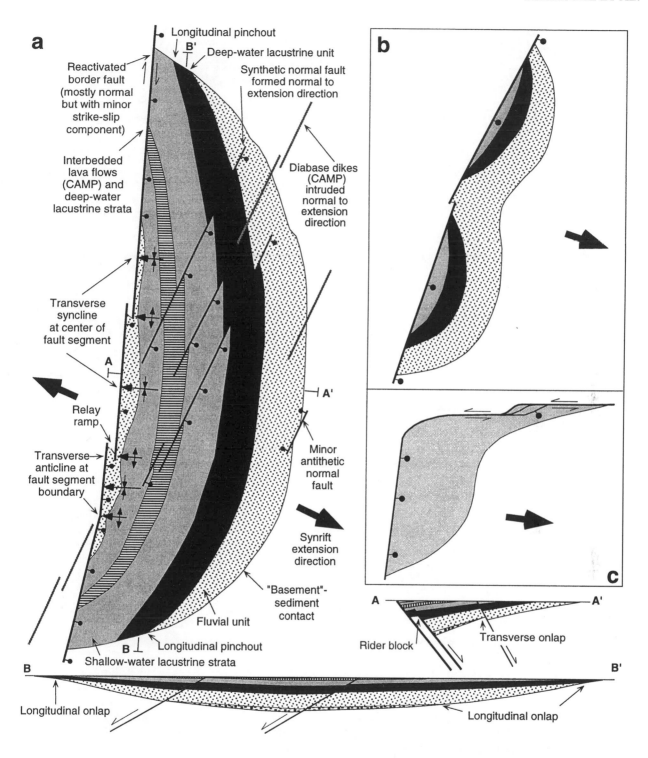

Figure 3. a) Geologic map and cross sections of an idealized, dip-slip dominated Mesozoic rift basin in eastern North America. Note that synrift units thicken toward the border fault in tranverse section and thicken toward the center of the basin in longitudinal section. **b)** Geologic map of idealized rift basin containing multiple subbasins related to large-scale segmentation of border-fault system. **c)** Geologic map of basin with both dip-slip and strike-slip-dominated margins. The basins shown in a) and b) apply strictly to the northern rift segment. The basins in the southern rift segment do not contain Jurassic lava flows or strata, and dikes are oriented at a high angle to the trend of the basin. The idealized basin geometry shown here does not include the effects of basin inversion. Modified from *Schlische* [1993].

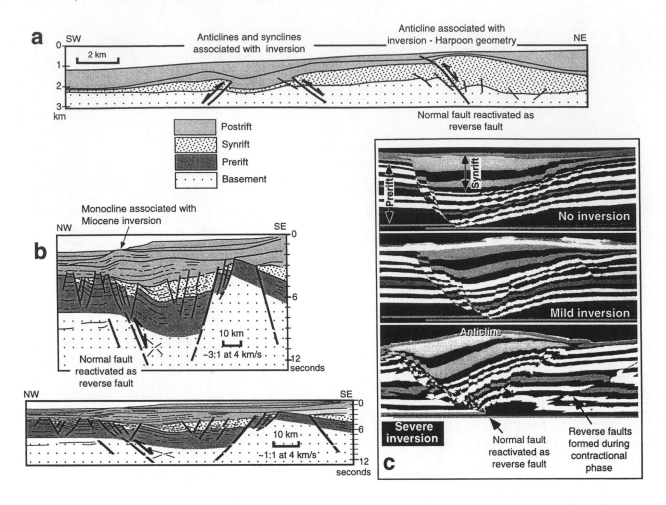

Figure 4. Examples of positive inversion structures. **a)** Cross section across part of Sunda arc. During inversion, normal faults became reverse faults, producing synclines and anticlines with harpoon geometries [after *Letouzey*, 1990]. **b)** Interpreted line drawings (with 3:1 and 1:1 vertical exaggeration) of AGSO Line 110-12 from Exmouth sub-basin, NW Shelf Australia [after *Withjack and Eisenstadt*, 1999]. During Miocene inversion, deep-seated normal faults became reverse faults. In response, gentle monoclines formed in the shallow, postrift strata. **c)** Cross sections through three clay models showing development of inversion structures [after *Eisenstadt and Withjack*, 1995]. Top section shows model with extension and no shortening; middle section shows model with extension followed by minor shortening; bottom section shows model with extension followed by major shortening.

followed by a contractional phase creates positive inversion structures. Basin inversion occurs on active margins [e.g., *de Graciansky et al.*, 1989; *Letouzey*, 1990; *Ginger et al.*, 1993] and passive margins [e.g., *Malcolm et al.*, 1991; *Hill et al.*, 1995; *Doré and Lundin*, 1996; *Vågnes et al.*, 1998; *Withjack and Eisenstadt*, 1999]. The above field and seismic studies, combined with experimental analysis [e.g., *Buchanan and McClay*, 1991; *Mitra and Islam*, 1994; *Eisenstadt and Withjack*, 1995; *Keller and McClay*, 1995; *Brun and Nalpas*, 1996], show that typical inversion structures include normal faults reactivated as reverse faults, newly formed reverse and thrust faults, monoclines, anticlines, and synclines (Figure 4). The term "postrift

contractional structure" refers to those structures that did not have an earlier extensional deformational phase or for which the pre-contractional history is unclear or unknown. For the sake of brevity, we refer to all contractional structures as inversion structures unless there is definitive evidence that the structure did *not* undergo older extensional deformation.

Because eastern North America has undergone at least three phases of deformation (Paleozoic orogeny, early Mesozoic rifting, and inversion), it is not always straightforward to constrain the timing of deformation. Cross-cutting relationships can be used to determine the relative sequence of deformation. In addition, the presence or absence of growth features can be used to determine the absolute age of deformation. For example, strata that thicken into the troughs of synclines and thin onto the crests of anticlines indicate that these folds formed at the same time as the growth strata. Alternatively, strata whose thickness is unaffected by folds indicate that the folds post-date the deposition of the strata, although the absolute age of the deformation is unconstrained unless the fold is overlapped by undeformed strata. Cross-cutting relationships between extensional and inversion structures and CAMP-related dikes as well as growth geometries involving CAMP are particularly useful because CAMP magmatism is well dated and of short duration.

3. CAMP AS A TEMPORAL BENCHMARK

Our understanding of the age and duration of rift-related magmatism has been continually refined over the last two decades [e.g., *Sutter*, 1988; *Olsen et al.*, 1989, 1996b; *Hames et al.*, 2000], as isotopic dating methods have improved and innovative stratigraphic methods have been employed. The igneous rocks of the Newark basin provide the best constraints on the age and duration of earliest Jurassic magmatism. Here, core and outcrop studies [*Fedosh and Smoot*, 1998; *Olsen et al.*, 1996a, b] show that three quartz-normative lava-flow sequences are interbedded with cyclical lacustrine strata. Based on Milankovitch cyclostratigraphy, the oldest lava is 20-40 kyr younger than the palynologically defined Triassic-Jurassic boundary [e.g., *Fowell et al.*, 1994]. Milankovitch cyclostratigraphy also constrains the duration of the igneous episode (base of the oldest lava flow to top of youngest flow) to 580±100 kyr [*Olsen et al.*, 1996b]. The absolute age of the igneous rocks, ~200 Ma, is based on isotopic dating of the lava flows [*Hames et al.*, 2000] and the Palisades intrusive sheet [*Sutter*, 1988; *Dunning and Hodych*, 1990; *Turrin*, 2000], which is physically connected to one of the flows [*Ratcliffe*, 1988]. The age and duration of the earliest Jurassic magmatic episode also apply to the Culpeper, Hartford, and Deerfield basins, based on almost exact match in cyclostratigraphy [e.g., *Olsen et al.*, 1989; 1996b], basalt geochemistry [e.g., *Tollo and Gottfried*, 1992] (Figure 2b), and paleomagnetic signature [*Prevot and McWilliams*, 1989; *Hozik*, 1992]. In the Fundy basin, the North Mountain Basalt is isotopically dated at ~200 Ma [*Hodych and Dunning*, 1992]. Furthermore, all extrusive lava flows and interbedded strata are normally magnetized [*Witte et al.*, 1991; *Kent et al.*, 1995].

The age and duration for the extrusive rocks discussed above also apply to most of the early Mesozoic intrusive rocks in the northern segment (see Figure 1) of the eastern

North American rift system [*Olsen et al.*, 1996b], for the following reasons: 1. Nearly all intrusive igneous rocks (sills, sheets, generally NE-trending dikes) (Figures 5a, 6) belong to the same geochemical categories (all quartz-normative) as the flows [e.g., *Puffer and Philpotts*, 1988; *McHone*, 1996]. 2. In some cases, intrusives served as feeders to the flows, as in the Newark basin [*Ratcliffe*, 1988]. In the Hartford basin, three NE-trending diabase dike swarms served as feeders for the three lava-flow sequences [*Philpotts and Martello*, 1986]. 3. Isotopic dates for diabase sheets in the Gettysburg basin are very similar to the Palisades sill [*Sutter*, 1988; *Dunning and Hodych*, 1990]. The Shelburne dike, located near the Fundy basin, is isotopically dated at ~200 Ma [*Dunn et al.*, 1998]. 4. No tholeiitic intrusions cut strata younger than the youngest flow in a basin [e.g., *Olsen et al.*, 1989]. Therefore, we are confident that the CAMP flows and intrusives in the northern segment of the rift system were emplaced in less than 1 million years at ~200 Ma.

The situation is not as well defined in the southern segment of the rift system (see Figure 1). There are no Early Jurassic flows in the exposed rift basins [e.g., *Olsen*, 1997] (Figure 2a). However, interbedded olivine- and quartz-normative flows are present in the subsurface of South Carolina and Georgia [*Behrendt et al.*, 1981; *Hamilton et al.*, 1983; *McBride et al.*, 1989]. The flows are flat lying whereas rift strata are dipping, suggesting that the flows developed *after* rifting. Isotopic ages for these postrift flows span the interval 97-217 Ma, but none are considered reliable [e.g., *Lamphere*, 1983; *Ragland*, 1991]. Although the wider compositional range of these flows may suggest a longer temporal duration of magmatic activity, the olivine-normative compositions may be due to alteration [P. Ragland, personal communication, in *Olsen et al.*, 1996b]. Alternating magnetic polarity zones for these flows [*Phillips*, 1983] also suggest a longer duration than for the northern-segment magmatism, but the paleomagnetic results may be unreliable [J. Phillips, personal communication, in *Olsen et al.*, 1996b]. Intrusions in the southern segment of the rift system include abundant dikes and subsurface sills. The dikes fall into two groups: NW-trending (mostly olivine normative) and N-trending dikes (mostly quartz-normative) [e.g., *Ragland et al.*, 1983, 1992] (Figure 6). All NW-trending dikes are of normal polarity, and at least one N-trending dike is of reversed polarity [*Smith*, 1987]. *Ragland et al.* [1992] believed that both dike sets are ~200 Ma, although the N-trending dikes appear to be slightly younger than the NW-trending dikes based on rare cross-cutting relationships. Recent Ar^{40}/Ar^{39} ages for two of the NW-trending dikes in South Carolina are ~200 Ma [*Hames et al.*, 2000]. In summary, the NW-trending dikes

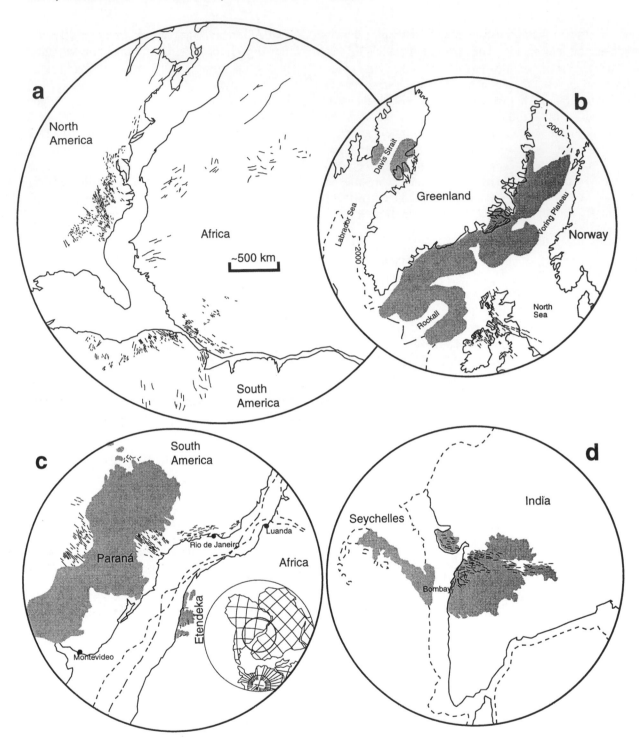

Figure 5. a) Pre-drift reconstruction of the Atlantic Ocean, showing the schematic geometry of CAMP-related dikes. Simplified from *McHone* [2000]. Diameter of circle is approximately 4000 km. **b)** Reconstruction of northern North Atlantic at 55 Ma, showing extent of Tertiary Igneous Province (gray) emplaced during breakup and known extent of dikes of same age. **c)** Reconstruction of the South Atlantic at 120 Ma, showing extent of Paraná flood basalt and Etendeka igneous province (gray) and known dikes of similar age. **d)** Reconstruction of the Indian-Seychelles region at ~65 Ma, showing extent of Deccan plateau basalts and related igneous activity (gray), and dikes of similar age. Parts (b), (c), and (d) are modified from *White* [1992], use a Lambert stereographic projection, and encompass areas with diameters of 3000 km, approximately the same scale as (a).

in the southern segment of the rift system appear to have the same age as the CAMP flows and intrusives in the northern segment of the rift system; however, the age of the N-trending dikes and the postrift basalt flows is not as well constrained.

4. STRUCTURAL GEOLOGY

4.1. Extensional Structures

Synrift extensional structures are reviewed by *Schlische* [1993, 2002] and *Withjack et al.* [1998]. Briefly, the largest-scale extensional structures are scoop-shaped half-graben produced by fault-displacement folding [*Schlische*, 1995; *Withjack et al.*, 2002] (Figure 3). In cross section perpendicular to the border fault, the basin consists of a broad downfold that results because displacement progressively decreases away from the fault. In cross section parallel to the border fault, the basin consists of a broad syncline that results because displacement is greatest at the center of the border fault and progressively decreases toward the fault tips. Seismic, outcrop, and drill hole data indicate that these large-scale folds are growth folds. Smaller-scale structures include mostly NE-striking intrabasinal normal faults and transverse fault-displacement folds related to segmentation and/or undulations on the border fault. At least in the Newark basin, these folds began to form during synrift sedimentation. Commonly, the idealized geometry shown in Figure 3 is modified by large-scale basin segmentation, significant strike slip along the border-fault system, and postrift inversion.

4.2. Inversion Structures

Almost all of the large exposed basins in eastern North America contain inversion structures (see *Withjack et al.* [1995], and reviews by *Withjack et al.* [1998], and *Schlische* [2002]). The most notable structures are: (1) broad NE-trending anticlines developed in the hanging walls of gently dipping, NE-striking border faults of the Fundy subbasin of the Fundy rift basin [*Withjack et al.*, 1995] (Figures 7b, 8a); (2) tight ENE- to E-trending synclines and anticlines formed in proximity to the steeply dipping ENE- to E-striking border faults of the Minas subbasin of the Fundy rift basin [*Withjack et al.*, 1995] (Figures 7b, 8c); (3) generally NE-striking basement-involved reverse faults and associated folds in the Richmond basin [*Shaler and Woodworth*, 1898; *Venkatakrishnan and Lutz*, 1988] (Figures 9a, b); (4) generally NE-trending inversion folds developed above NE-striking intrabasinal faults of the Taylorsville basin [*LeTourneau*, 1999] (Figure 9c); and

(5) WNW-striking axial planar cleavage in some WNW-trending folds in the Newark basin [e.g., *Lucas et al.*, 1988].

Inversion, by elevating the center of a rift basin relative to its margins, may have contributed to substantial erosion of these rift basins. For example, the center of the Minas subbasin has been uplifted relative to its margins [*Withjack et al.*, 1995] (Figure 8d). The Taylorsville basin has undergone 0.9-2.6 km of erosion, with the largest amount occurring over inversion-related anticlines [*Malinconico*, 2002] (Figure 9c). The Newark basin has also undergone 2 to 5+ km of erosion [e.g., *Pratt et al.*, 1988; *Steckler et al.*, 1993; *Malinconico*, 1999] that may, in part, be related to inversion. Inversion may also be responsible for some of the anomalously high stratal dips recorded in many of the exposed rift basins. Experimental clay models [*Eisenstadt and Withjack*, 1995] indicate that low (<10°) synrift stratal dips develop during low to moderate amounts of extension, but that these dips may steepen appreciably during inversion (Figure 4c). The Danville rift basin contains steeply dipping beds (~45°) that may have resulted from inversion (Figure 10b). The basin also contains small-scale reverse faults [*Ackermann et al.*, 2002]. The Danville basin is also exceptionally narrow relative to its length. The narrow width is probably related to erosion associated with uplift.

5. RELATIONSHIPS AMONG RIFTING, INITIATION OF INVERSION, AND CAMP

5.1. CAMP Basalts, Rift Basins, and the Diachronous End of Rifting

Synrift CAMP basalt flows and related Early Jurassic-age strata are found only in rift basins in the northern segment of the central Atlantic margin [e.g., *Olsen et al.*, 1989; *Olsen*, 1997] (Figures 1, 2a). Structural and stratigraphic data indicate that the northern-segment rift basins were active during this time (Early Jurassic). In the Newark basin, alluvial-fan conglomeratic facies are present in strata interbedded with and succeeding the lava flows, and some of these units contain clasts of basalt [e.g., *Manspeizer*, 1980; *Parker et al.*, 1988]. In the Connecticut Valley basin (Figure 10a), the basalt flows and interbedded strata are thickest adjacent to the border-fault system and thin toward the intrabasin high between the Hartford and Deerfield subbasins [e.g., *LeTourneau and McDonald*, 1988]. In the Fundy basin, outcrop, drill-hole, and seismic data indicate that the Late Triassic- and Early Jurassic-age synrift units generally thicken from the hinged margin of the basin toward the border-fault system [e.g., *Olsen et al.*, 1989; *Withjack et al.*, 1995] (Figure 7d, 8a, 8c, 8d). In ad-

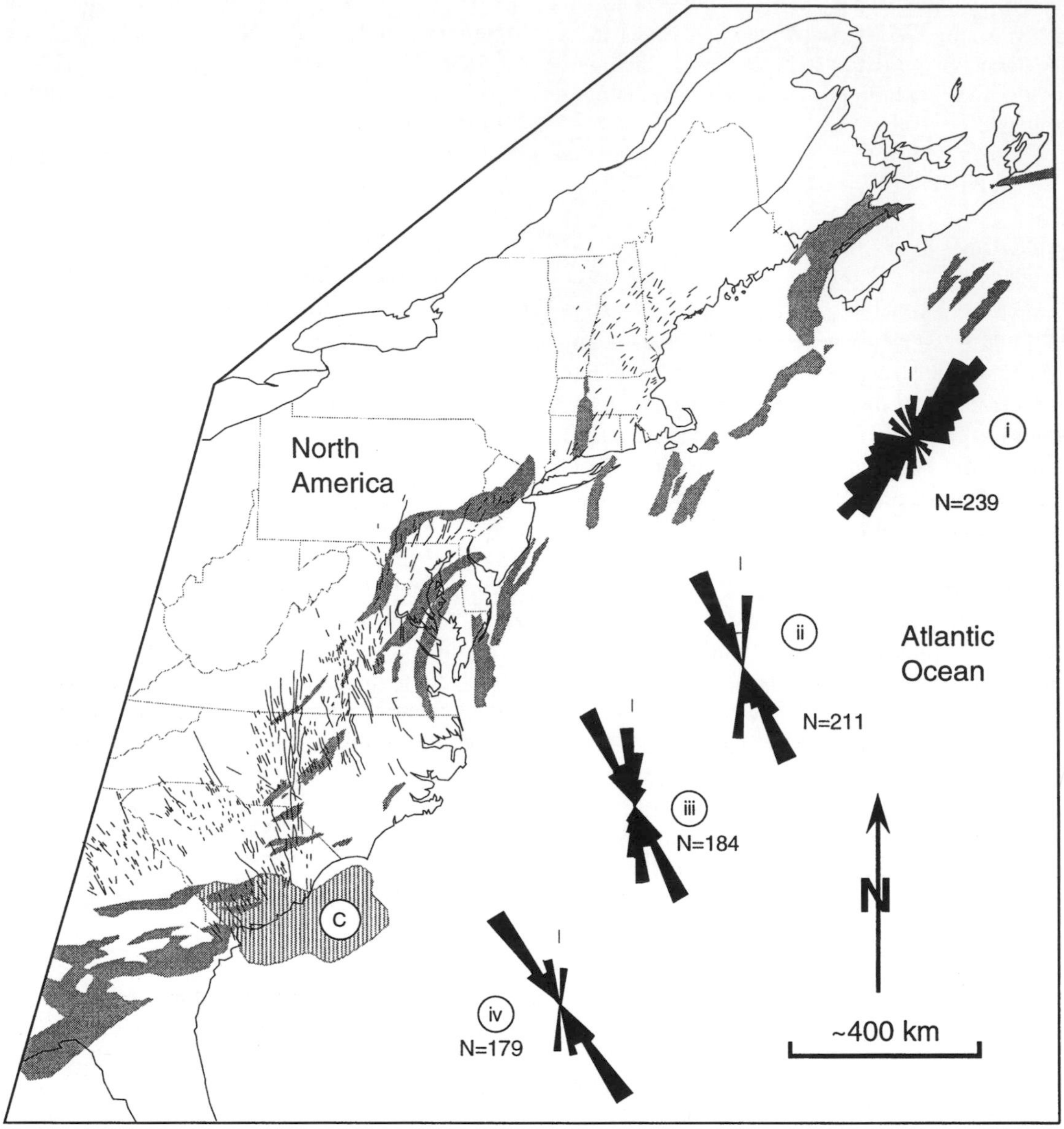

Figure 6. Early Jurassic-age diabase dikes (thin black lines) in eastern North America. Rift basins are stippled. C = possible extent of Clubhouse Crossroads Basalt [*Oh et al.*, 1995]. Rosettes indicate dike orientations (small tick marks indicate north) for the following regions: (i) Maritime Canada, New England, and New Jersey; (ii) Pennsylvania, Maryland and Virginia; (iii) North Carolina; (iv) South Carolina and Georgia. Modified from *McHone* [1988].

dition, the sedimentary unit above the North Mountain Basalt, the McCoy Brook Formation, contains paleo-talus slope deposits composed predominantly of basalt clasts, indicating active fault-controlled subsidence along the oblique-slip border-fault system [*Olsen and Schlische*, 1990; *Tanner and Hubert*, 1991] (Figure 8d). In the Newark basin, accumulation rates for earliest Jurassic-age strata are 2-5 times higher than for latest Triassic-age strata [e.g., *Olsen et al.*, 1989; *Schlische and Olsen*, 1990]. A plot of

cumulative stratigraphic thickness versus age for the northern-segment rift basins shows a pronounced steepening in the curves in earliest Jurassic time (Figure 2c).

The conditions in the southern-segment rift basins (Figure 1) were notably different. Early Jurassic-age strata are absent from all of the southern basins (Figure 2a). In fact, the youngest preserved Late Triassic-age strata in any of these basins (found in the Taylorsville basin) are ~10 Myr older than the Triassic-Jurassic boundary. Erosion and/or

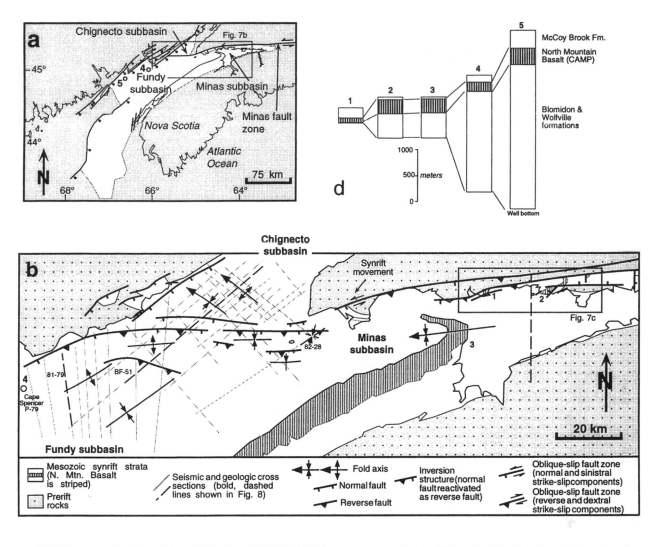

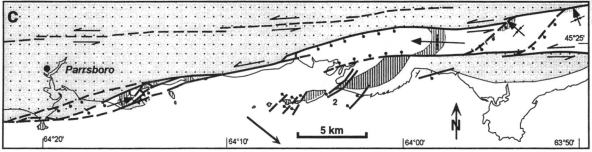

Figure 7. Maps and stratigraphic sections of the Fundy basin, Canadian Maritimes [from *Olsen and Schlische*, 1990; *Withjack et al.*, 1995]. **a)** The Fundy basin consists of the NE-trending Fundy and Chignecto subbasins and the E-trending Minas subbasin. **b)** Geologic map of part of the Fundy basin showing faults and folds related to extension and inversion and locations of seismic lines and cross section in Figure 8. **c)** Geologic map of the part of the northern margin of the Minas subbasin showing two sets of faults: ENE- to E-striking faults with oblique motion and NE-striking faults with mostly dip-slip motion. Arrow indicates extension direction. Only fault motions associated with extensional phase are shown in (a) and (c). **d)** Stratigraphic sections from five locations in the Fundy basin; see numbers in (a), (b), and (c).

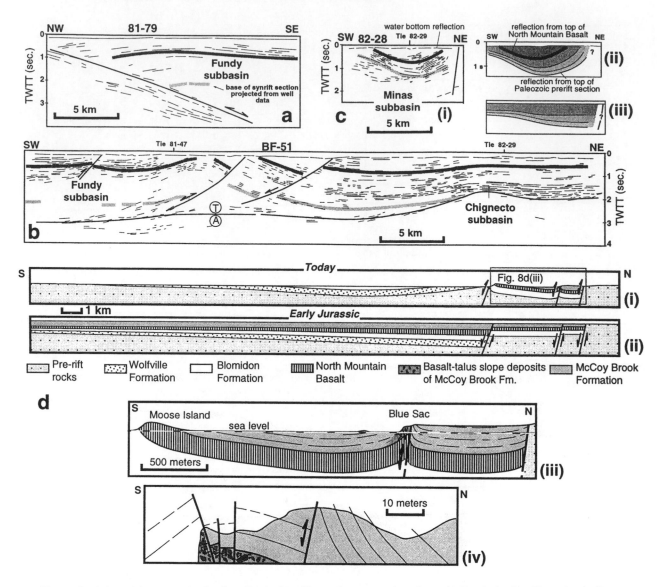

Figure 8. Selected interpreted seismic reflection profiles and cross sections from the Fundy basin. The seismic lines show the variation in the geometry of inversion-related folds: **a)** broad anticlines above gently dipping faults (81-79), **b)** tighter anticlines above moderately dipping faults (BF-51), and **c)** tight synclines adjacent to steeply dipping faults (82-28). Light-shaded line marks base of synrift section; dark-shaded line marks reflection from North Mountain Basalt. T, movement toward viewer; A, movement away. For 82-28 (c), (i) shows the line drawing of the seismic data, (ii) shows the geological interpretation, and (iii) shows the pre-inversion geometry. Note that units thicken toward the boundary fault but do not thicken toward the axis of the inversion syncline. **d)** Regional geologic cross section through the Minas subbasin today (i), showing inversion-related folds, and during Early Jurassic deposition (ii). (iii) Cross section at Blue Sac with central anticline and surrounding synclines. Basalt-talus slope deposits occur only in hanging wall, suggesting that the hanging wall was downthrown during deposition and upthrown after deposition. (iv) Sketch of outcrop at Blue Sac showing high-angle fault with reverse separation and steeply dipping beds of McCoy Brook Formation. Most faults have strike-slip as well as dip-slip displacement. Modified from *Withjack et al.* [1995].

non-deposition may account for the absence of these deposits. Modeling studies based on thermal maturation indices [*Malinconico*, 2002] and fission-track analyses [*Tseng et al.*, 1996] in the Taylorsville basin indicate that, although more strata accumulated in this basin than is cur-

rently preserved, synrift subsidence most likely ceased close to the Triassic-Jurassic boundary. If these results are applicable to other southern-segment basins, then synrift subsidence ceased in the south prior to earliest Jurassic time while it continued in the north into Early Jurassic

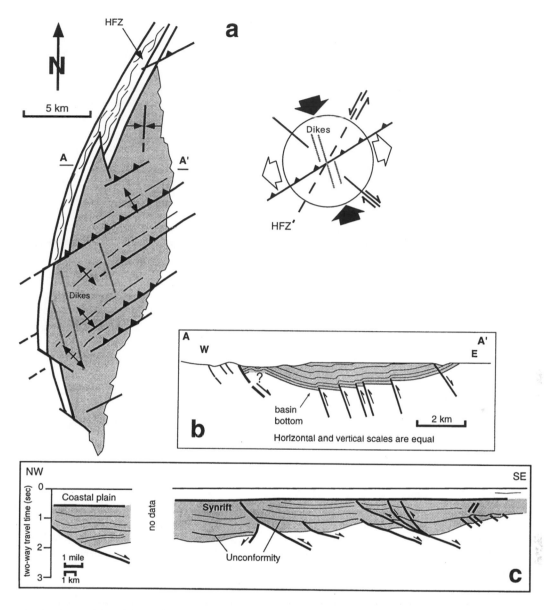

Figure 9. Geology of the Richmond and Taylorsville basins. **a)** Structural sketch map of the Richmond basin showing NNW-trending dikes cross-cutting NE-trending folds. Inferred shortening direction (black arrows) is based on contractional structures. HFZ is the Hylas fault zone. Modified from *Venkatakrishnan and Lutz* [1988]. **b)** Cross section of the Richmond basin showing basement-involved reverse faults and fault-propagation folds. Modified from *Shaler and Woodworth* [1899] and *Withjack et al.* [1998]. **c)** Interpreted seismic line across the Taylorsville basin, showing inverted faults and inversion-related folds. Simplified from *LeTourneau* [1999].

time. In fact, the northern-segment basins underwent accelerated subsidence during this time.

5.2. Different State of Stress in Southern and Northern Rift Segments During CAMP

The Mesozoic dikes in eastern North America do not have a uniform orientation [e.g., *King*, 1971; *May*, 1971]

(Figures 5a, 6). An interpreted radial pattern [*May*, 1971] has been attributed to doming related to a mantle plume [e.g., *Hill*, 1991]. However, as discussed by *McHone* [1988, 2000], the dikes in eastern North America fall into several partially overlapping sets that have a fairly uniform orientation within each set. In the northern rift segment, the dikes are generally NE trending. In the southern rift segment, the dikes fall into two sets, a NW-trending set and slightly younger N-trending set. These dikes cut uniformly across synrift and prerift rocks, indicating little or no influence of preexisting structures on dike orientations. Thus, the northern and southern rift segments experienced different states of stress during CAMP magmatism. In the northern segment, the maximum horizontal stress (S_{Hmax}) was oriented NE-SW, and the minimum horizontal stress (S_{Hmin}) was oriented NW-SE. In the southern segment, S_{Hmax} was NW-SE and S_{Hmin} was NE-SW. The younger N-trending dike set indicates that S_{Hmax} was N-S and S_{Hmin} was E-W. Thus, the state of stress was different for the northern and southern rift segments during CAMP magmatism, and it changed between the emplacement of the NW-trending set and the younger N-trending set in the southern rift segment. Because the boundary between the northern and southern segments probably was diffuse (see section 2), the state of stress should have been transitional between NE-SW-oriented S_{Hmin} in the south and NW-SE-oriented S_{Hmin} in the north. In fact, dikes in the transitional region (the area near the Culpeper and Gettysburg basins) trend ~N-S, indicating that S_{Hmin} was oriented ~E-W.

The Early Jurassic stress state in the northern rift segment (S_{Hmin} oriented NW-SE) is consistent with continued NW-SE extension, normal faulting, and basin subsidence in this region (Figures 3a, 6, 10a). However, the stress state in the southern rift segment (S_{Hmin} oriented NE-SW) is incompatible with NW-SE extension, normal faulting, and basin subsidence (Figures 6, 9a, 10b). Both dike sets in the southern rift segment cut across the rift basins and their bounding faults at a high angle [e.g., *Ragland et al.*, 1992]. These results corroborate the major conclusion from Section 5.1 that NW-SE extension, faulting, and subsidence had ceased in the southern segment prior to earliest Jurassic time while it continued in the northern segment [*Withjack et al.*, 1998].

5.3. Diachronous Initiation of Inversion

All results for the southern segment of the rift system indicate that NW-SE extension had ceased prior to CAMP time, and that NW-SE to N-S shortening was taking place before, during, and after CAMP time [*Withjack et al.*, 1998]. In the Richmond basin, NW-trending dikes cut a series of NNE-trending folds and reverse faults [*Venkatakrishnan and Lutz*, 1988], suggesting that NW-SE shortening began *before* and continued *during* CAMP time (Figure 9a). As mentioned in Section 3, postrift basalts are present in the subsurface of the southeastern United States [e.g., *Behrendt et al.*, 1981; *Hamilton et al.*, 1983; *McBride et al.*, 1989]. Locally, these postrift basalts are deformed by contractional structures. For example, the NE-striking Cooke fault in South Carolina underwent at least 140 m of

reverse displacement prior to the emplacement of the postrift basalt and had continued reverse displacement after the emplacement of the postrift basalt [*Behrendt et al.*, 1981; *Hamilton et al.*, 1983] (Figure 11e). If these basalts are CAMP basalts, then they provide additional evidence that rifting ended and inversion began prior to earliest Jurassic time along the southern segment of the central Atlantic margin.

All inversion structures in the northern segment rift basins are post-CAMP and also post-date the youngest preserved synrift deposits [*Withjack et al.*, 1995, 1998]. For example, in the Fundy basin, seismic reflection profiles show that Triassic and Early Jurassic-age units thicken toward border faults but exhibit no thickening or thinning toward the hinges of inversion-related folds (Figure 8c). Inversion therefore post-dates the Early Jurassic-age North Mountain Basalt (CAMP) and overlying McCoy Brook Formation. Without the presence of younger Jurassic, Cretaceous, and Tertiary strata, it is impossible to further constrain the timing of the inversion in the Fundy basin. Information from the adjacent Orpheus graben, however, can provide this information. The Cobequid-Chedabucto fault system bounds the Minas subbasin and its offshore continuation, the Orpheus graben, on the north (Figure 1). If the faults bounding the Minas subbasin had a component of reverse movement during inversion, then the faults bounding the kinematically linked Orpheus graben likely experienced similar movements. Previous studies have shown that, except for regional subsidence and minor salt movement, structural activity in the Orpheus graben ceased during Early Cretaceous time [*Tankard and Welsink*, 1989; *Wade and MacLean*, 1990; *MacLean and Wade*, 1992]. Thus, any inversion in the Orpheus graben and, by inference, in the Fundy basin occurred before or during Early Cretaceous time.

Although the duration of inversion is not known, it is clear that inversion began earlier in the southern rift segment than for the northern rift segment. In the southern segment, inversion occurred prior to and continued through CAMP time. In the northern segment, inversion began not only after CAMP time but after the deposition of all Early Jurassic-age synrift strata. Thus, not only is the end of rifting diachronous from south to north but the initiation of inversion is diachronous from south to north.

5.4. Initiation of Seafloor Spreading and Formation of SDR's

Conventionally, continental breakup and the onset of seafloor spreading are thought to have begun in Middle Jurassic time along the entire central Atlantic margin [e.g., *Klitgord and Schouten*, 1986]. However, this estimate is

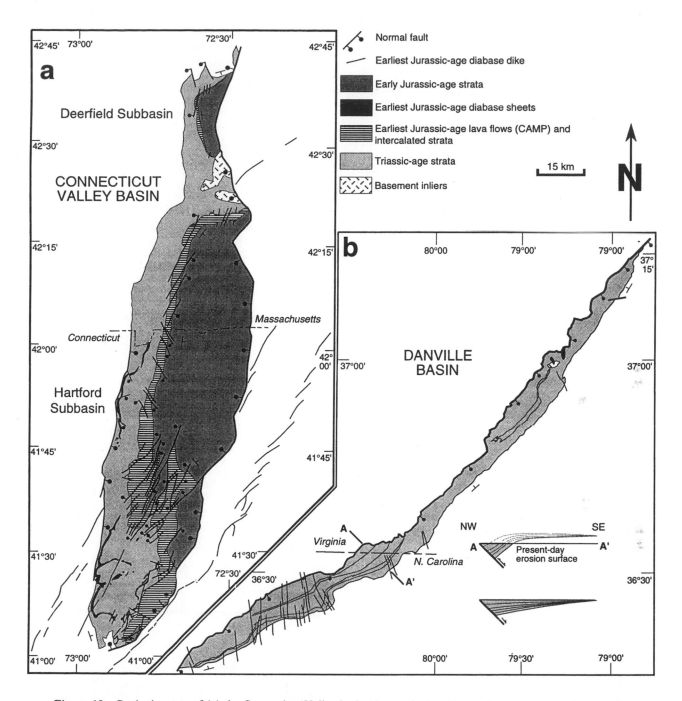

Figure 10. Geologic maps of (**a**) the Connecticut Valley basin (Connecticut and Massachusetts) and (**b**) the Danville basin (North Carolina and Virginia). Near the Connecticut Valley basin, earliest Jurassic diabase dikes are northeast trending and subparallel to the regional trend of the basin. Near the Danville basin, earliest Jurassic diabase dikes are north and northwest trending and are subperpendicular to the regional trend of the basin. Early Jurassic lava flows and sedimentary rocks are present in the Connecticut Valley basin, but are absent in the Danville basin. Danville basin cross section (top) shows inferred inversion-related geometry (see Figure 4c for experimental analog) and restored basin geometry prior to inversion (bottom) (assuming that the width of the Danville basin was approximately the same as the present-day width of the Connecticut Valley basin). Modified from *Schlische* [1993] and *Ackermann et al.* [2002].

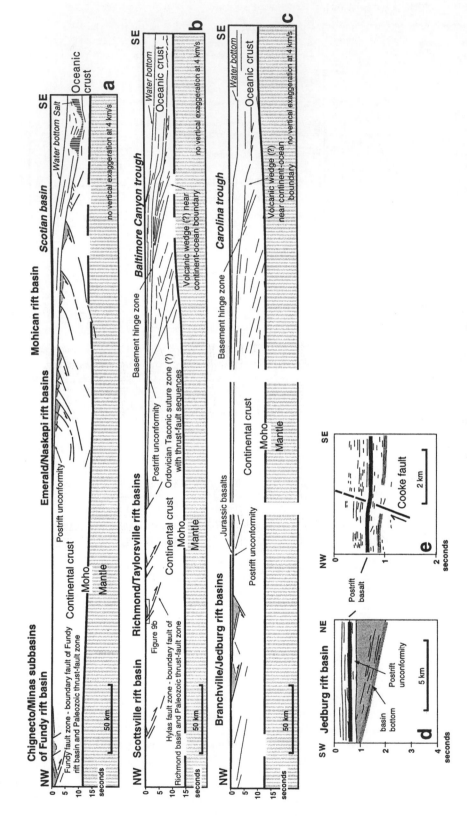

Figure 11. Regional and local cross sections through the passive margin of eastern North America. Regional sections show Paleozoic structures, early Mesozoic rift basins, and Mesozoic-Cenozoic postrift basins. Vertical axes are in two-way travel time. Section locations are shown in Figure 1. See *Withjack et al.* [1998] for data sources. **a**) Transect through southeastern Canada. **b**) Transect through the eastern United States is based on geologic data. Onshore geology was converted to two-way travel time assuming a velocity of 4000 m/s. **c**) Transect through the southeastern United States. **d**) Line drawing of seismic line VT-5 through the Jedburg rift basin showing postrift basalts. Modified from *Costain and Coruh* [1989]. **e**) Line drawing of segment of seismic line SC10 from onshore South Carolina showing that the Cooke fault was active before and during the emplacement of the postrift basalt. Modified from *Hamilton et al.* [1983].

subject to considerable uncertainty because the oldest oceanic crust does not contain identifiable and/or dated magnetic anomalies. In their calculations of the breakup age, *Klitgord and Schouten* [1986] assumed that the half-spreading rates of the Late Jurassic (19 mm/yr) also applied to earlier times. If spreading rates during earlier times were slower, however, then seafloor spreading would have initiated earlier. In fact, a half-spreading rate of 4 mm/yr yields a breakup age of ~200 Ma.

A massive wedge, presumably composed of volcanic or volcaniclastic rocks, is present along the edge of much of the central Atlantic margin [*Hinz*, 1981; *Benson and Doyle*, 1988; *Klitgord et al.*, 1988; *Austin et al.*, 1990; *Holbrook and Keleman*, 1993, *Sheridan et al.*, 1993; *Keleman and Holbrook*, 1995; *Oh et al.*, 1995] (Figures 11a, b, c). The wedge lies near the continent-ocean boundary, and formed during the transition from rifting to drifting [*Hinz*, 1981; *Benson and Doyle*, 1988; *Austin et al.*, 1990]. A wedge is not observed on the passive margin of southeastern Canada [*Keen and Potter*, 1995]. The exact age of the wedge (and, by inference, the onset of seafloor spreading) is unknown and, in fact, may vary along the margin [e.g., *Oh et al.*, 1995]. Beneath the Georges Bank basin, SDR's within the wedge underlie a relatively flat-lying Middle Jurassic postrift sequence [*Schlee and Klitgord*, 1988]. Thus, the wedge beneath the Georges Bank basin formed before the Middle Jurassic deposition of these postrift strata. Beneath the Baltimore Canyon trough, SDR's overlap rift basins [*Benson and Doyle*, 1988] and underlie the postrift unconformity [*Sheridan et al.*, 1993]. Thus, the wedge beneath the Baltimore Canyon trough formed after the deposition of the Late Triassic to Early Jurassic-age synrift strata and before the deposition of the thick package of postrift strata beneath the Late Jurassic-age rocks sampled in offshore wells [*Poag and Valentine*, 1988]. Beneath the Carolina trough, SDR's appear to underlie the postrift basalts [*Austin et al.*, 1990; *Oh et al.*, 1995]. If so, then the wedge beneath the Carolina trough formed before the eruption of the postrift basalts. If these postrift basalts belong to CAMP, then it is possible that the SDR's in this region are also associated with CAMP.

6. DISCUSSION

6.1. Summary of Geological and Geophysical Data

1. The CAMP flows and intrusives in the northern segment of the rift system were emplaced in less than 1 million years at ~200 Ma (earliest Jurassic time). The NW-trending dikes in the southern segment of the rift system have the same age as the CAMP flows and intrusives in the northern segment of the rift system. It is possible, however, that the emplacement of the postrift basalts and some of the other intrusives in the southern segment of the rift system occurred shortly before, during, or shortly after CAMP magmatism.

2. Synrift subsidence ceased in the southern segment of the rift system prior to earliest Jurassic time. It continued, however, in the northern segment of the rift system into Early Jurassic time. In fact, the rift basins in the northern segment underwent accelerated subsidence during earliest Jurassic time.

3. During earliest Jurassic time, the stress state indicated by dike trends (S_{Hmin} oriented NW-SE) in the northern segment of the rift system is compatible with continued NW-SE extension, normal faulting, and basin subsidence. The stress state indicated by dike trends (S_{Hmin} oriented NE-SW) in the southern segment of the rift system, however, is incompatible with NW-SE extension, normal faulting, and basin subsidence.

4. Inversion began earlier in the southern segment of the rift system than in the northern segment. In the southern segment, inversion occurred prior to and continued through CAMP time. In the northern segment, inversion began not only after CAMP time but after the deposition of all Early Jurassic-age synrift strata.

5. The formation of SDR's and the onset of seafloor spreading in the northern segment of the rift system occurred after the deposition of the Early Jurassic-age synrift strata and before the deposition of the Early to Middle Jurassic-age postrift strata. The formation of the SDR's and the onset of seafloor spreading in the southern segment of the rift system occurred after the deposition of the Late Triassic-age synrift strata and before the deposition of the thick sedimentary wedge below sampled Late Jurassic-age postrift strata.

6.2. Interpretations

Two end-member scenarios can honor these available geological and geophysical observations. In scenario 1 (traditional scenario), the initiation of seafloor spreading was synchronous along the entire margin of the central North Atlantic Ocean during the late Early Jurassic to early Middle Jurassic. With this scenario, all of the SDR's on the margin of eastern North America are the same age (i.e., late Early Jurassic to early Middle Jurassic). Thus, with this scenario, two distinct episodes of magmatism affected eastern North America: CAMP (earliest Jurassic) and a later episode associated with the emplacement of all of the SDR's and possibly some of the igneous activity in the southern segment of the rift system. This scenario also implies that rifting ceased and inversion began in the southern segment of the rift system before the onset of sea-floor spreading. Presumably, localized extensional deformation continued at the eventual site of continental breakup. In this scenario, the cause of the NW-SE compression responsible for the NW-trending dikes of CAMP age is unexplained. In scenario 2, the initiation of seafloor spreading was diachronous, beginning in the southern segment of the rift system during latest Triassic to earliest Jurassic time and beginning in the northern segment during late Early Jurassic to early Middle Jurassic time. With this scenario, the emplacement of the SDR's on the margin of eastern North America was diachronous (latest Triassic/earliest Jurassic in the south and late Early Jurassic to early Middle Jurassic in the north). As in scenario 1, two distinct episodes of magmatism affected eastern North America: CAMP (including the southern SDR's) and a later episode associated with the emplacement of the northern SDR's (late Early Jurassic to early Middle Jurassic). This scenario also implies that the cessation of rifting and the onset of inversion occurred during the transition from rifting to drifting on both the northern and southern segments of the rift system. Coring and dating the postrift basalts and the SDR's is critical to deciding which scenario is most viable. This is a key scientific goal of the Pole-to-Pole Pangea Coring Project [*McHone et al.*, 2002; *Schlische et al.*, 2002b].

We believe that several lines of evidence support scenario 2. 1. Numerical simulations by Bott [1992] suggest that a change in the stress state is likely to occur during the rift-drift transition. 2. On many volcanic passive margins, the timing of the rift-drift transition and LIP magmatism are synchronous (e.g., the passive margins of the North Atlantic Ocean) [e.g., *White*, 1992]. 3. By analyzing subsidence curves, *Dunbar and Sawyer* [1989] proposed that spreading first began between the Blake Spur and Delaware Bay fracture zones and then subsequently initiated between the Delaware Bay and Newfoundland fracture zones (Figure 1). These and other fracture zones may have accommodated some of the differences in strain between the area that was undergoing seafloor spreading and the area to the north that was still undergoing continental extension. 4. The inferred diachronous initiation of spreading for the central North Atlantic Ocean is part of regional trend that reflects the progressive breakup of Pangea. Seafloor spreading between the Grand Banks and southwestern Europe began during the Early Cretaceous [e.g., *Srivastava and Tapscott*, 1986]; seafloor spreading between Labrador and western Greenland began during the early Tertiary (anomaly 27N) [e.g., *Chalmers et al.*, 1993]; seafloor spreading between eastern Greenland and northwestern Europe began slightly later during the early Tertiary

(anomaly 24R) [e.g., *Talwani and Eldholm*, 1977; *Hinz et al.*, 1993].

6.3. Relationship to Middle Jurassic Hydrothermal Event

Inversion in the southern rift segment is temporally associated with CAMP and possibly with the emplacement of SDR's at the continent-ocean boundary. Inversion in the northern basins is broadly contemporaneous with an inferred hydrothermal event (~175-178 Ma) that reset many isotopic clocks [*Sutter*, 1988], that remagnetized many rocks [*Witte and Kent*, 1991; *Witte et al.*, 1991], and that may be temporally associated with the emplacement of SDR's during the initiation of seafloor spreading along the northern segment of the central North Atlantic margin. *Sutter* [1988] hypothesized that the hydrothermal fluids migrated in response to regional tilting and/or uplift. *Schlische and Olsen* [1990] attributed this fluid flow to a period of accelerated normal faulting immediately prior to breakup. However, faulting, tilting, and uplift during inversion is equally plausible. For example, *Schlische* [2002] described a NE-striking reverse fault from the Newark basin that exhibits evidence of extensive fluid flow. Paleomagnetic data from the Jacksonwald and Sassamansville synclines of the Newark basin show that the magnetization component related to the hydrothermal event was acquired during folding [*Witte and Kent*, 1991; *Kodama et al.*, 1994]. Both synclines probably initiated as synrift fault-displacement folds [*Schlische*, 1992, 1995], but the presence of very steep dips and axial planar cleavage suggest that the folds were amplified during inversion. The paleomagnetic data suggest that the inversion occurred penecontemporaneously with the hydrothermal event. Interestingly, *Sibson* [1995] argued that high fluid pressures are required to reverse-reactivate normal faults that are not optimally oriented (i.e., those dipping steeper than 50°) and that fluid pressures increase during inversion as open subvertical fractures close during regional subhorizontal compression. Thus, inversion, fluid flow, and hydrothermal activity are all related processes.

6.4. Uncertainties in Inversion-Related Shortening Direction and Magnitude

The shortening direction for the southern rift segment is relatively well constrained because it is likely subparallel to S_{Hmax} inferred from the CAMP dikes, i.e., NW-SE for the NW-trending set and N-S for the N-trending set (Figure 6). However, it is not known why the S_{Hmax} and the shortening direction rotated from NW-SE to N-S between the emplacement of the two sets of dikes.

The shortening direction for the northern rift segment is not as well constrained because it post-dates the CAMP dikes. *Withjack et al.* [1995] inferred that the shortening direction was NW-SE for the Fundy basin, based principally on the mean orientation of inversion-related folds and reverse- and oblique-slip faults associated with inversion structures (Figure 7b, c). *DeBoer* [1992] used small-scale faults to infer a NE-SW shortening direction for the Fundy basin. *DeBoer and Clifton* [1988] also used small-scale faults to infer a NE-SW shortening direction for the Hartford basin. In the Newark basin, *Lomando and Engelder* [1984] inferred a N-S shortening direction based on calcite twins, whereas *Lucas et al.* [1988] inferred a NNE-SSW shortening direction based on the geometry of the Jacksonwald syncline and its axial planar cleavage.

Part of the uncertainty is related to the fact that inversion structures are inexact recorders of the shortening direction. By definition, inversion structures are two-phase structures that must have undergone an earlier extensional phase of deformation. Thus, the attitude of an inversion structure is, at least in part, controlled by the original extensional geometry. In addition, experimental modeling of reactivated structures shows that both extensional and contractional forced folds form subparallel to the trend of a preexisting fault, even where the shortening direction is oblique to the trend of the fault [*Schlische et al.*, 1999a, b, 2002a; *Tindall et al.*, 1999a, b]. Postrift contractional structures that did not have an earlier extensional phase are more likely to reflect the true shortening direction. Regional analyses of inversion structures may be useful in constraining the shortening direction. As shown in Figure 12, different shortening directions result in different senses of displacement on the differently oriented boundary faults of various eastern North American rift basins.

Experimental clay models [*Eisenstadt and Withjack*, 1995] indicate that the basin-scale inversion structures may be difficult to detect. Obvious basin-scale inversion folds only develop in the models when the amount of shortening is equal to or greater than the amount of extension (Figure 4c). However, if erosion removes the upper part of the model, the majority of the inversion-related regional fold disappears, leaving a classical half-graben basin with somewhat steeper dips than normal (which may be misinterpreted to indicate higher amounts of regional extension). The foregoing discussion suggests that inversion in eastern North America may be even more widespread than currently appreciated.

6.5. Applications to Other Passive Margins

In addition to the central North Atlantic passive margin, inversion in a passive-margin setting has occurred along the Norwegian margin of the North Atlantic Ocean [e.g., *Doré and Lundin*, 1996; *Vågnes et al.*, 1998], the southeastern Australian margin [e.g., *Hill et al.*, 1995], and the Northwest Shelf of Australia [e.g., *Malcolm et al.*, 1991; *Withjack and Eisenstadt*, 1999] (Figure 4b). Like the central North Atlantic margin, the Norwegian margin is a volcanic passive margin [e.g., *Hinz*, 1981; *Hinz et al.*, 1993], and parts of the Northwest Shelf of Australia are volcanic [*Hopper et al.*, 1992]. On the Norwegian margin, inversion began soon after the initiation of seafloor spreading [*Vågnes et al.*, 1998] and continued into Miocene time. Inversion may have occurred on the Australia margin soon after the initiation of seafloor spreading [*Withjack and Eisenstadt*, 1999]. Significant inversion occurred during late Miocene time [*Malcolm et al.*, 1991; *Withjack and Eisenstadt*, 1999].

A key feature related to inversion during the rift-drift transition is the presence of dikes trending at a high angle to the continental margin. A number of other rifted margins have dikes oriented at a high angle to the rifted margin [e.g., *White*, 1992]. These include the east Greenland and United Kingdom margins of the northern North Atlantic (breakup age: 55 Ma); the southwest India margin (breakup age: 65 Ma); and the southeastern margin of South America of the southern South Atlantic (breakup age: 130 Ma) (Figures 5b, c, d). Interestingly, in the last example, both margin-perpendicular and margin-parallel dike sets are present, and, like the central North Atlantic region, is associated with south-to-north diachronous initiation of seafloor spreading [*White*, 1992]. Although all three examples discussed here are associated with mantle plumes and *White* [1992] related the dikes to lateral transport away from the plume head, we speculate that the dike orientations may be more indicative of a change in stress state (from rifting to inversion) that is broadly contemporaneous with the rift-drift transition on volcanic passive margins.

6.6. Causes of Inversion on Passive Margins

The causes of shortening on passive margins include plate collision or subduction, ridge-push forces, continental resistance to plate motion, and active asthenospheric upwelling [e.g., *Dewey*, 1988; *Bott*, 1992; *Boldreel and Anderson*, 1993; *Withjack et al.*, 1998; *Vågnes et al.*, 1998]. The last two are likely to be most intense during the rift-drift transition and the earliest stages of drifting and to act in concert to produce shortening on the passive margin. *Holbrook and Kelemen* [1993] and *Kelemen and Holbrook* [1995] argued that active asthenospheric upwelling (in which the upwelling rate is greater than the lithospheric extension rate) is necessary to produce the volume and seismic velocities of the SDR's along the central North

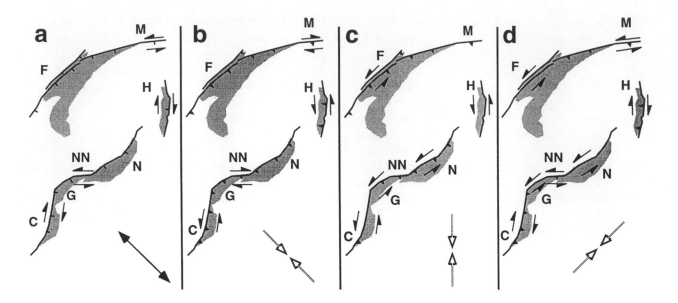

Figure 12. The extension direction and shortening direction acting on variably oriented preexisting zones of weakness affect the type of fault reactivation. Each panel shows the present-day configuration of the boundary faults and synrift strata for the Culpeper (C), Gettysburg (G), Narrow Neck (NN), Newark (N), Hartford-Deerfield (H), Fundy subbasin (F) and Minas subbasin (M) of the Fundy basin, and the inferred type of fault reactivation for **a**) NW-SE extension, **b**) NW-SE shortening, **c**) N-S shortening, and **d**) NE-SW shortening. Because the shortening direction associated with basin inversion is not well constrained, three possibilities are shown here. Each possible shortening direction predicts different types of faulting along the boundary faults of the rift basins. For example, NE-SW shortening predicts reverse slip along the Newark basin and Fundy subbasin border faults and reverse-right-oblique slip along the Minas subbasin and Narrow Neck border faults. In contrast, NW-SE shortening predicts reverse-left-oblique slip along the Newark and Fundy subbasin border faults and reverse-slip along the Minas subbasin and Narrow Neck border faults.

Atlantic margin. The active asthenospheric upwelling is a type of convection driven in part by the change in lithospheric thickness across the rifted zone [e.g., *Mutter et al.*, 1988] and possibly aided by the presence of a thermal anomaly associated with the thermal-blanketing effect of the Pangean supercontinent [e.g., *Anderson*, 1982]. The active asthenospheric upwelling would occur along a laterally extensive, linear zone rather than the domal or radial zone associated with a mantle plume. (Additional evidence against a mantle-plume origin for the eastern North American Jurassic magmatism is discussed in *Kelemen and Holbrook* [1993], *McHone* [2000], and *Puffer* [2001].)

In the geodynamic model of *Withjack et al.* [1998], the lithospheric mantle and the asthenosphere are initially partially decoupled (Figure 13). This means that the displacement rates of the lithosphere and the asthenosphere are slightly different. Prior to plate rupture, this leads to widespread intraplate extension (Figure 13a). Following plate rupture, the mantle lithosphere is likely to become coupled to the asthenosphere near the spreading center, whereas they remain uncoupled farther away from the spreading center. Because the portion of the plate adjacent to the spreading center is moving at a faster rate than the

portion of the plate farther from the spreading center, compensating crustal shortening occurs in this region (Figure 13b). In this model, the shortening is expected to decrease in magnitude through time because (1) continental resistance to plate motion decreases as the lithosphere and asthenosphere become increasingly coupled, and (2) active asthenospheric upwelling gives rise to passive upwelling associated with normal seafloor-spreading processes (Figure 13c). Compressional stresses will not disappear entirely, as ridge-push forces will always be present. These ridge-push forces may be responsible, at least in part, for the reverse faults that have offset coastal plain deposits [e.g., *Prowell*, 1988] and may contribute to the present-day state of stress [e.g., *Zoback and Zoback*, 1989]. However, the present-day S_{Hmax} is oriented NE-SW, which is not parallel to the present-day spreading direction (WNW-ESE) and the presumable direction of ridge push.

The diachronous rift-drift transition (scenario 2) has additional (albeit speculative) implications for this geodynamic model involving active asthenospheric upwelling. Along the southern segment of the margin, both CAMP and the SDR's are related to the active asthenospheric upwelling that culminated during the rift-drift transition.

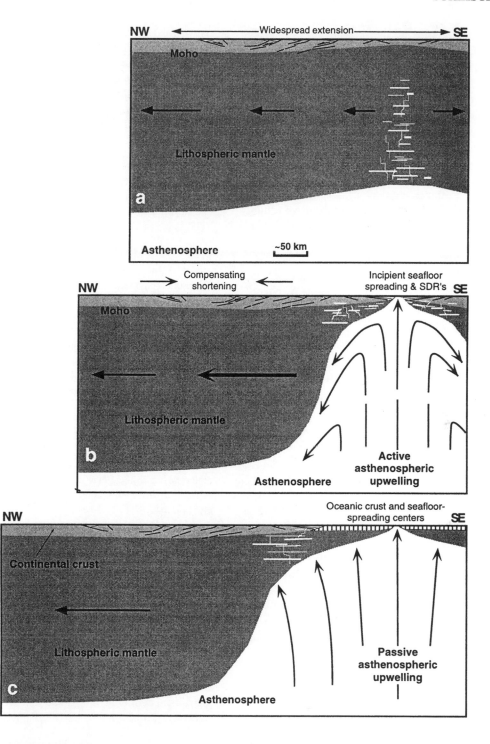

Figure 13. Tectonic model for central eastern North America. **a)** Early rifting. Distant plate-tectonic forces produce divergent lithospheric displacements. **b)** Late rifting. Lithosphere is substantially thinned. Gravitational-body forces and traction forces associated with the hot, low-density asthenospheric upwelling increase substantially. In response, lithospheric displacements near the upwelling exceed those far from the upwelling, causing shortening (inversion) in the intervening zone. **c)** Early drifting. Lithospheric displacements far from upwelling increase, eventually equaling those near the upwelling. Most shortening/inversion ceases, and the asthenospheric upwelling becomes passive. From *Withjack et al.* [1998].

CAMP magmatism along the northern segment of the margin may reflect "leakage" or lateral transport of the mostly southern magmas into the region that was still undergoing continental extension. In this case, the magmatism ceased abruptly (while continental extension continued) because the formation of the SDR's and the transition to normal seafloor spreading along the southern segment of the margin largely dissipated the original thermal anomaly in the source region for much of the magma. This thermal anomaly would not have been dissipated (or not dissipated sufficiently) along the northern segment of the margin, thus leading to the formation of SDR's during the active asthenospheric upwelling that culminated during the younger rift-drift transition. This geodynamic model of active asthenospheric upwelling coupled with a diachronous rift-drift transition therefore predicts that a given rifted margin may experience multiple magmatic pulses (some of which are synrift, some of which are postrift) and that the SDR's are diachronous along the margin, as observed on many volcanic margins [*Menzies et al.*, 2000].

7. SUMMARY AND CONCLUSIONS

1. CAMP igneous activity is an excellent temporal benchmark for assessing the timing of tectonic activity because it is well dated and of short duration.

2. Many of the rift basins in the eastern North American rift system have undergone basin inversion.

3. Along the southern segment of the central North Atlantic margin, rifting ended prior to CAMP igneous activity, while rifting continued during and after CAMP along the northern segment.

4. Along the southern segment of the central North Atlantic margin, basin inversion began prior to CAMP igneous activity. Along the northern segment, inversion began after CAMP igneous activity (i.e, after synrift deposition in Early Jurassic time) and ended before or during the Early Cretaceous.

5. The end of rifting and the initiation of inversion were diachronous along the central North Atlantic margin.

6. It is likely that the initiation of seafloor spreading was diachronous along the central North Atlantic margin. If so, then the rift-drift transition is associated with a change in the strain state from NW-SE extension to NW-SE to N-S shortening along the southern segment of the central North Atlantic margin. Also, SDR's, emplaced during the rift-drift transition, have different ages along the length of the margin.

7. Widespread magmatism and inversion in eastern North America are related to active asthenospheric upwelling that peaked during the rift-drift transition.

Acknowledgments We thank the editors of this volume for encouraging this contribution; Mark Baum, Jennifer Elder Brady, Dennis Kent, Peter LeTourneau, Maryann Malinconico, and Peter Rona for useful discussions; and Richard Benson, Georgia Pe-Piper, and Robert Sheridan for their helpful reviews.

REFERENCES

Ackermann, R. V., R. W. Schlische, L. C. Patiño, and L. A. Johnson, A Lagerstätte of rift-related tectonic structures from the Solite Quarry, Dan River/Danville rift basin, in *Aspects of Triassic-Jurassic Rift Basin Geoscience*, edited by P. M. LeTourneau, and P. E. Olsen, Columbia University Press, New York, in press, 2002.

Anderson, D. L., Hotspots, polar wander, Mesozic convection, and the geoid, *Nature*, *297*, 391-393, 1982.

Austin, J. A., and seven others, Crustal structure of the Southeast Georgia embayment-Carolina trough: preliminary results of a composite seismic image of a continental suture(?) and a volcanic passive margin, *Geology*, *18*, 1023-1027, 1990.

Behrendt, J. C., R. M. Hamilton, H. D. Ackermann, and V. J. Henry, Cenozoic faulting in the vicinity of the Charleston, South Carolina, 1886 earthquake, *Geology*, *9*, 117-122, 1981.

Benson, R. N., and R. G. Doyle, Early Mesozoic rift basins and the development of the United States middle Atlantic continental margin, in *Triassic-Jurassic Rifting, Continental Breakup and the Origin of the Atlantic Ocean and Passive Margins*, edited by W. Manspeizer, Elsevier, New York, pp. 99-127, 1988.

Boldreel, L. O., and M. S. Andersen, Late Paleocene to Miocene compression in the Faero-Rockall area, in *Petroleum Geology of Northwest Europe*, edited by J. R. Parker, Geological Society, London, pp. 1025-1034, 1993.

Bott, M. H. P., The stress regime associated with continental break-up, in *Magmatism and the Causes of Continental Breakup*, edited by B. C. Storey, et al., *Geol. Soc. Spec. Publ.*, *68*, 125-136, 1992.

Brun, J.-P., and T. Nalpas, Graben inversion in nature and experiments, *Tectonics*, *15*, 677-687, 1996.

Buchanan, P. G., and K. R. McClay, Sandbox experiments of inverted listric and planar fault systems, *Tectonophysics*, *188*, 97-115, 1991.

Chalmers, J. A., C. R. Pulvertaft, F. G. Christiansen, H. C. Laresen, K. H. Laursen, and T. G. Ottesen, The southern West Greenland continental margin: Rifting history, basin development, and petroleum potential, in *Petroleum Geology of Northwest Europe*, edited by J. R. Parker, Geological Society, London, pp. 915-931, 1993.

Costain, J. K., and C. Coruh, Tectonic setting of Triassic half-grabens in the Appalachians: Seismic data acquisition, processing, and results, in *Extensional Tectonics and Stratigraphy of the North Atlantic Margins*, edited by A. J. Tankard and H. R. Balkwill, *AAPG Mem.*, *46*, 155-174, 1989.

deBoer, J. Z., Stress configurations during and following emplacement of ENA basalts in the northern Appalachians, in *Eastern North American Mesozoic Magmatism*, edited by J. H.

Puffer and P. C. Ragland, *Geol. Soc. Am. Spec. Paper*, *268*, 361-378, 1992.

deBoer, J. Z., and A. E. Clifton, Mesozoic tectogenesis: Development and deformation of 'Newark' rift zones in the Appalachians (with special emphasis on the Hartford basin, Connecticut), in *Triassic-Jurassic Rifting, Continental Breakup and the Origin of the Atlantic Ocean and Passive Margins*, edited by W. Manspeizer, Elsevier, New York, pp. 275-306, 1988.

de Graciansky, P. C., G. Dardeau, M. Lemoine, and P. Tricart, The inverted margin of the French Alps and foreland basin inversion, in *Inversion Tectonics*, edited by M. A. Cooper and G. D. William, *Geol. Soc. Spec. Publ.*, *44*, 87-104, 1989.

Dewey, J. F., Lithospheric stress, deformation, and tectonic cycles: The disruption of Pangaea and the closure of the Tethys, in *Gondwana and Tethys*, edited by M. G. Audley-Charles and A. Hallam, *Geol. Soc. Spec. Publ.*, *37*, 23-40, 1988.

Dietz, R. S., and J. C. Holden, Reconstruction of Pangaea: breakup and dispersion of continents, Permian to present: *J. Geophy. Res.*, *75*, 4939-4956, 1970.

Dore, A. G., and E. R. Lundin, Cenozoic compressional structures on the NE Atlantic margin: Nature, origin, and potential significance for hydrocarbon exploration, *Petrol. Geoscience*, *2*, 299-311, 1996.

Dunbar, J. A., and D. S. Sawyer, Patterns of continental extension along the conjugate margins of the central and North Atlantic Oceans and Labrador Sea, *Tectonics*, *8*, 1059-1077, 1989.

Dunn, A. M., P. H. Reynolds, D. B. Clarke, J. M. Ugidos, A comparison of the age and composition of the Shelburne dyke, Nova Scotia, and the Messejana dyke, Spain, *Can. J. Earth Sci.*, *35*, 1110-1115, 1998.

Dunning, G. R., and J. D. Hodych, U-Pb zircon and baddeleyite age for the Palisade and Gettysburg sills of northeast United States: Implications for the age of the Triassic-Jurassic boundary, *Geology*, *18*, 795-798, 1990.

Eisenstadt, G., and M. O. Withjack, Estimating inversion: results from clay models, in *Basin Inversion*, edited by J. G. Buchanan and P. G. Buchanan, *Geol. Soc. Spec. Publ.*, *88*, 119-136, 1995.

Fedosh, M. S., and J. P. Smoot, A cored stratigraphic section through the northern Newark basin, New Jersey, in *Studies of the Early Mesozoic Basins of the Eastern United States*, edited by A. J. Froelich and G. R. Robinson, Jr., *U.S. Geol. Surv. Bull.*, *1776*, 19-24, 1988.

Fowell, S. J., B. Cornet, and P. E. Olsen, Geologically rapid Late Triassic extinctions: Palynological evidence from the Newark Supergroup, *Geol. Soc. Am. Spec. Paper*, *288*, 197-206, 1994.

Froelich, A. J., and P. E. Olsen, Newark Supergroup, a revision of the Newark Group in eastern North America, *U.S. Geol. Surv. Bull.*, *1537A*, A55-A58, 1984.

Ginger, D. C., W. O. Ardjakusumah, R. J. Hedley, and J. Pothecary, Inversion history of the West Natuna basin: Examples from the Cumi-Cumi PSC, in *Proceedings Indonesian Petrol. Assoc.*, *22ⁿᵈ Ann. Convention*, *IPA93-1.1-171*, pp. 635-657, 1993.

Hames, W. E., P. R. Renne, and C. Ruppel, New evidence for geologically instantaneous emplacement of earliest Jurassic Central Atlantic magmatic province basalts on the North American margin, *Geology*, *28*, 859-862, 2000.

Hamilton, R. M., J. C. Behrendt, and H. D. Ackermann, Land multichannel seismic-reflection evidence for tectonic features near Charleston, South Carolina, in *Studies Related to the Charleston, South Carolina, Earthquake of 1886 — Tectonics and Seismicity*, edited by G. S. Gohn, *U.S. Geol. Surv. Prof. Paper*, *1313*, I1-I18, 1983.

Hill, K. C., K. A. Hill, G. T. Cooper, A. J. O'Sullivan, P. B. O'Sullivan, and M. J. Richardson, Inversion around the Bass basin, SE Australia, in *Basin Inversion*, edited by J. G. Buchanan and P. G. Buchanan, *Geol. Soc. Spec. Publ.*, *88*, 525-548, 1995.

Hill, R. I., Starting plumes and continental breakup, *Earth Planet. Sci. Lett.*, *87*, 398-416, 1991.

Hinz, K., A hypothesis on terrestrial catastrophes; wedges of very thick, oceanward-dipping layers beneath passive continental margins, *Geol. Jahrbuch*, *E22*, 3-38, 1981.

Hinz, K., O. Eldholm, M. Block, and J. Skogseid, Evolution of North Atlantic volcanic continental margins, in *Petroleum Geology of Northwest Europe*, edited by J. R. Parker, Geological Society, London, pp. 901-913, 1993.

Hodych, J. P., and G. R. Dunning, Did the Manicouagan impact trigger end-of-Triassic mass extinction?, *Geology*, *20*, 51-54, 1992.

Holbrook, W. S., and P. B. Kelemen, Large igneous province on the US Atlantic margin and implications for magmatism during continental breakup, *Nature*, *364*, 433-436, 1993.

Hopper, J. R., J. C. Mutter, R. L. Larson, C. Z. Mutter, and Northwest Australia Study Group, Magmatism and rift margin evolution: Evidence from northwest Australia, *Geology*, *20*, 853-857, 1992.

Hozik, M. J., Paleomagnetism of igneous rocks in the Culpeper, Newark, and Hartford/Deerfield basins, in *Eastern North American Mesozoic Magmatism*, edited by J. H. Puffer and P. C. Ragland, *Geol. Soc. Am. Spec. Paper*, *268*, 279-308, 1992.

Keen, C. E., and D. P. Potter, The transition from a volcanic to a nonvolcanic rifted margin off eastern Canada, *Tectonics*, *14*, 359-371, 1995.

Kelemen, P. B., and W. S. Holbrook, Origin of thick, high-velocity igneous crust along the U.S. east coast margin, *J. Geophys. Res.*, *100*, 10,077-10,094, 1995.

Keller, J. V. A., and K. R. McClay, 3D sandbox models of positive inversion, in *Basin Inversion*, edited by J. G. Buchanan and P. G. Buchanan, *Geol. Soc. Spec. Publ.*, *88*, 137-146, 1995.

Kent, D. V., P. E. Olsen, and W. K. Witte, Late Triassic-earliest Jurassic geomagnetic polarity sequence and paleolatitudes from drill cores in the Newark rift basin, eastern North America, *J. Geophys. Res.*, *100*, 14,965-14,998, 1995.

King, P. B., Systematic pattern of Triassic dikes in the Appalachian region, second report, *U.S. Geol. Surv. Prof. Paper*, *750D*, D84-D88, 1971.

King, S. D., and D. L. Anderson, Edge-driven convection, *Earth Planet. Sci. Lett.*, *160*, 289-296, 1998.

Klitgord, K. D., and H. Schouten, Plate kinematics of the central Atlantic, in *The Geology of North America, v. M, The Western North Atlantic Region*, edited by P. R. Vogt and B. E. Tucholke, Geological Society of America, Boulder, pp. 351-378, 1986.

Klitgord, K. D., D. R. Hutchinson, and H. Schouten, U.S. Atlantic continental margin; structural and tectonic framework, in *The Geology of North America, v. I-2, The Atlantic Continental Margin, U.S.*, edited by R. E. Sheridan and J. A. Grow, Geological Society of America, Boulder, pp. 19-56, 1988.

Kodama, K. P., C. Hedlund, J. Gosse, and J. Strasser, Rotated paleomagnetic poles from the Sassamansville syncline, Newark basin, southeastern Pennsylvania, *J. Geophys. Res.*, *99*, 4643-4653, 1994.

Lanphere, M. A., $^{40}Ar/^{39}Ar$ ages of basalt from Clubhouse Crossroads test hole #2, near Charleston, South Carolina, in *Studies Related to the Charleston, South Carolina, Earthquake of 1886 — Tectonics and Seismicity*, edited by G. S. Gohn, *U.S. Geol. Surv. Prof. Paper*, *1313*, B1-B8, 1983.

LeTourneau, P. M., Depositional history and tectonic evolution of Late Triassic age rifts of the U.S. Central Atlantic margin: Results of an integrated stratigraphic, structural, and paleomagnetic analysis of the Taylorsville and Richmond basins, Ph.D. thesis, Columbia University in the City of New York, 1999.

LeTourneau, P. M., and N. G. McDonald, Facies analysis of Early Jurassic lacustrine sequences, Hartford basin, Connecticut and Massachusetts, *Geol. Soc. Am. Abstr. Programs*, *20*, 32, 1988.

Letouzey, J., Fault reactivation, inversion, and fold-thrust belt, in *Petroleum and Tectonics in Mobile Belts*, edited by J. Letouzey, IFP Editions Technip, Paris, pp. 101-128, 1990.

Lindholm, R. C., Triassic-Jurassic faulting in eastern North America--a model based on pre-Triassic structures, *Geology*, *6*, 365-368, 1978.

Lomando, A. J., and T. Engelder, Strain within the rocks of the Newark basin, New York, *Geol. Soc. Am. Abstr. Programs*, *12:2*, 70-71, 1980.

Lucas, M., J. Hull, and W. Manspeizer, A foreland-type fold and related structures of the Newark rift basin, in *Triassic-Jurassic Rifting, Continental Breakup and the Origin of the Atlantic Ocean and Passive Margins*, edited by W. Manspeizer, Elsevier, New York, pp. 307-332, 1988.

MacLean, B. C., and J. A. Wade, Petroleum geology of the continental margin south of the islands of St. Pierre and Miquelon, offshore eastern Canada, *Bull. Can. Petrol. Geol.*, *40*, 222-253, 1992.

Malcolm, R. J., M. C. Pott, E. Delfos, A new tectonostratigraphic synthesis of the north west Cape area, *APEA Journal*, *31*, 154-176, 1991.

Malinconico, M. L., Thermal history of the Early Mesozoic Newark (NJ/PA) and Taylorsville (VA) basins using borehole vitrinite reflectance: conductive and advective effects, *Geol. Soc. Am. Abstr. Programs*, *31:2*, A-31, 1999.

Malinconico, M. L., Estimates of eroded strata using borehole vitrinite reflectance data, Triassic Taylorsville rift basin, Virginia: Implications for duration of syn-rift sedimentation and evidence of structural inversion, in *Aspects of Triassic-Jurassic Rift Basin Geoscience*, edited by P. M. LeTourneau and P. E. Olsen, Columbia University Press, New York, in press, 2002.

Manspeizer, W., Rift tectonics inferred from volcanic and clastic structures, in *Field Studies of New Jersey Geology and Guide to Field Trips*, edited by W. Manspeizer, 52nd Annual Meeting of the New York State Geological Association, Rutgers University, Newark, N.J., pp. 314-350, 1980.

Manspeizer, W., Triassic-Jurassic rifting and opening of the Atlantic: An overview, in *Triassic-Jurassic Rifting, Continental Breakup and the Origin of the Atlantic Ocean and Passive Margins*, edited by W. Manspeizer, Elsevier, New York, pp. 41-79, 1988.

Marzolli, A., P. R. Renne, E. M. Piccirillo, M. Ernesto, G. Bellieni, and A. De Min, Extensive 200-million-year-old continental flood basalts of the Central Atlantic Magmatic Province, *Science*, *284*, 616-618, 1999.

May, P. R., Pattern of Triassic-Jurassic diabase dikes around the North Atlantic in the context of the predrift configuration of the continents, *Geol. Soc. Am. Bull.*, *82*, 1285-1292, 1971.

McBride, J. H., K. D. Nelson, and L. D. Brown, Evidence and implications of an extensive early Mesozoic rift basin and basalt/diabase sequence beneath the southeast Coastal Plain, *Geol. Soc. Am. Bull.*, *101*, 512-520, 1989.

McHone, J. G., Tectonic and paleostress patterns of Mesozoic intrusions in eastern North America, in *Triassic-Jurassic Rifting, Continental Breakup and the Origin of the Atlantic Ocean and Passive Margins*, edited by W. Manspeizer, Elsevier, New York, pp. 607-620, 1988.

McHone, J. G., Broad-terrane Jurassic flood basalts across northeastern North America, *Geology*, *24*, 319-322, 1996.

McHone, J. G., Non-plume magmatism and rifting during the opening of the central Atlantic Ocean, *Tectonophysics*, *316*, 287-296, 2000.

McHone, J. G., M. Talwani, W. Hames, A. Marzoli, P. Ragland, B. Turrin, M. Coffin, S. Barr, C. White, and D. Morris, Large igneous provinces, seaward dipping reflectors, and Central Atlantic Magmatic Province, in *Climatic, Biotic, and Tectonic Pole-to-Pole Coring Transect of Triassic-Jurassic Pangea*, edited by P. E. Olsen and D. V. Kent, International Continental Drilling Program, in press, 2002.

Menzies, M. A., C. Ebinger, and S. Klemperer, Penrose Conference report: Volcanic rifted margins, *GSA Today, August 2000*, 8-11, 2000.

Mitra, S., and Q. T. Islam, Experimental (clay) models of inversion structures, *Tectonophysics*, *230*, 211-222, 1994.

Mutter, J. C., W. R. Buck, and C. M. Zehnder, Convective partial melting I: A model for formation of thick igneous sequences during initiation of spreading, *J. Geophys. Res.*, *93*, 1031-1048, 1988.

Oh, J., J. A. Austin, Jr., J. D. Phillips, M. F. Coffin, and P. L. Stoffa, Seaward-dipping reflectors offshore the southeastern United States: Seismic evidence for extensive volcanism accompanying sequential formation of the Carolina trough and Blake Plateau basin, *Geology*, *23*, 9-12, 1995.

Olsen, P. E., Stratigraphic record of the early Mesozoic breakup of Pangea in the Laurasia-Gondwana rift system, *Ann. Rev. Earth Planet. Sci.*, *25*, 337-401, 1997.

Olsen, P. E., Giant lava flows, mass extinctions, and mantle plumes, *Science, 284*, 604-605, 1999.

Olsen, P. E., and R.W. Schlische, Transtensional arm of the early Mesozoic Fundy rift basin: Penecontemporaneous faulting and sedimentation, *Geology, 18*, 695-698, 1990.

Olsen, P. E., R. W. Schlische, and P. J. W. Gore, *Tectonic, Depositional, and Paleoecological History of Early Mesozoic Rift Basins of Eastern North America*, International Geological Congress Field Trip T-351, AGU, Washington, 174 pp., 1989.

Olsen, P. E., D. V. Kent, B. Cornet, W. K. Witte, and R. W. Schlische, High-resolution stratigraphy of the Newark rift basin (early Mesozoic, eastern North America), *Geol. Soc. Am. Bull., 108*, 40-77, 1996a.

Olsen, P. E., R. W. Schlische, and M. S. Fedosh, 580 kyr duration of the Early Jurassic flood basalt event in eastern North America estimated using Milankovitch cyclostratigraphy, in *The Continental Jurassic*, edited by M. Morales, *Museum of Northern Arizona Bull., 60*, 11-22, 1996b.

Parker, R. A., H. F. Houghton, and R. C. McDowell, Stratigraphic framework and distribution of early Mesozoic rocks of the northern Newark basin, New Jersey and New York, in *Studies of the Early Mesozoic Basins of the Eastern United States*, edited by A. J. Froelich and G. R. Robinson, Jr., *U.S. Geol. Surv. Bull., 1776*, 31-39, 1988.

Phillips, J. D., Paleomagnetic investigations of the Clubhouse Crossroads basalt, *U.S. Geol. Surv. Prof. Paper, 1313-C*, 1-18, 1983.

Philpotts, A. R., and A. Martello, Diabase feeder dikes for the Mesozoic basalts in southern New England, *Am. J. Sci., 286*, 105-126, 1986.

Poag, C. W., and P. C. Valentine, Mesozoic and Cenozoic stratigraphy of the United States Atlantic continental shelf and slope, in *The Geology of North America, v. I-2, The Atlantic Continental Margin, U.S.*, edited by R. E. Sheridan and J. A. Grow, Geological Society of America, Boulder, pp. 67-85, 1988.

Pratt, L. M., C. A. Shaw, and R. C. Burruss, Thermal histories of the Hartford and Newark Basins inferred from maturation indices of organic matter, in *Studies of the Early Mesozoic Basins of the Eastern United States*, edited by A. J. Froelich and G. R. Robinson, Jr., *U.S. Geol. Surv. Bull., 1776*, 58-62, 1988.

Prevot, M., and M. McWilliams, Paleomagnetic correlation of the Newark Supergroup volcanics, *Geology, 17*, 1007-1010, 1989.

Prowell, D. C., Cretaceous and Cenozoic tectonism on the Atlantic coastal margin, in *The Geology of North America, v. I-2, The Atlantic Continental Margin, U.S.*, edited by R. E. Sheridan and J. A. Grow, Geological Society of America, Boulder, pp. 557-564, 1988.

Puffer, J. H., Contrasting high field strength element contents of continental flood basalts from plume versus reactivated-arc sources, *Geology, 29*, 675-678, 2001.

Puffer, J. H., and A. R. Philpotts, Eastern North American quartz tholeiites: geochemistry and petrology, in *Triassic-Jurassic Rifting, Continental Breakup and the Origin of the Atlantic Ocean and Passive Margins*, edited by W. Manspeizer, Elsevier, New York, pp. 579-605, 1988.

Ragland, P. C., Mesozoic igneous rocks, in *Geology of the Carolinas*, edited by J. W. Horton, and V. A. Zullo, University of Tennessee Press, Knoxville, pp. 171-190, 1991.

Ragland, P. C., R. D. Hatcher, Jr., and D. Whittington, Juxtaposed Mesozoic diabase dike sets from the Carolinas: a preliminary assessment, *Geology, 11*, 394-399, 1983.

Ragland, P. C., L. E. Cummins, and J. D. Arthur, Compositional patterns for early Mesozoic diabases from South Carolina to central Virginia, in *Eastern North American Mesozoic Magmatism*, edited by J. H. Puffer and P. C. Ragland, *Geol. Soc. Am. Spec. Paper, 268*, 301-331, 1992.

Ratcliffe, N. M., and W. C. Burton, Fault reactivation models for the origin of the Newark basin and studies related to U.S. eastern seismicity, *U.S. Geol. Surv. Circular, 946*, 36-45, 1985.

Ratcliffe, N. M., W. C. Burton, R. M. D'Angelo, and J. K. Costain, Low-angle extensional faulting, reactivated mylonites, and seismic reflection geometry of the Newark Basin margin in eastern Pennsylvania, *Geology, 14*, 766-770, 1986.

Ratcliffe, N. M., Structural analysis of the Furlong fault and the relationship of mineralization to faulting and diabase intrusion, Newark basin, Pennsylvania, in *Studies of the Early Mesozoic Basins of the Eastern United States*, edited by A. J. Froelich and G. R. Robinson, Jr., *U.S. Geol. Surv. Bull., 1776*, 176-193, 1988.

Sanders, J. E., Late Triassic tectonic history of northeastern United States, *Am. J. Sci., 261*, 501-524, 1963.

Schlee, J. S., and K. D. Klitgord, Georges Bank basin: A regional synthesis, in *The Geology of North America, v. I-2, The Atlantic Continental Margin, U.S.*, edited by R. E. Sheridan and J. A. Grow, Geological Society of America, Boulder, pp. 243-268, 1988.

Schlische, R. W., Structural and stratigraphic development of the Newark extensional basin, eastern North America; Implications for the growth of the basin and its bounding structures, *Geol. Soc. Am. Bull., 104*, 1246-1263, 1992.

· Schlische, R. W., Anatomy and evolution of the Triassic-Jurassic continental rift system, eastern North America, *Tectonics, 12*, 1026-1042, 1993.

Schlische, R. W., Geometry and origin of fault-related folds in extensional settings, *AAPG Bull., 79*, 1661-1678, 1995.

Schlische, R. W., Progress in understanding the structural geology, basin evolution, and tectonic history of the eastern North American rift system, in *Aspects of Triassic-Jurassic Rift Basin Geoscience*, edited by P. M. LeTourneau and P. E. Olsen, Columbia University Press, New York, in press, 2002.

Schlische, R. W., and M. H. Anders, Stratigraphic effects and tectonic implications of the growth of normal faults and extensional basins, in *Reconstructing the Structural History of Basin and Range Extension Using Sedimentology and Stratigraphy*, edited by K. K. Beratan, *Geol. Soc. Am. Spec. Paper, 303*, 183-203, 1996.

Schlische, R. W., and P. E. Olsen, Quantitative filling model for continental extensional basins with applications to early Mesozoic rifts of eastern North America, *J. Geol., 98*, 135-155, 1990.

Schlische, R. W., M. O. Withjack, and G. Eisenstadt, Using

scaled physical models to study the deformation pattern produced by basement-involved oblique extension, *AAPG Ann. Meeting Abstr.*, *8*, A-125, 1999a.

Schlische, R. W., M. O. Withjack, and G. Eisenstadt, An experimental study of the secondary fault patterns produced by oblique-slip normal faulting, *Eos Trans. AGU*, *80 (17)*, S339, 1999b.

Schlische, R. W., M. O. Withjack, and G. Eisenstadt, An experimental study of the secondary deformation produced by oblique-slip normal faulting, *AAPG Bull.*, in press, 2002a.

Schlische, R. W., M. O. Withjack, J. A. Austin, D. E. Brown, J. Contreras, E. Gierlowski-Kordesch, L. F. Jansa., M. L. Malinconico, J. P. Smoot, and R. P. Wintsch, Basin Evolution, in *Climatic, Biotic, and Tectonic Pole-to-Pole Coring Transect of Triassic-Jurassic Pangea*, edited by P. E. Olsen and D. V. Kent, International Continental Drilling Program, in press, 2002b.

Sepkowski, J. J., Biodiversity: past, present and future, *J. Paleont.*, *71*, 533-539, 1997.

Shaler, N. S., and J. B. Woodworth, Geology of the Richmond basin, Virginia, *U.S. Geol. Surv. Ann. Report*, *19*, pp. 1246-1263, 1899.

Sheridan, R. E., and J. A. Grow, editors, 1988, *The Geology of North America, v. I-2, The Atlantic Continental Margin, U.S.*, Geological Society of America, Boulder, 1988.

Sheridan, R. E., D. L. Musser, L. Glover III., M. Talwani, J. I. Ewing, W. S. Holbrook, G. M. Purdy, R. Hawman, and S. Smithson, Deep seismic reflection data of EDGE U.S. mid-Atlantic continental-margin experiment: Implications for Appalachian sutures and Mesozoic rifting and magmatic underplating, *Geology*, *21*, 563-567, 1993.

Sibson, R. H., Selective fault reactivation during basin inversion: potential for fluid redistribution through fault-valve action, in *Basin Inversion*, edited by J. G. Buchanan and P. G. Buchanan, *Geol. Soc. Spec. Publ.*, *88*, 3-19, 1995.

Smith, W. A., Paleomagnetic results from a crosscutting system of northwest and north-south trending diabase dikes in the North Carolina Piedmont, *Tectonophysics*, *136*, 137-150, 1987.

Srivastava, S. P., and C. R. Tapscott, Plate kinematics of the North Atlantic, in *The Geology of North America, v. M, The Western North Atlantic Region*, edited by P. R. Vogt and B. E. Tucholke, Geological Society of America, Boulder, pp. 379-404, 1986.

Steckler, M. S., G. I. Omar, G. D. Karner, and B. P. Kohn, Pattern of hydrothermal circulation with the Newark basin from fission-track analysis, *Geology*, *21*, 735-738, 1993.

Sutter, J. F., Innovative approaches to the dating of igneous events in the early Mesozoic basins of the Eastern United States, in *Studies of the Early Mesozoic Basins of the Eastern United States*, edited by A. J. Froelich and G. R. Robinson, Jr., *U.S. Geol. Surv. Bull.*, *1776*, 194-200, 1988.

Swanson, M. T., Preexisting fault control for Mesozoic basin formation in eastern North America, *Geology*, *14*, 419-422, 1986.

Talwani, M., and O. Eldholm, Evolution of the Norwegian-Greenland Sea, *Geol. Soc. Am. Bull.*, *88*, 969-999, 1977.

Talwani, M., J. Ewing, R. E. Sheridan, W. S. Holbrook, and L. Glover III, The EDGE experiment and the U.S. East Coast magnetic anomaly, in *Rifted Ocean-Continent Boundaries*, edited by E. Banda, et al., Kluwer Academic Publishers, Dordrecht, The Netherlands, pp. 155-181, 1995.

Tankard, A. J., and H. J. Welsink, Mesozoic extension and styles of basin formation in Atlantic Canada, in *Extensional Tectonics and Stratigraphy of the North Atlantic Margins*, edited by A. J. Tankard and H. R. Balkwill, *AAPG Mem.*, *46*, 175-195, 1989.

Tanner, L. H., and J. F. Hubert, Basalt breccias and conglomerates in the Lower Jurassic McCoy Brook Formation, Fundy basin, Nova Scotia: Differentiation of talus and debris-flow deposits, *J. Sed. Petrol.*, *61*, 15-27, 1991.

Tindall, S. E., G. Eisenstadt, M. O. Withjack, and R. W. Schlische, Interpretation of oblique compressional structures -- comparisons of field examples and physical models, *AAPG Ann. Meeting Abstr.*, *8*, A139, 1999a.

Tindall, S. E., G. Eisenstadt, M. O. Withjack, and R. W. Schlische, Characteristic features of oblique basement-cored uplifts, *Geol. Soc. Am., Abstr. Programs*, *31*, A-236, 1999b.

Tollo, R. P., and D. Gottfried, Petrochemistry of Jurassic basalt from eight cores, Newark basin, New Jersey; implications for volcanic petrogenesis of the Newark Supergroup, in *Eastern North American Mesozoic Magmatism*, edited by J. H. Puffer and P. C. Ragland, *Geol. Soc. Am. Spec. Paper*, *268*, 233-259, 1992.

Tseng, H-Y., T. C. Onstott, R. C. Burruss, and D. S. Miller, Constraints on the thermal history of the Taylorsville basin, Virginia, U.S.A., from fluid-inclusion and fission-track analyses: implications for subsurface geomicrobiology experiments, *Chem. Geol.*, *127*, 297-311, 1996.

Turrin, B. D., $^{40}Ar/^{39}Ar$ mineral ages and potassium and argon systematics from the Palisade Sill, New York, *Eos Trans. AGU*, *81 (48)*, Fall Meet. Suppl., Abstract V72E-13, 2000.

Vågnes, E., R. H. Gabrielsen, and P. Haremo, Late Cretaceous-Cenozoic intraplate contractional deformation at the Norwegian continental shelf: timing, magnitude and regional implications, *Tectonophysics*, *300*, 29-46, 1998.

Venkatakrishnan, R., and R. Lutz, A kinematic model for the evolution of the Richmond basin, in *Triassic-Jurassic Rifting, Continental Breakup and the Origin of the Atlantic Ocean and Passive Margins*, edited by W. Manspeizer, Elsevier, New York, pp. 445-462, 1988.

Wade, J. A., and B. C. MacLean, The geology of the southeastern margin of Canada, in *Geology of the Continental Margin of Eastern Canada*, edited by M. J. Keen and G. L. Williams, Geol. Surv. Canada, *Geology of Canada No. 2*, 167-238, 1990.

White, R. S., Magmatism during and after continental break-up, in *Magmatism and the Causes of Continental Break-up*, edited by B. C. Storey, et al., *Geol. Soc. Spec. Publ.*, *68*, 1-16, 1992.

Williams, G. D., C. M. Powell, and M. A. Cooper, Geometry and kinematics of inversion tectonics, in *Inversion Tectonics*, edited by M. A. Cooper and G. D. William, *Geol. Soc. Spec. Publ.*, *44*, 3-15, 1989.

Wise, D. U., Dip domain method applied to the Mesozoic Connecticut Valley rift basins, *Tectonics*, *11*, 1357-1368, 1993.

Withjack, M. O., and G. Eisenstadt, Structural history of the Northwest Shelf, Australia -- an integrated geological, geophysical and experimental approach, *AAPG Ann. Meeting Abstr.*, *8*, A151, 1999.

Withjack, M. O., P. E. Olsen, and R. W. Schlische, Tectonic evolution of the Fundy basin, Canada: Evidence for extension and shortening during passive margin formation, *Tectonics*, *14*, 390-405, 1995.

Withjack, M. O., R. W. Schlische, and P. E. Olsen, Diachronous rifting, drifting, and inversion on the passive margin of central eastern North America: An analog for other passive margins, *AAPG Bull.*, 82, 817-835, 1998.

Withjack, M. O., R. W. Schlische, and P. E. Olsen, Influence of rift basin structural geology on sedimentology and stratigraphy: a review, in *Continental Rift Basin Sedimentology*, edited by R. Renaut and G. M. Ashley, *SEPM Spec. Publ.*, in press, 2002.

Witte, W. K., and D. V. Kent, Tectonic implications of a remagnetization event in the Newark Basin, *J. Geophys. Res.*, *96*, 19,569-19,582, 1991.

Witte, W. K., D. V. Kent, and P. E. Olsen, Magnetostratigraphy and paleomagnetic poles from Late Triassic-earliest Jurassic strata of the Newark Basin, *Geol. Soc. Am. Bull.*, *103*, 1648-1662, 1991.

Zoback, M. L., and M. D. Zoback, Tectonic stress field of the conterminous United States, *Geol. Soc. Am. Mem.*, *172*, 523-539, 1989.

Paul E. Olsen, Department of Earth and Environmental Sciences and Lamont-Doherty Earth Observatory of Columbia University, Rt 9W, Palisades, New York 10964.

Roy W. Schlische and Martha Oliver Withjack, Department of Geological Sciences, Rutgers University, 610 Taylor Road, Piscataway, New Jersey 08854-8066.

Age Estimates of the Seaward-Dipping Volcanic Wedge, Earliest Oceanic Crust, and Earliest Drift-Stage Sediments Along the North American Atlantic Continental Margin

Richard N. Benson

Delaware Geological Survey, University of Delaware, Newark Delaware

Owing to their depths of burial along and adjacent to the North American continental margin, there is no direct evidence obtained from boreholes for the ages of the seaward-dipping volcanic wedge, earliest drift-stage sediments overlying the wedge, and the earliest Atlantic oceanic crust between the East Coast (ECMA) and Blake Spur (BSMA) magnetic anomalies. Maximum ages of late Sinemurian for drift-stage sediments have been determined from exploration wells in the Scotian Basin. A similar age is postulated for those sediments in the Georges Bank Basin, but palynomorphs from exploration wells may indicate that earliest drift-stage sediments, in places associated with volcanic rocks, are of Bajocian age and occur higher in the section above the postrift unconformity as recognized on seismic lines. In the Southeast Georgia Embayment of the Blake Plateau Basin, the oldest drift-stage sediments overlying the postrift unconformity that were drilled are of Kimmeridgian-Tithonian age. In the Baltimore Canyon Trough, the volcanic wedge overlies the postrift unconformity which truncates buried synrift rocks that may be as young as Sinemurian. In the Carolina Trough and Blake Plateau Basin, a possible offshore flood basalt marking the postrift unconformity and traced as a reflector to the volcanic wedge may correspond to a subsurface flood basalt onshore that may be part of CAMP (Hettangian). Alternatively, its magmatic source may have been that of the possibly younger volcanic wedge. Sea-floor-spreading-rate lines based on the latest Jurassic time scales and extended to the BSMA and ECMA indicate ages of 166 and 171 Ma for the BSMA and 172 and 179 Ma for the ECMA. An alternative model suggests a middle Pliensbachian/early Toarcian age (188 –190 Ma) for the igneous activity that produced the volcanic wedge and earliest oceanic crust.

INTRODUCTION

Seaward-dipping reflectors (SDRs) on offshore seismic reflection profiles of the Atlantic continental margin of North America [*Hinz*, 1981; *Benson and Doyle*, 1988; *Austin et al.*, 1990; *Sheridan et al.*, 1993; *Oh et al.*, 1995] are assumed to represent a seaward-thickening, wedge-shaped sequence of volcanic and interbedded sedimentary rocks that accumulated within a few million years. These rocks and underlying intrusive and underplated rocks are considered to be the major source of the East Coast Magnetic Anomaly (ECMA) [e.g., *Kelemen and Holbrook*, 1995; *Talwani et al.*, 1995; *Lizarralde and Holbrook*, 1997; *Talwani and Abreu*, 2000].

The Central Atlantic Magmatic Province:
Insights from Fragments of Pangea
Geophysical Monograph 136
Copyright 2003 by the American Geophysical Union
10.1029/136GM04

The volcanic wedge is associated with the final breakup of Pangea and is considered nearly contemporaneous with the formation of the earliest Atlantic oceanic crust as sea-floor spreading began. Its seismic image is similar to that of the early Tertiary volcanic wedge that accumulated during the initial opening of the Norwegian-Greenland Sea and was drilled off the Norwegian continental margin [*Talwani and Udintsev et al.*, 1976; *Mutter et al.*, 1982; *Eldholm et al.*, 1989].

The rocks represented by the SDRs along the North American margin have not been sampled by drilling as they are too deeply buried by sedimentary rocks of the drift stage. As their chemistry, petrology, and age are unknown, the rocks of the volcanic wedge cannot be directly related to the igneous rocks of the Central Atlantic Magmatic Province (CAMP). *McHone* [2000] summarizes that most dates of the CAMP period of magmatism are between 196 and 202 Ma and cites *Olsen*'s [1997] stratigraphic studies of basalts of the Triassic-Jurassic rift basins that indicate most or all of the magmatism occurred close to 201 Ma during early Hettangian time. *McHone and Puffer* [2001, in press] estimated the volume of the volcanic wedge as more than 1 million km³. They included this figure in arriving at their estimate of over 2 million km³ for all CAMP basalts, a figure that would be halved if the volcanic wedge was found to be significantly younger than the CAMP basalts.

Sediments of the drift stage would not have begun to accumulate until sufficient accommodation space was provided by the relatively rapid subsidence of the wedge and adjacent earliest oceanic crust by loading and thermal contraction [*Eldholm et al.*, 1989]. Given this assumption, the age of the oldest drift-stage sediments should be close to the minimum age of the magmatism that produced the volcanic wedge and earliest oceanic crust. Knowledge of the beginning and rates of subsidence is important in modeling the petroleum systems of the offshore sedimentary basins. In this paper I review the interpretations of the ages of the oldest drift-stage rocks recovered from exploration wells in the offshore sedimentary basins of the North American Atlantic continental margin (see *Wade and MacLean* [1990] and *Mattick and Libby-French* [1988] for well locations).

Neither the oldest Atlantic oceanic crust, the overlying sediments, nor acoustic horizons correlated to those sediments have been drilled. *Klitgord and Schouten* [1986, Fig. 5] extrapolated western Atlantic sea-floor-spreading rates to the ECMA to estimate the age of the earliest oceanic crust as 175 Ma. I revise their figure using more recent Jurassic time scales, an older age for basement at DSDP Site 534, and results of the 1990 EDGE mid-Atlantic onshore/offshore seismic experiment [*Lizarralde and Holbrook*, 1997].

OFFSHORE BASIN DRILLING RESULTS

The maximum thicknesses of drift-stage (postrift) sediments of the North American continental margin occur in the offshore basins shown in Figure 1. Seismic reflectors representing these sediments onlap a major regional unconformity in a landward direction. For trailing continental margins, *Falvey* [1974] considered the unconformity to represent the "breakup unconformity" and defined it as marking the onset of sea-floor spreading [*Wade and MacLean*, 1990]. *Schlee and Klitgord* [1988] refer to this seismic marker as the "postrift unconformity." *Wade and MacLean* [1990] refer to it as the "base event." Because of its genetic implications the term "breakup unconformity" should not be applied to the entire landward extent of the unconformity. The unconformity represents continuous erosion through time after breakup until it is preserved beneath progressively younger onlapping sediments of the drift stage. In Delaware, for example, the oldest drift-stage sediments at the entrance to Delaware Bay are represented by Kimmeridgian-age seismic reflectors that onlap the unconformity [*Benson et al.*, 1986]; 90 km updip, the oldest Atlantic Coastal Plain sediments above the unconformity are of Barremian?-Aptian? age [*Doyle and Robbins*, 1977].

The oldest marine to marginal marine drift-stage sediments should occur at the edge of the continental margin as this was where subsidence of the margin was sufficient to accommodate the first incursion of the Tethyan sea. For the North American margin these should be the first sediments to accumulate above the volcanic wedge; most likely they were evaporites and carbonates deposited in a shallow ocean with limited water circulation [*Talwani et al.*, 1995]. Only in the Scotian and Georges Bank basins and the Southeast Georgia Embayment of the Blake Plateau Basin did exploration wells penetrate the unconformity, either into underlying pre-Mesozoic basement or Mesozoic rift basin (synrift) rocks. These wells, however, are updip from the volcanic wedge; therefore, ages of the oldest drift-stage sediments drilled may be younger than the earliest sediments overlying the wedge.

Figure 2 summarizes the published ages of the oldest drift-stage rocks that were drilled in each of the offshore basins and the rocks underlying the postrift (PRU) or breakup unconformity. The 1994 time scale for the Triassic and Jurassic is that of *Gradstein et al.* [1994], and the 2000 time scale for the Jurassic is that of *Pálfy et al.* [2000].

Scotian Basin

Wade and MacLean [1990] describe the geology of the Scotian Basin from seismic and well data. The oldest dated (by palynomorphs) Mesozoic rocks in the Scotian Basin are Rhaetian to Hettangian-Sinemurian. They are the Eurydice red beds and Argo evaporites which occupy rift grabens. Along the margin of the basin they are truncated by the breakup unconformity. The authors interpret the breakup unconformity as the upper eroded surface of the volcanic rocks in the offshore Glooscap C-63 well (Fig. 1). They correlate the "Glooscap volcanics," a 150-m thick basalt sequence between the Argo and Eurydice formations and the overlying Mohican Formation, with the North Mountain Basalt, which is part of CAMP, and date them as early Sinemurian. The Mohican Formation completed the filling of the rift grabens and overlapped the basement highs. Together with the Iroquois Formation the two formations overlie the breakup unconformity in the Scotian Basin. *Wade and MacLean* [1990] cite personal communications from E.H. Davies and G.L. Williams [1986] that the overall age of these formations is late Sinemurian? to early Bajocian. Thus, the earliest drift-stage sediments were deposited during late Sinemurian time.

Biostratigraphic indications of the first marine conditions in the drift-stage sediments are summarized by *Williams et al.* [1990]. Whereas Liassic biozones are based on nonmarine spores and pollen, predominantly marine dinoflagellate assemblages first appeared in the Toarcian to Aalenian. The oldest Jurassic sediments in which foraminifers and ostracodes have been found are of Middle Jurassic (Bathonian) age in the Scotian Basin; the earliest occurrence of foraminifers on the Grand Banks was during the Pliensbachian.

The Scatarie Member of the Abenaki Formation overlies the Mohican Formation. It is predominantly an oolitic limestone that in cross section is a seaward-thickening wedge. The unit is a strong reflector on seismic profiles. *Wade and MacLean* [1990. Fig. 5.21] show a Bathonian-age unconformity for the western Nova Scotian shelf that separates the Scatarie Member from overlying units.

Georges Bank Basin

There are different interpretations of the stratigraphy of the Georges Bank Basin that are based on data from two Continental Offshore Stratigraphic Test (COST) wells G-1 and G-2, eight exploration wells, and seismic reflection profiles. For the oldest Mesozoic rocks drilled, only palynomorphs yield biostratigraphic data as calcareous fossils are absent.

Although seismic profiles indicate the presence of rift basins beneath the breakup unconformity (base event), according to *Wade* [1990], no synrift rocks were drilled.

The COST G-1 well penetrated the unconformity at 4737 m subsea into early Paleozoic metamorphic rocks. The oldest datable rocks occur 250 m above the unconformity and are of Bajocian age. *Wade* [1990, Fig. 5.7] states that the Eurydice Formation and the thick salt sequence of the Argo Formation do not extend west of the Scotian Basin. In Georges Bank Basin, coeval continental redbeds may have filled the rift basins, but that area was above the level of the invading Tethys Sea. The minimum age of the Argo is Sinemurian; therefore, if synrift rocks had been drilled in Georges Bank Basin, the assumption is that they would be no younger than this age and that this is the maximum age of the breakup unconformity.

Wade [1990] subdivides the Mesozoic-Cenozoic drift-stage rocks into four informal sequences. Sequence 1 extends from the unconformity to a distinctive seismic reflection that, on the basis of correlation to the COST and other wells, is at or near the base of Bathonian-age sediments. The maximum age of the base of sequence 1 is late Sinemurian to Pliensbachian based on the age of the breakup unconformity and associated volcanic rocks in the Scotian Basin. The sequence is 900 m thick in the COST G-1 well. The lower part is nonmarine and consists of coarse-grained to conglomeratic sandstones overlain by shales succeeded by a mixed dolostone-anhydrite facies likely deposited in a shallow marine to tidal flat environment. The COST G-2 well penetrated approximately 2200 m of sequence 1 but did not reach the postrift unconformity ("base event"). The lowermost 13 m is halite and anhydrite, apparently conformable with the overlying carbonate rocks at 6654 m subsea. The seismic character of this evaporite facies is mappable over an extensive area and averages about 300 m thick but may be twice this thickness, locally [*Ward*, 1990, Fig. 5.8]. As this salt body occurs above the unconformity, *Ward* [1990] states that it is not equivalent to the Argo Formation salt of the Scotian Basin. It probably represents an evaporite facies , possibly part of a sabkha, associated with the initial incursion of the Tethys Sea into Georges Bank Basin during Sinemurian-Pliensbachian time.

Schlee and Klitgord [1988] present an alternative interpretation of the COST well data. They place the postrift (breakup) unconformity as the acoustic reflector correlated to the top of a conglomerate layer at ~4544 m subsea in the COST G-1 well. They trace the reflector to a thin red siltstone-sandstone layer at ~6220 m in the COST G-2 well. This layer caps a 300-m-thick dolomite which overlies salt at 6654 m, the same salt that *Ward* [1990] assigns to sequence 1 of the drift-stage rocks. Although not dated, the age of the salt may be as old as Late Triassic. The authors cite no age data for the lowermost section of postrift (drift-stage) rocks

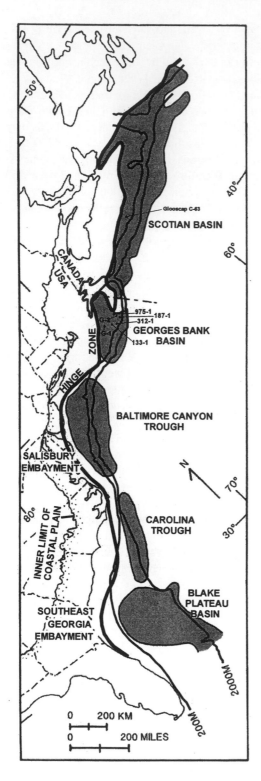

Figure 1. Offshore sedimentary basins of the Atlantic continental margin of North America [after *Benson*, 1984; *Grant et al.*, 1986; and *Klitgord* 1988] with locations of exploration wells discussed in the text. Water depths are indicated by the 200- and 2000-m bathymetric contours. Hinge zone marks the position of the abrupt increase in thickness of the drift-stage (postrift) rocks.

but indicate a possible range of Sinemurian? to early Pliensbachian?.

The palynomorph biostratigraphy determined for the COST G-2 well by *Cousminer and Steinkraus* [1988] differs significantly from the above interpretations. Those authors recovered palynomorphs of Carnian-Norian age (Late Triassic) between 4435 m and TD in the COST G-2 well. Palynomorphs from drill cuttings above this indicate a 183-m-thick attenuated Liassic? (Hettangian–Toarcian) section. The next dated interval above this is of Bajocian marine limestones between 3340 and 4050 m. They place the postrift unconformity (PRU) at 4050 m, between the dolomite-anhydrite sequence and the overlying limestone unit. This is within the drift-stage sedimentary section as interpreted on Georges Bank seismic reflection profiles [see *Manspeizer and Cousminer*, 1988, Fig. 2., and *Manspeizer et al.*, 1989, Fig. 5]. *Schlee et al.* [1988] state that the unconformity in the COST G-2 well as recognized by *Cousminer and Steinkraus* [1988] covers a significant part of Early Jurassic time and is an important tectono-stratigraphic boundary separating Upper Triassic [and Early Jurassic?] synrift clastic-evaporite facies from the overlying Middle Jurassic drift-stage carbonate-clastic facies.

Additional palynomorph data from three exploration wells on Georges Bank (Fig. 1) appear to support a major unconformity at the base of Bajocian rocks, or at least suggest that the oldest drift-stage rocks in this part of the basin are of Bajocian age. In the Exxon 975-1 well, Bajocian-age sandstones overly a Norian-age interbedded sequence of limestone, salt, and anhydrite at 4252 m subsea [*Edson et al.*, 2000a]. The interval 3356–3410 m in the Mobil 312-1 well [*Edson et al.* 2000b] is highly oxidized with residual detritus and indicates probable regression and erosion. It occurs at the base of a marginal marine unit with no palynomorph species diagnostic of either Aalenian or Toarcian age. Above this undated interval is a Bajocian–Kimmeridgian marginal marine to inner shelf depositional sequence beginning at 3191 m. Below the oxidized interval are rocks of early Pliensbachian to as old as Norian-Rhaetian age, thus indicating that an unconformity separates Bajocian from Early Jurassic sediments. Palynomorphs from the Tenneco 187-1 well [*Edson et al.* 2000c] indicate a relatively thin Bajocian interval between 3066 and 3203 m above a barren interval from 3203 to 3532 m consisting of drill cuttings of mostly micritic limestone together with oolitic limestone and shale. Below the barren interval are rocks of Pliensbachian age, also of dominant micritic limestone lithology, and below this are rocks dated as Hettangian-Sinemurian (3560–3795 m), Rhaetian-Norian (3795–4441 m), and Carnian (5087–5114 m).

Manspeizer and Cousminer [1988] claim that their paleontologic data from the COST G-2 well indicate uplift and

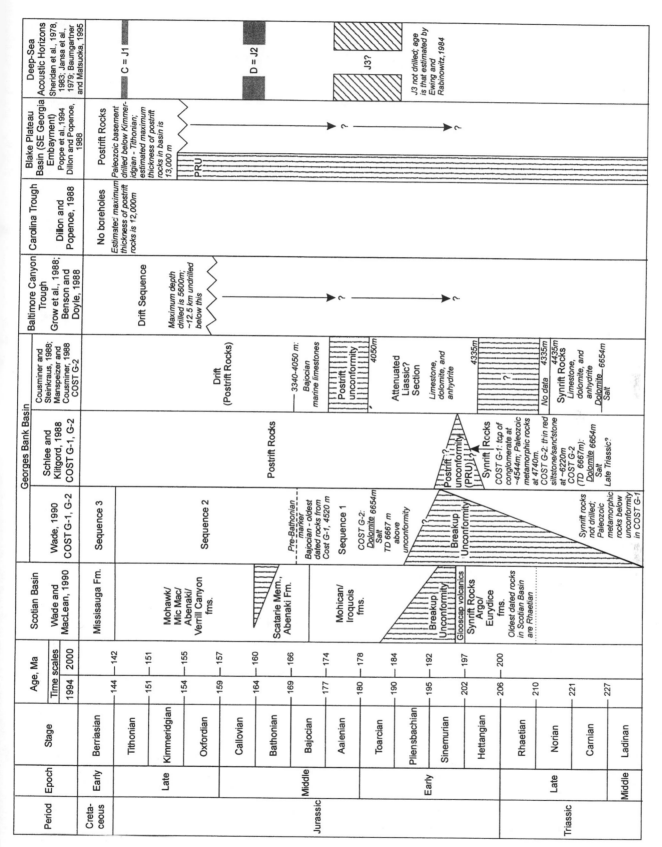

Figure 2. Stratigraphic chart showing interpreted ages of rocks drilled in the offshore basins of the Atlantic continental margin of North America and adjacent Atlantic Ocean basin. See text for explanation and discussion. Time scales are those of *Gradstein et al.* [1994] and *Pálfy et al.* [2000].

erosion after a long period of deposition of Late Triassic–Early Liassic? shallow, restricted-marine evaporites and dolomites. They propose a late rifting event, accompanied by uplift and erosion, during late Liassic–early Middle Jurassic time and that this led to Middle Jurassic vulcanism accompanied by subsidence and the onset of marine sedimentation of the earliest drift stage (Bajocian). Evidence for Bajocian-age vulcanism in the COST G-2 well is given by *Simonis* [1980] who observed that green tuff comprises as much as five percent of each sample collected between 3572 and 3627 m. In the Exxon 133-1 well (Fig. 1), *Hurtubise et al.*[1987] describe a 225-m thick unit of volcaniclastic rock, basalt flows, and minor limestone interbeds. Core samples from above the igneous unit as well as from drill cuttings just below it yielded palynomorphs of Bajocian-Bathonian age. Are these Bajocian volcanic rocks coeval with the seaward-dipping volcanic wedge?

Pe-Piper et al. [1992, Fig. 2] recognize three stages of tectonism during continental breakup of the eastern Canadian margin that accords with the above interpretation for the Georges Bank Basin by *Manspeizer and Cousminer* [1988]. (1) The first phase (Anisian–Hettangian) was rifting with the formation of grabens and their filling with sediments followed by evaporite deposition. (2) After this was a post-rifting phase of flexural uplift along and landward of the hinge zone and an increase in subsidence seaward. This was the time of emplacement of the CAMP flood basalts that overlie the postrift unconformity observed on seismic reflection profiles that had been referred to as the breakup unconformity. The Mohican Formation plus the first open-marine deposits of the Iroquois Formation belong to this phase. (3) The drift stage began near the end of the Bajocian with the formation of the first oceanic crust and the development of SDRs. Rocks of this phase, Abenaki and younger formations, are separated from those of the postrift stage by what *Pe-Piper et al.* [1992] call the breakup unconformity.

Baltimore Canyon Trough

Exploration wells in the Baltimore Canyon Trough penetrated only about a third of the entire nearly 18-km-thick sedimentary section overlying the postrift unconformity, an erosional unconformity that truncates buried Mesozoic rift basins and intervening pre-Mesozoic basement rocks. Industry paleontology reports (National Geophysical Data Center, Boulder, Colo.) indicate that the oldest rocks drilled are Late Jurassic (Oxfordian); therefore, the age of the earliest drift-stage rocks is unknown. *Grow et al.*[1988] estimate that approximately 70 percent of the section is Middle and Upper Jurassic and 15 percent each is Cretaceous and Cenozoic.

Figure 3 shows the landward limit of seaward-dipping reflectors imaged on seismic reflection profiles across the Baltimore Canyon Trough [after *Benson and Doyle*, 1988].

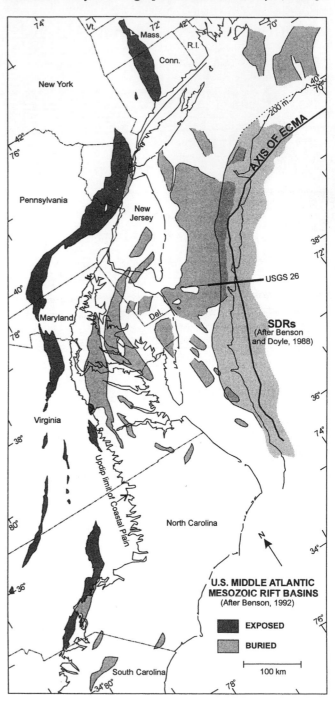

Figure 3. Map of landward limit of seaward dipping reflectors (SDRs) showing their overlap of offshore buried early Mesozoic rift basins. Bold portion of USGS seismic reflection line 26 is that shown in Figure 4.

In this interpretation, the SDRs overlap the postrift (=breakup?) unconformity that truncates reflectors interpreted as representing synrift rocks of offshore buried rift basins as mapped by *Benson* [1992]. Figure 4 [after *Benson and Doyle*, 1988] illustrates this interpretation for seismic profile USGS 26. If the synrift rocks of the offshore buried basins are of the same Carnian–Sinemurian age as some of the northern rift basins (Hartford, Fundy, Orpheus, Jeanne d'Arc, and Carson) [*Olsen*, 1997, Fig. 3], the seaward-dipping volcanic wedge, must be no older than Sinemurian, therefore younger than the CAMP volcanic rocks of Hettangian age. Reflectors representing the earliest drift-stage sediments onlap and downlap an unconformity that truncates the SDRs. This unconformity merges landward with the postrift unconformity.

Carolina Trough and Blake Plateau Basin

Dillon and Popenoe [1988] estimate maximum thicknesses of 12 and 13 km of postrift rocks, respectively, in the Carolina Trough and Blake Plateau Basin. No exploration wells were drilled in the Carolina Trough. In the Southeast Georgia Embayment of the Blake Plateau Basin, the oldest postrift rocks drilled above early Paleozoic basement at subsea depths of about 2100–3100 m are of Kimmeridgian-Tithonian age [*Poppe et al.*, 1994]. The embayment is landward of the hinge zone of the Blake Plateau Basin. On the basis of seismic profiles *Dillon and Popenoe* [1988] infer the presence of synrift rocks in rift basins and older Jurassic rocks in the overlying postrift deposits in the deeper parts of both basins.

Dillon and Popenoe [1988, Fig. 8] conclude that the onset of drifting and oceanic crust generation occurred much earlier at the Carolina Trough than at the Blake Plateau Basin. They based this on the presence of a much broader zone of oceanic crust between the Carolina Trough and the Blake Spur Magnetic Anomaly (BSMA) than there is between the Blake Plateau Basin and the BSMA. The two basins are separated by the Blake Spur Fracture Zone. The fracture zone also marks the approximate southwestern terminus of the ECMA, (therefore, of the volcanic wedge and associated intrusive rocks characteristic of the margin to the northeast?)

Oh et al. [1995] also recognize a time-transgressive breakup of northwest Africa and North America on the basis of their interpretation of the relationship between SDRs and a prominent reflector named J on seismic reflection profiles. SDRs along the hinge zone of the Carolina Trough lie beneath a reflector they identify as "J?." Along the hinge zone of the Blake Plateau Basin they also recognize a presumably younger set of SDRs but in this case above the J reflector.

The J reflection was identified by *Hamilton et al.* [1983] on land seismic reflection profiles for the contact between Upper Cretaceous sediments and Jurassic basalt recovered from the Clubhouse Crossroads core holes drilled beneath the Atlantic Coastal Plain near Charleston, South Carolina. Where the basalt may be missing the J reflection would represent the contact with pre-Mesozoic basement or presumably Mesozoic synrift redbeds. *Gottfried et al.* [1983] described the J basalts recovered from the three Clubhouse Crossroads core holes and concluded that they are strikingly similar to the exposed lower Mesozoic high-Ti, quartz-normative tholeiites and olivine-normative tholeiites in eastern North America, i.e., the 201-Ma CAMP basalts. Test hole no. 3 recovered a minimum of 256 m of basalt above a sequence of lower Mesozoic? sedimentary red beds (assumed to be synrift rocks). *Lanphere* [1983] radiometrically dated three basalt samples from core hole no. 2 with results that range between 182 and 236 Ma. He determined that only one sample met the criteria for a reliable crystallization age of 184±3.3 Ma.

McBride et al. [1989] mapped the postulated extent of the J reflector (=basalt/diabase) from Florida and Georgia through South Carolina and offshore. Several Coastal Plain boreholes recovered basalt at the depth of the reflector. Those authors conclude that the basalt marks the postrift unconformity and that the extrusive/intrusive episode across the region began as sea-floor spreading commenced in the North Atlantic. The J basalt, therefore, would be approximately contemporaneous with the volcanic wedge sequence offshore.

On seismic profiles offshore South Carolina, *Dillon et al.* [1979, p. 31] describe a high amplitude reflector marking the postrift unconformity as "A very strong and smooth reflector... that ... correlates with a high-velocity refractor of refraction velocity about 5.8 to 6.2 km/sec... that ... represents basaltic flows and pyroclastic deposits because it can be projected landward to volcanic rocks (basalts) of the ...test wells at Clubhouse Crossroads..." *Behrendt et al.* [1983] labeled this strong reflection "J" on the marine seismic profiles, although they note that the marine data are not tied to the land data. From land seismic-reflection evidence, *Hamilton et al.* [1983, Fig. 1] show a zone of "missing J" between the Clubhouse Crossroads test holes and the coast of South Carolina; therefore, it is possible that the basalts recovered from the core holes may not extend offshore and therefore would not be correlative with the J reflector offshore.

Dillon et al. [1983] inferred that the offshore presumably volcanic layer is slightly younger than the postrift unconformity. According to *Behrendt et al.* [1983], this indicates that the offshore basalt layer postdates the early opening of

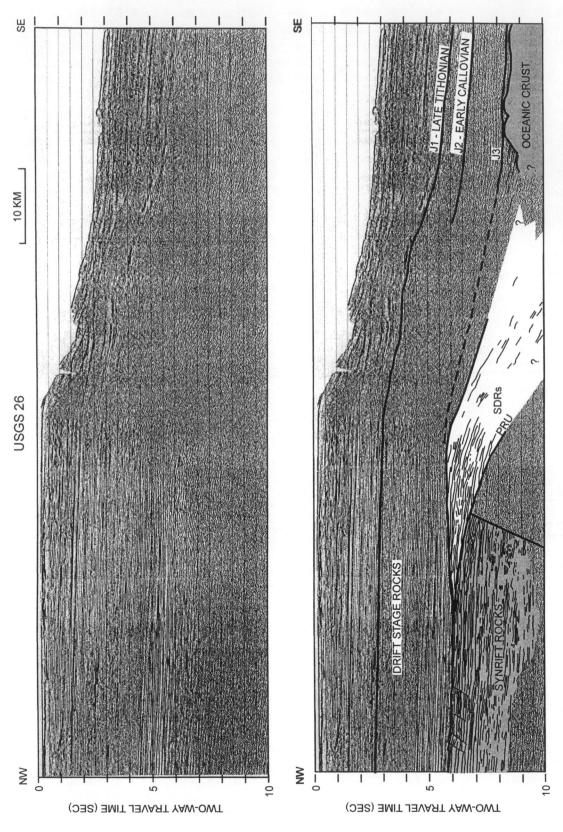

Figure 4. Part of USGS line 26 across the outer Baltimore Canyon Trough showing the overlap of the wedge of seaward-dipping reflectors (SDRs) on the postrift unconformity (PRU). The PRU truncates continuous reflectors interpreted as representing synrift rocks (as young as Sinemurian?) downfaulted (note diffractions) against pre-Mesozoic continental crust. Deep-sea acoustic horizons J1–J3 are discussed in the text. Drift-stage rocks beneath the continental shelf onlap and downlap on the PRU and the volcanic wedge. Figure is modified after *Benson and Doyle* [1988]. See Figure 3 for location of USGS 26.

the Atlantic. They also note that the seaward increase in velocity of "J" probably indicates a decrease in thickness of interbedded sedimentary rock, and they also suggest that the basalt sequence thickens seaward based on the observation by *Dillon et al.* [1979] that they were rarely able to observe reflections below "J." This suggests that these presumed volcanic rocks may be a landward extension of the seaward-dipping volcanic wedge farther offshore. An alternative is that the volcanic layer is older than the postrift unconformity which is the upper eroded surface of the volcanic rocks where present, similar to the interpretation for the Glooscap volcanics of the Scotian Basin. In this case, the layer might be older than the volcanic wedge, and its source that of the flood basalt beneath the southeastern Atlantic Coastal Plain. The offshore extent of the J reflector as mapped by *Dillon et al.* [1983] may not everywhere represent a volcanic layer but instead the postrift unconformity. If so, the J reflector and its relationship to the SDRs as described by *Oh et al.* [1995] is similar to the interpretation shown in Figure 4 for the Baltimore Canyon Trough—an unconformity (PRU) below the wedge and another above it.

Deep-Sea Acoustic Horizons

Three prominent Jurassic-age deep-sea acoustic horizons within the sedimentary prism of the western Atlantic were identified and named by *Klitgord and Grow* [1980] J1, J2, and J3 (Fig. 2). The youngest one, J1, marks the top of the Cat Gap Formation of *Jansa et al.* [1979]. It is equivalent to Horizon C in the Blake-Bahama Basin, a reflector that corresponds to the top of a red argillaceous Tithonian limestone that was drilled at DSDP Site 391[*Sheridan et al.*, 1978].

J2 terminates seaward against the oceanic basement scarp associated with the Blake Spur Magnetic Anomaly [*Klitgord and Grow*, 1980]. At DSDP Site 534 in the Blake-Bahama Basin, *Sheridan et al.* [1983] drilled through the equivalent of J2, their horizon D, and identified it as a lower Oxfordian turbiditic limestone interbedded with dark green and maroon claystone at 1550 m sub-bottom depth. Later radiolarian studies of Site 534 by *Baumgartner and Matsuoka* [1995, Fig. 4] indicate the age of Horizon D at 1550 m as early Callovian.

The oldest horizon, J3, is present over the upper continental rise where it can be traced seaward to where it onlaps the oldest (closest to the continental margin) oceanic basement that has the typical hyperbolic acoustic character [*Klitgord and Grow,* 1980]. The reflector has not been drilled as it lies near the base of the sedimentary prism. The Toarcian–Aalenian estimate of its age shown in Figure 2 is that of *Ewing and Rabinowitz* [1984].

On the USGS 26 seismic reflection profile of Figure 4, the J1 (late Tithonian) reflector can be traced to the reflector that corresponds to the top of Tithonian-age rocks in Baltimore Canyon Trough exploration wells. J2 (early Callovian) and J3 cannot be traced landward through the zone where coherent reflections are lacking. J3 projected landward more or less parallel to the J1 horizon appears to onlap or lie just a short distance above the SDRs, thus indicating that the oldest sediments of the drift stage overlying the SDRs are nearly the same age as J3. On the EDGE deep seismic reflection line MA 801 across the Virginia continental margin, *Sheridan et al.* [1993] trace J2 to the top of the Jurassic carbonate platform which overlies the Jurassic volcanic wedge. On the depth section of USGS 25 in *Ewing and Rabinowitz* [1984], there is ~1.4 km of sediment between J3 and oceanic basement rocks.

AGES FROM SEA-FLOOR-SPREADING RATES

By extending the spreading-rate line between sea-floor magnetic anomalies M-21 and M-25 across the Jurassic Magnetic Quiet Zone along the IPOD/USGS seismic reflection profile, *Klitgord and Schouten* [1986, Fig. 5] estimated the ages of the Blake Spur Magnetic Anomaly (BSMA) and the East Coast Magnetic Anomaly (ECMA) on a plot of distance from Mid-Atlantic Ridge vs. age by using two different time scales, *Harland et al.* [1982] and *Kent and Gradstein* [1986]. The line based on the latter time scale gives an early Bathonian age of ~175 Ma for the ECMA, thus the presumed age of the earliest Atlantic oceanic crust. For control between M-25 and the BSMA and ECMA, they used the middle Callovian age determination by *Sheridan et al.* [1983] for the oldest sediments on oceanic basalt at DSDP Site 534.

Figure 5 is adapted from their plot. I used the 1994 time scale of *Gradstein et al.* [1994] and, for comparison, the 2000 time scale of *Pálfy et al.* [2000]. For age control on the oldest sediments on oceanic basalt at DSDP Site 534, I used the age determination of middle Bathonian based on *Baumgartner and Matsuoka*'s [1995] study of radiolarians recovered from those sediments. The 2000 and 1994 spreading-rate lines extended to the BSMA indicate its age as 166 Ma (early Bathonian) and 171 Ma (late Bajocian), respectively. The lines extended to the ECMA without change in slope give ages of 172 Ma (early Bajocian) for the 2000 time scale and 179 Ma (early Aalenian) for the 1994 time scale. *Vogt* [1973] noted the absence on the conjugate African margin of a corridor of oceanic crust comparable in width to that between the ECMA and BSMA. This corridor, therefore, should be the oldest Atlantic Ocean crust. *Vogt*

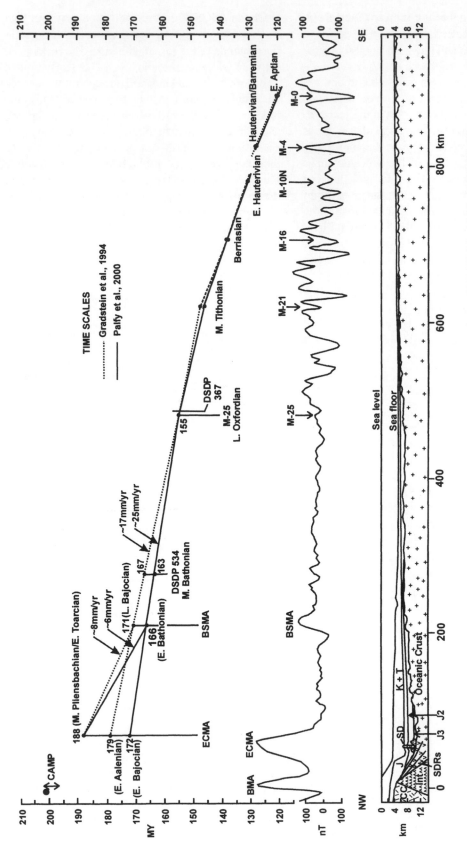

Figure 5. Sea-floor-spreading rate lines across the western Atlantic basin along IPOD-USGS seismic reflection line modified after *Klitgord and Schouten* [1986, Fig. 5] and based on the time scales of *Gradstein et al.* [1994] and *Pálfy et al.* [2000]. Ages of sea-floor magnetic anomalies M-0 through M-25 are those of *Gradstein et al.* [1994]. Bottom panel is a composite interpretation after *Klitgord and Schouten* [1986], *Lizarralde and Holbrook* [1997], and *Talwani and Abreu* [2000]. CAMP–Central Atlantic Magmatic Province; BMA–Brunswick magnetic anomaly; ECMA–East Coast Magnetic Anomaly; BSMA–Blake Spur Magnetic Anomaly; CC–continental crust; INT–intrusives; SDRs–seaward-dipping reflectors; SD–salt domes; J2, J3–deep-sea acoustic horizons; J–Jurassic, K–Cretaceous, T–Tertiary.

[1973] postulated a spreading center jump to the BSMA following the generation of the earliest crust. Extrapolation of spreading-rate lines between M-25 and DSDP 534 to the BSMA, therefore, may be justified. Maintaining the same slope of the spreading-rate lines between the BSMA and ECMA is not justified without direct age data for this corridor.

Figure 5 shows another set of spreading-rate lines between the BSMA and ECMA that are based on the pattern of continental margin subsidence during the accumulation of the seaward-dipping volcanic wedge as modeled by *Lizarralde and Holbrook* [1997]. Their analysis is based on data from the 1990 EDGE mid-Atlantic onshore/offshore seismic experiment. They assume (1) the wedge basalts (SDRs) represent the extrusive component of "subaerial sea-floor spreading;" (2) the seaward termination of the SDRs, which coincides with the ECMA [*Sheridan et al.*, 1993, Fig. 3], represents the submergence of the spreading center; and (3) a "somewhat arbitrary" age of 190 Ma for the initiation of subaerial sea-floor spreading that corresponds to the end of onshore igneous activity according to *Manspeizer et al.* [1989]. *Lizarralde and Holbrook* [1997] calculate a time of 2.3 m.y. for the accumulation of the volcanic wedge and subaerial spreading, after which submarine sea-floor spreading began as the spreading center subsided below sea level, at approximately 188 Ma. This age estimate corresponds to middle Pliensbachian on the 2000 time scale and early Toarcian on the 1994 time scale. Assuming the spreading center remained at the position of the ECMA until the jump to the BSMA position, average spreading rates for this corridor of the earliest Atlantic oceanic crust calculate as ~6–8 mm/yr for the two time scales shown. According to this model, if submarine sea-floor spreading began instead shortly after the CAMP flood basalts at ~201 Ma, the average spreading rate would be about 2 mm/yr.

At the conjugate continental margin of Africa, *Roeser* [1982] mapped two linear, subparallel, closely spaced (~50 km separation) magnetic anomalies, S1 (seaward) and S2 (landward). He considered S1 to be the counterpart of the ECMA, marking the approximate landward edge of oceanic crust. If the conclusion by *Roest et al.* [1992] is correct that S1, instead, is a sea-floor spreading lineation and the counterpart of the BSMA, *Steiner et al.* [1998] state that this would imply that the continent-ocean boundary is situated east of S1, just before the S2 anomaly. This has importance with regard to the age of the BSMA and earliest oceanic crust. *Steiner et al.* [1998] propose a Toarcian? age for the oldest sedimentary rocks unconformably overlying the uplifted Atlantic oceanic crust (N-MORB flows and breccias) at the base of the exposed Fuerteventura Island (Canary Islands) Jurassic sedimentary sequence. The site is located approximately 50 km <u>west</u> of S1. If S1 is the counterpart of the BSMA, the BSMA must be older than its extrapolated ages shown in Figure 5. *Steiner et al.*'s [1998] age determination, however, is based on indirect evidence as they do not mention recovering any fossils from the 130- to 150-m-thick "sediment-dominated" subunit B of their basal unit, the subunit that overlies the basalts. It is in the overlying pelagic bivalve limestone unit where they note the first occurrence of fossils, that of *Bositra buchi*, a biostratigraphic marker they cite as indicating a Toarcian to Oxfordian age. Because of its long geologic range, the first occurrence of this fossil does not provide definitive evidence for the age of the underlying subunit B rocks; they could be older than Toarcian but as young as Oxfordian, depending on when *Bositra* first appeared at the site.

SUMMARY AND CONCLUSIONS

There is no direct evidence available to determine the ages of emplacement of the seaward-dipping volcanic wedge, the formation of the earliest Atlantic oceanic crust, and when the earliest sediments of the drift stage were deposited. Owing to their depth of burial, neither the wedge sequence nor the earliest drift-stage sediments overlying the wedge have been drilled.

Offshore basin drilling results are the most extensive for the Scotian Basin. Earliest sediments above the postrift unconformity are of late Sinemurian age based on nonmarine spores and pollen. Predominantly marine dinoflagellate assemblages first appeared in the Toarcian to Aalenian.

Authors of studies of the Georges Bank Basin sedimentary record do not agree on definition of the postrift unconformity, nor on the age of breakup or of the earliest drift-stage sediments. On seismic reflection profiles the postrift unconformity has been identified (1) as the "base event" overlying pre-Mesozoic basement and the inset sedimentary rocks of Mesozoic rift basins, (2) a few hundred meters above the "base event," and (3) on the basis of a major unconformity defined by palynomorphs, >2600 m above the "base event." For the first and second definitions, the maximum age of the drift-stage sediments is late Sinemurian to Pliensbachian based on comparison with the Scotian Basin section. For the third, which is based on palynomorph data, the oldest drift-stage sediments are of Bajocian age. Volcanic rocks of Bajocian-Bathonian age have been recovered from two Georges Bank Basin exploration wells. This supports the interpretation of *Pe-Piper et al.* [1992] that for the eastern Canadian margin, the drift stage began near the end of the Bajocian with the formation of the first oceanic crust and the development of SDRs. They interpret the CAMP flood basalts as overlying the "base event" which they define as the postrift unconformity; the breakup unconformity occurring higher in the section is overlain by Bajocian-age sedimentary rocks.

No drill holes reached the base of drift-stage sediments in the Baltimore Canyon Trough. No exploration wells have been drilled in the Carolina Trough. Two exploration wells in the Southeast Georgia Embayment, the updip part of the Blake Plateau Basin, reached Paleozoic basement beneath the oldest drift-stage rocks there of Kimmeridgian-Tithonian age.

On seismic reflection profiles across the Baltimore Canyon Trough, the seaward-dipping volcanic wedge overlaps the postrift unconformity which truncates reflectors interpreted as representing synrift rocks of buried Mesozoic rift basins. If, as in some of the northern rift basins, the synrift rocks are as young as Sinemurian, the volcanic wedge must be no older than Sinemurian, therefore, younger than the CAMP basalts of Hettangian age. The volcanic wedge is onlapped and downlapped by the earliest drift-stage sediments.

Seismic reflection profiles of the Carolina Trough and Blake Plateau Basin reveal a strong reflector named "J" marking the postrift unconformity and interpreted as a flood basalt. Its age is unknown. It may correspond to the subsurface J basalt drilled in the adjacent Atlantic Coastal Plain that is chemically and mineralogically similar to the 210-Ma CAMP basalts, but the published radiometric age of the J basalt is ~184 Ma. Offshore, the reflector, identified as "J?," lies above the SDRs along the hinge zone of the Carolina Trough but below the SDRs along the hinge zone of the Blake Plateau Basin. If the J reflector represents a basalt at these locations and is part of CAMP, the SDRs would be of Hettangian age. Alternatively, if the J reflector represents a flood basalt, its magmatic source may be the same as that of the presumably younger volcanic wedge basalts represented by the SDRs. If this were case, the basalt extended landward over the postrift unconformity, as interpreted for the SDRs overlying the postrift unconformity in the Baltimore Canyon Trough.

Three prominent Jurassic-age deep-sea acoustic horizons are identified within the sedimentary prism overlying the oldest oceanic crust of the western Atlantic basin. Two of these have been drilled and dated. J1 is equivalent to Horizon C in the Blake-Bahama Basin where it was drilled and dated as late Tithonian. J2 is equivalent to Horizon D in the same basin where it was drilled and recently given a revised age of early Callovian. The oldest horizon, J3, has not been drilled; it lies near the base of the sedimentary prism.

According to *Vogt* [1973], the earliest Atlantic oceanic crust lies in the corridor off North America between the East Coast Magnetic Anomaly (ECMA) and the Blake Spur Magnetic Anomaly (BSMA). The oldest crust should be found at the position of the spreading center that generated the subaerial extrusive volcanic rocks comprising the seaward-dipping volcanic wedge. According to the model of *Lizarralde and Holbrook* [1997], the spreading center was located at the place where the seaward-dipping reflectors (SDRs) representing the wedge volcanic rocks on seismic reflection profiles give way to the hyperbolic reflectors typical of oceanic crust. This is at the approximate location of the North American continent-ocean boundary marked by the ECMA. Neither the oldest oceanic crust nor the overlying sediments including those marked by the J3 deep-sea acoustic horizon have been drilled. The oldest oceanic crust that has been drilled is east of the Blake Spur Magnetic Anomaly (BSMA) and represents crust produced after the spreading-center jump to the BSMA [*Vogt*, 1973]; at DSDP Site 534, sediments overlying basalt are dated as middle Bathonian [*Baumgartner and Matsuoka*, 1995]. Off the conjugate African margin ~50 km west of the S1 magnetic anomaly, the presumed counterpart of the BSMA, the oldest sedimentary rocks overlying uplifted oceanic crust exposed in the Canary Islands are interpreted by indirect means to be of Toarcian? age [*Steiner et al*, 1998], but no fossils giving definitive age information have been recovered from those rocks.

A modification of *Klitgord and Schouten*'s [1986, Fig. 5] plot of sea-floor-spreading-rate lines for the western Atlantic based on the time scales of *Gradstein et al.* [1994] and *Pálfy et al.* [2000] plus the middle Bathonian age for basalts at Site 534 give revised extrapolated ages for the BSMA and ECMA. Extension of the slope of the lines between sea-floor magnetic anomaly M-25 and Site 534 to the BSMA and ECMA give ages for the BSMA of ~171 Ma (late Bajocian, spreading rate of 17mm/yr for the 1994 time scale) and ~166 Ma (early Bathonian, spreading rate of 25 mm/yr for the 2000 time scale). The same spreading-rate lines give extrapolated ages for the ECMA of ~179 Ma (early Aalenian, 1994 time scale) and ~172 Ma (early Bajocian, 2000 time scale). An alternative plot shows sea-floor-spreading rates of ~ 6 to 8 mm/yr between the BSMA and the ECMA according to the model of *Lizarralde and Holbrook* [1997]. They indicate that earliest production of oceanic crust began when the "subaerial spreading center" that produced the volcanic wedge subsided below sea level at~188 Ma (middle Pliensbachian/early Toarcian), approximately 2.3 m.y. after 190 Ma when they assumed wedge volcanism began. If submarine sea-floor spreading began instead shortly after the CAMP flood basalts at ~201 Ma, the average spreading rate between the ECMA and BSMA would be about 2 mm/yr.

In order to obtain direct evidence to answer some of the unresolved questions this review of the literature has revealed, deep drilling and sampling of the seaward-dipping volcanic wedge and overlying sediments are required. This is unlikely as they are too deeply buried; however, drilling of less deeply buried targets may be feasible. The J reflector offshore the southeastern United States is one. Given the feasibility of deep-sea drilling with a marine riser system, the J3 acoustic horizon and even the underlying oldest oceanic crust could be future targets.

Acknowledgments. I thank James A. Austin, Jr., Daniel Lizarralde, and Peter P. McLaughlin, Jr., for their critical reviews of the manuscript and helpful suggestions for its improvement.

REFERENCES

Austin, J. A., Jr., P. L Stoffa, J. D. Phillips, J. Oh, D. S. Sawyer, G. M. Purdy, E. Reiter, and J. Makris, Crustal structure of the Southeast Georgia Embayment-Carolina trough: Preliminary results of a composite seismic image of a continental suture(?) and a volcanic passive margin, *Geology, 18,* 1023-1027, 1990.

Baumgartner, P. O., and A. Matsuoka, New radiolarian data from DSDP Site 534A, Blake Bahama Basin, central Northern Atlantic, in *Middle Jurassic to Lower Cretaceous Radiolaria of Tethys: Occurrences, Systematics, Biochronology,* edited by P. O. Baumgartner, L. O'Dogherty, S. Gorican, E. Urquhart, and P. De Wever, pp. 709-715, Mémoires de Géologie (Lausanne), 23, 1995.

Behrendt, J. C., R. M. Hamilton, H. D. Ackermann, V. J. Henry, and K .C. Bayer, Marine multichannel seismic-reflection evidence for Cenozoic faulting and deep crustal structure near Charleston, South Carolina, in *Studies Related to the Charleston, South Carolina, Earthquake of 1886–Tectonics and Seismicity* edited by G. S. Gohn, pp. J1-J29, U. S. Geological Survey Professional Paper 1313, 1983.

Benson, R. N., Structure contour map of pre-Mesozoic basement, landward margin of Baltimore Canyon Trough, scale 1:500,000, *Delaware Geological Survey Miscellaneous Map Series No. 2,* 1984.

Benson, R. N., Map of exposed and buried early Mesozoic rift basins of the U. S. middle Atlantic continental margin, scale 1:1,000,000, *Delaware Geological Survey Miscellaneous Map Series No. 5,* 1992.

Benson, R. N., A. S. Andres, J. H. Roberts, and K. D. Woodruff, Seismic stratigraphy along three multichannel seismic reflection profiles off Delaware's coast, *Delaware Geological Survey Miscellaneous Map Series No. 4,* 1986.

Benson, R. N., and R. G. Doyle, Early Mesozoic rift basins and the development of the United States middle Atlantic continental margin, in *Triassic-Jurassic Rifting,* edited by W. Manspeizer, pp. 99-127, Elsevier Science Publishers B. V., Amsterdam, 1988.

Cousminer, H. L., and W. E. Steinkraus, Biostratigraphy of the COST G-2 well (Georges Bank): a record of Late Triassic synrift evaporite deposition; Liassic doming; and mid-Jurassic to Mio-

cene postrift marine sedimentation, in *Triassic-Jurassic Rifting,* edited by W. Manspeizer, pp. 168-183, Elsevier Science Publishers B. V., Amsterdam, 1988.

Dillon, W. P., K. M. Klitgord, and C. K. Paull, Mesozoic development and structure of the continental margin off South Carolina, in *Studies Related to the Charleston, South Carolina, Earthquake of 1886–Tectonics and Seismicity* edited by G. S. Gohn, pp. N1-N16, U. S. Geological Survey Professional Paper 1313, 1983.

Dillon, W. P., C. K. Paull, R. T. Buffler, and J.-P. Fail, Structure and development of the Southeast Georgia Embayment and northern Blake Plateau: Preliminary analysis, in *Geological and Geophysical Investigations of Continental Margins,* edited by J. S. Watkins, L. Montadert, and P. W. Dickerson, pp. 27-41, American Association of Petroleum Geologists Memoir 29, 1979.

Dillon, W. P., and P. Popenoe, The Blake Platcau Basin and Carolina Trough, in *The Atlantic Continental Margin: U. S.,* edited by R. E. Sheridan and J. A. Grow, pp. 291-328, Geological Society of America, The Geology of North America, I-2, 1988.

Doyle, J. A., and E. I. Robbins, Angiosperm pollen zonation of the continental Cretaceous of the Atlantic Coastal Plain and its application to deep wells in the Salisbury Embayment, *Palynology, 1,* 43-78, 1977.

Edson, G. M., D. L. Olson, and A. J. Petty, Exxon Corsair Canyon Block 975 No. 1 well, geological and operational summary, *OCS Report MMS-2000-032,* 43 pp., U. S. Department of the Interior, Minerals Management Service, New Orleans, La.,2000a.

Edson, G. M., D. L. Olson, and A. J. Petty, Mobil Lydonia Canyon Block 312 No. 1 well, geological and operational summary, *OCS Report MMS-2000-037,* 48 pp., U. S. Department of the Interior, Minerals Management Service, New Orleans, La.,2000b.

Edson, G. M., D. L. Olson, and A. J. Petty, Tenneco Lydonia Canyon Block 187 No. 1 well, geological and operational summary, *OCS Report MMS-2000-035,* 53 pp., U. S. Department of the Interior, Minerals Management Service, New Orleans, La.,2000c.

Eldholm, O., J. Thiede, and E. Taylor, Evolution of the Vøring volcanic margin, in *Proceedings of the Ocean Drilling Program, Scientific Results, 104,* 1033-1065, 1989.

Ewing, J. I., and P. D. Rabinowitz (Eds.), *Eastern North American Continental Margin and Adjacent Ocean Floor, 34 ° to 41 °N and 68 ° to 78 °W, Ocean Margin Drilling Program Atlas 4,* 40 pp., Marine Science International, Woods Hole, Mass., 1983.

Falvey, D. A., The development of continental margins in plate tectonic theory, *Australian Petroleum Exploration Association Journal, 14,* 95-106, 1974.

Gottfried, D., C. S. Annell, and G. R Byerly, Geochemistry and tectonic significance of subsurface basalts from Charleston, South Carolina: Clubhouse Crossroads test holes #2 and #3, in *Studies Related to the Charleston, South Carolina, Earthquake of 1886–Tectonics and Seismicity,* edited by G. S. Gohn, pp. A1-A19, U. S. Geological Survey Professional Paper 1313, 1983.

Gradstein, F. M., F. P. Agterberg, J. G. Ogg, J. Hardenbol, P. van Veen, J. Thierry, and Z. Huang, A Mesozoic time scale, *Journal of Geophysical Research, 99 (B12),* 24,051-24,074, 1994.

Grant, A. C., K. D. McAlpine, and J. A. Wade, The continental margin of eastern Canada: Geological framework and petroleum potential, in *Future Petroleum Provinces of the World,* edited by

M. T. Halbouty, pp. 177-205, American Association of Petroleum Geologists Memoir 40, 1986.

Grow, J. A., K. D. Klitgord, and J. S. Schlee, Structure and evolution of Baltimore Canyon Trough, in *The Atlantic Continental Margin: U. S.*, edited by R. E. Sheridan and J. A. Grow, pp. 269-290, Geological Society of America, The Geology of North America, I-2, 1988.

Hamilton, R. M., J. C. Behrendt, and H. D. Ackermann, Land multichannel seismic-reflection evidence for tectonic features near Charleston, South Carolina, in *Studies Related to the Charleston, South Carolina, Earthquake of 1886–Tectonics and Seismicity*, edited by G. S. Gohn, pp. I1-I18, U. S. Geological Survey Professional Paper 1313, 1983.

Harland, W. B., A. V. Cox, P. G Llewellyn, C. A. G. Pickton, A. G. Smith, and R. Walters, *A Geologic Time Scale*, Cambridge University Press, 131 pp., 1982.

Hinz, K., A hypothesis on terrestrial catastrophes: Wedges of very thick oceanward dipping layers beneath passive continental margins—their origin and paleoenvironmental significance, *Geologisches Jahrbuch, Reihe E, Geophysik, 22*, 3-28, 1981.

Hurtubise, D. O., J. H. Puffer, and H. L. Cousminer, An offshore Mesozoic igneous sequence, Georges Bank basin, North Atlantic, *Geological Society of America Bulletin, 98*, 430-438, 1987.

Jansa, L. F., P. Enos, B. E. Tucholke, F. M. Gradstein, and R. E. Sheridan, Mesozoic-Cenozoic sedimentary formations of the North American basin; western North Atlantic, in *Deep Drilling Results in the Atlantic Ocean: Continental Margins and Paleoenvironments*, edited by M. Talwani, W. Hay, and W. B. F. Ryan, pp. 1-57, American Geophysical Union Maurice Ewing Series 3, 1979

Kelemen, P. B., and W. S. Holbrook, Origin of thick, high-velocity igneous crust along the U. S. East Coast Margin, *Journal of Geophysical Research, 100 (B7)*, 10,077-10,094, 1995.

Kent, D. V., and F. M. Gradstein, A Jurassic to recent chronology, in *The Western North Atlantic Region*, edited by P. R. Vogt and B. E. Tucholke, pp. 45-50, Geological Society of America, The Geology of North America, M, 1986.

Klitgord, K. D., Base of postrift sedimentary deposits, in *The Atlantic Continental Margin: U. S.*, edited by R. E. Sheridan and J. A. Grow, pl. 2D, Geological Society of America, The Geology of North America, I-2, 1988.

Klitgord, K. D., and J. A. Grow, Jurassic seismic stratigraphy and basement structure of western Atlantic magnetic quiet zone, *American Association of Petroleum Geologists Bulletin, 64*, 1658-1680, 1980.

Klitgord, K. D., and H. Schouten, Plate kinematics of the central Atlantic, in *The Western North Atlantic Region*, edited by P. R. Vogt and B. E. Tucholke, pp. 351-378, Geological Society of America, The Geology of North America, M, 1986.

Lanphere, M. A., ^{40}Ar/^{39}Ar ages of basalt from Clubhouse Crossroads test hole #2, near Charleston, South Carolina, in *Studies Related to the Charleston, South Carolina, Earthquake of 1886–Tectonics and Seismicity*, edited by G .S. Gohn, pp. B1-B8, U. S. Geological Survey Professional Paper 1313, 1983.

Lizarralde, D., and W. S. Holbrook, U. S. mid-Atlantic margin structure and early thermal evolution, *Journal of Geophysical Research, 102 (B10)*, 22,855-22,875, 1997.

Manspeizer, W., and H. L. Cousminer, 1988, Late Triassic–Early Jurassic synrift basins of the U. S. Atlantic margin, in *The Atlantic Continental Margin: U. S.*, edited by R. E. Sheridan and J. A. Grow, pp. 197-216, Geological Society of America, The Geology of North America, I-2, 1988.

Manspeizer, W., J. DeBoer, J. K. Costain, A. J. Froelich, C. Çoruh, P. E. Olsen, G. J. McHone, J. H. Puffer, and D. C. Prowell, Post-Paleozoic activity, in *The Appalachian-Ouachita Orogen in the United States*, edited by R. D. Hatcher, Jr., W. A. Thomas, and G. W. Viele, pp. 319-374, Geological Society of America, The Geology of North America, F-2, 1989.

Mattick, R. E., and J. Libby-French, Petroleum geology of the United States Atlantic continental margin, in *The Atlantic Continental Margin: U. S.*, edited by R. E. Sheridan and J. A. Grow, pp. 445-462, Geological Society of America, The Geology of North America, I-2, 1988.

McBride, J. H., K. D. Nelson, and L. D. Brown, Evidence and implications of an extensive early Mesozoic rift basin and basalt/diabase sequence beneath the southeast Coastal Plain, *Geological Society of America Bulletin, 101*, 512-520, 1989.

McHone, J. G., Non-plume magmatism and rifting during opening of the central Atlantic Ocean, *Tectonophysics, 316*, 287-296, 2000.

McHone, J. G., and J. H. Puffer, Flood basalt provinces of the Pangaean Atlantic rift: Regional extent and environmental significance, in *Aspects of Triassic-Jurassic Geoscience*, edited by P. M. LeTourneau and P. E. Olsen, Columbia University Press, New York, 2001(in press).

Mutter, J. C., M. Talwani, and P. L. Stoffa, Origin of seaward-dipping reflectors in oceanic crust off the Norwegian margin by 'subaerial sea-floor spreading,' *Geology, 10*, 353-357, 1982.

Oh, J., J. A. Austin, Jr., J. D. Phillips, M. F. Coffin, and P. L. Stoffa, Seaward-dipping reflectors offshore the southeastern United States: Seismic evidence for extensive volcanism accompanying sequential formation of the Carolina trough and Blake Plateau basin, *Geology, 23*, 9-12, 1995.

Olsen, P. E., Stratigraphic record of the early Mesozoic breakup of Pangea in the Laurasia-Gondwana rift system, *Annual Review of Earth and Planetary Sciences, 25*, 337-401, 1997.

Pálfy, J., P. L. Smith, and J. K. Mortensen, A U-Pb and ^{40}Ar/^{39}Ar time scale for the Jurassic, *Canadian Journal of Earth Sciences, 37*, 923-944, 2000.

Pe-Piper, G., L. F. Jansa, and R. St. J. Lambert, Early Mesozoic magmatism on the eastern Canadian margin: Petrogenetic and tectonic significance, in *Eastern North American Mesozoic Magmatism*, edited by J. H. Puffer and P. C. Ragland, pp. 13-36, Geological Society of America Special Paper 268, 1992.

Poppe, L. J., P. Popenoe, C. W. Poag, and B. A. Swift, Correlations of subsurface rocks in the Southeast Georgia Embayment, U. S. South Atlantic Continental Shelf, *U. S. Geological Survey Miscellaneous Field Studies Map MF-2251*, 1994.

Roeser, H. A., Magnetic anomalies in the magnetic quiet zone off Morocco, in *Geology of the Northwest African Continental Margin*, edited by U. von Rad, K. Hinz, M. Sarnthein, and E. Seibold, pp .61-68, Springer-Verlag, Berlin, 1982.

Roest, W. R., J. J. Dañobeitia, J. Verhoef, and B. J. Collette, Magnetic anomalies in the Canary Basin and the Mesozoic evolution of the central North Atlantic, *Marine Geophysical Researches, 14*, 1-24, 1992.

Schlee, J. S., and K. D. Klitgord, Georges Bank Basin: A regional synthesis, in *The Atlantic Continental Margin: U. S.*, edited by R. E. Sheridan and J. A. Grow, pp. 243-268, Geological Society of America, The Geology of North America, I-2, 1988.

Schlee, J. S., W. Manspeizer, and S. R. Riggs, Paleoenvironments: Offshore Atlantic U. S. margin, in *The Atlantic Continental Margin: U. S.*, edited by R. E. Sheridan and J. A. Grow, pp. 365-385, Geological Society of America, The Geology of North America, I-2, 1988.

Sheridan, R. E., L. G. Bates, T. H. Shipley, and J. T. Crosby, Seismic stratigraphy in the Blake-Bahama Basin and the origin of Horizon D, in *Initial Reports of the Deep Sea Drilling Project, 76*, pp. 667-683, U. S. Government Printing Office, Washington, D. C., 1983.

Sheridan, R. E., D. L. Musser, L. Glover, III, M. Talwani, J. I. Ewing, W. S. Holbrook, G. M. Purdy, R. Hawman, and S. Smithson, Deep seismic reflection data of EDGE U. S. mid-Atlantic continental-margin experiment: Implications for Appalachian sutures and Mesozoic rifting and magmatic underplating, *Geology, 21*,. 563-567, 1993.

Sheridan, R. E., L. Pastouret, and G. Mosditchian, Seismic stratigraphy and related lithofacies of the Blake-Bahama Basin, in *Initial Reports of the Deep Sea Drilling Project, 44*, pp. 529-546, U. S. Government Printing Office, Washington, D. C., 1978.

Simonis, E. K., Lithologic description, in *Geologic and Operational Summary, COST No. G-2 well*, edited by R. V. Amato and E. K. Simonis, pp. 14-19, U. S. Department of the Interior Geological Survey Open-File Report No. 80-269, 1980.

Steiner, C., A. Hobson, P. Favre, G. M. Stampfli, and J. Hernandez, Mesozoic sequence of Fuerteventura (Canary Islands): Witness of Early Jurassic sea-floor spreading in the central Atlantic, *Geological Society of America Bulletin, 110*, 1304-1317, 1998.

Talwani, M., and V. Abreu, Inferences regarding initiation of oceanic crust formation from the U. S. East Coast margin and conjugate South Atlantic margins, in *Atlantic Rifts and Continental Margins*, edited by W. Mohriak and M. Talwani, pp. 211-233, American Geophysical Union, Geophysical Monograph 115, Washington, D. C., 2000.

Talwani, M., J. Ewing, R. E. Sheridan, W. S. Holbrook, and L. Glover, III, The EDGE experiment and the U. S. East Coast Magnetic Anomaly, in *Rifted Ocean-Continent Boundaries*, edited by E. Banda, M. Torné, and M. Talwani, pp. 155-181, Kluwer Academic Publishers, Norwell, Mass., 1995.

Talwani, M., and G. Udintsev, Tectonic synthesis, in *Initial Reports of the Deep Sea Drilling Project, 38*, pp. 1213-1242, U. S. Government Printing Office, Washington, D. C., 1976.

Vogt, P. R., Early events in the opening of the North Atlantic, in *Implications of Continental Drift to the Earth Sciences, Volume 2*, edited by D. H. Tarling and S. K. Runcorn, pp. 693-712, Academic Press, New York, 1973.

Wade, J. A., The stratigraphy of Georges Bank Basin and relationships to the Scotian Basin, in *Geology of the Continental Margin of Eastern Canada*, edited by M. J. Keen and G. L. Williams, pp.171-190, Geological Society of America, The Geology of North America, I-1, 1990.

Wade, J. A., and B. C. MacLean, The geology of the southeastern margin of Canada, in *Geology of the Continental Margin of Eastern Canada*, edited by M. J. Keen and G. L. Williams, pp.169-238, Geological Society of America, The Geology of North America, I-1, 1990.

Williams, G. L., P. Ascoli, M. S. Barss, J. P. Bujak, E. H. Davies, R. A. Fensome, and M. A. Williamson, Biostratigraphy and related studies, in *Geology of the Continental Margin of Eastern Canada*, edited by M. J. Keen and G. L. Williams, pp. 89-137, Geological Society of America, The Geology of North America, I-1, 1990.

Richard N. Benson, Delaware Geological Survey, University of Delaware, Newark, Delaware 19716
e-mail: rnbenson@udel.edu

Critical Evaluation Of ^{40}Ar/^{39}Ar Ages for the Central Atlantic Magmatic Province: Timing, Duration and Possible Migration of Magmatic Centers

Ajoy K. Baksi

Louisiana State University, Baton Rouge, LA

Recent reports suggest the Central Atlantic Magmatic Province (CAMP), initially covering sections of North America, Africa and South America, was formed within a relatively short period of time around 200 Ma. All relevant ^{40}Ar/^{39}Ar ages are corrected for interlaboratory differences, mini or marginal plateau values are rejected as accurate estimates of crystallization age, and results scrutinized for the presence of excess argon and/or alteration effects. Twenty-one accurate ages span ~230 to 175 Ma, with a marked concentration at ~200 Ma. Magmatism in North America and contiguous parts of Africa in Pangaea (N = 7) average 199.7 Ma; in South America, seven samples yield an average age of 198.3 Ma. The age difference appears to be statistically significant, and in the southern areas, a second phase of magmatism (N = 3) centers around 192 Ma. From a temporal viewpoint, a genetic link between CAMP and the Triassic-Jurassic boundary extinction event appears likely.

1. INTRODUCTION

Magmatism at ~200 Ma, related to the breakup of Pangaea and the opening of the central Atlantic Ocean, produced sills/dikes and lava flows over an area > 4 x 10^6 km^2 (Marzoli et al., 1999; Hames et al., 2000, and references therein). These rocks are said to belong to the Central Atlantic Magmatic Province (CAMP), and are now found in (south)eastern USA, (north)western Africa and Brazil (see Figure 1). This magmatic event has been compared to the formation of flood basalt provinces such as the Siberian and Deccan Traps, in that each may be genetically linked to a global faunal extinction. For the CAMP this is the event recorded at the Triassic-Jurassic boundary.

The Central Atlantic Magmatic Province:
Insights from Fragments of Pangea
Geophysical Monograph 136

The age data relevant to the CAMP has been gathered by a number of different laboratories and by different dating techniques. Herein, I critically examine all relevant age data to evaluate (a) over what duration of time the magmatism occurred (b) how many of the bodies currently grouped with CAMP have ages significantly different from 200 Ma and (c) whether any progression of ages can be detected within CAMP. Specifically, I look to data in Marzoli et al. (1999) and Hames et al. (2000), and references therein, and to geochronological data for the Gettysburg and Palisades sills, USA, and sills in Liberia.

2. EVALUATION OF AGES

Data obtained from ^{40}Ar/^{39}Ar stepheating studies will be utilized. All ages will be (re)calculated using the decay constants recommended by Steiger and Jager (1977). Guidelines for evaluation of such data have been set out elsewhere (Baksi, 1999). These include the fact that plateau sections of age spectra should be made up of at least three

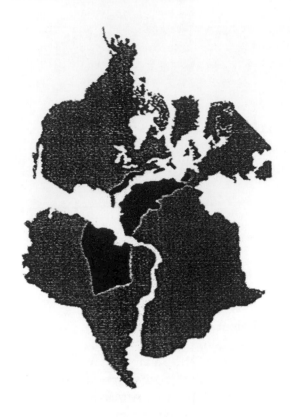

Figure 1. Simplified location map of CAMP (in black) in a Pangaea reconstruction at 200 Ma. Modified from Marzoli et al. (1999).

contiguous steps carrying > 50% of the total ^{39}Ar, that do not differ in age at the 95% confidence interval,. The term marginal plateau will be used for sections of age spectra where consecutive steps carrying (a) 30-50% of the total ^{39}Ar overlap at the 95% confidence interval or (b) >50% of the total ^{39}Ar marginally fail to overlap at the 95% confidence interval. Such sections of age spectra may, *occasionally,* approach crystallization values, but will not be used herein as *accurate* estimates thereof. In assessment of both age spectra and isochron plots, acceptable values of the parameter MSWD - hereinafter referred to as F – varies with the number of points used for analysis. Herein, the probability (p) of obtaining the given value of F for the straight line fit on the isochron will be listed, using χ^2 tables (e.g. DeVor et al., 1992).

Min et al. (2000) noted that ages determined by the ^{40}Ar/^{39}Ar and the U/Pb methods on undisturbed rocks may differ, in part, due to possible errors in the constants used for ^{40}K decay. The scenario for rocks of Early Mesozoic age (such as CAMP), is not entirely clear, but suggest ^{40}Ar/^{39}Ar ages are ~0.5-1.5% younger than U/Pb ages.

For the Siberian Traps, ages are, 250.0±0.8 Ma (^{40}Ar/^{39}Ar - Renne et al., 1995) and 251.2±0.2 Ma (U/Pb - Kamo et al., 1996). For the Permo-Triassic boundary ages are, 250.0±0.8 Ma (^{40}Ar/^{39}Ar - Renne et al., 1995) and 251.4±0.2 or slightly older than 253 Ma (U/Pb - Bowring et al., 1998; Mundil et al., 2001). For the Gettys-burg and Palisades sills ages are, 200.3±0.7 Ma (^{40}Ar/^{39}Ar - this work), and 201.1±0.5 Ma (U/Pb - Dunning and Hodych, 1990). Since it is currently not possible to *precisely* relate U/Pb and ^{40}Ar/^{39}Ar ages, I will examine only the latter set for CAMP rocks. Errors in ^{40}Ar/^{39}Ar ages will be quoted at the 1σ level, including a term (generally σJ = 0.4%) for uncertainty in determination of the irradiation parameter, but without allowing for possible errors in the constants for ^{40}K decay.

3. REDUCTION OF ^{40}Ar/^{39}Ar AGES TO A COMMON BASE

It is necessary to compare ages obtained at various laboratories using different monitor samples. Ages must be reduced to a common set of standards and I utilize the ages reported by Renne et al. (1998) as a large number of ages evaluated herein were generated at Berkeley. Renne et al. (1998) analyzed a limited number of standards; my methodology for reducing ages to a common set is outlined below. Ages on standards determined by Baksi et al. (1996) and Renne et al. (1998) show internally consistency, with the former yielding values 1.0% younger than the latter for ages of ~30-100 Ma (see results in these papers for TCR San and GA1550 Bio). The results for MMhb-1 in these two studies are not consistent. For splits of 3-5 grains, Renne et al. (1998) report an age of 523.1±1.3 Ma, whereas larger subsamples (~20 mg each) yield an average age of 516.4±1.7 (Baksi et al., 1996). In both of these studies, the error on each individual determination is 0.7-1.0 m.y.. The average age in each case yields a higher error, suggesting inhomogeneity within MMhb-1.

For results reported with respect to LP-6 Bio 40-60#, ages were first recalculated with an age of 128.1 Ma (Baksi et al., 1996) and subsequently increased by 1% to bring them to the values preferred by Renne et al. (1998). Results in the literature reported relative to MMhb-1, were obtained on subsamples of at least a few milligrams. These ages were adjusted to an age of 516.4 Ma (Baksi et al., 1996) and subsequently raised by 1.0% to bring them into agreement with the values preferred by Renne et al. (1998).

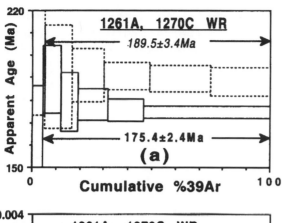

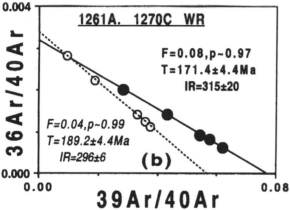

Figure 2. (a) Age spectra and (b) isochron plot for whole-rocks from the Gettysburg sill (recalculated from Sutter and Smith, 1979). Results for 1261A and 1270C are shown in solid lines (bold letters) and dotted lines (italics), respectively. Isochron plots follow York (1969) and utilize plateau steps only; F = goodness of fit parameter (p = probability of obtaining this value), T = age and IR = $(^{40}Ar/^{36}Ar)_i$. All errors listed and shown at the 1σ level. The "ages" for two samples collected from a single outcrop, ~35 m apart, differ significantly. Both ages are younger than the crystallization age of 201 Ma (Dunning and Hodych, 1990), and result from the use of altered specimens (see text).

Where FCT-3 Bio was used as the monitor, its age was adjusted to 28.23 Ma, i.e. 1.0% older than in Baksi et al. (1996).

4. CAMP SUBPROVINCES

4.1 Dikes/Sills/Lavas in the (North) Eastern United States

Accurate ages for Dallmeyer's (1975) $^{40}Ar/^{39}Ar$ work on the Palisades sill cannot be listed herein, as the monitor

sample used remains unknown. Sutter and Smith (1979) dated two specimens from the Gettysburg sill. Though the monitor used is unknown, examination of the results is instructive. Plateau ages of 175 and 190 Ma (Figure 2) differ at >99% confidence interval, for samples (1261A and 1270C) collected from a single outcrop ~35 m apart. Isochron plots yield very low F values (Figure 2), suggesting overestimation of analytical (mass spectrometric) measurements (Baksi, 1991). Biotite (PAL1) from a recrystallized xenolith of the Stockton Formation close to the Palisades sill, was analyzed. Sutter's (1988) age, based on *two* high temperature steps, carrying ~44% of the total ^{39}Ar, was recalculated herein to 203.0±1.2 Ma. Since lower temperature steps yield higher ages, the sample contains excess argon. An isochron plot utilizing 4 steps carrying 95% of the total ^{39}Ar, yields an age of 200.4±2.6 Ma (Figure 3). New whole-rock ages for the Gettysburg and Palisades sills are reported herein. GS14 (York Haven type composition) and PS6 were analyzed using sieved 10-30 mesh fractions. Spectra are inverted U-shaped (Figure 4 and Table 1), showing partial loss of $^{40}Ar^*$, and yield plateau ages of 200±1 Ma for intermediate temperature steps. These ages may be compared to the U/Pb determinations of Dunning and Hodych (1990), namely 201.2±0.5 Ma. The incorrect (low) ages for the Gettysburg sill obtained by Sutter and Smith (1979) resulted from the use of altered rocks that contain high quantities of ^{36}Ar (Baksi, 1987). Plateau steps for the rocks analyzed by Sutter and Smith (1979) show 15-60% $^{40}Ar^*$, whereas those of Baksi (this work) show >90%. A similar test can be used to distinguish incorrect (young) plateau ages on feldspar separates as well. This will be addressed in detail elsewhere (A.K. Baksi, in prep.).

Ages for other rocks from this area are available in the literature. A hornblende separate (GUP84-1) from a granophyric differentiate of the Germana Bridge diabase sheet (Sutter, 1988) gives an age of 200.9±1.0 Ma (plateau age) - the recalculated age of 203 Ma listed by Hames et al. (2000) is an overestimate. A whole-rock diabase sample (MS1A) from the Gainesville sheet of the Culpeper basin (Sutter, 1988) gives a low-intermediate temperature plateau age of 197.7±1.0 Ma (Figure 3), marginally younger than the Germana Bridge diabase sheet. (Step ages listed in Sutter, 1988, Table 1, have minor errors). The plateau steps show relatively low amounts of $^{40}Ar^*$ (45-80%) and the age will be treated as a minimum value. Hames et al. (2000) reported plateau ages of 198.8±1.0 and 201.0±1.1 Ma for the Wachtung flows III and I of the Newark Basin,

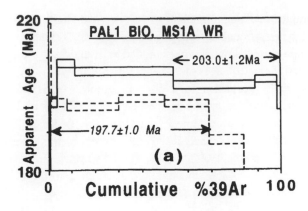

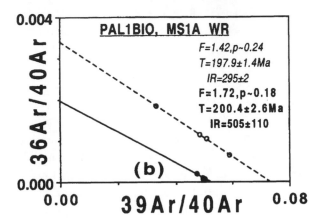

Figure 3. (a) Age spectra and (b) isochron plot for two samples from the Newark and Culpeper Basins, USA. Results for PAL1 biotite and MS1A whole-rock are shown in solid lines (bold letters) and dotted lines (italics) respectively. Errors and notations as in Figure 2. The recrystallized biotite specimen contains excess argon and yields an isochron age of 200 Ma. The whole-rock specimen is altered, showing relatively high amounts of ^{36}Ar; its age is regarded as a minimum value (see text).

respectively. Utilizing the guidelines herein, the latter is not usable as a crystallization age as it entails only 41% of the total ^{39}Ar released. These authors also obtained ages for two dikes from South Carolina. Crystallization values are 199.7±0.8 Ma and 199.2±0.7 Ma (average of two determinations).

4.2 Lava Flows in Central Morocco

Fiechtner et al. (1992) analyzed six plagioclase samples and reported five ages in the range ~195-205 Ma. Their isochron analyses are faulty, since *non-consecutive steps* were utilized in many cases, and ages were accepted where the initial argon ratio was lower than the atmospheric value. Recalculation yields acceptable isochron ages for

three specimens that contain excess argon (Figure 5 and Table 2); one of these (111/88 -175 Ma) is markedly younger than generally accepted for CAMP.

4.3 Rocks From Iberia, Morocco and Mali

A large number of samples were studied by Sebai et al. (1991). Few accurate ages resulted, since "mini plateaus" involve <50% of the total ^{39}Ar released, and are not acceptable herein as estimates of the crystallization ages. As noted by the authors, many plagioclase samples showed considerable sericitization - denoted by a sharp fall in Ca/K values for intermediate heating steps. There are difficulties in interpretation of complex age spectra. For the Messejana dike in Iberia, cloudy (altered) plagioclase grains gave a plateau age of 190 Ma, whereas transparent grains give higher step ages (Figure 6). The latter, is a better *estimate* for the crystallization age. For plagioclase from the Ksi Ksou dike in Algeria, a low temperature marginal plateau yields an

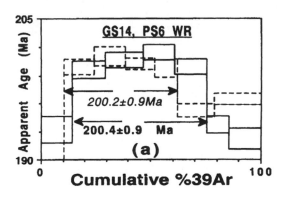

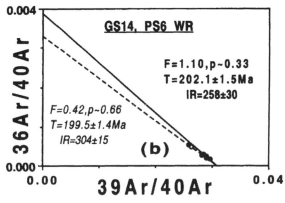

Figure 4. (a) Age spectra and (b) isochron plot for whole-rocks from the Gettysburg (solid lines - bold letters) and Palisades sills (dotted lines - italics). Data listed in Table 1. Errors and notations as in Figure 2. These ^{40}Ar/^{39}Ar ages are in good agreement with the U/Pb age (201 Ma) of Dunning and Hodych (1990).

Table 1. Analytical data for $^{40}Ar/^{39}Ar$ dating of whole-rock diabase from the Gettysburg and Palisades sill.

Step (^{o}C)	Cumulative % ^{39}Ar	$^{40}Ar/^{39}Ar$	$^{36}Ar/^{39}Ar$	$^{37}Ar/^{39}Ar$	$^{40}Ar*$ (%)	Age±1σ (Ma)
		GS14, 510 mg, J = 0.003646				
1 (700)	14.2	46.81	0.05540	6.119	61.2	193.1±1.3
2 (800)	29.8	34.99	0.01162	5.913	91.8	199.6±0.9#
3 (880)	47.1	32.26	0.00647	4.462	95.6	200.6±0.8#
4 (950)	61.1	34.38	0.00776	3.459	94.3	201.4±0.8#
5 (1020)	75.5	34.54	0.00887	2.598	93.1	199.9±0.8#
6 (1100)	85.7	33.78	0.01027	3.845	92.1	193.7±0.8
7 (1180)	100.0	33.21	0.01299	15.75	92.9	192.2±1.1
		PS6, 610 mg, J = 0.003596, degassed at 500°C for 15 min.				
1 (700)	10.4	36.48	0.02057	3.406	84.2	188.9±0.8
2 (750)	24.4	37.91	0.01935	4.424	86.5	199.9±0.8#
3 (800)	38.9	36.87	0.02131	4.473	84.8	201.0±1.0#
4 (850)	52.3	35.48	0.01048	2.940	92.1	200.2±0.6#
5 (900)	62.2	32.55	0.01070	2.543	91.8	199.7±0.9#
6 (1000)	78.7	33.63	0.00696	2.392	94.6	195.2±0.7
7 (1180)	100.0	35.23	0.01592	15.47	90.8	196.4±0.6

Isotopic ratios corrected for decay. Ca/K = 1.82 x $(^{37}Ar/^{39}Ar)$. Monitor sample, FCT-3 Bio, age adjusted to 28.23 Ma (see text). Analyzed at Queen's University following techniques described elsewhere (Baksi et al., 1996). Internal precision errors listed (σJ =0). # indicates steps used for plateau/isochron age calculations.

older age than a higher temperature one (Figure 6); this is anomalous, since argon released at low temperatures - derived from the less retentive sites - should yield younger ages. Table 3 lists the results of my evaluation of the results of Sebai et al. (1991). Only two crystallization ages result, average ages for a dike in North Mali (203.4±0.9 Ma) and flows in central Morocco (200.0±1.1 Ma).

4.4 Tholeiites from French Guyana, Surinam, Guinea and Liberia

Deckart et al. (1997) reported results for minerals separated from these rocks. The analytical data was obtained for examination (G. Feraud, written communication, 1999). Many samples contain excess argon; "mini plateaus" are re-

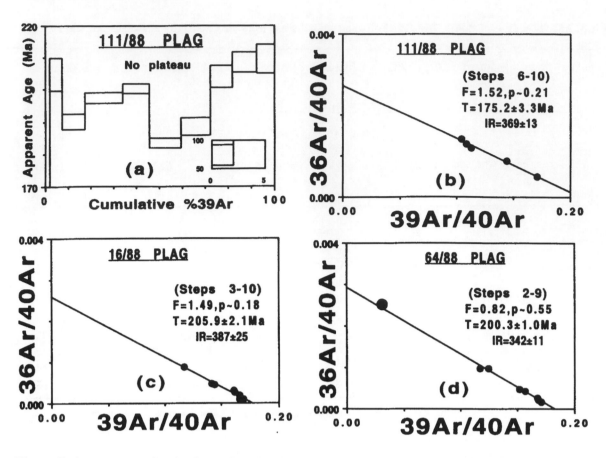

Figure 5. Age spectra and/or isochron plots for plagioclase separates from three tholeiitic flows, central Morocco (recalculated from Fiechtner et al., 1992). Errors and symbols as in Figure 2. (a) and (b)111/88 (c) and (d) 16/88 and 64/88, respectively. All specimens contain excess argon. 111/88 shows no plateau; the high temperature steps yield an isochron age of 175 Ma. 16/88 and 64/88 yield plateau ages (see Table 2) in general agreement with the isochron values; the latter are preferred as crystallization ages as the initial argon ratios are higher than the atmospheric argon value (295.5).

Table 2. Ages of flows in central Morocco. Plagioclase separates of Fiechtner et al. (1992).

Sample	Plateau Age (Ma) (Steps, % ^{39}Ar)	Isochron steps (% ^{39}Ar)	Age (Ma)	MSWD	$(^{40}Ar/^{36}Ar)_i$
16/88	209.1±1.1 (4 - 8, 76)	4 - 8 (76)	205.9±2.3	1.49 (p~0.20)	387±25
64/88	202.8±1.1 (4 - 9, 86)	4 - 9 (86)	200.3±1.3	0.82 (p~0.55)	342±11
111/88	No plateau	6 - 10 (54)	175.2±3.4	1.52 (p~0.20)	369±3

Isochron ages preferred as specimens contain excess argon. Errors listed at 1σ level. p = probability of obtaining the observed MSWD value from χ^2 tables.

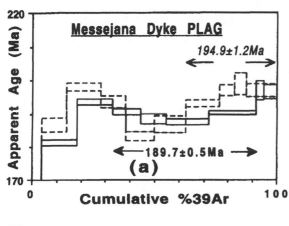

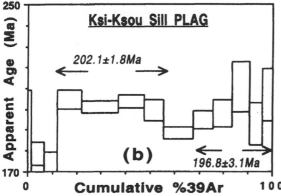

Figure 6. Age spectra for plagioclase separates of rocks from west Africa (data from Sebai et al., 1991). Errors and symbols as in Figure 2. (a) For the Messejana dike (C1), cloudy (sericitised) grains (solid line, bold letters) define a plateau (189.7±1.0 Ma) significantly younger than the marginal plateau (194.9±1.2 Ma) shown by transparent grains (dotted lines, italics) (b) Specimen from the Ksi Ksou sill yields two marginal plateaus that barely overlap (see text).

jected herein as *accurate* estimates of crystallization ages. Amphibole from G1 and G10 yield low precision plateau ages that average 195.9±4.4 Ma (see Table 4). Biotite fractions from Bourda-2 give low precision plateau ages of ~224-233 Ma; the best estimate of its age would appear to be 227.0±2.9 Ma, considerably older than the ~200 Ma generally accepted for CAMP. For the Fouta Djalon sill of Guinea, plagioclase separates (GUI35, 70a and 92) give plateau ages of 190.5±1.0, 195.0±1.0 and 198.9±1.1 Ma, respectively. These ages do *not* overlap; Ca/K ratio of steps used for plateau calculation vary by a factor of ~2, indicating alteration. For the Kakoulima layered intrusion, numerous plagioclase separates were studied. GUI115 gave a plateau age of 212.5±2.2 Ma. GUI110 and GUI76 both show two marginal plateaus (Figure 7) whose ages do not

overlap. Three plagioclase specimens from GUI118 show further problems in interpretation of ages. The coarsest fraction (125-160 μm) gives a marginal plateau of 198.9±1.8 Ma; replicate analyses of a finer fraction (160-200 μm), gives a plateau age of 191.8±2.2 Ma, and a marginal plateau of 200.3±1.7 Ma. For GUI141 plagioclase, a coarse fraction gives a lower marginal plateau than a finer fraction (see Table 4). The crystallization age of the Kakoulima intrusion remains unknown. Overall, only three acceptable ages were recovered, namely for G1/G10 amphibole, Bourda biotite, and GUI115 plagioclase. A sill (L21) in Liberia (Dalrymple and Lanphere, 1974; Dalrymple et al., 1975) gives an age of 191.5±1.5 Ma.

4.5. Dikes and Lava Flows in Brazil

Marzoli et al. (1999) reported ages for eleven samples. Guidelines for acceptance of crystallization ages are: (a) all isochrons must exhibit acceptably low F values and (b) for dikes and lavas where excess argon is present, isochron ages are preferred over plateau values. Six rocks yield acceptable crystallization ages (see Table 5); three ages listed by Marzoli et al. (1999) have very small error estimates (e.g. ±0.2 m.y. for 8820 PL). For the purposes of frequency age spectra and average calculations, these were increased to ±0.8 m.y. (see Table 6). Two whole-rock basalts from the Maranhao province gave crystallization ages (Table 5). BR-10 shows considerable loss of ^{40}Ar*, and an inverted U-shaped age spectrum; however, its plateau age is thought to accurately reflect the crystallization value, since the constituent steps show high amounts (>90%) of ^{40}Ar* (Baksi and Archibald, 1997).

5. DISCUSSION

Table 6 lists accurate crystallization ages for CAMP and these will be utilized for further discussion. Many of the rejected "ages" (see Tables 3-5), fall in the range expected for CAMP; however, based on the criteria adopted herein, these cannot be treated as accurate estimates of crystallization ages. Figure 8a shows an age probability spectra for all acceptable ages. The total range of ages is greater than reported by Marzoli et al. (1999), and includes recalculated ages (e.g. Fiechtner et al., 1992) from the same sources as utilized by these authors. The frequency of ages outside 205-190 Ma is low, and the main phase of CAMP igneous activity is centered at 200 Ma. These latter ages are exam-

Table 3. Ages from Iberia, Morocco, Algeria and Mali. Mineral separates and whole rocks of Sebai et al. (1991).

Sample number	Plateau Age (Ma) (Steps, %^{39}Ar)	Comments
	IBERIA - MESSEJANA DIKE	
C1 PL (cloudy)	*189.7±1.0* (5 - 8, 59)	Altered sample - age unacceptable
CI PL (transparent)	*194.9±1.2* (8 - 11, 37)	Marginal plateau
P1A PYX	*203.4±1.2* (8 - 11, 32)	Marginal plateau, forms bottom of U-shaped spectrum - excess argon present.
P3 WR	*196.9±0.9* (6 - 11, 59)	Marginal plateau, steps 6 and 9 do not overlap
	MOROCCO - HIGH ATLAS LAVAS	
M1216 PL	199.3±1.2 (4 - 11, 79)	Highest flow.
M1125 PL	202.4±2.3 (3 - 7, 57)	Middle flow.
M1208 PL	*201.8±0.9* (7 - 12, 43)	Bottom flow. Marginal plateau.
	ALGERIA - KSI-KSOU DIKE	
K-K PL	*202.1±1.9* (4 - 6, 35)	Low temperature marginal plateau.
	196.8±3.1 (9 - 13, 32)	High temperature marginal plateau.
	MALI - TAOUDENNI DIKE SWARM	
TA8 PL	202.1±1.3 (3 - 10, 95)	Two different
TA32 PL	204.6±1.3 (2 - 10, 99.5)	dikes.

PL, PYX and WR denote plagioclase, pyroxene and whole-rock, respectively. "Ages" in italics are unacceptable as crystallization values. Errors listed at 1σ level. The Mali dike swarm is taken to (geographically) lie outside CAMP.

ined more critically in Figure 8b. Ages from the northern areas (North America/northwest Africa) that do not differ from 200 Ma at the 2σ level, show a symmetrical distribution. (Ten) ages in the range 190-200 Ma from southern areas (South America/Liberia), are skewed to the younger side; the main peak clusters at ~198 Ma. Based on ages then available, Baksi and Archibald (1997) suggested "CAMP" ages in North America were marginally older than those from South America. Herein, the average ages (with SEM errors) are 199.7±0.3 Ma (northern sections, N = 7) and 198.3±0.4 Ma (southern sections, N = 7). These

ages differ at the >99% confidence level, suggesting the main phase of magmatism in the south took place ~1.5 m.y. later than in the north. In North America, ages from South Carolina average 199.4±0.5 Ma (Hames et al., 2000), whereas those further north average 200.4±0.3 Ma. These ages overlap at the ~10% confidence level; future work will help establish whether there is also a latitudinal progression of ages for the North American section of CAMP. Marzoli et al. (1999) suggested magmatism in Brazil continued after 200 Ma towards the future rifted margin of the continent. The scenario appears more complex,

Table 4. Ages for rocks from French Guyana, Surinam, Guinea and Liberia. All samples studied by Deckart et al. (1997), except L21 (Dalrymple and Lanphere,1974; Dalrymple et al., 1975).

Sample Number	Age (Ma)	Comments
		FRENCH GUYANA - DIKES
G1 AMP	194.6±6.1	Five step plateau (99.6% ^{39}Ar).
G10 AMP	197.3±6.3	Five step plateau (99.0% ^{39}Ar).
Bourda-2 BIO	227.0±2.9	Weighted plateau average of two runs on 160 μm sized subsamples.
67SUR BIO	*220.9±1.4*	Three step marginal plateau (45% ^{39}Ar).
G3 PL	*196.3±1.9*	Five step marginal plateau (49% ^{39}Ar).
G12 PL	*191.9±1.6*	Five step marginal plateau (41% ^{39}Ar).
G50 PL	*189.5±1.4*	Four step marginal plateau (37% ^{39}Ar).
Tortue PL	*198.7±1.8*	Five step marginal plateau (33% ^{39}Ar).
		GUINEA - FOUTA DJALON SILLS AND DIKES
GUI35 PL	*190.5±1.0*	Seven step plateau (51% ^{39}Ar) on altered sample (see text).
GUI70a PL	*195.0±1.0*	Six step plateau (57% ^{39}Ar) on altered sample (see text).
GUI92 PL	*198.9±1.1*	Six step plateau (52% ^{39}Ar) on altered sample (see text).
GUI18 PL	*190.2±1.0*	Five step marginal plateau (48% ^{39}Ar).
GUI87 PL	*193.8±1.9*	Four step marginal plateau (43% ^{39}Ar).
		GUINEA - KAKOULIMA LAYERED INTRUSION
GUI115 PL	212.5±2.2	Six step plateau (52% ^{39}Ar).
GUI110 PL	*194.2±3.6*	Two marginal plateaus on one sample (Figure 6). Both unacceptable
	205.9±3.1	(see text).
GUI76 PL	*194.6±3.1*	Two marginal plateaus on one sample (Figure 6). Both unacceptable
	205.5±2.9	(see text).
GUI118 PL	*198.9±1.8*	Seven step marginal plateau (49.5% ^{39}Ar) for 125-160 μm fraction.
	191.8±2.2	Six step plateau (80% ^{39}Ar) for 160-200 μm fraction
	200.3±1.7	Nine step marginal plateau (35% ^{39}Ar) for 160-200 μm fraction.
GUI141 PL	*195.6±1.3*	Seven step marginal plateau (48% ^{39}Ar) for 125-160 μm fraction.
	201.7±1.8	Four step plateau (63% ^{39}Ar) on 160-200 μm fraction.
	201.5±1.5	Five step marginal plateau (33% ^{39}Ar) on 160-200 μm fraction.
		LIBERIA - SILL
L21 PL	191.5±1.5	Average of plateaus for whole-rock and plagioclase separate.
		Recalculated to current decay constants.

AMP, BIO and PL denote amphibole, biotite and plagioclase, respectively. "Ages" in italics are unaccceptable as crystallization values (see text). Errors listed at 1σ level.

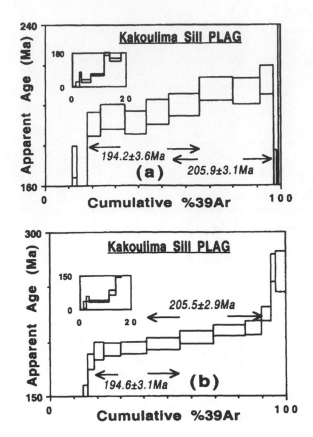

Figure 7. Age spectra for rocks from the Kakoulima layered intrusion, Guinea (data from G. Feraud, written communication, 1999). (a) and (b) GUI110 and GUI76 plagioclase, respectively. Errors and symbols as in Figure 2. Each specimen exhibits discordant age spectra, in which two marginal plateaus can be defined. In each case, these ages do not overlap; this highlights problems in use of marginal plateau ages as estimates of crystallization values.

since BR-4A (200 Ma) lies ~100 km *east* of BR-10 (193 Ma). - see Baksi and Archibald (1997), Figure 1. Magmatism in South America (Brazil) and Africa (Liberia), is confirmed at 192 Ma. Marginal plateau ages for the oldest tholeiitic flows in Morocco (M1208PL - Sebai et al., 1991) and for the Wachtung I flow in the Newark Basin (Hames et al., 2000) are >200 Ma. This may indicate the earliest phase of magmatism in North America predated the Triassic-Jurassic boundary, with the bulk of the material formed at 200 Ma.

McHone (2000) and Hames et al. (2000) suggested that CAMP rocks were formed by lithospheric processes that led to upper mantle melting, rather than by hotspot derivation. Herein, I utilize published trace element data and the ΔNb approach (Fitton et al., 1997), in seeking a possible

plume connection to CAMP rocks. A corollary to be noted is that though the method appears to have merit for lava flows (Baksi, 2001), sills and dikes frequently yield spurious positive ΔNb values (A.K. Baksi, unpubl. data). Movement of volatile complexes in differentiated sills may lead to enrichment of Nb and other elements (Greenough and Papezik, 1986). Differentiated sections of the Palisades and Gettysburg sills contain minor amounts of zircon/baddeleyite, which have been used for U-Pb dating (Dunning and Hodych, 1990). The effect of fractionation of a minor amount of these minerals on the (Nb – Y - Zr) chemistry of the residual melt is problematic, since values for the requisite partition coefficients, are uncertain (see Green, 1994). For extrusive rocks belonging to CAMP, the North Mountain basalt, Nova Scotia, (Dostal and Greenough, 1992) and the Holyoke basalt, Connecticut (Phillpotts et al., 1996) exhibit ΔNb values of +0.1 to +0.4, whereas the Orange Mountain and Hook Mountain Basalts, New Jersey (Tollo and Gottfried, 1992), yield marginally negative (+0.03 to -0.10) values. Early Mesozoic tholeiites from Morocco (Bertrand et al., 1982) and the western Maranhao lava flows (Fodor et al., 1990), with ages of ~193-200 Ma (Baksi and Archibald, 1997), yield values of +0.1 to +0.3. A deep mantle component is indicated in some of these rocks. High quality trace element data are required for other *extrusive* CAMP rocks, to critically test for a genetic connection to plumes, and to elucidate the possible role of the latter in continental breakup (cf. Storey, 1995).

A possible genetic link between formation of the CAMP and the faunal extinction event seen at the Triassic-Jurassic boundary has been suggested (Marzoli et al., 1999). The timing of the latter event, determined by U/Pb dating, suggest it was centred around 200 Ma and occurred ~0.5-0.7 m.y. earlier on continents than in the sea (Palfy et al., 2000). Ages currently available for the northern sections of CAMP span this age range (Table 6), whereas those from Brazil appear to be marginally younger. The early Hettangian lag in biodiversity (a post extinction lag - Palfy et al., 2000), may reflect the effects of extensive basaltic magmatism in Brazil. Based on organic carbon data from UK and Greenland (Hesselbo et al., 2002) and Re-Os abundances and Os isotopic composition of marine mudrock samples in the UK (Cohen et al., 2002), it has been argued that isotopic excursions (shifts) took place in the latest Triassic. Both sets of authors favor a genetic link between the faunal extinction event and the initiation of igneous activity form-

Table 5. ^{40}Ar /^{39}Ar ages for rocks from Brazil. Plagioclase separates of Marzoli et al. (1999) and whole-rock basalts of Baksi and Archibald (1997).

Sample Number	Age (Ma)	Comments
		RORAIMA DIKES
8820 PL	197.9±0.2*	Plateau age (200.3±0.3 Ma) older than isochron value - excess argon present. Isochron age listed.
8804 PL	*199.0±1.2*	Marginal plateau (<50% of ^{39}Ar)
8818 PL	197.4±1.9	Plateau age (203.2±0.9 Ma) older than isochron value - excess argon present. Isochron age listed.
		CASSIPORE DIKES
8026 PL	199.0±2.3	Plateau (202.0±1.0 Ma) and listed isochron ages overlap.
8034 PL	*192.7±0.9*	Plateau age listed, but not acceptable, as isochron shows high F value#.
8041 PL	*197.1±0.9*	Plateau age listed, but not acceptable, as isochron shows high F value#.
		CEARA LAVA FLOW
8232 PL	*198.4±0.7*	Plateau age < isochron age (203.7±0.8 Ma), denotes (^{40}Ar/^{36}Ar)$_i$ < 295.5.
		ANARI LAVA FLOW
ANG-6 PL	198.0±0.4*	Listed plateau age overlaps isochron value (196.8±0.6 Ma).
		TAPIRAPUA LAVA FLOW
TRG-10 PL	*196.6±0.9*	Listed plateau age unacceptable; isochron shows high F value#.
		MARANHAO LAVA FLOWS
5013 PL	190.5±0.8	Plateau and isochron age (189.0±0.7 Ma) overlap - former listed.
5042 PL	198.5±0.4*	Plateau and isochron age (197.3±0.6) overlap - former listed.
BR-10 WR	193.0±0.9	Plateau age: isochron values, age = 193.1±0.9 Ma, F = 0.85, (^{40}Ar/^{36}Ar)$_i$ = 284±13
BR-4A WR	200.3±1.3	Plateau age: isochron values, age = 202.1±0.9 Ma, F = 0.83, (^{40}Ar/^{36}Ar)$_i$ = 240±35

PL and WR denote plagioclase and whole-rock, respectively. "Ages" in italics not used as crystallization values. Errors listed at 1σ level. F = MSWD. *Listed errors adjusted to ±0.8 m.y. in Table 6 (see text). # F values > 10 (P.R. Renne, written comm., 1999).

ing CAMP. However, Tanner et al. (2001) suggest there was only a modest increase in atmospheric carbon dioxide levels across the Triassic-Jurassic boundary, and do not support sudden global warming caused by carbon dioxide emissions from CAMP volcanism. They prefer short term cooling or acidification of oceans and atmosphere from sulfur dioxide aerosols (Olsen, 1999), or rapid sea-level change resulting from thermal doming linked to a plume (Hallam, 2000). Future work should focus on other rocks from North and South America to delineate the timing and total areal extent of CAMP magmatism. It is probable that CAMP magmatism began at ~205 Ma, and peaked at 198-200 Ma.

Acknowledgments. ^{40}Ar/^{39}Ar dating of the Gettysburg and Palisades sill samples, generously provided by Joe Hodych, were carried out at Queen's University; Ed Farrar is thanked for access to his dating laboratory. Gilbert Feraud supplied supporting analytical data for the work carried out by Deckart et al. (1997). Pat Smith kindly plotted the age frequency spectra. Paul Renne and John Sutter made numerous perceptive com-

Table 6. Accurate ^{40}Ar /^{39}Ar ages for CAMP rocks.

Age±1σ (Ma)	Reference.	Age±1σ (Ma)	Reference
North America, northwest Africa.		South America, west Africa	
200.3±0.9	Gettysburg sill [1]	197.9±0.8	Roraima dike [2]
200.4±0.9	Palisades sill [1]	197.4±1.9	Roraima dike [2]
198.8±1.0	Wachtung flow [3]	199.0±2.3	Cassipore dike [2]
199.7±0.7	South Carolina dike [3]	198.0±0.8	Anari lava flow [2]
199.2±0.7	South Carolina dike [3]	190.5±0.8	Maranhao lava flow [2]
200.3±1.3	Morocco lava flow [4]	198.5±0.8	Maranhao lava flow [2]
175.2±3.4	Morocco lava flow [4]	193.0±0.9	Maranhao lava flow [5]
205.9±2.3	Morocco lava flow [4]	200.3±1.3	Maranhao lava flow [5]
200.0±1.1	Morocco lava flows [6]	195.9±4.4	French Guyana dike [7]
		227.0±2.9	French Guyana dike [7]
		212.5±2.2	Guinea layered intrusion [7]
		191.5±1.5	Liberian dike [8]

Errors listed at the 1σ level. References: [1] = this work: [2] = Marzoli et al. (1999): [3] = Hames et al. (2000): [4] = Fiechtner et al. (1992): [5] = Baksi and Archibald (1997): [6] = Sebai et al. (1991): [7] = Deckart et al. (1997): [8] = Dalrymple and Lanphere (1974) and Dalrymple et al. (1975). U/Pb ages for CAMP rocks are: 201.8±0.5 Ma for the North Mountain Basalt (Hodych and Dunning,1992); 201.5±0.5 Ma for the Gettysburg and Palisades Sills (Dunning and Hodych, 1990).

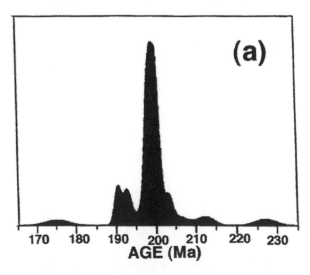 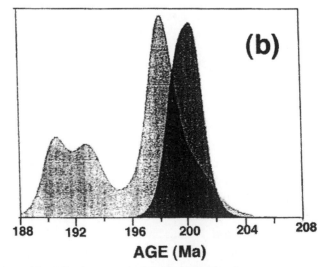

Figure 8. Frequency spectra of CAMP ages with 1σ errors. (a) Ages (N = 21), show a wider range than noted by Marzoli et al. (1999); the main peak supports their conclusion of a marked magmatic event at ~200 Ma. (b) Ages in the range 200 -190 Ma, for the northern sections (N = 8, dark gray) and southern section (N = 10, light gray). The two plots show different skewness and the main peak of magmatic activity in the north apparently preceded that in the south (see text).

ments during formal review of this manuscript. This work was supported by a grant from the Ardhendu-Roma Foundation.

REFERENCES

Baksi, A.K., Critical evaluation of the age of the Deccan Traps, India: Implications for flood-basalt volcanism and faunal extinctions, *Geology, 15*, 147-150, 1987.

Baksi, A.K., Comment on "U/Pb zircon and baddeleyite ages for the Palisades and Gettysburg sills of the northeastern United States: Implications for the age of the Triassic/Jurassic boundary", *Geology, 19*, 860-861, 1991.

Baksi, A.K., Reevaluation of plate motion models based on hotspot tracks in the Atlantic and Indian Oceans, *J. Geol., 107*, 13-26, 1999.

Baksi, A.K., Search for a deep mantle component in mafic rocks using a Nb-Y-Zr plot, *Can. Jour. Earth. Sci., 38*, 813-824, 2001.

Baksi, A.K. and D.A. Archibald, Mesozoic igneous activity in the Maranhao province, northern Brazil: $^{40}Ar/^{39}Ar$ evidence for separate episodes of basaltic magmatism, *Earth Planet. Sci. Lett., 151*, 139-153, 1997.

Baksi, A.K., D.A. Archibald and E. Farrar, Intercalibration of $^{40}Ar/^{39}Ar$ dating standards, *Chem. Geol., 129*, 307-324, 1996.

Bertrand, H., J. Dostal and C. Dupuy, Geochemistry of Early Mesozoic tholeiites from Morocco, *Earth Planet. Sci. Lett., 58*, 225-239, 1982.

Bowring, S.A., D.H. Erwin, Y.G. Jin, M.W. Martin, K. Davidek and W. Wang, U/Pb zircon geochronology and tempo of the end-Permian mass extinction, *Science, 280*, 1039-1045, 1998.

Cohen, A.S. and A.L. Coe, New geochemical evidence for the onset of volcanism in the Central Atlantic magmatic province and environmental change at the Triassic-Jurassic boundary, *Geology, 30*, 267-270.

Dallmeyer, R.D.,The Palisades sill: A Jurassic intrusion? Evidence from $^{40}Ar/^{39}Ar$ incremental release ages, *Geology, 3*, 243-245, 1975.

Dalrymple, G.B. and M.A. Lanphere, $^{40}Ar/^{39}Ar$ age spectra of some undisturbed terrestrial samples, *Geochim. Cosmochim. Acta, 38*, 715-738, 1974.

Dalrymple, G.B., C.S. Gromme and R.S. White, Potassiumargon age and paleomagnetism of diabase dikes in Liberia: Initiation of central Atlantic rifting, *Geol. Soc. Am Bull., 86*, 399-411, 1975.

Deckart, K., G. Feraud and H. Bertrand, Age of Jurassic continental tholeiites of French Guyana/ Suriname and Guinea: Implications to the opening of the central Atlantic Ocean, *Earth Planet. Sci . Lett., 150*, 205-220, 1997.

DeVor, R.E., T-H. Chang and J.W. Sutherland, *Statistical quality design·an d control*, Prentice Hall, Upper Saddle River, New Jersey, 1992, pp. 787-797.

Dostal, J. and J.D. Greenough, Geochemistry and petrogenesis of the Early Mesozoic North Mountain Basalts of Nova Scotia, Canada, in *Eastern North American Mesozoic magmatism* (J.H. Puffer and P.C. Ragland, Eds.), Geol. Soc. Am. Spec. Paper 268, 149-160, 1992.

Dunning, G.R. and J.P. Hodych, U/Pb zircon and baddeleyite ages for the Palisades and Gettysburg sills of the northeastern United States: Implications for the age of the Triassic/Jurassic boundary, *Geology, 18*, 795-798, 1990.

Fiechtner, L., H. Friedrichsen and K. Hammerschmidt, Geochemistry and geochronology of Early Mesozoic tholeiites from central Morocco, *Geol. Ru ndschau, 81*, 45- 62, 1992.

Fitton, J.G., A.D. Saunders, M.J. Norry, B.S. Hardarson a nd R.N. Taylor, Thermal and chemical structure of the Iceland plume, *Earth Planet. Sci. Lett., 153*, 197-208, 1997.

Fodor, R.V., A.N. Sial, S.B. Mukasa and E.H. McKee, 1990, Petrology, isotope characteristics and K-Ar ages of the Maranhao northern Brazil Mesozoic basalt province, *Contrib. Mineral. Petrol., 104*, 555-567, 1990.

Green, T.H., Experimental studies of trace-element partitioning applicable to igneous petrogenesis - Sedona 16 years later, *Chem. Geol., 117*, 1-36, 1994.

Greenough, J.D. and V.S. Papezik, Volatile control of differentiation in sills from the Avalon Peninsula, Newfoundland, Canada, *Chem. Geol., 54*, 217-236, 1986.

Hallam, A., The end-Triassic extinction in relation to a superplume event (abstract), *Geol. Soc. Am Abs. Prog. 37*, 380, 2000.

Hames, W.E., P.R. Renne and C. Ruppel, New evidence for geologically instantaneous emplacement of earliest Jurassic Central Atlantic magmatic province basalts on the North American margin, *Geology 28*, 859-862, 2000.

Hesselbo, S.P., S.A. Robinson, F. Surlyk and S. Piasecki, Terrestrial and marine extinction at the Triassic-Jurassic boundary synchronized with major carbon-cycle perturbation; A link to initiation of massive volcanism? *Geology, 30*, 251-254.

Hodych, J.P. and G.R. Dunning, Did the Manicouagan impact trigger end-of-Triassic mass extinction? *Geology, 20*, 51-54, 1992.

Kamo, S.L., G.K. Czamanske and T.E. Krogh, A minimum U-Pb age for Siberian flood-basalt volcanism, *Geochim. Cosmochim. .Acta, 60* , 3505-3511, 1996.

Marzoli, A., P.R. Renne, E.M. Piccirillo, M. Ernesto, G. Bellieni and A. De Min, Extensive 200-million-year-old continental flood basalts of the Central Atlantic Magmatic Province, *Science, 284,* 616-618, 1999.

McHone, J.G., Non-plume magmatism and rifting during the opening of the Central Atlantic Ocean, *Tectonophys., 316,* 287-296, 2000.

Min, K.W., R. Mundil, P.R. Renne and K.R. Ludwig, A test for systematic errors in $^{40}Ar/^{39}Ar$ geochronology through comparison with U/Pb analysis of a 1.1Ga rhyolite, *Geochim. Cosmochim. Acta, 64,* 73-98, 2000.

Mundil, R., I. Metcalfe, K.R. Ludwig, P.R. Renne, F. Oberli and R.S. Nicoll, Timing of the Permian-Triassic biotic crisis: implications from new zircon U/Pb age data (and their limitations), *Earth Planet. Sci. Lett., 187,* 131-145, 2001.

Olsen, P.E., Giant lava flows, mass extinctions, and mantle plumes, *Science, 284,* 604-605, 1999.

Palfy, J., J.K. Mortensen, E.S. Carter, P.L. Smith, R. M. Friedman and H.W. Tipper, Timing the end-Permian mass extinction: First on land, then in the sea? *Geology, 28,* 39-42, 2000.

Philpotts, A.R., M. Carroll and J.M. Hill, Crystal-mush compaction and the origin of pegmatitic segregation sheets in a thick flood-basalt flow in the Mesozoic Hartford Basin, Connecticut, *Jour. Petrol., 37,* 811-836, 1996.

Renne, P.R., C.C. Swisher, A.L. Deino, D.B. Karner, T. Owens and D.J. DePaolo, Intercalibration of standards, absolute ages and uncertainties in $^{40}Ar/^{39}Ar$ dating, *Chem. Geol., 145,* 117-152, 1998.

Renne, P.R., Z. Zichao, M.A. Richards, M.T. Black and A.R. Basu, Synchrony and causal relations between Permian-Triassic boundary crises and Siberian flood volcanism, *Science, 269,* 1413-1416, 1995.

Sebai, A., G. Feraud, H. Bertrand and J. Hanes, $^{40}Ar/^{39}Ar$ dating and geochemistry of tholeiitic magmatism related to the early opening of the Central Atlantic rift, *Earth Planet. Sci. Lett., 104,* 455-472, 1991.

Steiger, R.H. and E. Jager, Subcommission on geochronology: Convention on the use of decay constants in geo- and cosmochronology, *Earth Planet. Sci. Lett., 36,* 359- 362, 1977.

Storey, B.C., The role of mantle plumes in continental breakup: case histories from Gondwanaland, *Nature, 377,* 301-308, 1995.

Sutter, J.F., Innovative approaches to the dating of igneous Events in the Early Mesozoic basins of the eastern United States, in A.J. Froelich and G.R. Robinson, Eds., Study of the Early Mesozoic basins of the eastern United States: *U.S. Geol. Surv. Bull. 1776,* 194- 198, 1988.

Sutter, J.F. and T.E. Smith, $^{40}Ar/^{39}Ar$ ages of diabase intrusions from Newark trend basins in Connecticut and Maryland: Initiation of central Atlantic rifting, *Am. J. Sci., 279,* 808-831, 1979.

Tanner, L.H., J.F. Hubert, B.P. Coffey and D.P. McInerney, Stability of atmospheric CO_2 levels across the Traissic/Jurassic boundary, *Nature, 411,* 675-677, 2001.

Tollo, R.P. and D. Gottfried, Petrochemistry of Jurassic basalt from eight cores, Newark Basin, New Jersey, in *Eastern North American Mesozoic magmatism* (J.H. Puffer and P.C. Ragland, Eds.), Geol. Soc. Am. Spec. Paper 268, 233-260, 1992.

York, D., Least squares fitting of a straight line using correlated errors, *Earth Planet. Sci. Lett., 5,* 320-324, 1969.

Ajoy Baksi, Department of Geology & Geophysics, Louisiana State University, Baton Rouge, LA 70803, USA. Email: abaksi@geol.lsu.edu

The Central Atlantic Magmatic Province (CAMP) in Brazil: Petrology, Geochemistry, $^{40}Ar/^{39}Ar$ Ages, Paleomagnetism and Geodynamic Implications

Angelo De Min[1], Enzo M. Piccirillo[1], Andrea Marzoli[2], Giuliano Bellieni[3], Paul R. Renne[4], Marcia Ernesto[5] and Leila S. Marques[5]

The CAMP tholeiitic magmatism in Brazil (mean $^{40}Ar/^{39}Ar$ age of 199.0±2.4 Ma) occurs on the continental margin to ca. 2,000 km into the South American platform, near the boundary between the ancient terrains of the Amazonia craton and Proterozoic/Brazilian-cycle related mobile belts. Geological evidence indicates that this magmatism was preceded, in Permo-Triassic times, by continental sedimentation, indicating a possible regional uplift. The Brazilian CAMP tholeiites are generally evolved and characterized by a low TiO_2 concentration (less than 2 wt%). The Cassiporé dykes, which are usually high in TiO_2 (more than 2wt%) are an exception. The Cassiporé low- and high-TiO_2 basalts are characterized by a positive Nb anomaly and Sr-Nd isotopes that are parallel to "typical" mantle array. Except for one sample, all the other Brazilian CAMP tholeiites that are low in TiO_2, show Sr-Nd isotopes trending towards crustal components. The latter isotopic characteristics could be related to "crustal recycling" ancient (Middle-Late Proterozoic) subductions, and/or low-pressure crustal interaction. All the Brazilian CAMP tholeiites show a decoupling between their Sr-Nd isotopic composition and Rb/Sr and Sm/Nd values, suggesting "mantle metasomatism", and/or subduction-related crustal interaction before mantle melting. Notably, the chemical data show that tholeiites from specific Brazilian regions are related to mantle sources that reflect compositional mantle heterogeneity, including the lower mantle of the lithospheric thermal boundary layer. In general, paleomagnetic poles for CAMP rocks from South America, Africa and North America match an age of ca. 200 Ma, but also show a distribution pattern trending to younger ages (e.g. 190 Ma), especially for the South American poles relative to the CAMP magmatism of the continental edge. The Brazilian CAMP magmatism cannot be easily explained through "plume head" (active) models, being instead consistent with mantle geodynamic processes where the unstable buoyancy of the Pangea "supercontinent" played an essential role to approach isostatic stabilization. Therefore, it is proposed that the Brazilian CAMP magmatism was related to hot "upper" mantle incubation under thick continental lithosphere, and to edge-driven convection between lithospheric domains with different thickness.

[1] Dipartimento di Scienze della Terra, University of Trieste, Trieste, Italy.

[2] Département de Minéralogie, Université de Genève, Genève, Switzerland.

[3] Dipartimento di Mineralogia e Petrologia, University of Padova, Padova, Italy.

[4] Berkeley Geochronology Center, Berkeley, California State, USA.

[5] Departamento de Geofisica; Instituto Astronomico e Geofisico, University of São Paulo (USP), São Paulo, SP, Brazil.

The Central Atlantic Magmatic Province:
Insights from Fragments of Pangea
Geophysical Monograph 136
Copyright 2003 by the American Geophysical Union
10.1029/136GM06

INTRODUCTION

The major rifting related to the Pangea break-up and the opening of the central Atlantic ocean was preceded by extensive tholeiitic basalt activity in once-contiguous parts of central-northern South America, northeastern America, western Africa and western Europe (Fig. 1). *Marzoli et al.* [1999] demonstrated that such tholeiitic magmatism (ca. 2-$3*10^6$ km^3) mainly occurred at 200±4 Ma ($^{40}Ar/^{39}Ar$ dates) and covered an area of ca. $7*10^6$ km^2; the largest known continental flood large igneous province (LIP). The Central Atlantic Magmatic Province (CAMP) is mainly represented by low-TiO_2 (<2 wt%; LTi) and high-TiO_2 (>2 wt%; HTi) evolved (MgO<10 wt%) tholeiites, similar to the Middle Jurassic continental tholeiitic magmatism of Karoo [*Cox,* 1988; ca. 180 Ma: *Encarnación et al.,* 1996] and Antarctica [*Antonini et al.,* 1999; ca. 177 Ma: *Heimann et al.,* 1994; *Encarnación et al.,* 1996; *Fleming et al.,* 1997], and the Early Cretaceous low- and high-TiO_2 analogue from the Paraná basin [SE-Brazil; cf. *Piccirillo et al.,* 1989; *Piccirillo and Melfi,* 1988; mainly ca. 132 Ma: e.g. *Renne et al.,* 1992; *Stewart et al.,* 1996].

In the South American platform, most of the Early Jurassic magmatism (Fig. 1) crops out from French Guyana to Venezuela and from northern (e.g. Cassiporé, Roraima, Amazonia basin *latu sensu* (l.s.), Maranhão basin, Lavras da Mangabeira) to central (e.g. Anarí and Tapirapuá) regions of Brazil. The Brazilian CAMP magmatism is represented by low- and minor high-TiO_2 tholeiites represented by dyke swarms (Cassiporé, Roraima) crosscutting Archean-Proterozoic crystalline terrains, sill-type intrusions in Palaeozoic sediments (Amazonia basin l.s.) and stratoid flows overlying Palaeozoic and/or Triassic sedimentary sequences (Maranhão basin, Lavras da Mangabeira, Anarí and Tapirapuá).

The present paper addresses the geology, petrology, elemental and Sr-Nd isotope geochemistry, $^{40}Ar/^{39}Ar$ dating and paleomagnetism of the Brazilian CAMP tholeiites. These data are essential to understand the relationships between the mantle geodynamic processes involved in the generation of the CAMP tholeiites at pre-Atlantic ocean crust formation.

GEOLOGICAL OUTLINES

The oldest terrains of the investigated areas in central-northern Brazil belong to the Amazonia craton l.s. which represents the extension of the West African craton l.s. in South America [*Trompette,* 1994]. The Amazonia craton is constituted by Archean-Early Proterozoic crystalline rocks (granulites, gneisses, migmatites and amphibolites) stabilized during the Transamazonian-Paraguayan Cycle [2.0-1.7 Ga; *Cordani et al.,* 1984; *Schobbenhaus et al.,* 1984]. This craton is subdivided by the ENE-trending

Palaeozoic sedimentary basin in the Guyana (north) and Guaporé (south) cratonic blocks. The Amazonia craton l.s. is bordered at the western and eastern sides by Proterozoic mobile belts [*Trompette,* 1994].

After the Laurasia-Gondwana collision, the northern part of South America underwent mainly E-W extensional tectonics which allowed the emplacement of voluminous tholeiitic magmasm Early Jurassic times. This magmatism is represented by generally N-S trending dyke swarms [e.g. Roraima and Cassiporé; *Almeida,* 1986; *Oliveira and Tarney,* 1990; *Menezez Leal et al.,* 2000 and unpublished data], mafic sills [e.g. Amazonia basin l.s.; *Almeida,* 1986], and basalt flows [e.g. Maranhão basin, Lavras da Mangabeira, Anarí and Tapirapuá; *Bellieni et al.,* 1990; *Fodor et al.,* 1990; *Montes Lauar et al.,* 1994 and unpublished data]. Field and petrological data support the view that the emplacement of this Early Jurassic magmatism in central-northern Brazil, and in the Paraná-Etendeka (and Angola) LIP [*Piccirillo and Melfi,* 1988], preceded the major rifting which lead to the central-northern Atlantic ocean crust formation.

Recent $^{40}Ar/^{39}Ar$ dates of the Jurassic tholeiites in central-northern Brazil (Fig. 1) range from 188 to 202 Ma (Early Jurassic) with a peak of activity at 200±2 Ma, including the tholeiitic dyke swarms of French Guyana, Surinam, West Africa and SE-USA [cf. *Deckart et al.,* 1997; *Baksi and Archibald,* 1997; *Marzoli et al.,* 1999, *Hames et al.* 2000 and present study]. It follows that the Early Jurassic tholeiitic magmatism of the central-northern Brazil is an important part of the CAMP and testifies that the tholeiitic magmatism extends into South America for about 2,000 km.

Roraima dykes

The Early Jurassic magmatism in NE-Roraima State (N-Brazil) is characterized by generally NE-trending tholeiitic dyke swarms (Fig. 1). The dykes are 5-50 m thick, may reach 100 km in length and crosscut Early Proterozoic terrains of the Sumuru Group (Guyana cratonic block). The Roraima dykes parallel the Proterozoic fault system (Tacutú lineament), restored in Gondwana break-up times. Towards the South, the dykes underlie the NE-trending Tacutú sedimentary basin, seat of an important Jurassic-Cretaceous volcano-sedimentary sequence which includes tholeiitic sills and flows (Apoteri Basalt Formation) of Middle Jurassic age, i.e. K/Ar=150-180 Ma [*Berrangé and Dearley,* 1975] and $^{40}Ar/^{39}Ar$=172 Ma [*Renne,* unpublished data].

Cassiporé dykes

Early Jurassic NNW to NS trending dyke swarms outcrop in the Cassiporé region (Amapá State, N-Brazil) and extend into French Guyana. These dykes are 10-100 m

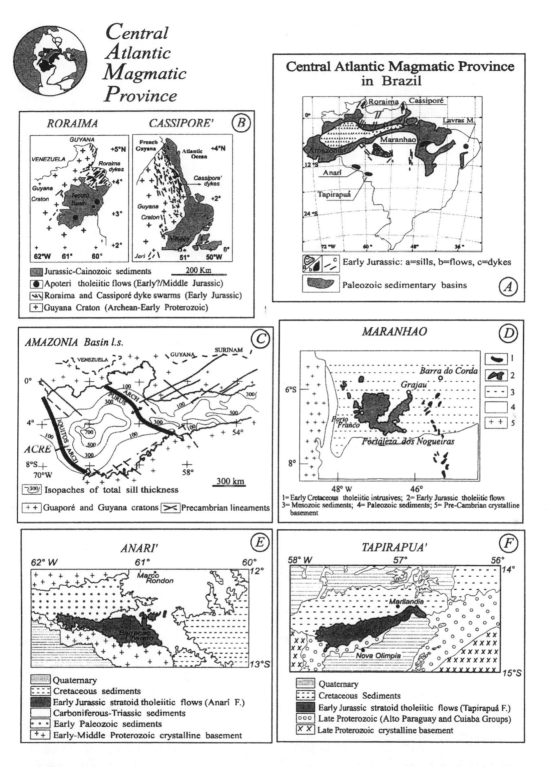

Figure 1. Geological sketch maps of the Early Jurassic (191-202 Ma) tholeiitic magmatism (sills, dykes and flows) of the central-northern Brazil belonging to the Central Atlantic Magmatic Province [CAMP; *Marzoli et al.*, 1999]. A= distribution of the Early Jurassic tholeiitic magmatism in Brazil and of the most important Palaeozoic sedimentary basins. B= Roraima and Cassiporé dykes. C= sills of the Amazonian basin. D, E, F= stratoid flow occurrences in the Marañhao, Anarí and Tapirapuá basins, respectively.

thick, up to 300 km in length and intrude Archean-Early Proterozoic terrains of the Guyana cratonic block. In the contiguous French Guyana (and Surinam) the Early Jurassic dykes are mainly 1-60 m thick and up to 20 km in length [*Deckart et al.*, 1997]. Southwest of Macapá city (Fig. 1), the NE- to NS-trending Early Jurassic (K/Ar=180-250 Ma; *Lima et al.*, 1974] tholeiitic dyke swarms of the Jari River intrude the Palaeozoic sediments of the Amazonia basin. The latter dykes may reach several hundred meters in thickness and tens kilometers in length [*Oliveira and Tarney*, 1990].

The Cassiporé and Guyana dyke swarms probably continue into the once contiguous regions of Western Africa, where they are represented by the coeval Early Jurassic tholeiitic dykes of Liberia and Guinea [*Dupuy et al.*, 1988; *Deckart et al.*, 1997].

Amazonia Sills

The Amazonia sills (Fig. 1) represent the most extensive Early Jurassic magmatism in the northern part of South America. They cover an area of about $1*10^6$ km^2 and have a total thickness that varies between 100 and 809 m (ca. $5*10^5$ km^3).

The Amazonia sills intrude the Palaeozoic sediments of three contiguous basins: Acre ($2.3*10^5$ km^2), Solimoes ($4*10^5$ km^2) and Amazonia s.s. ($5*10^5$ km^2) [*Petri and Fulfaro*, 1983; *Schobbenhaus et al.*, 1984]. These basins are separated by the Iquitos and Purus structural highs (arches) formed since Silurian-Devonian times.

The stratigraphy of the Amazonia basin l.s. is based on bore-hole investigation [e.g. *Almeida*, 1986]. The Palaeozoic sedimentary rocks range in thickness from 2,300 to 5,500 m and are divided into three main sequences (1) Ordovician-Devonian (480-400 Ma; Trombetas Group), (2) Devonian-Carboniferous (390-340 Ma; Urupadi, Curuá and Marimari Groups) and (3) Carboniferous-Permian (330-250 Ma; Tapajós and Tefé Groups). The Mesozoic sediments are of Cretaceous age, as Triassic and Jurassic strata are absent. The maximum total thickness (809 m) of the sills decreases from the Solimoes towards both the Amazonia s.s. (east) and Acre (west) basins [*Almeida*, 1986].

Maranhão (Parnaíba) and Lavras da Mangabeira Flows

The Maranhão sedimentary basin (Fig. 1) extends for ca. $6*10^5$ km^2 and is characterized by uplifted structures (Ferre-Urbano Santos and Tocantin arches) at its NNE-margin. The sedimentary rocks are mainly related to Palaeozoic subsidence [*Petri and Fulfaro*, 1983; *Almeida*, 1986; *Arora et al.*, 1999] and are of Silurian-Late Permian age, while the Triassic sediments are scarce. The Early Jurassic sediments are constituted by scarce aeolian sandstones intercalated within the Early Jurassic tholeiitic flows which locally may be covered by Middle-Late Jurassic sediments [*Petri and Fulfaro*, 1983]. The maximum thickness of the sediments attains ca. 2,000 m [*Almeida*, 1986]. In Early Jurassic times, the Maranhão basin was the seat of a widespread volcanic activity (presently preserved ca. 40,000 km^2; *Bellieni et al.*, 1990] represented by the stratoid tholeiites of the Mosquito Formation. The basalt sequence has a maximum preserved thickness of 175 m in the western side of the basin [*Almeida*, 1986].

The scarce and strongly vesiculated two-pyroxene tholeiites of Early Jurassic age from Lavras da Mangabeira (Fig. 1) are substantially altered showing widespread celadonitic glass, secondary carbonates and zeolites (see below).

Anari and Tapirapuá Flows

The Early Jurassic stratoid tholeiites of Anari and Tapirapuá [^{40}Ar/^{39}Ar=196-198 Ma; *Marzoli et al.*, 1999] flooded SE-trending structural lows [*Drueker and Gay*, 1985]. These tholeiites (Fig. 1) represent the southernmost occurrence of the CAMP magmatism in South America.

The Anari tholeiites (presently ca. 1,000 km^2) outcrop in SE-Rondonia State (W-Brazil), have a total thickness of 50-80 m and extend for ca. 100 km. The stratoid flows overlie Middle-Late Proterozoic terrains (Xingú Complex, Guaporé cratonic block) and Palaeozoic to Mesozoic sediments at the southern and northern sides, respectively. Marine-fluvial to aeolian sediments of Carboniferous to Triassic age, respectively, preceded the basalt extrusion.

The Tapirapuá tholeiites (Mato Grosso State, W-Brazil; presently ca. 1,000 km^2) extend for ca. 115 km and may reach a total thickness of 310 m. These tholeiites overlie the Late Proterozoic terrains of the Alto Paraguay Group (Guaporé cratonic block) and are covered by the Cretaceous continental sediments of the Parecis Group at the northern side [*Montes Lauar*, 1993].

We emphasize that, in general, the transition from Permo-Triassic to Early Jurassic in northern-central Brazil is characterized by shallow-water to aeolian sedimentation preceding the Early Jurassic CAMP magmatism [*Petri and Fulfaro*, 1983].

^{40}Ar/^{39}Ar AGES OF THE BRAZILIAN CAMP THOLEIITES

Tholeiitic dykes and lava flows from Roraima, Cassiporé, Maranhão, Lavras da Mangabeira, Anari and Tapirapuá regions (Fig. 1) were dated via ^{40}Ar/^{39}Ar incremental heating of either plagioclase separates [*Marzoli et al.*, 1999, and present data] or whole rocks [*Baksi and Archibald*, 1997]. Three plagioclase separates of tholeiites from the Amazonia basin were dated for this study, adopting the analytical procedures described in

Renne et al. [1996]. All the ages were calculated relative to an age of 28.02 Ma for the Fish Canyon sanidine (FCs) neutron fluence monitor [*Renne et al.*, 1998] in order to allow comparison. The analytical errors are quoted at the 2σ confidence level. Whole-rock compositions of the dated samples are reported in Tables 1 and 6.

Roraima Dikes

Three Roraima dykes from the SW (sample R/8804) and NE (samples R/8818 and R/8820) portions of the dyke swarms were $^{40}Ar/^{39}Ar$ dated. A detailed step heating analysis (32-84 steps) allowed ages ranging from 203.2±1.8 to 199.0±2.4 Ma [*Marzoli et al.*, 1999]. Isochron ages for the same samples are slightly younger and range from 201.1±1.4 to 197.4±3.8. In view of the excess argon which is present in the analyzed plagioclase separates (i.e. initial $^{40}Ar/^{36}Ar$ up to 390), the isochron ages are favored. In summary, no significant age difference between the analyzed dykes is evident, and a mean age of 198.8 ± 2.0 Ma is obtained for the Roraima dyke swarms.

Cassiporé Dykes

Two high-TiO_2 (samples C/8034 and C/8041) and one low-TiO_2 (sample C/8026) tholeiitic dykes from Cassiporè were $^{40}Ar/^{39}Ar$ dated [*Marzoli et al.*, 1999]. The plagioclase separates from these samples yielded plateau ages of 202.0±2.0 Ma (C/8026), 197±1.8 Ma (C/8041) and 192.7±1.8 Ma (C/8034). As for the Roraima dykes, one Cassiporé dyke (8026) yielded an isochron age of 199.0±1.6 Ma which is slightly younger than its plateau age, and is favored here. Nonetheless, the apparent age variation of the Cassiporé dykes (199.0±1.6-192.7±1.8 Ma) is larger than that for the Roraima dykes, and consistent with $^{40}Ar/^{39}Ar$ ages of the French Guyana dykes [*Deckart et al.*, 1997], which represent the northern extension of the Cassiporé dyke swarms. No evident age difference exists between high- and low-TiO_2 Cassiporé tholeiites.

Maranhão, Lavras da Mangabeira, Anarí and Tapirapuá Flows

Plateau ages obtained from plagioclase separates (samples M/5013 and M/5042: eastern and western portions of the volcanic plateau, respectively; *Marzoli et al.*, 1999] and whole rocks [*Baksi and Archibald*, 1997] of the Maranhão (Mosquito Formation tholeiites) range from 190.5±1.6 to 198.5±0.8 Ma. The age relationships with the samples dated by *Marzoli et al.* [1999] are consistent with the stratigraphy. Fresh plagioclase separates of a vesiculated tholeiitic flow from Lavras da Mangabeira (sample L/8232) yielded a plateau age of 198.4±1.4 Ma

Fresh plagioclase separates of a vesiculated tholeiitic flow from Lavras da Mangabeira (sample L/8232) yielded a plateau age of 198.4±1.4 Ma.

Two basalts from and Anarí (sample AN/A6) and Tapirapuá (sample TR/A10) yielded plateau ages of 198.06±0.8 and 196.0±1.8 Ma, respectively.

Amazonia Sills

Fine-grained plagioclase separates of three samples from the Amazonian tholeiitic sills (AM/PB1, AM/PB9 and AM/PB30) yielded discordant apparent age spectra, with integrated ages of 181±2, 188±2 and 193±2 Ma, respectively (Fig. 2). The sinusoidal age and correlated Ca/K (deduced from corrected $^{37}Ar/^{39}Ar$ ratios) spectra in all three of these samples is striking. The range in apparent Ca/K in one sample (PB-30) is much larger than

TABLE 1. Representative major (wt%) and trace (ppm) element contents of 5 samples of the 14 Brazilian CAMP tholeiites used for the $^{40}Ar/^{39}Ar$ high-precision geochronologic analysis; the other dated samples (i.e. R/8804, R/8818, C/8026, M/5042, L/8232, AN/A6, PB/1, PB/9 and PB/30) are reported in table 6. Major elements recalculated to 100% on volatile-free basis.

| Sample | R/8820 | C/8034 | C/8041 | M/5013 | TR/A10 |
Rock-type	AB	ThB	AB	AB	AB
SiO_2 (wt%)	53.56	49.16	51.55	53.90	52.62
TiO_2	2.05	2.05	3.15	1.02	1.37
Al_2O_3	14.52	16.80	14.41	15.51	15.30
FeO_{total}	12.63	12.18	14.70	9.58	11.40
MnO	0.19	0.19	0.20	0.15	0.19
MgO	4.00	5.73	3.66	6.01	5.31
CaO	9.51	10.65	7.90	9.49	10.58
Na_2O	2.34	2.52	3.16	2.44	2.35
K_2O	0.94	0.48	0.82	1.75	0.40
P_2O_5	0.26	0.24	0.45	0.15	0.48
L.O.I.	1.15	1.47	1.55	2.44	1.83
Cr (ppm)	57	185	61	166	85
Ni	47	91	48	61	71
Ba	266	95	299	978	151
Rb	31	19	20	29	15
Sr	229	241	451	281	196
Zr	183	135	261	100	115
Y	29	35	42	34	41
Nb	14	14	28	7	11
La	14	10	28	13	19
Ce	33	31	70	22	25
Nd	25	19	34	17	7

Sample labels: R= Roraima, C= Cassiporé, M= Maranhão and TR= Tapirapuá. AB= andesitic basalt, ThB= tholeiitic basalt. L.O.I.= loss on ignition.

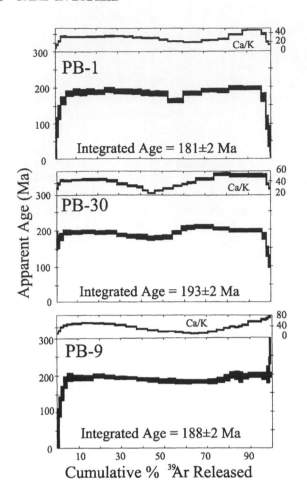

Figure 2. Apparent age and Ca/K spectra of plagioclase separates of samples PB1, PB9 and PB30 (Amazonia sills).

determined with the electron microprobe, and we tentatively conclude that significant recoil redistribution of ^{37}Ar and ^{39}Ar between fine-scale (i.e., much less than the 1 micrometer electron beam) exsolved phases is manifest. The net effect on integrated ages is unclear, but it seems likely that low-temperature discordance in these samples reflects alteration, and thus the integrated ages are minimum ages.

In general, the Brazilian Early Jurassic tholeiites yielded ^{40}Ar/^{39}Ar dates (199.0±2.4 Ma) which are indistinguishable from the ^{40}Ar/^{39}Ar ages of CAMP basalts from North America and West Africa. The youngest ages obtained for the CAMP tholeiites of the South American platform relate to the northernmost regions of Brazil [e.g. Cassiporé and Maranhão and Amazonia; *Baksi and Archibald*, 1997; *Marzoli et al.*, 1999; present study] and in French Guyana [*Deckart et al.*, 1997], i.e. the regions near the continental margin. Notably, a few samples of the once contiguous African continental margin in Guinea also yielded similar

younger ages, i.e. as low as 189.8 ± 1.9 Ma [*Deckart et al.*, 1997].

CLASSIFICATION AND PETROGRAPHY

The Brazilian CAMP magmatism is represented by basalts and basaltic andesites according to TAS [*Le Bas et al.*, 1986; Fig. 3) and by tholeiitic basalts, andesitic basalts, and scarce transitional basalts and latibasalts, respectively, following our preferred classification of *De La Roche et al.* [1980] and *Bellieni et al.* [1981]. Most basalts are quartz (Q) CIPW-normative (Q=0.1-11.6wt%), except a few samples of Cassiporé, Jari, Amazonia and Tapirapuá that show CIPW-normative olivine and have a olivine/hyperstene ratios (OL/HY) as high as 1.8 (cf. Table 6).

Roraima Dykes

The samples plot in the tholeiite and andesitic basalt fields (Q=4-6wt%). They are holocrystalline and have intergranular or slightly porphyritic texture. Augite ($Wo_{39-33}En_{49-45}Fs_{12-21}$), pigeonite ($Wo_{7-10}En_{63-48}Fs_{30-42}$), plagioclase ($An_{72-47}$) and Ti-magnetite are the common minerals, while orthopyroxene ($Wo_{4.5-4.4}En_{74.8-69.6}Fs_{20.7-26.0}$) microphenocrysts are rare. The groundmass is characterized by quartz-feldspar intergrowths, plagioclase, pyroxenes, Ti-magnetite (ulvöspinel=61-75%) and ilmenite (R_2O_3=2.3-4.7%). Apatite and zircon are accessory phases. Biotite and amphibole may be present as very thin rims around augite, or as groundmass microlites.

Cassiporé Dykes

Tholeiitic and andesitic basalts, sometimes straddling the transitional basalt and latibasalt fields, represent the common rock-types (Q=0.1-4.6wt%; Ol/Hy=0.3-1.1) of the Cassiporé dykes which are holocrystalline and have ophitic to subophitic texture. They are made up of augite ($Wo_{40-23}En_{42-39}Fs_{18-38}$), pigeonite ($Wo_{11-14}En_{55-56}Fs_{33-30}$), plagioclase ($An_{81-44}$) and Ti-magnetite. Olivine (Fo_{73-23}) is rare and often strongly altered. The groundmass is characterized by plagioclase, pyroxenes, quartz-feldspar intergrowths, Ti-magnetite (ulvöspinel=36-75%) and ilmenite (R_2O_3=1.8-7.2%). Apatite and zircon are common accessory minerals, sometimes associated with small patches of amphibole and biotite.

According to *Oliveira and Tarney* [1990] the Jarí dykes are mainly composed of augite, plagioclase, opaques and rare olivine (Fo_{70-51}).

Amazonia Sills

The tholeiitic and andesitic basalts (Q=0-6.5wt%, Ol/Hy=0.6) of the Amazonia sills are holo- to micro-crystalline and have subophitic to intergranular texture. They are characterized by phenocrysts and

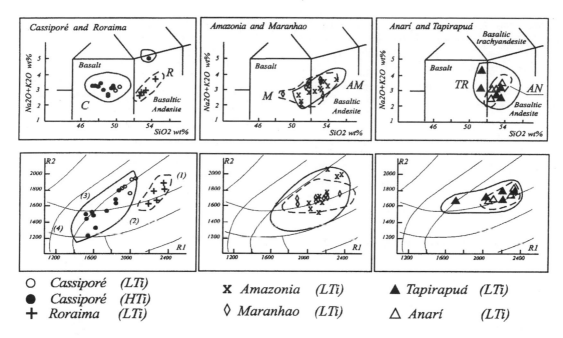

Figure 3. Total Alkali Silica [TAS; *Le Bas et al.* 1986] and R1-R2 [*De La Roche et al.* 1980; *Bellieni et al.*, 1981; R1 = 4Si − 11(Na+K) − 2(Fe²⁺+Fe³⁺+Ti), R2 = 6Ca + 2Mg + Al; oxide percentages converted to millications] classificative diagrams for representative Brazilian CAMP basalts; solid and dashed fields include all the analyzed samples. Abbreviations: AM= Amazonia sills, AN= Anarí flows, C= Cassiporé dykes, M= Marañhao flows and TR= Tapirapuá flows. 1= tholeiitic basalt, 2= andesitic basalt, 3= transitional basalt, 4= latibasalt; LTi and HTi = low (<2wt%) and high (>2wt%) TiO₂, respectively.

microphenocrysts of augite ($Wo_{40-23}En_{50-48}Fs_{10-29}$), pigeonite ($Wo_8En_{69}Fs_{23}$), plagioclase ($An_{81-51}$), and rare orthopyroxene ($Wo_{4.7}En_{69.2}Fs_{26.1}$) and olivine ($Fo_{65-59}$). In general, olivine is altered in iddingsitic products and opaques, augite may have thin rims of chlorite and/or amphibole, while plagioclase may be partly replaced by smectitic products or, more rarely, by carbonates. The groundmass is made up of augite, plagioclase, Ti-magnetite (ulvöspinel=7-53%), ilmenite (R_2O_3=4-12%) and quartz-feldspar intergrowths; apatite and zircon are common accessory minerals. The matrix is sometimes altered in chloritic and/or smectitic and/or celadonitic products, and may contain biotite and amphibole microlites.

Maranhão (Parnaíba) and Lavras da Mangabeira Flows

The stratoid tholeiites and andesitic basalts (Q=5-12wt%) of the Maranhão basin are subaphyric or slightly porphyritic [maximum phenocryst content=7vol%; *Bellieni et al.*, 1990]. Phenocrysts and microphenocrysts are represented by augite ($Wo_{35-30}En_{51-40}Fs_{13-30}$), pigeonite ($Wo_{8-9}En_{63-55}Fs_{29-36}$), plagioclase ($An_{80-58}$), Ti-magnetite and completely altered olivine. The groundmass is constituted by pyroxenes, plagioclase, Ti-magnetite

(ulvöspinel=61-74%) and ilmenite (R_2O_3=3.5-20.1%). Apatite and zircon are common accessory minerals. The vesiculated samples usually contain secondary carbonates and zeolites.

East of the Maranhão basin, towards the continental margin, sparsely Early Jurassic tholeiitic basalts outcrop at Lavras da Mangabeira. They are characterized by abundant zeolite- and carbonate-bearing vesicles and amygdales. Their tholeiitic nature is testified by the occurrence of phenocrysts and/or microphenocrysts of augite ($Wo_{36-31}En_{50-45}Fs_{14-24}$), pigeonite ($Wo_{9.8-8.7}En_{60.1-46.1}Fs_{10.1-45.2}$), fresh plagioclase ($An_{67-68}$) and Ti-magnetite. The matrix is constituted by pyroxenes, plagioclase, Ti-magnetite (ulvöspinel=34-80%), ilmenite, and apatite and zircon as accessory phases. Carbonates, zeolites and clay minerals are secondary and/or alteration products.

Anari and Tapirapuá Flows

The samples correspond to tholeiitic basalts (Q=0-7wt%; Ol/Hy=1.8). They vary from aphyric to slightly porphyritic types or are holocrystalline with moderately porphyritic texture. The vesiculated samples often contain zeolites infilling. These tholeiites are made up of phenocrysts and microphenocrysts of augite ($Wo_{41-20}En_{47-49}Fs_{12-31}$), pigeonite (Wo$_{8-9}En_{64-36}Fs_{28-55}$), plagioclase An$_{84-46}$), Ti-magnetite and completely altered olivine. The groundmass is given by pyroxenes, plagioclase, Ti-magnetite (ulvöspinel=49%), ilmenite (R$_2$O$_3$=0.1%) and accessory apatite and glass, the latter sometimes replaced by zeolitic products and/or clay minerals.

MINERAL COMPOSITIONS

Microprobe analyses were carried out with a Cameca-Camebax operating at 15 kV and 15 nA. A PAP-Cameca program was used to convert X-ray counts into weight percent of the corresponding oxides. Results are considered accurate to within 2-3% for major elements and better than 10% for minor elements. The compositions of pyroxenes, olivine, plagioclase and opaques of representative Brazilian CAMP tholeiites are in the Tables 2 to 5. With "early crystallization" are indicate cores of phenocrysts and/or microphenocrysts, while with "late crystallization" are indicate rims of phenocrysts and/or rims of microphenocrysts and/or groundmass.

Pyroxenes

In general, the tholeiites and andesitic basalts (dykes, sills and flows) are characterized (Table 2) by augite, pigeonite and rare orthopyroxene (Roraima and Amazonia) and olivine (Cassiporé and Amazonia).

In the conventional Ca-Mg-Fe*(Fe^{2+}+Mn+Fe^{3+}) diagram [*Deer et al.*, 1978] of Fig. 4, the Cassiporé and Amazonia augites straddle the Skaergaard tholeiitic pattern and then trend to subcalcic augite compositions, as those of late crystallization from the Tapirapuá tholeiites. The augite-subcalcic augite path is similar to the "quench trend" of *Muir and Tilley* [1964] and *Muir and Mattey* [1982] which, as documented by *Mellini et al.* [1988], is due to spinodal exsolutions that occur in the C2/c pyroxenes with Ca<0.5 atoms per formula unit, a.p.f.u. Ca-rich pyroxenes of the Roraima dykes are represented by augite and Fe-augite with low and virtually constant or slightly decreasing Ca content. The augite to Fe-augite variation of the basalt flows (Fig. 4) is similar to that of Skaergaard, but at a lower Ca content. The TiO$_2$ of early-crystallized augites correlates with TiO$_2$ of the host-rocks (Fig. 4). Finally, the pigeonites fit the Skaergaard trend.

A general characteristics of the early-crystallized augites is the low Al^{VI} content (max. 0.041 a.p.f.u.) which reflects low crystallization pressure [max. 1-3 kbar; *Nimis*, 1995]. The application of Kretz's [1982] geothermometer to the augites coexisting with low-Ca pyroxenes, yielded mean temperatures (°C) of early and late crystallization ranging from 1190±13 (1178-1209; Roraima dykes) to 1130±29 (1103-1164; Cassiporé dykes), respectively.

Olivine

Rare microphenocrysts of olivine are present in some Cassiporé dykes and one Amazonia sill. The forsterite content (Table 3) of the Cassiporé samples decreases from tholeiites (73-64%) to andesitic basalts (23-24%), while that of the Amazonia tholeiites ranges from 65 (core) to 59% (rim). In general, Mg/(Mg+Fe^{2+}) of olivines are not in equilibrium with mg# (Mg/(Mg+Fe^{2+}); Fe$_2$O$_3$/FeO= 0.15) values of the host-rocks [*Jaques and Green, 1980*], being lower than the expected ones (e.g. 0.24 vs. 0.60; 0.65 vs. 0.87). This suggests that olivines crystallized from melts more evolved than those represented by the host-rocks.

Plagioclase

Plagioclase compositions are given in Table 4. In general, the tholeiites contain plagioclases with a mean anorthite (An) content ranging from 73±5% (66-81%: early crystallization) to 55±9% (46-67%: late crystallization), while the andesitic tholeiites have a mean An content ranging from 69±7% (57-78%: early crystallization) to 54±10% (44-68%: late crystallization).

The crystallization temperatures (°C) of plagioclase [*Kudo and Weill*, 1970; *Mathez*, 1973] for the tholeiites and andesitic basalts under dry conditions range from 1225±44 (1158-1336: early crystallization) to 1150±56 (1094-1242: late crystallization). At P$_{H2O}$=0.5 kbar, estimated temperatures vary from 1180±42 (1150-1282: early crystallization) to 1105±54 (1023-1193: late crystallization). These temperatures are similar to those calculated for Ca-rich pyroxenes, indicating near anhydrous crystallization conditions.

Fe-Ti oxides

Ti-magnetite and ilmenite are common groundmass phases. Homogeneous magnetite (ulvöspinel=7-75%)-ilmenite (R$_2$O$_3$=0.1-12.0%) pairs of the matrix (Table 5) yielded a mean subsolidus temperature (°C) of 772±134 (600-1029) and NNO and QFM buffer conditions [*Buddington and Lindsley*, 1964; Fig. 5].

WHOLE-ROCK COMPOSITIONS

Major and Trace Elements

Major and trace element concentrations of the Brazilian CAMP tholeiites (Table 6) have been determined with a PW1404-XRF spectrometer, following the procedure of *Philips Procedure* [1994] for the correction of matrix effects. Results are considered accurate to within 2-5% for major elements and better than 10% for minor and trace elements. Rare Earth Elements (REE) were measured by

TABLE 2. Microprobe compositions of Ca-rich and Ca-poor pyroxenes of the Brazilian CAMP tholeiites.

	Roraima (R) dykes									Cassiporé (C) dykes					
Sample	R/8821	R/8821	R/8803	R/8803	R/8821	R/8821	R/8824	R/8818	R/8818	C/8002	C/3009	C/8002	C/8009	C/8003	C/8002
rock-type	ThB	ThB	AB	AB	ThB	ThB	AB	AB	AB	AB/HTi	AB/HTi	AB/HTi	AB/HTi	AB/HTi	AB/HTi
	Early	Late	Early	Late	Early	Late	Late	Early	Late	Early	Early	Late	Late	Late	Late
SiO$_2$ (wt%)	53.36	51.60	52.46	51.40	53.57	52.60	50.98	54.85	53.74	51.32	48.85	52.19	49.12	51.52	49.15
TiO$_2$	0.38	0.45	0.33	0.49	0.19	0.31	0.46	0.15	0.25	1.16	1.58	0.37	1.41	0.84	1.05
Al$_2$O$_3$	1.70	1.96	2.11	1.94	0.91	1.01	0.84	0.85	1.11	2.60	3.93	1.07	4.48	1.38	4.35
FeO$_{total}$	7.37	14.22	7.96	12.88	18.77	20.29	25.30	13.50	16.34	10.96	12.31	14.80	14.43	15.99	20.98
MnO	0.20	0.34	0.19	0.27	0.35	0.48	0.59	0.22	0.30	0.23	0.30	0.52	0.29	0.49	0.31
MgO	17.51	14.54	17.53	15.83	22.43	20.12	16.34	27.81	25.02	14.31	13.68	12.66	13.69	16.50	12.17
CaO	19.34	16.71	18.35	16.17	3.24	5.04	4.83	2.32	2.22	19.26	18.94	18.68	16.00	13.20	10.32
Na$_2$O	0.20	0.14	0.20	0.21	0.03	0.04	0.04	0.04	0.01	0.29	0.36	0.21	0.52	0.13	1.60
Cr$_2$O$_3$	0.05	0.00	0.24	0.01	0.01	0.00	0.00	0.09	0.01	0.00	0.08	0.00	0.03	0.02	0.04
Sum	100.11	99.96	99.37	99.20	99.50	99.89	99.38	99.83	99.00	100.13	100.03	100.50	99.97	100.07	99.97
Si	1.954	1.942	1.936	1.932	1.984	1.969	1.968	1.967	1.973	1.916	1.832	1.972	1.847	1.935	1.869
Al(IV)	0.046	0.058	0.064	0.068	0.016	0.031	0.032	0.033	0.027	0.084	0.168	0.028	0.153	0.061	0.131
Sum	2.000	2.000	2.000	2.000	2.000	2.000	2.000	2.000	2.000	2.000	2.000	2.000	2.000	1.996	2.000
Al(VI)	0.027	0.029	0.028	0.018	0.024	0.014	0.006	0.003	0.021	0.031	0.007	0.020	0.046	0.000	0.064
Fe^{2+}	0.216	0.433	0.221	0.369	0.582	0.632	0.815	0.383	0.502	0.332	0.288	0.466	0.390	0.480	0.543
Fe^{3+}	0.010	0.000	0.025	0.036	0.000	0.003	0.001	0.022	0.000	0.010	0.098	0.002	0.064	0.023	0.124
Cr	0.001	0.000	0.007	0.000	0.000	0.000	0.000	0.002	0.000	0.000	0.002	0.000	0.001	0.001	0.001
Mg	0.956	0.816	0.965	0.887	1.238	1.123	0.941	1.487	1.369	0.796	0.765	0.713	0.767	0.924	0.689
Mn	0.006	0.011	0.006	0.009	0.011	0.015	0.019	0.007	0.009	0.007	0.010	0.017	0.009	0.016	0.010
Ti	0.010	0.013	0.009	0.014	0.005	0.009	0.013	0.004	0.007	0.033	0.044	0.011	0.040	0.024	0.030
Ca	0.759	0.676	0.726	0.651	0.129	0.202	0.200	0.089	0.087	0.770	0.761	0.756	0.645	0.531	0.420
Na	0.014	0.011	0.014	0.015	0.002	0.003	0.003	0.003	0.000	0.021	0.026	0.016	0.038	0.009	0.118
Sum	1.999	1.987	2.001	1.999	1.991	2.001	1.998	2.000	1.995	2.000	2.001	2.001	2.000	2.008	1.999
Ca	38.98	34.85	37.36	33.35	6.58	10.23	10.12	4.48	4.42	40.21	39.59	38.69	34.40	26.90	23.52
Mg	49.10	42.19	49.67	45.44	63.16	56.86	47.62	74.80	69.60	41.57	39.81	36.49	40.91	46.81	38.58
Fe*	11.92	22.96	12.97	21.21	30.26	32.91	42.26	20.72	25.98	18.22	20.60	24.82	24.69	26.29	37.91

ThB= tholeiitic basalt, AB= andesitic basalt; HTi and LTi= high (>2wt%) and low (<2wt%) TiO$_2$, respectively. Fe^{3+} calculated according to *Papike et al.* [1974]; Fe*= Fe^{2+}+Mn+Fe^{3+}. Early and Late= early- and late-crystallized pyroxenes. Mean compositions of Maranhao pyroxenes after *Bellieni et al.* [1990].

Table 2. (continued).

Sample	Cassiporé (C) dykes							Amazonia (AM) sills						
	C/8019	C/8009	C/8037	C/8026	C/8037	C/8026	C/8037	AM/PB25	AM/PB25	AM/PB30	AM/PB18	AM/PB25	AM/PB2A	AM/PB30
rock-type	AB/HTi	AB/HTi	ThB/LTi	ThB/LTi	ThB/LTi	ThB/LTi	ThB/LTi	ThB	ThB	ThB	ThB	ThB	ThB	ThB
	Late	Late	Early	Early	Late	Late	Late	Early	Late	Late	Late	Late	Late	Late
SiO_2 (wt%)	51.70	51.11	50.63	51.67	50.36	52.30	51.11	52.58	53.12	51.57	52.89	54.49	53.44	54.48
TiO_2	0.53	0.97	0.95	0.87	0.86	0.62	0.52	0.36	0.38	0.55	0.49	0.23	0.24	0.15
Al_2O_3	0.72	1.77	2.97	2.51	1.93	1.51	0.78	2.68	1.92	1.48	0.93	0.97	0.77	0.83
FeO_{total}	25.20	20.41	9.86	11.33	15.36	18.45	25.92	6.38	7.72	18.26	18.24	14.52	20.27	16.83
MnO	0.66	0.40	0.28	0.23	0.40	0.44	0.70	0.14	0.12	0.49	0.43	0.38	0.36	0.33
MgO	15.65	19.60	15.17	15.34	13.29	19.44	15.75	17.34	17.47	16.40	19.63	25.03	21.76	25.01
CaO	5.51	5.56	19.44	17.83	17.31	7.05	5.10	19.62	18.93	11.12	7.34	4.23	3.11	2.37
Na_2O	0.08	0.08	0.35	0.27	0.23	0.08	0.06	0.17	0.21	0.13	0.03	0.03	0.05	0.00
Cr_2O_3	0.00	0.05	0.27	0.03	0.00	0.07	0.01	0.72	0.11	0.01	0.02	0.12	0.00	0.00
Sum	100.05	99.95	99.92	100.08	99.74	99.96	99.95	99.99	99.98	100.01	100.00	100.00	100.00	100.00
Si	1.983	1.918	1.881	1.923	1.914	1.953	1.970	1.926	1.948	1.946	1.972	1.975	1.983	1.988
Al(IV)	0.017	0.078	0.119	0.077	0.086	0.047	0.030	0.074	0.052	0.054	0.028	0.025	0.017	0.012
Sum	2.000	1.996	2.000	2.000	2.000	2.000	2.000	2.000	2.000	2.000	2.000	2.000	2.000	2.000
Al(VI)	0.016	0.000	0.011	0.033	0.000	0.019	0.005	0.041	0.031	0.011	0.012	0.017	0.017	0.024
Fe^{2+}	0.808	0.613	0.234	0.339	0.434	0.576	0.836	0.191	0.226	0.555	0.569	0.440	0.629	0.512
Fe^{3+}	0.000	0.028	0.072	0.013	0.054	0.000	0.000	0.005	0.011	0.021	0.000	0.000	0.000	0.000
Cr	0.000	0.001	0.008	0.001	0.000	0.002	0.000	0.021	0.003	0.000	0.001	0.003	0.000	0.000
Mg	0.894	1.096	0.840	0.851	0.753	1.082	0.905	0.946	0.955	0.922	1.091	1.353	1.204	1.357
Mn	0.021	0.013	0.009	0.007	0.013	0.014	0.023	0.004	0.004	0.016	0.014	0.012	0.011	0.010
Ti	0.015	0.028	0.027	0.024	0.025	0.017	0.015	0.010	0.011	0.008	0.014	0.006	0.007	0.004
Ca	0.227	0.224	0.774	0.711	0.705	0.282	0.211	0.770	0.744	0.450	0.293	0.164	0.124	0.092
Na	0.006	0.006	0.025	0.020	0.017	0.006	0.004	0.012	0.015	0.010	0.002	0.002	0.004	0.000
Sum	1.987	2.009	2.000	1.999	2.001	1.998	1.999	2.000	2.000	1.993	1.996	1.997	1.996	1.999
Ca	11.64	11.35	40.12	37.01	35.99	14.43	10.68	40.36	38.65	23.34	15.02	8.39	6.32	4.71
Mg	45.85	55.52	43.55	44.30	38.44	55.37	45.82	49.63	49.62	47.85	55.86	69.12	61.53	69.17
Fe*	42.51	33.13	16.33	18.69	25.57	30.19	43.49	10.01	11.73	28.81	29.12	22.50	32.16	26.12

Table 2. (continued).

	Lavras (L) flows				Anarí (AN) and Tapirapuá (TR) flows						
Sample	L/8232	L/8231	L/8231	L/8232	AN/A11	TR/A1	TR/A2	TR/A16	TR/A23	TR/A16	TR/A24
rock-type					ThB	ThB	ThB	ThB	ThB	ThB	ThB
	Late	Late	Late	Late	Late	Early	Late	Late	Late	Late	Late
SiO_2 (wt%)	53.55	51.83	53.27	51.20	54.19	53.57	53.50	52.58	50.87	53.79	50.19
TiO_2	0.36	0.52	0.34	0.39	0.27	0.30	0.44	0.68	0.70	0.44	0.48
Al_2O_3	1.76	1.23	0.86	1.07	1.01	2.19	1.51	2.24	1.67	1.62	0.50
FeO_{total}	8.48	14.76	18.78	26.88	17.23	7.47	9.19	10.80	18.78	15.37	32.28
MnO	0.11	0.28	0.40	0.53	0.31	0.21	0.26	0.26	0.48	0.34	0.65
MgO	17.33	15.96	21.49	15.69	22.52	16.23	16.92	17.23	17.44	20.81	12.23
CaO	17.53	15.20	4.86	4.12	4.37	19.77	18.05	15.94	9.92	7.46	3.65
Na_2O	0.54	0.16	0.00	0.08	0.09	0.17	0.01	0.22	0.13	0.10	0.00
Cr_2O_3	0.35	0.06	0.00	0.04	0.02	0.09	0.03	0.04	0.00	0.07	0.03
Sum	100.01	100.00	100.00	100.00	100.01	100.00	99.91	99.99	99.99	100.00	100.01
Si	1.965	1.944	1.975	1.973	1.987	1.966	1.970	1.942	1.917	1.975	1.985
Al(IV)	0.035	0.055	0.025	0.027	0.013	0.034	0.030	0.058	0.074	0.025	0.015
Sum	2.000	1.999	2.000	2.000	2.000	2.000	2.000	2.000	1.991	2.000	2.000
Al(VI)	0.041	0.000	0.012	0.022	0.031	0.060	0.036	0.040	0.000	0.045	0.008
Fe^{2+}	0.258	0.428	0.582	0.866	0.528	0.229	0.283	0.334	0.548	0.472	1.068
Fe^{3+}	0.002	0.035	0.000	0.000	0.000	0.000	0.000	0.000	0.044	0.000	0.000
Cr	0.010	0.002	0.000	0.001	0.001	0.003	0.001	0.001	0.000	0.002	0.001
Mg	0.948	0.892	1.187	0.901	1.231	0.887	0.929	0.949	0.980	1.139	0.721
Mn	0.003	0.009	0.013	0.017	0.010	0.006	0.008	0.008	0.015	0.011	0.022
Ti	0.010	0.015	0.009	0.011	0.007	0.008	0.012	0.019	0.020	0.012	0.014
Ca	0.689	0.611	0.193	0.170	0.172	0.777	0.712	0.631	0.401	0.293	0.155
Na	0.038	0.012	0.000	0.006	0.007	0.012	0.007	0.016	0.010	0.007	0.000
Sum	1.999	2.004	1.996	1.994	1.987	1.982	1.988	1.998	2.018	1.981	1.988
Ca	36.26	30.94	9.77	8.70	8.86	40.92	36.85	32.84	20.17	15.30	7.88
Mg	49.89	45.16	60.10	46.11	63.42	46.71	48.08	49.38	49.30	59.48	36.67
Fe*	13.84	23.90	30.13	45.19	27.72	12.37	15.06	17.79	30.53	25.22	55.44

using ICP-MS technique [CNRS, Vandoeuvre, France; *Govindarajau and Mevelle*, 1987]. FeO was determined by titration and loss on ignition, corrected for Fe^{2+} oxidation, by gravimetry after 1100 °C (12 hours) heating.

Variation Diagrams and Trace Element Relationships

The Brazilian CAMP tholeiites are evolved, having mg# values mainly in the range 0.30-0.60 (mean 0.50±0.06). Only a few Amazonia sills trend to primary basalt compositions (mg# = 0.66-0.65; Cr and Ni = 693 and 443 ppm, respectively). It should be noted that in general the Brazilian CAMP basalts have low TiO_2 (mean 1.3±0.3wt%) and relatively higher mg# (0.46-0.66); the scarce high TiO_2 (mean 3.2±0.7wt%) analogues show instead a lover mg# value (0.31-0.52). The HTi-thoeiites prevail in the Cassiporé region, including Jarí (J/ML.47,

J/ML49, J/ML39 and J/ML55; Table 6) and neighboring areas of French Guyana (samples C/GF38 and C/GF49c; Table 6).

The variation diagrams for mg# vs. major and trace elements (Figs. 6, 7) show that the LTi-tholeiites of Roraima, Amazonia, Maranhão, Anarí and Tapirapuá have similar trends and plot in narrow compositional fields, except for K_2O, Al_2O_3 and Sr (Fig. 6), and Rb and Ba (Fig. 7). In general, the HTi-tholeiites from Cassiporé in general display well defined variation trends: mg# decreases to 0.35, there is a significant decrease in SiO_2, CaO, Al_2O_3, Sr and (Cr+Ni), and an increase in K_2O, TiO_2, FeO_t, P_2O_5, Ba, Rb, Zr, Y, Nb, La and Sm (Figs. 6, 7).

The relationships among the incompatible elements (IE) reveal that LTi- and HTi-tholeiites (excluding C/8048) from Cassiporé yield good linear correlation coefficient

TABLE 3. Microprobe olivine compositions of Cassiporé (C) dykes and one Amazonia (AM) sill.

	Cassiporé (C) dykes					Amazonia (AM) sill	
Sample	C/8048	C/8048	C/8003	C/8037	C/8037	AM/PB2	AM/PB2
Rock-type	AB/HTi	AB/HTi	AB/HTi	ThB/LTi	ThB/LTi	ThB/LTi	ThB/LTi
	Early	Late	Late	Early	Late	Early	Late
SiO_2 (wt%)	31.79	31.75	36.73	37.98	37.73	37.01	35.72
FeO_{total}	56.34	57.12	30.66	24.22	25.43	30.26	34.66
MnO	0.83	0.83	0.50	0.36	0.42	0.27	0.44
MgO	10.23	9.71	31.69	36.80	35.98	32.10	28.05
CaO	0.43	0.32	0.28	0.27	0.17	0.22	0.24
Sum	99.62	99.73	99.86	99.63	99.73	99.86	99.11
Si	1.000	1.002	0.999	1.002	1.000	1.002	1.000
Fe^{2+}	1.482	1.507	0.697	0.534	0.564	0.685	0.811
Mn	0.022	0.022	0.012	0.008	0.009	0.006	0.010
Mg	0.480	0.456	1.284	1.446	1.421	1.296	1.170
Ca	0.014	0.011	0.008	0.008	0.005	0.006	0.007
Sum	2.998	2.998	3.000	2.998	2.999	2.995	2.998
Fo	24.00	22.85	64.18	72.46	71.09	65.19	58.74
Fa	74.18	75.49	34.84	26.76	28.20	34.49	40.73
Tph	1.10	1.12	0.57	0.40	0.47	0.31	0.52
Lar	0.72	0.54	0.41	0.38	0.24	0.01	0.01
mg (ol-eq)	0.60		0.67	0.78		0.87	
mg# (wr)	0.31		0.38	0.52		0.66	

mg# of the whole rock (wr) = $Mg/(Mg+Fe^{2+}$; $Fe_2O_3/FeO = 0.15$); mg(ol-eq)= $Mg(Mg+Fe^{2+})$ of olivine in equilibrium with the mg# value of whole rock, assuming Kd Fe-Mg (ol/melt)= 0.30. ThB= tholeiitic basalts, AB= andesitic basalt; HTi and LTi= high (>2wt%) and low (<2wt%) TiO_2 respectively. Fo= forsterite, Fa= fayalite, Tph= tephroite and Lar= larnite. Early and Late = early- and late-crystallized olivines.

("r"), usually higher than 0.95 for REE, Zr, Nb and Y (e.g. r: La-Nb=0.985, Zr-Dy=0.991), but lower than 0.70 for Ba and Rb (e.g. Zr-Ba, Zr-Rb=0.643 and 0.646, respectively). Similar relationships for all the other investigated Brazilian CAMP tholeiites yielded lower "r" values (e.g. La-Nb, Zr-Ba and Zr-Rb: 0.031, 0.042 and 0.242, respectively).

As illustrated in Fig. 8, the Cassiporé tholeiites are characterized by Zr, La, Yb and Y straight line correlations (r: 0.953 to 0.996) that intersect the origin for Yb, Y and Zr at 0.86-0.96, 8-13 and 24 ppm, respectively. This (and major element variations) implies that the Cassiporé HTi-tholeiites cannot be derived from LTi-tholeiites through simple fractional crystallization (cf. also Zr vs. Cr+Ni log-log relationship). All the other Brazilian CAMP tholeiites have IE ratios that may be similar to those of the Cassiporé dykes (e.g. La/Yb=3.6-7.1; La/Y=0.26-0.51) or substantially different (e.g. Zr/Yb=32-56 vs. 56-95; Zr/La=5.2-13.3 vs. 16.8) (Fig. 6). It follows that different parental melts (and mantle sources) for the Brazilian CAMP tholeiites should be considered.

It should be noted that the Brazilian CAMP tholeiites from specific regions are characterized by different La/Nb and/or Zr/La (Fig. 9). The Cassiporé, Anarí-Tapirapuá and part of the Amazonia samples, for example, have low La/Nb (0.6-1.1), while those from Roraima, Maranhão (M/5040 excepted) and part of the Amazonia sills have high La/Nb (1.2-2.5) values. Another important difference between these two groups of tholeiites is apparent in terms of Zr/La (Fig. 9) with the Cassiporé samples having the highest (13-17) and the Maranhão ones the lowest (6-10) ratios.

REE and multi-elemental Diagrams

The Brazilian CAMP tholeiites show important variations in chondrite-normalized [cn; Boynton, 1984] La/Sm_cn (1.1 to 2.4) and Sm/Yb_cn (1.6-2.2) ratios, for similar mg# value (Fig. 10A). This cannot be easily accounted by variable degrees of differentiation, and probably reflects different parental magma characteristics. (La/Sm)_cn-(Sm/Yb)_cn relationships (Fig. 10B) show that the

TABLE 4. Microprobe plagioclase compositions of the Brazilian CAMP tholeiites.

	Roraima (R) dykes						Cassiporé (C) dykes								
Sample	R/8822	R/8822	R/8802	R/8802	R/8818	R/8818	C/8009	C/8009	C/8003	C/8003	C/8001	C/8001	C/8026	C/8026	C/8037
rock-type	ThB	ThB	AB	AB	AB	AB	AB/HTi	AB/HTi	AB/HTi	AB/HTi	AB/HTi	AB/HTi	ThB/LTi	ThB/LTi	ThB/LTi
	Early	Late	Early	Late	Early	Late	Early	Late	Early	Late	Early	Late	Early	Early	Late
SiO_2 (wt%)	51.26	56.24	49.88	55.73	50.31	55.93	53.44	54.54	50.78	50.74	50.45	57.16	47.59	50.72	53.71
TiO_2	0.02	0.01	0.05	0.10	0.05	0.07	0.20	0.14	0.07	0.07	0.01	0.05	0.08	0.01	0.12
Al_2O_3	30.85	27.18	31.67	27.52	30.88	27.26	28.79	28.08	30.97	30.92	31.17	26.86	33.08	31.15	29.12
FeO_{total}	0.59	0.77	0.72	0.66	0.65	0.62	1.06	1.10	0.70	0.77	0.82	0.57	0.77	0.69	0.60
CaO	13.50	9.30	14.49	9.70	13.74	9.44	11.27	10.44	13.70	13.67	13.94	8.83	16.19	13.86	11.49
Na_2O	3.73	6.01	3.17	5.74	3.48	5.88	4.79	5.22	3.61	3.58	3.49	6.39	2.17	3.53	4.90
K_2O	0.18	0.37	0.11	0.41	0.17	0.40	0.40	0.48	0.12	0.18	0.08	0.29	0.08	0.13	0.17
Sum	100.13	99.88	100.09	99.86	99.28	99.60	99.95	100.00	99.95	99.93	99.96	100.15	99.96	100.09	100.11
Or (wt%)	1.05	2.19	0.64	2.47	1.02	2.39	2.38	2.88	0.71	1.07	0.48	1.74	0.47	0.76	1.01
Ab	31.70	51.27	27.00	48.97	29.85	50.27	40.98	44.70	30.76	30.51	29.79	54.27	18.52	30.07	41.62
An	67.25	46.54	72.36	48.56	69.13	47.34	56.64	52.42	68.53	68.42	69.73	43.99	81.01	69.17	57.37

	Amazonia (AM) sills						Lavras (L) flows		Anari (AN) flows				Tapirapuá (TR) flows		
Sample	AM/PB17	AM/PB2A	AM/PB2A	AM/PB9	AM/PB9	AM/PB25	L/8231	L/8232	AN/A11	AN/A11	AN/A9	AN/A9	TR/A16	TR/A16	TR/A1
rock-type	ThB	ThB	ThB	ThB	ThB	ThB	ThB	ThB	ThB	ThB	AB	AB	ThB	ThB	AB
	Early	Early	Late	Early	Late	Late	Late	Late	Early	Late	Early	Late	Early	Late	Early
SiO_2 (wt%)	47.40	49.50	55.22	50.81	55.11	50.990	51.28	51.14	51.25	56.54	48.44	51.24	49.75	51.29	46.82
TiO_2	0.07	0.03	0.05	0.01	0.03	0.070	0.09	0.09	0.10	0.06	0.00	0.09	0.03	0.03	0.05
Al_2O_3	33.08	31.89	28.05	30.93	27.95	30.630	30.37	30.70	30.52	27.18	32.56	30.46	31.77	30.41	33.78
FeO_{total}	1.01	0.71	0.58	0.84	0.89	1.030	0.54	0.75	0.91	0.54	0.80	1.06	0.63	1.14	0.58
CaO	16.24	14.76	10.25	13.66	10.20	13.370	13.39	13.39	13.22	9.23	15.56	13.18	14.16	13.12	16.95
Na_2O	2.12	3.00	5.47	3.63	5.52	3.720	3.75	3.74	3.86	6.09	2.53	3.89	3.11	3.94	1.79
K_2O	0.08	0.11	0.38	0.12	0.30	0.190	0.22	0.19	0.12	0.36	0.10	0.10	0.10	0.08	0.03
Sum	100.00	100.00	100.00	100.00	100.00	100.00	99.64	100.00	99.98	100.00	99.99	100.02	99.55	100.01	100.00
Or (wt%)	0.47	0.65	2.24	0.71	1.80	1.13	1.28	1.11	0.74	2.11	0.62	0.62	0.61	0.48	0.19
Ab	18.14	25.60	46.59	30.97	47.14	31.81	31.92	31.92	33.00	51.82	21.57	33.28	26.47	33.69	15.21
An	81.39	73.75	51.17	68.32	51.06	67.06	66.80	66.97	66.26	46.07	77.81	66.10	72.92	65.83	84.60

ThB = tholeiitic basalt, AB = andesitic basalt, HTi and LTi= high (>2wt%) and low (<2wt%) TiO_2, respectively. Or, Ab and An = orthoclase, albite and anorthite, respectively. Early and Late = early- and late-crystallized plagioclases.

TABLE 5. Microprobe compositions of homogeneous groundmass magnetite-ilmenite pairs of the Brazilian CAMP tholeiites.

Roraima (R) and   Cassiporé (C) dykes

Sample	R/8821	R/8822	C/8003	C/8009	C/8048	C/8019	C/8002	C/8001	C/8026
Magnetites rock-type	ThB	ThB	AB/HTi	AB/HTi	AB/HTi	AB/HTi	AB/HTi	AB/HTi	ThB/LTi
SiO_2 (wt%)	0.26	0.13	0.26	0.26	0.10	0.26	0.26	0.10	0.08
TiO_2	17.83	21.43	19.05	21.68	17.71	21.45	9.53	26.00	12.85
Al_2O_3	1.74	1.14	1.17	2.95	1.34	1.63	1.27	0.57	4.29
FeO_{total}	75.88	73.20	75.24	72.10	77.38	70.04	78.98	69.75	78.06
MnO	1.25	1.14	0.58	0.29	0.83	0.34	0.05	0.82	0.37
MgO	0.19	0.22	0.73	0.04	0.02	0.51	0.36	0.04	1.47
CaO	0.40	0.19	0.27	0.20	0.08	3.11	7.33	0.02	0.00
Cr_2O_3	0.05	0.05	0.02	0.07	0.08	0.05	0.02	0.04	0.14
Sum	97.60	97.50	97.32	97.59	97.54	97.39	97.80	97.34	97.26
FeO	46.51	49.65	47.40	51.43	47.19	46.76	32.04	54.25	41.62
Fe_2O_3	32.64	26.16	30.92	22.97	33.54	25.86	52.15	17.23	40.48
Sum	100.87	100.11	100.40	99.89	100.89	99.97	103.01	99.07	101.30
Ulvöspinel(%)	74.52	61.17	54.65	62.86	50.47	61.20	26.94	74.52	36.49
T°C	820	730	806	1039	600	822	721	929	714
Log $f(O_2)$	-14.70	-18.30	-15.40	-10.50	-22.60	-15.30	-16.70	-13.00	-17.90
Ilmenites									
SiO_2 (wt%)	0.04	0.03	0.12	0.23	0.00	0.05	0.02	0.00	0.04
TiO_2	49.66	50.97	50.32	48.05	51.66	50.64	49.26	50.50	51.17
Al_2O_3	0.03	0.00	0.00	0.21	0.05	0.01	0.00	0.00	0.00
FeO_{total}	48.06	46.62	46.53	48.83	45.23	46.60	47.20	46.74	44.86
MnO	0.79	0.59	1.20	0.59	0.55	0.58	2.03	1.35	0.73
MgO	0.11	0.37	0.13	0.34	1.25	0.50	0.05	0.05	1.72
CaO	0.03	0.13	0.29	0.23	0.03	0.11	0.11	0.03	0.04
Cr_2O_3	0.01	0.02	0.00	0.00	0.00	0.00	0.03	0.03	0.04
Sum	98.73	98.73	98.59	98.48	98.77	98.49	98.70	98.70	98.60
FeO	43.67	44.44	43.57	41.99	43.63	43.98	42.03	43.91	42.19
Fe_2O_3	4.88	2.42	3.30	7.60	1.78	2.91	5.74	3.14	2.96
Sum	99.22	98.97	98.93	99.24	98.95	98.78	99.27	99.01	98.89
R_2O_3(%)	4.73	2.34	3.17	7.20	1.77	2.81	5.54	3.05	2.85

ThB= tholeiitic basalt, AB= andesitic basalt, HTi and LTi= high (>2wt%) and low (<2wt%) TiO_2 respectively. Fe_2O_3 calculated according to *Carmichael* [1967]; T(°C) and $-Log fO2$ calculated according to *Buddington and Lindsley* [1964].

LTi and HTi Cassiporé samples (i.e. the samples with low La/Sm_{cn} and high Sm/Yb_{cn} and the Maranhão sample M/5040) are distinct from the Maranhão ones (i.e. the samples with high La/Sm_{cn} and low Sm/Yb_{cn}, M/5040 excepted), while the other tholeiites mainly have intermediate $(La/Sm)_{cn}$ and variable $(Sm/Yb)_{cn}$ values. It is notable that the LTi-Maranhão sample M/5040 is similar to the LTi-tholeiites from Cassiporé (cf. also major and trace elements; cf. Table 6).

REE patterns (Fig. 11) of the Brazilian CAMP LTi-tholeiites, Cassiporé excepted, have a moderate LREE enrichment, mean $(La/Sm)_{cn}$ ranging from 1.55±0.12 (part

Table 5. (continued).

Magnetites	Amazonia (AM) sills					Lavras (L) and				Tapirapuá (TR) flows
Sample	AM/PB25	AM/PB18	AM/PB17	AM/PB9	AM/PB1	L/8231	L/8231	L/8231	L/8232	TR/A24
rock-type	ThB	ThB	ThB	ThB	AB					ThB
SiO_2 (wt%)	0.04	0.25	0.06	0.11	0.20	0.26	0.27	0.26	0.18	0.26
TiO_2	4.39	18.41	5.95	9.41	2.17	27.70	22.81	27.54	23.86	17.10
Al_2O_3	2.25	1.27	2.44	1.09	0.17	1.96	1.01	1.68	1.41	4.98
FeO_{total}	87.25	76.64	84.64	83.46	91.88	65.77	71.98	67.03	70.65	72.21
MnO	0.24	0.37	0.82	0.72	0.06	0.34	0.17	0.12	0.58	0.33
MgO	0.52	0.59	0.67	0.02	0.01	0.60	0.37	0.14	0.18	1.45
CaO	0.07	0.04	0.02	0.00	0.04	0.19	0.42	0.31	0.27	1.06
Cr_2O_3	0.04	0.04	0.07	0.06	0.00	0.00	0.00	0.06	0.00	0.01
Sum	94.80	97.61	94.67	94.87	94.53	96.82	97.03	97.14	97.13	97.40
FeO	34.65	47.63	35.24	39.30	33.56	55.35	51.32	56.09	52.27	44.41
Fe_2O_3	58.44	32.23	54.88	49.06	64.80	11.57	22.92	12.15	20.42	30.88
Sum	100.64	100.83	100.15	99.77	101.01	97.97	99.29	98.35	99.17	100.48
Ulvöspinel(%)	12.74	52.71	17.32	27.54	6.99	80.45	65.84	79.97	68.74	49.14
T°C	631	805	670	703	600					991
Log f(O_2)	-17.90	-15.30	-16.90	-17.50	-18.60					-10.70

Ilmenites

	AM/PB25	AM/PB18	AM/PB17	AM/PB9	AM/PB1	L/8231	L/8231	L/8231	L/8232	TR/A24
SiO_2 (wt%)	0.04	0.01	0.02	0.02	0.02	-	-	-	-	0.26
TiO_2	47.41	50.42	47.32	50.74	45.74	-	-	-	-	46.90
Al_2O_3	0.00	0.01	0.03	0.00	0.01	-	-	-	-	0.79
FeO_{total}	50.56	46.85	50.82	47.61	49.95	-	-	-	-	44.01
MnO	0.73	1.95	0.40	0.53	2.06	-	-	-	-	0.64
MgO	0.29	0.17	0.63	0.97	0.08	-	-	-	-	1.24
CaO	0.12	0.06	0.01	0.00	0.14	-	-	-	-	3.62
Cr_2O_3	0.00	0.01	0.03	0.03	0.01	-	-	-	-	0.00
Sum	99.15	99.48	99.26	99.90	98.01					97.46
FeO	41.26	43.00	41.03	43.38	38.74					34.95
Fe_2O_3	10.33	4.28	10.87	4.70	12.46					10.06
Sum	100.18	99.91	100.34	100.37	99.26					10.75
R_2O_3(%)	9.82	4.10	10.38	4.46	12.01					0.06

of Amazonia) to 2.20±0.12 (Maranhão). In general, REE show U-shape patterns characterized by negative to slightly positive Eu-anomaly (mean Eu/Eu*: Anarí+Tapirapuá= 0.89±0.03, Roraima= 1.01±0.04, Amazonia= 0.95±0.06 and Maranhão= 0.91±0.03). These patterns are consistent with that of E-MORB of *Sun and McDonough* [1989].

It should be noted that the Maranhão sample M/5040 is characterized by a bell-shape La-Nd REE pattern having $(La/Sm)_{cn}$ and $(La/Ce)_{cn}$ values of 1.13 and 0.98, respectively. Similar bell-shape patterns are typical of

both LTi- and HTi-tholeiites from Cassiporé (Fig. 11), all characterized by low mean values of $(La/Sm)_{cn}$ (LTi= 1.17±0.08 and HTi=1.32±0.14) and $(La/Ce)_{cn}$ (LTi= 0.98±0.03 and HTi=1.03±0.05). The Cassiporé REE patterns are subparallel and have Eu/Eu* values ranging from 0.97 to 1.10 (LTi samples) and from 0.91 to 1.05 (HTi samples).

The multi-elemental diagrams of the Brazilian CAMP tholeiites, normalized to the primitive mantle (PM) of *Sun and McDonough* [1989], are shown in Fig. 12. In general, all the samples are characterized by negative anomalies of

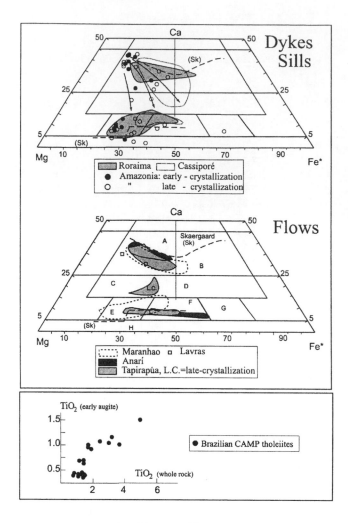

Figure 4. Ca-Mg-Fe* (Fe^{2+}+Mn+Fe^{3+}) plot [*Deer et al.*, 1978] for the Brazilian CAMP tholeiites. A= augite, B= Fe-augite, C= subcalcic augite, D= subcalcic Fe-augite, E= Mg-pigeonite, F= transitional pigeonite, G= Fe-pigeonite and H= orthopyroxene. Inset: relationships between TiO$_2$ (wt%) of early-crystallized augites and corresponding host-rocks.

Sr, P and Ti, the latter only slightly appreciable for the Cassiporé tholeiites (except C/8048). The positive and negative Ba-spikes are another compositional feature common to the Brazilian CAMP basalts from specific localities. In addition, it is notable that the tholeiites from Roraima, Maranhão and part of Amazonia (samples with La/Sm$_{cn}$> 1.8) have a distinct negative Nb-anomaly, while both the LTi- and HTi-basalts from Cassiporé and those from Anarí show a positive Nb-spike. On the other hand, the tholeiites of Tapirapuá and part of Amazonia (samples with La/Sm$_{cn}$<1.7) are characterized by (K/Nb)$_{pm}$>1 and (La/Nb)$_{pm}$<1. Finally, the Maranhão sample M/5040,

compositionally similar to the LTi-toleiites of Cassiporé (see above), shows a negative Nb-anomaly.

In summary, the Brazilian CAMP magmatism is dominated by low-TiO$_2$ tholeiites, while high-TiO$_2$ tholeiites are scarce (i.e. Cassiporé region). Variation diagrams illustrate that the evolved character and the compositional differences of the basalts require gabbro fractionation, starting from different parental melts. This is supported for the LTi- and HTi-tholeiites from Cassiporé (cf. Fig. 8) by the incompatible element relationships. The latter relationships also allowed to ascertain that the LTi-basalts of specific regions have quite different incompatible element ratios. It follows that the low-TiO$_2$ CAMP tholeiites from different regions, in spite of their similar compositions, may be related to parental magmas with different compositions (cf. Figs. 9 to 11). Finally, we stress that some LTi-tholeiites from the Maranhão region

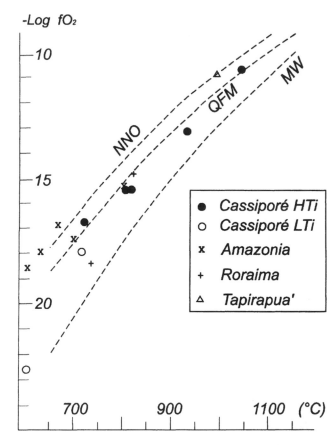

Figure 5. Temperature (°C) vs. –Log fO$_2$ [*Buddington and Lindsley*, 1964] for groundmass homogeneous magnetite-ilmenite pairs of the Brazilian CAMP tholeiites. NNO, QFM and MW = nickel-nickel oxide, quartz-fayalite-magnetite and magnetite-wustite buffers.

TABLE 6. Major (wt%) and trace (ppm) element contents of the Brazilian CAMP tholeiites.

	Roraima (R) dykes					Cassiporé (C) dykes				
Sample	R/8802	R/8804	R/8818	R/8822	R/8826	C/8026	C/8037	C/8043	C/8001	C/8003
Rock-type	AB	AB	AB	ThB	ThB	ThB	ThB	ThB	AB	AB
SiO_2 (wt%)	52.66	53.16	54.37	53.09	52.59	48.98	50.36	48.97	49.42	48.41
TiO_2	1.24	1.26	1.16	0.95	0.99	1.72	1.69	1.87	2.39	3.66
Al_2O_3	15.36	15.37	15.56	16.91	16.57	16.88	18.51	16.77	16.66	15.23
FeO_{total}	11.01	11.16	9.90	9.34	9.45	11.48	9.96	11.87	12.63	14.77
MnO	0.19	0.19	0.17	0.16	0.16	0.18	0.15	0.19	0.20	0.20
MgO	5.58	5.31	5.06	5.24	5.86	6.86	5.31	6.13	5.14	4.48
CaO	10.97	10.45	9.95	11.39	11.62	10.73	10.74	11.20	10.20	9.16
Na_2O	2.26	2.24	2.48	2.29	2.18	2.69	2.69	2.46	2.69	2.78
K_2O	0.56	0.67	1.16	0.49	0.44	0.28	0.40	0.33	0.39	0.79
P_2O_5	0.17	0.19	0.19	0.14	0.14	0.20	0.19	0.21	0.28	0.52
mg#	0.50	0.49	0.51	0.53	0.56	0.55	0.52	0.51	0.45	0.38
Fe_2O_3 (a)	4.30	1.68	1.90	2.27	1.74	1.08	1.93	1.86	2.90	2.64
FeO (a)	7.14	9.65	8.19	7.30	7.88	10.51	8.22	10.20	10.02	12.39
L.O.I. (a)	0.69	0.96	1.36	0.96	1.28	1.76	1.22	1.30	1.37	13.42
Q	4.24	5.38	5.47	4.01	3.98		0.28		0.07	
Ol/Hy						1.11		0.29		0.03
Cr (ppm) *	84	92	56	61	85	183	157	215	92	61
Ni	68	69	73	54	68	141	121	99	76	71
Ba	150	139	348	167	120	47	72	94	114	176
Rb	16	20	30	13	12	8	13	10	11	23
Sr	192	209	234	212	205	206	272	250	267	272
Zr	116	112	121	88	84	118	128	122	163	286
Y	22	23	24	18	19	29	31	32	35	57
Nb	7	7	9	4	5	11	12	11	15	23
La (ppm) #	9.76	9.63	12.75	8.44	8.87	6.83	8.19	8.30	11.02	19.30
Ce	22.11	22.45	28.35	17.68	18.55	18.09	21.13	21.57	27.79	49.88
Pr	-	-	-	-	-	-	-	-	-	-
Nd	13.76	13.40	16.01	10.18	9.92	13.56	15.01	15.03	19.05	34.71
Sm	3.55	3.37	3.73	2.84	2.91	4.12	4.34	4.41	5.30	9.48
Eu	1.29	1.23	1.25	1.02	1.07	1.43	1.54	1.54	1.75	3.02
Gd	4.00	3.64	4.24	3.49	3.64	4.01	4.08	4.22	4.76	8.57
Tb	0.59	0.59	0.65	0.48	0.56	-	-	-	-	-
Dy	3.79	3.70	4.26	3.02	3.15	4.37	4.48	4.69	5.24	8.76
Er	2.30	2.07	2.19	1.69	1.79	2.28	2.36	2.53	2.74	4.45
Yb	2.27	1.92	2.17	2.06	2.10	2.10	2.15	2.47	2.69	4.16
Lu	0.32	0.31	0.36	0.30	0.31	0.32	0.33	0.40	0.41	0.62
$(La/Sm)_{cn}$	1.73	1.80	2.15	1.87	1.92	1.04	1.19	1.18	1.31	1.28
$(Sm/Yb)_{cn}$	1.68	1.88	1.84	1.48	1.49	2.10	2.16	1.91	2.11	2.44
$(Eu/Eu^*)_{cn}$	1.04	1.07	0.96	0.99	1.00	1.06	1.10	1.08	1.05	1.01
$(La/Ce)_{cn}$	1.15	1.12	1.17	1.24	1.25	0.98	1.01	1.54	1.03	1.00
$(La/Nb)_{pm}$	1.45	1.43	1.47	2.19	1.84	0.64	0.71	0.78	0.76	0.87

Major elements recalculated to 100% on a volatile free basis. mg# = $Mg/(Mg+Fe^{2+})$ ($Fe_2O_3/FeO=0.15$); a= original values; Q, Ol and Hy = CIPW-normative quartz, olivine and hyperstene, respectively, assuming $Fe_2O_3/FeO=0.15$. (*) and (#) correspond to trace elements measured by XRF and ICP-MS techniques, respectively; cn=chondrite normalized [Boynton,1984]; PM= primitive mantle normalized [Sun and McDonough, 1989]. Jarí and part of Marañhao data after Oliveira and Turney [1990], and Bellieni et al. [1990], respectively. Abbreviations: R= Roraima, C= Cassiporé, J= Jarí, AM= Amazonia, M= Marañhao, L= Lavras da Mangabeira, AN= Anarí, TR= Tapirapuá, AB= andesitic basalt, ThB= tholeiitic basalt, OB= transitional basalt.

Table 6. (continued).

	Cassiporé (C) dykes						Jarí (J) dykes			
Sample	C/8009	C/8019	C/8048	C/GF38	C/GF49c	C/GF9B	J/ML.47	J/ML.49	J/ML.39	J/ML.55
Rock-type	AB	AB	AB	AB	AB	Pl-cumulus	ThB	ThB	ThB	OB
SiO_2 (wt%)	47.43	49.66	53.55	47.94	49.66	49.39	50.28	49.66	48.98	48.72
TiO_2	4.41	2.91	2.93	3.80	2.70	0.61	1.59	1.99	2.36	3.37
Al_2O_3	13.62	15.35	15.27	14.99	14.80	15.18	16.18	14.29	14.78	12.52
FeO_{total}	16.85	14.07	12.66	15.90	14.06	9.43	9.92	11.86	12.94	15.89
MnO	0.22	0.21	0.18	0.22	0.22	0.16	0.19	0.20	0.20	0.25
MgO	4.32	4.53	2.85	4.37	4.99	9.61	6.47	7.40	6.88	5.85
CaO	8.96	9.59	7.13	9.04	10.14	13.61	12.04	11.75	10.63	9.65
Na_2O	2.37	2.71	3.71	2.65	2.67	1.76	2.83	2.33	2.53	2.98
K_2O	1.02	0.60	1.27	0.60	0.38	0.18	0.34	0.34	0.44	0.46
P_2O_5	0.80	0.37	0.45	0.49	0.38	0.07	0.16	0.18	0.26	0.31
mg#	0.34	0.39	0.31	0.36	0.42	-	0.57	0.56	0.52	0.43
Fe_2O_3 (a)	3.97	2.96	1.72	2.99	2.56	2.60	-	-	-	-
FeO (a)	13.28	11.41	11.11	13.21	11.76	7.09	-	-	-	-
L.O.I. (a)	1.35	1.16	1.19	1.04	0.94	2.63	-	-	-	-
Q	1.30	1.40	4.61	0.58	0.91					
Ol/Hy							0.63	0.22	0.32	0.33
Cr (ppm) *	79	93	42	69	121	281	224	352	147	92
Ni	71	74	38	82	71	101	71	93	112	80
Ba	263	152	415	162	112	66	141	105	183	136
Rb	45	16	31	14	9	5	9	10	9	9
Sr	179	288	492	262	203	314	234	200	217	290
Zr	516	213	275	300	210	31	104	141	167	227
Y	98	46	37	63	52	20	26	33	35	40
Nb	41	17	33	24	16	6	8	9	11	19
La (ppm) #	39.05	14.49	35.69	20.41	13.45	3.08	7.5	8.7	12.0	16.4
Ce	94.90	36.78	80.69	52.53	34.74	7.60	19.8	24.0	31.2	42.3
Pr	-	-	-	-	-	-	2.6	3.3	4.2	5.5
Nd	59.20	25.38	45.08	36.26	24.72	5.20	12.6	15.7	20.6	26.5
Sm	15.56	7.15	10.04	9.99	7.27	1.64	3.9	4.4	6.0	7.5
Eu	3.88	2.36	2.92	3.05	2.20	0.73	1.4	1.5	1.9	2.5
Gd	16.68	6.66	8.25	9.00	6.71	1.75	4.6	5.1	6.7	8.3
Tb	-	-	-	-	-	-	-	-	-	-
Dy	14.48	6.89	6.71	9.20	7.10	2.21	4.3	5.0	6.2	7.1
Er	7.66	3.52	3.01	-	-	-	2.2	2.9	3.0	3.4
Yb	7.35	3.38	2.64	4.23	3.81	1.35	2.0	2.4	2.9	3.0
Lu	1.14	0.51	0.42	0.66	0.53	0.22	0.29	0.36	0.40	0.43
$(La/Sm)_{cn}$	1.58	1.27	2.24	1.29	1.16	1.18	1.21	1.24	1.26	1.38
$(Sm/Yb)_{cn}$	2.27	2.27	4.08	2.53	2.05	1.30	2.09	1.96	2.22	2.68
$(Eu/Eu^*)_{cn}$	0.73	1.03	0.95	0.97	0.95	1.31	1.01	0.97	0.91	0.96
$(La/Ce)_{cn}$	1.07	1.03	1.15	1.01	1.01	1.14	0.99	0.95	1.00	1.01
$(La/Nb)_{pm}$	0.99	0.88	1.12	0.88	0.87	-	1.00	0.96	1.11	0.92

are compositionally very similar to those of the Cassiporé region.

Sr-Nd Isotopes

Sr-Nd isotopic ratios in the Brazilian CAMP tholeiites (Table 7) were measured using the procedure in *Petrini et al.* [1987]. Repeated analyses of NBS987 and La Jolla standards gave averages of 0.710239(9) and 0.511855(4), respectively, and no corrections were applied to the measured ratios for instrumental bias. The reported uncertainties on Sr-Nd isotopic compositions are at the 2σ confidence level. The carbonate- and zeolite-bearing

Table 6. (continued).

Amazonia (AM) sills

Sample	AM/PB1	AM/PB2A	AM/PB9	AM/PB17	AM/PB18	AM/PB25	AM/PB30	AM/PB37	AM/PB40	AM/7AM
Rock-type	AB	ThB	ThB	ThB	ThB	ThB	ThB	AB	ThB	ThB
SiO_2 (wt%)	53.40	53.83	51.70	50.67	50.73	50.35	52.96	54.77	53.07	52.34
TiO_2	1.43	1.20	1.41	1.02	0.97	0.91	1.48	1.32	1.20	1.20
Al_2O_3	13.38	13.84	15.18	14.08	15.74	15.10	14.93	14.75	13.68	15.00
FeO_{total}	12.09	10.82	11.70	10.76	9.66	9.22	11.62	11.13	10.48	10.40
MnO	0.20	0.16	0.19	0.20	0.16	0.16	0.20	0.18	0.17	0.20
MgO	6.60	6.92	6.50	9.75	8.63	8.78	5.51	4.84	7.60	6.96
CaO	9.21	9.79	10.34	11.20	11.87	12.75	10.00	9.16	10.66	10.86
Na_2O	2.68	2.53	2.28	1.80	1.82	1.96	2.72	2.53	2.16	2.31
K_2O	0.88	0.79	0.56	0.43	0.33	0.68	0.43	1.15	0.84	0.60
P_2O_5	0.13	0.12	0.14	0.09	0.09	0.09	0.15	0.17	0.14	0.13
mg#	0.52	0.56	0.53	0.65	0.64	0.66	0.49	0.47	0.59	0.57
Fe_2O_3 (a)	2.87	1.80	2.49	3.17	2.79	2.29	2.68	1.24	2.07	2.19
FeO (a)	9.51	9.20	9.46	7.91	7.15	7.16	9.21	10.01	8.62	8.43
L.O.I. (a)	1.74	1.80	1.35	2.18	1.82	1.10	0.87	1.42	1.48	1.28
Q	2.79	3.59	2.20		0.07		3.48	6.54	2.86	2.15
Ol/Hy				0.01		0.57				
Cr (ppm) *	38	23	63	234	308	523	63	30	180	151
Ni	41	39	81	209	217	170	58	36	55	106
Ba	238	128	132	152	81	105	122	320	148	133
Rb	22	25	11	9	6	14	10	30	26	17
Sr	224	244	196	212	189	200	180	238	201	202
Zr	123	110	111	89	85	76	121	144	111	92
Y	31	28	30	26	24	21	31	31	28	28
Nb	11	9	8	8	8	7	9	10	9	10
La (ppm) #	*12	10.70	8.28	7.03	*6	5.59	8.99	15.42	11.61	8.58
Ce	*33	23.51	19.41	15.55	*17	12.82	21.39	33.65	24.95	19.90
Pr	-	-	-	-	-	-	-	-	-	-
Nd	*17	13.96	12.52	9.92	*11	8.66	13.96	17.17	12.84	11.89
Sm	-	3.66	3.39	2.77	-	2.26	4.16	4.31	3.23	3.23
Eu	-	1.15	1.15	1.00	-	0.73	1.30	1.33	1.14	1.01
Gd	-	3.69	3.73	2.94	-	2.34	4.17	5.29	4.42	3.26
Tb	-	0.63	0.68	0.51	-	0.44	0.72	0.81	0.66	-
Dy	-	4.14	4.52	3.35	-	2.61	4.83	4.80	3.89	3.55
Er	-	2.37	2.40	1.95	-	1.51	2.71	-	-	-
Yb	-	2.31	2.50	1.94	-	1.46	2.59	2.56	2.24	2.03
Lu	-	0.35	0.36	0.30	-	0.23	0.42	0.41	0.35	0.33
$(La/Sm)_{cn}$	-	1.84	1.54	1.60	-	1.56	1.36	2.25	2.26	1.67
$(Sm/Yb)_{cn}$	-	1.70	1.45	1.53	-	1.66	1.72	1.80	1.55	1.71
$(Eu/Eu*)_{cn}$	-	0.95	0.98	1.06	-	0.96	0.94	0.85	0.92	0.94
$(La/Ce)_{cn}$	0.95	1.19	1.11	1.18	0.92	1.14	1.10	1.19	1.21	1.12
$(La/Nb)_{pm}$	1.13	1.23	1.07	0.91	0.78	0.83	1.04	1.60	1.34	0.89

samples from Lavras da Mangabeira were leached by several washes in 2.5N HCl at 60 °C, until the solution was clear.

As shown by the histograms in Fig. 13, the LTi- and HTi-tholeiitic dykes from Cassiporé and the LTi-tholeiitic flow M/5040 from the Maranhão basin are characterized by lower εSr (-12 to 39) and higher εNd (5.5 to -1.8)

values with respect to those of the other Brazilian CAMP LTi-tholeiites (εSr=7 to 91 and εNd=2.2 to -3.7).

In general, the important variations in the initial (200 Ma) $^{87}Sr/^{86}Sr$ (0.7031-0.7107) and $^{143}Nd/^{144}Nd$ (0.51269-0.51223) are broadly positively (εSr) and negatively (εNd) correlated with SiO_2, K_2O, Rb, Ba, La/Nb and $(La/Sm)_{cn}$ (Fig. 14), suggesting that the Brazilian CAMP tholeiites

Table 6. (continued).

Marañhao (M) flows

Sample	M/5007	M/5015	M/5020	M/5022	M/5023	M/5033	M/5039	M/5040	M/5041	M/5042	M/5044
Rock-type	ThB	ThB	ThB	ThB	ThB	ThB	ThB	ThB	ThB	ThB	ThB
SiO_2 (wt%)	53.12	52.80	53.41	53.97	53.82	51.72	53.98	49.95	54.19	52.71	53.56
TiO_2	0.99	1.05	0.95	1.06	0.94	0.94	0.82	1.88	1.01	1.05	1.12
Al_2O_3	16.27	15.83	16.80	15.81	14.66	15.07	15.81	17.10	15.32	15.91	16.20
FeO_{total}	8.76	10.33	9.52	9.82	9.76	9.73	9.42	12.94	10.10	9.60	10.07
MnO	0.16	0.17	0.17	0.15	0.14	0.16	0.14	0.22	0.16	0.16	0.19
MgO	7.40	5.91	5.11	5.09	7.35	8.11	6.04	5.13	5.70	5.06	4.64
CaO	9.93	10.41	10.40	10.45	10.59	10.74	10.31	9.92	10.15	11.06	10.58
Na_2O	2.32	2.22	2.10	2.35	1.93	2.24	2.34	2.30	2.17	2.39	2.20
K_2O	0.90	1.08	1.38	1.14	0.67	1.16	0.96	0.32	1.00	1.90	1.29
P_2O_5	0.15	0.18	0.18	0.15	0.14	0.13	0.17	0.24	0.19	0.16	0.17
mg#	0.64	0.54	0.52	0.51	0.60	0.63	0.56	0.47	0.53	0.52	0.48
Fe_2O_3 (a)	2.91	4.01	3.33	3.72	4.00	5.59	4.22	4.72	3.68	3.56	3.57
FeO (a)	6.14	6.72	6.52	6.47	6.16	4.70	5.62	8.69	6.79	6.40	6.86
L.O.I. (a)	1.69	2.15	3.69	1.39	2.59	2.84	1.81	1.99	2.06	1.77	3.13
Q	7.85	9.55	10.42	11.05	11.58	4.86	5.97	9.76	12.39	6.52	11.22
Ol/Hy											
Cr (ppm) *	461	183	246	170	190	314	206	64	148	272	204
Ni	114	66	67	56	63	86	62	86	51	55	67
Ba	162	266	548	163	494	142	174	50	185	291	171
Rb	25	17	22	36	9	23	36	7	41	41	43
Sr	193	200	204	190	218	195	183	213	188	210	229
Zr	115	108	68	112	104	95	104	120	120	110	88
Y	28	31	33	30	49	25	30	32	33	30	27
Nb	8	7	5	7	7	5	8	#6.2	8	6	5
La (ppm) #	13.71	12.16	12.42	12.48	12.87	10.06	13.03	7.77	12.82	10.87	10.96
Ce	29.93	25.93	24.53	27.35	25.77	21.80	26.39	20.63	26.84	23.56	23.81
Pr	3.54	3.22	3.21	3.40	3.13	2.73	3.31	2.90	3.34	2.86	2.89
Nd	15.56	13.38	13.17	13.89	13.21	11.52	14.75	14.49	14.71	12.66	13.65
Sm	3.89	3.44	3.22	3.57	3.60	3.19	3.71	4.33	3.52	3.25	3.26
Eu	1.19	1.03	1.13	1.04	1.13	1.00	1.17	1.43	1.11	1.05	1.05
Gd	4.06	3.77	3.86	3.67	4.07	3.36	3.89	4.98	4.00	3.43	3.63
Tb	0.65	0.58	0.63	0.65	0.59	0.54	0.62	0.78	0.59	0.52	0.52
Dy	3.81	3.52	3.81	3.47	3.88	3.45	3.97	5.00	3.98	3.43	3.77
Er	-	-	-	-	-	-	-	-	-	-	-
Yb	2.08	1.97	2.16	2.11	1.87	1.79	2.17	2.38	2.13	1.90	2.02
Lu	0.31	0.33	0.34	0.33	0.29	0.29	0.33	0.34	0.41	0.33	0.33
$(La/Sm)_{cn}$	2.22	2.22	2.43	2.20	2.25	1.98	2.21	1.13	2.29	2.10	2.11
$(Sm/Yb)_{cn}$	2.00	1.87	1.60	1.81	2.06	1.91	1.83	1.95	1.77	1.83	1.73
$(Eu/Eu*)_{cn}$	0.91	0.87	0.98	0.87	0.90	0.93	0.94	0.94	0.90	0.96	0.93
$(La/Ce)_{cn}$	1.19	1.22	1.32	1.19	1.30	1.20	1.29	0.98	1.24	1.20	1.20
$(La/Nb)_{pm}$	1.78	1.80	2.58	1.85	1.91	2.09	1.69	1.30	1.66	1.88	2.27

may reflect interaction with SiO_2-rich crustal components [e.g. upper crust of *Taylor and McLennan*, 1985; terrigenous sediments: EMII of *Weaver,* 1991; Fig. 15). It should be noted, however, that many samples of the Cassiporé dykes show important εSr and εNd variations in spite of their similar and low SiO_2, K_2O, Rb, Ba and La/Nb and $(La/Sm)_{cn}$ values (Fig. 14). Therefore, the Sr-Nd isotopic and elemental variability of the Cassiporé LTi- and HTi-tholeiites does not support appreciable interaction with SiO_2-rich crustal components during basalt uprising to the surface, but point to different mantle sources.

The relationships between εSr and εNd (Fig. 15) show that many Cassiporé low- and high-TiO_2 tholeiites, and the Maranhão sample M/5040, trend from the depleted to the

Table 6. (continued).

Sample Rock-type	Lavras (L) flows		Anarí (AN) flows				Tapirapuá (TR) flows				
	L/8231	L/8232	AN/A2 AB	AN/A6 ThB	AN/A9 ThB	AN/A11 ThB	TR/A1 AB	TR/A2 AB	TR/A16 ThB	TR/A23 ThB	TR/A24 OB
SiO_2 (wt%)	50.55	51.01	52.96	53.54	52.58	52.76	53.46	53.84	52.72	51.29	51.69
TiO_2	1.06	1.09	1.36	1.21	1.39	1.10	1.32	1.27	1.34	1.42	1.13
Al_2O_3	14.30	14.30	15.26	14.98	16.13	15.38	15.67	16.15	15.22	16.84	16.60
FeO_{total}	10.04	10.20	11.21	10.54	11.37	10.53	10.97	10.64	11.23	11.45	10.16
MnO	0.17	0.17	0.19	0.17	0.19	0.18	0.18	0.17	0.18	0.20	0.17
MgO	5.89	6.17	5.66	5.95	5.07	6.63	5.40	4.73	5.70	4.90	5.89
CaO	11.43	11.21	10.19	9.93	10.39	10.41	10.36	9.94	10.68	10.69	9.91
Na_2O	5.45	4.64	2.52	2.78	2.24	2.30	2.30	2.61	2.46	2.72	3.79
K_2O	1.03	1.12	0.54	0.75	0.53	0.57	0.23	0.55	0.37	0.38	0.54
P_2O_5	0.08	0.09	0.11	0.15	0.11	0.14	0.11	0.10	0.10	0.11	0.12
mg#	0.55	0.55	0.51	0.53	0.47	0.56	0.50	0.47	0.51	0.46	0.54
Fe_2O_3 (a)	3.72	3.29	5.01	5.53	4.79	4.44	4.94	4.96	4.17	5.27	3.62
FeO (a)	6.69	7.24	6.70	5.56	7.06	6.53	6.52	6.18	7.48	6.71	6.90
L.O.I. (a)	4.29	3.74	1.54	1.82	1.93	1.63	1.97	1.00	1.63	2.06	1.79
Q			3.98	3.05	5.05	3.31	6.65	5.86	3.99	1.40	
Ol/Hy											1.81
Cr (ppm) *	354	368	71	71	76	111	75	74	86	72	105
Ni	80	80	72	62	71	90	66	72	80	66	72
Ba	214	206	146	171	477	128	95	130	108	101	123
Rb	13	12	14	22	18	16	6	14	14	14	14
Sr	274	279	169	165	194	159	189	163	174	177	164
Zr	103	93	111	111	116	101	112	107	111	114	96
Y	25	26	35	34	35	30	29	28	29	30	30
Nb	7	7	10	11	10	10	9	10	10	10	10
La (ppm) #	*10	*13	10.02	9.54	9.89	10.80	9.42	8.89	9.65	9.79	8.60
Ce	*30	*28	26.73	19.11	26.86	22.76	23.73	23.27	28.92	28.59	22.97
Pr	-	-	-	-	-	-	-	-	-	-	-
Nd	*13	*14	13.20	13.30	13.65	13.84	13.64	12.06	13.36	14.19	12.51
Sm	-	-	3.74	3.62	3.90	3.95	3.57	3.45	3.86	3.83	3.16
Eu	-	-	1.14	1.18	1.17	1.15	1.14	1.04	1.23	1.13	1.11
Gd	-	-	4.32	3.92	4.49	3.84	4.49	3.67	4.42	4.18	3.66
Tb	-	-	-	-	-	-	-	-	-	-	-
Dy	-	-	4.43	4.37	4.49	4.25	4.28	3.85	4.28	4.26	3.76
Er	-	-	2.46	2.49	2.52	2.43	2.30	2.16	2.38	2.39	2.09
Yb	-	-	2.19	2.36	2.29	2.17	2.09	1.94	2.06	2.06	1.86
Lu	-	-	0.35	0.42	0.42	0.35	0.36	0.33	0.40	0.36	0.32
$(La/Sm)_{cn}$	-	-	1.69	1.66	1.60	1.72	1.66	1.62	1.57	1.61	1.71
$(Sm/Yb)_{cn}$	-	-	1.83	1.64	1.83	1.95	1.83	1.91	2.01	1.99	1.82
$(Eu/Eu^*)_{cn}$	-	-	0.87	0.95	0.85	0.89	0.87	0.89	0.91	0.86	1.00
$(La/Ce)_{cn}$	0.87	1.21	0.98	1.30	0.96	1.24	1.03	1.00	0.87	0.89	0.98
$(La/Nb)_{pm}$	1.48	1.93	1.04	0.90	1.03	1.12	1.09	0.92	1.00	1.02	0.89

enriched quadrant (Tristan da Cunha and Gough islands). The Cassiporé tholeiites mainly plot in or approach the fields of the volcanics from Fernando de Noronha and Abrolhos Atlantic islands. Note that the sample C/GF9B plots near the lowest εNd-tip of the Cape Verde islands, while the Cassiporé sample C/8043 plots in the field of the NE-Brazil Tertiary alkali basalts containing spinel-peridotite xenoliths [*Novello*, 1995; *Fodor et al.*, 1998]. All the other Brazilian CAMP tholeiites define a general pattern trending to upper crustal components. The latter pattern fits quite well that of the Paraná-Etendeka, low-TiO_2 tholeiites [e.g. *Erlank et al.*, 1984; *Piccirillo et al.*, 1988; *Milner and Le Roex*, 1996] which might result from (1) MORB-type melts contaminated by SiO_2-rich crustal

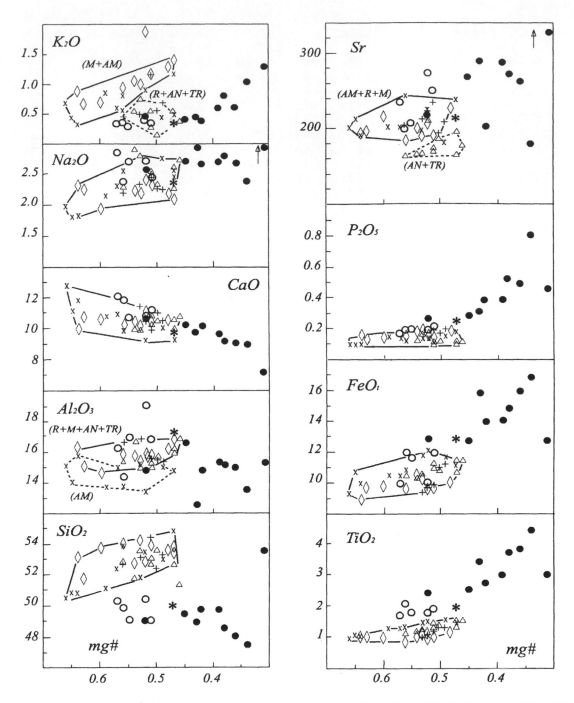

Figure 6. mg# (Mg/(Mg+Fe^{2+}); Fe$_2$O$_3$/FeO=0.15) vs. SiO$_2$, Al$_2$O$_3$, CaO, Na$_2$O, K$_2$O, TiO$_2$, FeO$_t$, P$_2$O$_5$ (wt%) and Sr (ppm) for the Brazilian CAMP tholeiites. Circles and solid circles= low- (<2wt%) and high- (>2wt%) TiO$_2$ tholeiitic dykes from Cassiporé; crosses= Roraima dykes; X= Amazonia sills; diamonds= Marañhao flows; asterisk= sample M/5040 from Marañhao; triangles= Anarí-Tapirapuá flows.   Solid field without labels includes Roraima (R), Amazonia (AM), Marañhao (M) and Anarí (AN)-Tapirapuá (TR) samples.

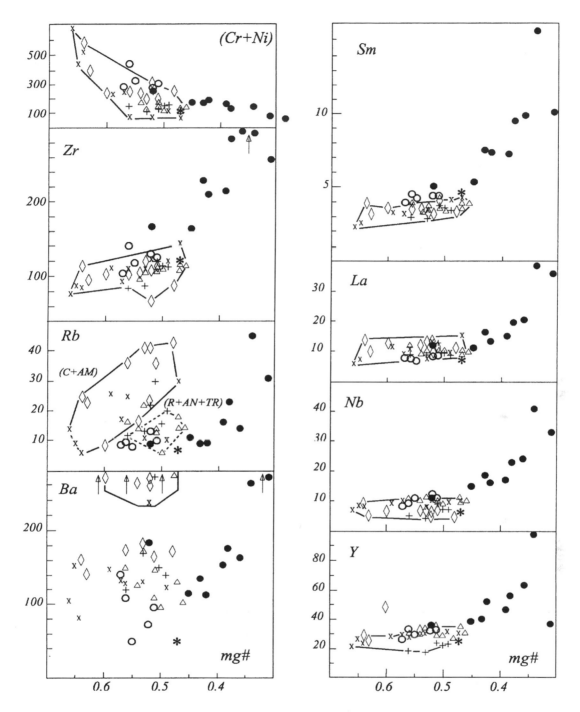

Figure 7. mg# vs. Ba, Rb, Zr, Cr+Ni, Y, Nb, La and Sm (ppm) for the Brazilian CAMP tholeiites. Symbols, fields, labels and other abbreviations as in Fig. 6.

components during magma uprising [*Petrini et al.*, 1987; *Piccirillo and Melfi*, 1988; *Piccirillo et al.*, 1989; *Comin-Chiaramonti et al.*, 1997; *Marques et al.*, 1999], or (2) mixing between "asthenospheric" and "lithospheric" magmas [Esmeralda- and Gramado-types, respectively;

e.g. *Peate and Hawkesworth*, 1996]. In general, it is difficult to model low-pressure crustal contamination for the Brazilian CAMP tholeiites for specific regions through simple mixing or AFC processes since, for example, an increase in εSr and decrease in εNd may be associated with

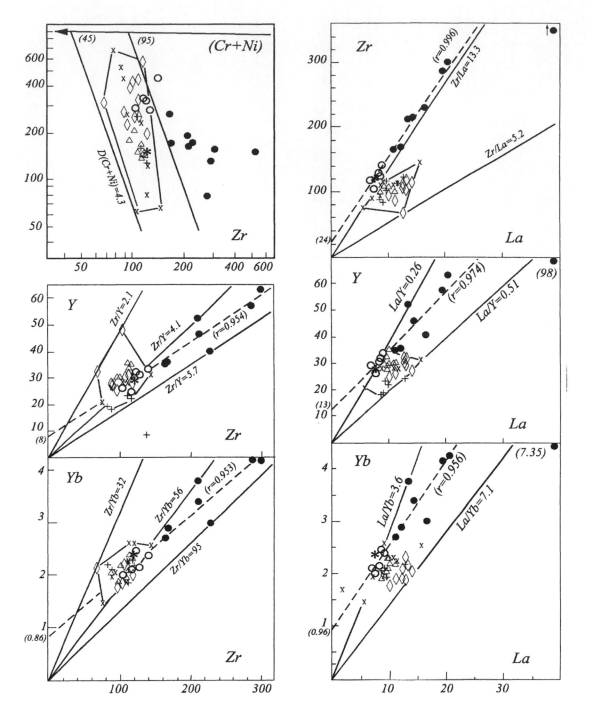

Figure 8. Zr vs. Yb, Y and Cr+Ni, and La vs. Yb, Y and Zr (ppm) plots for the Brazilian CAMP tholeiites. Symbols and fields as in figure 6. In the Log Zr vs. Log (Cr+Ni) plot the number in parentheses (45 and 95), indicate the maximum estimated Zr (ppm) content for the melts parental to the evolved low TiO$_2$ Brazilian Early Jurassic tholeiites. Zr content of the parental magmas of the evolved high-TiO$_2$ basalts is higher than 95 ppm; D= bulk distribution coefficient for (Cr+Ni). r = linear correlation coefficient for the Cassiporé straight lines; Yb, Y and Zr (ppm) intercepts are reported in parentheses.

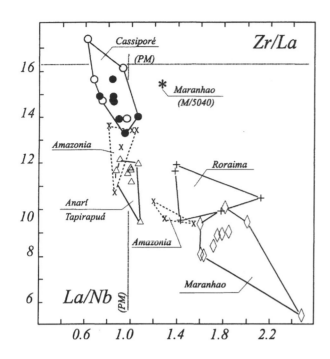

Figure 9. La/Nb vs. Zr/La plot for the Brazilian CAMP tholeiites. Symbols as in Fig. 6. PM= Primitive Mantle of *Sun and McDonough* [1989]: La/Nb=0.96 and Zr/La=16.3.

SiO$_2$ decrease (e.g. Amazonia sills), or other not expected elemental variations (e.g. Anarí and Tapirapuá flows) (Fig. 14).

Inset A of Fig. 15 shows that the low- and high-TiO$_2$ CAMP tholeiites from Liberia [*Dupuy et al.*, 1988] and those from North Florida [*Heatherington and Mueller*, 1999] have εSr (-11 to 53, and 22 to 93, respectively) and εNd (5.2 to –9.4, and 0.1 to –4.5, respectively) values that fit quite well the field of the Brazilian CAMP basalts. According to the above authors, the Liberia magmatism would be related to "asthenospheric" and "lithospheric" mantle sources, while that from North Florida would be derived from "lithospheric" mantle variably contaminated by incompatible-element rich crust.

Inset B of Fig. 15 shows that, in general, the Brazilian (and Liberian, not shown) CAMP tholeiites with negative εSr and/or positive εNd have Rb/Sr and Sm/Nd values higher and lower, respectively, relative to Bulk Earth. This suggests that Rb vs. Sr, and Nd vs. Sm enrichments occurred before mantle melting. Assuming that these incompatible-element enrichments substantially reflect "mantle metasomatism" l.s., including small-volume melts, [*Menzies and Hawkesworth*, 1987], or "subduction-related" crustal contamination, we tentatively use T$_{DM}$(Nd) model ages to infer the time when these processes may have occurred. The histograms of Fig. 16A show that Nd-

model ages of the Cassiporé dykes (M/5040 sample included) which conform to a "typical" mantle array (Fig. 15) are in general lower (range=0.7-1.5 Ga, mean=1.0±0.3 Ga) than those for the other Brazilian CAMP tholeiites (range=1.0-1.9 Ga, mean=1.4±0.3 Ga). The Liberian samples (Fig. 16B) yielded Nd-model ages ranging from 0.7 to 2.1 Ga (mean=1.2±0.5 Ga), with the "asthenospheric" tholeiites (Groups I and II) having T$_{DM}$(Nd) model ages lower (range=0.7-1.0 Ga, mean=0.8±0.1 Ga) than those of the "lithospheric" ones (Group V: range=1.8-2.3 Ga, mean=2.1±0.3 Ga). Finally, Nd-model ages of the "lithospheric" tholeiites from North Florida (Fig. 16B) range from 1.4 to 2.1 Ga (mean=1.6±0.2 Ga) and cluster between those of the "asthenospheric" and "lithospheric" tholeiites from Liberia.

In summary, if low pressure crustal contamination did not play a significant role, Nd-model ages indicate that the CAMP basalts with "typical" mantle array (most Cassiporé dykes, the Maranhão flow M/5040 and the Group I and II samples from Liberia) may have been affected by "mantle metasomatism" in Middle-Late Proterozoic times (range=0.7-1.5 Ga, mean=0.9±0.3 Ga). The other CAMP tholeiites (Brazil, Group V samples from Liberia and north Florida) may instead reflect older processes of subduction-related "crustal contamination" which occurred in Early-Middle Proterozoic times (range=2.3-1.0 Ga, mean=1.5±0.3 Ga).

PALEOMAGNETISM

Paleomagnetic data from most of the CAMP-related magmatic occurrences in South America are now available. Previous published paleomagnetic poles from the Guyana Shield (Surinam, Guyana, French Guyana and the Bolivar dykes in Venezuela) and the Maranhão basalts are listed in *Rapalini et al.* [1993]. Data from central Brazil (the Anarí-Tapirapuá flows and Amazonian Penatecaua intrusives) were reported by *Guerreiro and Schult* [1986] and *Montes-Lauar et al.* [1994]. Recently new data from the Maranhão basin, combined with data from the Lavras da Mangabeira flows and some dykes from the northeastern Brazil (Ceará-Mirim) allowed the calculation of a paleomagnetic pole [*Ernesto et al.*, this volume] that supersedes the already existent Maranhão Jurassic (Mosquito Formation) pole. The Cassiporé and Roraima dykes in Northern Brazil were also investigated and paleomagnetic poles based on 17 and 7 sites, respectively, were calculated [*Ernesto et al.*, in preparation]. All these poles are displayed in Fig. 17 as north poles rotated to Africa.

The South American CAMP poles tend to concentrate in two groups, one of them tightly defined by the poles from central Brazil (Anari-Tapirapuá and Amazonian

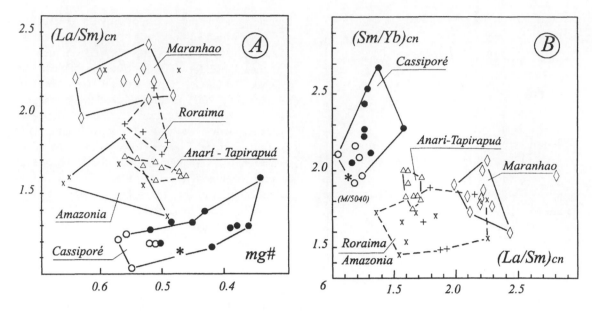

Figure 10. A) mg# vs. chondrite-normalized [cn; *Boynton*, 1984] (La/Sm)$_{cn}$ and B) (La/Sm)$_{cn}$ vs. (Sm/Yb)$_{cn}$ for the Brazilian CAMP tholeiites. Symbols and other notations as in Fig. 6.

Penatecaua) and the Bolivar dikes from Venezuela. The Cassiporé and Roraima dykes, and rocks from northeastern Brazil (Maranhão and Lavras da Mangabeira and Ceará-Mirim) yeld poles that belong to the second group. Poles from Surinam and Guyana are more scattered, and it is not clear in which group they should be included. Tentatively, they were added to the second group due to the geographic proximity, and on the basis that they pertain to the same dyke systems as Cassiporé and Roraima. The mean poles for the two groups are shown in Fig. 17 by shaded circles, which do not overlap indicating that the two groups are statistically independent. This difference could be explained by age difference but also by the effects of the variations of the geomagnetic field on the magnetic remanence of the rocks, especially on those based on few sites. However, the observation that in Brazil younger CAMP rocks are located towards the continental margin suggests that the hypothesis of two paleomagnetic age groups is not unrealistic, although this issue needs a more careful examination. It is interesting to note that the same pattern is observed for the African CAMP poles (inset of Fig. 17). The North American CAMP poles are also very consistent with data from the other two continents, mainly for those grouped closer to 200 Ma segment of the apparent polar wander path (APWP) for North America [*Besse and Courtillot*, 1991]; another group of poles (2, 3 and 4) seems to be displaced toward younger ages of the APWP, but trending distinctly from the South American and African poles.

GEODYNAMIC IMPLICATIONS

The areal extent and location of the Brazilian CAMP tholeiites span from the continental margin (Cassiporé) to ca. 2,000 km inside the South American platform. According to *Veevers* [1989], *Ernst et al.* [1995], *Hames et al.* [2000] and *McHone* [2000] the continental rifting in the central Pangean mountain belt (Appalachian Palaeozoic orogens and Roraima plateau) (Fig. 18A) was active ca. 25-35 My before and after the CAMP magmatic event, which evolved into a spreading ocean crust. In terms of geodynamic processes *Hames et al.* [2000] and *McHone* [2000] demonstrated that the North American/European CAMP magmatism cannot be easily related to plume (active) models [e.g. *Hill et al.*, 1992; *Campbell and Griffiths*, 1990; *Farnetani and Richards*, 1994] due to the (1) lack of important regional uplift, (2) the time of crustal extension which preceded the tectonic uplift and (3) the short duration (ca. 1-2 My) of the main magmatism emplacement. Note also that the models that require massive horizontal translation of deep mantle plume material along the incipient Atlantic major rifting [e.g. *Oyarzum et al.*, 1997; *Wilson*, 1997; *Thompson*, 1998] cannot account for the geometry of the magmatic margin wedge [*McHone*, 2000]. Therefore, passive geodynamic models are favoured [e.g. *Marzoli et al.*, 1999; *Hames et al.*, 2000; *McHone*, 2000] to explain the genesis of the "Appalachian"-related CAMP magmatism. These passive models require hot mantle incubation [e.g. *Anderson*,

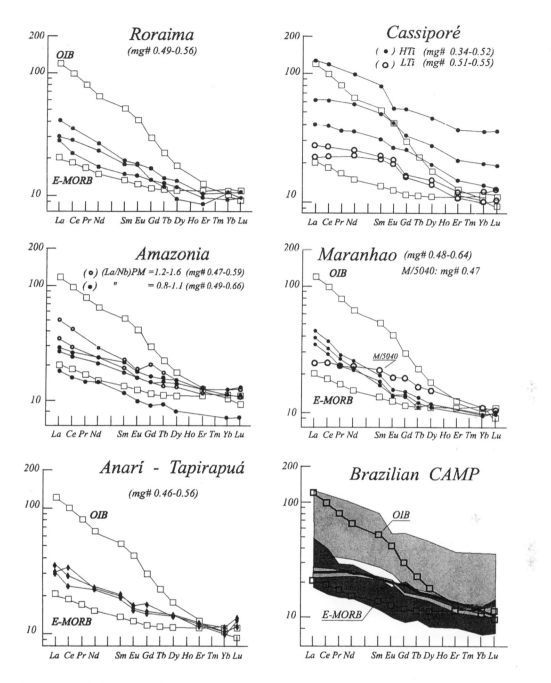

Figure 11. REE patterns for the Brazilian CAMP tholeiites. OIB and E-MORB [*Sun and McDonough*, 1989] patterns are shown for comparison. "Brazilian CAMP" plot: light-grey fields = Cassiporé tholeiites; dark-grey field = all the other Brazilian CAMP tholeiites; heavy solid pattern = M/5040 Marañhao sample.

1994a], as well as edge-driven convection [e.g. *Buck*, 1986; *King and Anderson*, 1998; *Boutilier and Keen*, 1999] related to important differences in the lithospheric thickness and associated extensional movements. *Tackley* [2000] suggested that upwelling mantle naturally develops

underneath a large continent and that the "subduction forces at its edges generate a stress sufficient to break up a supercontinent in about 200 Ma".

Figure 18A shows that the "Appalachian" rift-system does not include important regions of the African (e.g.

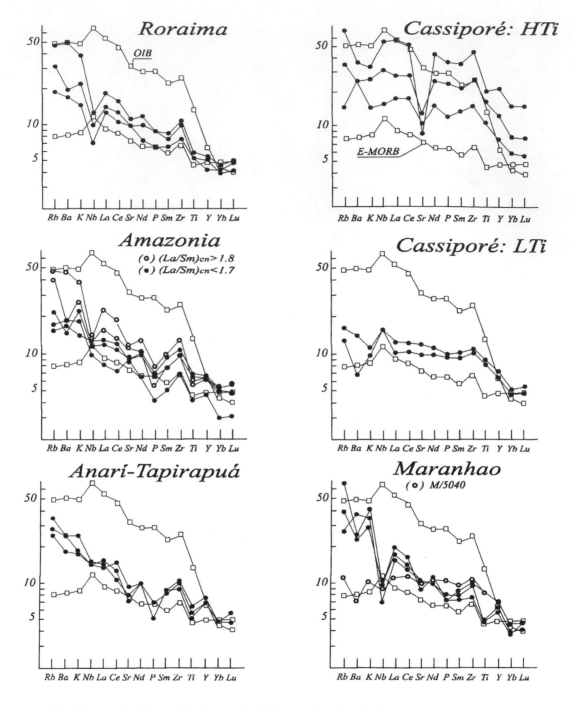

Figure 12. Primitive mantle [*Sun and McDonough*, 1989] normalized multi-elemental plots for the Brazilian CAMP tholeiites. OIB and E-MORB after *Sun and McDonough* [1989]; LTi and HTi= low- and high-TiO$_2$ tholeiitic basalts, respectively.

Guinea, Mali, Mauritania, Algeria) and Brazilian CAMP tholeiites. It is apparent from Fig. 18B that most Brazilian CAMP magmatism is concentrated along Proterozoic mobile belts and at their boundary with the Amazonia l.s. craton (i.e. Solimoes basin sills, part of the Central Brazil dykes and the Anarí and Tapirapuá flows), or at the boundary between the Amazonia l.s. craton and the Trans-Amazonian and/or Brazilian mobile belts (i.e. part of the Central Brazil dykes and the Maranhão flows). Therefore, the Brazilian CAMP magmatism, associated with generally

TABLE 7. Sr and Nd isotopic data of the Brazilian CAMP tholeiites.

Sample	Rock-type	Rb/Sr	Sm/Nd	$(^{87}Sr/^{86}Sr)_m$	$(^{143}Nd/^{144}Nd)_m$	$(^{87}Sr/^{86}Sr)_i$	$(^{143}Nd/^{144}Nd)_i$	ε Sr	ε Nd
R/8802	AB	0.083	0.258	0.70692±1	0.51255±1	0.70623	0.51235	28.03	-0.65
R/8804	AB	0.096	0.251	0.71148±1	0.51255±2	0.71069	0.51235	91.34	-0.55
R/8818	AB	0.128	0.233	0.70855±1	0.51242±1	0.70749	0.51224	45.93	-2.80
R/8822	ThB	0.061	0.279	0.70758±1	0.51246±1	0.70708	0.51224	39.98	-2.73
R/8826	ThB	0.059	-	0.70760±2	-	0.70712	-	40.59	-
C/8026	ThB	0.039	0.304	0.70462±2	0.51282±2	0.70430	0.51258	0.57	3.91
C/8037	ThB	0.048	0.289	0.70564±3	0.51272±2	0.70525	0.51249	14.01	2.19
C/8043	ThB	0.040	0.293	0.70483±2	0.51267±2	0.70450	0.51244	3.42	1.15
C/8001	AB	0.041	0.278	0.70405±2	0.51288±2	0.70371	0.51266	-7.80	5.48
C/8003	AB	0.085	0.273	0.70592±2	0.51288±2	0.70522	0.51266	13.69	5.56
C/8009	AB	0.251	0.263	0.70608±3	0.51278±2	0.70401	0.51257	-3.53	3.76
C/8019	AB	0.056	0.282	0.70746±1	0.51251±2	0.70700	0.51229	38.95	-1.80
C/8048	AB	0.063	0.223	0.70520±2	0.51251±2	0.70468	0.51233	5.98	-0.89
C/GF38	AB	0.053	0.276	0.70369±1	-	0.70325	-	-14.34	-
C/GF49c	AB	0.044	0.294	0.70379±2	0.51285±2	0.70343	0.51262	-11.85	4.65
C/GF9B	Pl-AB	0.016	0.315	0.70292±1	0.51255±1	0.70279	0.51230	-20.80	-1.50
AM/PB 1	AB	0.098	0.270	0.70696±1	0.512590±1	0.70616	0.51238	26.96	0.51
AM/PB 2A	ThB	0.102	0.262	0.70798±1	0.51250±1	0.70714	0.51230	40.93	-1.64
AM/PB 9	ThB	0.056	0.271	0.705934±	0.51264±1	0.70547	0.51242	17.23	0.85
AM/PB 17	ThB	0.042	0.279	0.70675±1	0.51264±1	0.70640	0.51242	30.42	0.78
AM/PB 18	ThB	0.032	0.270	0.70608±1	0.51263±1	0.70582	0.51244	22.22	1.08
AM/PB 25	ThB	0.070	0.261	0.70678±1	0.51256±1	0.70621	0.51235	27.69	-0.56
AM/PB30	ThB	0.056	0.298	0.70560±1	0.51267±1	0.70514	0.51244	12.56	1.17
AM/PB37	AB	0.126	0.251	0.70578±1	0.51269±1	0.70474	0.51249	6.85	2.17
AM/PB40	ThB	0.129	-	0.70747±7	-	0.70641	-	30.46	-
AM/7AM	ThB	0.084	-	0.70729±4	-	0.70660	-	33.19	-
M/5007	ThB	0.130	0.250	0.70731±2	0.51253±3	0.70624	0.51233	28.17	-0.92
M/5015	ThB	0.085	-	0.70724±2	-	0.70654	-	32.38	-
M/5020	ThB	0.108	0.244	0.70821±2	0.51248±2	0.70732	0.51229	43.49	-1.81
M/5022	ThB	0.189	0.257	0.70823±2	0.51253±4	0.70667	0.51233	34.23	-1.03
M/5023	ThB	0.041	0.273	0.70701±1	0.51263±4	0.70667	0.51242	34.22	0.69
M/5039	ThB	0.197	-	0.70778±2	-	0.70616	-	26.99	-
M/5040	ThB	0.033	0.299	0.70336±3	0.51293±3	0.70309	0.51269	-16.62	6.14
M/5044	ThB	0.188	0.239	0.70815±3	0.51249±3	0.70660	0.51230	33.29	-1.53
L/8231		0.047	0.270	0.70849±2	0.51240±2	0.70810	0.51219	56.60	-3.77
L/8232		0.043	0.270	0.70812 ±2	0.51252±1	0.70776	0.51230	51.40	-1.43
AN/A2	AB	0.083	0.283	0.70625±2	0.51245±3	0.70557	0.51223	18.58	-2.99
AN/A6	ThB	0.133	0.272	0.70692±3	0.51266±8	0.70582	0.51245	22.19	1.28
AN/A9	ThB	0.093	0.286	0.70703±4	0.51260±3	0.70627	0.51237	28.49	-0.10
AN/A11	ThB	0.101	0.285	0.70652±4	0.51256±4	0.70569	0.51234	20.33	-0.88
TR/A1	AB	0.032	-	0.70640±7	-	0.70614	-	26.68	-
TR/A2	AB	0.086	-	0.70652±2	-	0.70581	-	22.05	-
TR/A23	ThB	0.079	0.270	0.70647±3	0.51255±3	0.70582	0.51234	22.14	-0.83

"m" and "i" = measured and initial (200 Ma) Sr and Nd isotopic ratios, respectively. εSr and εNd calculated assuming present day $^{143}Nd/^{144}Nd = 0.512638$ and $^{87}Sr/^{86}Sr = 0.7045$, respectively. Pl= plagioclase cumulus; other abbreviations as in Table 6.

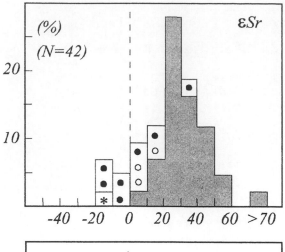

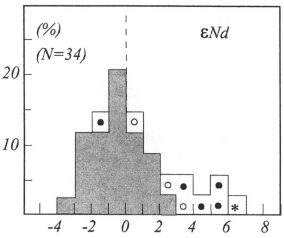

trending E-W extensional tectonics, was probably controlled by significative difference in the lithospheric thickness between the Amazonia craton and the adjacent mobile belts. This is consistent with the distribution of the CAMP magmatism in the western Africa [WA; Fig. 19; *Trompette*, 1994; *McHone*, 2000]. It is notable that WA-CAMP magmatism, outcropping east of the Hercynian (Mauritanides) fold belts, mainly occurs at the borders 1) between the Man (Leo) cratonic block (MCB) and Pan-African fold belts, 2) between MCB and SW-margins of the Taoudeni Late Proterozoic basin and 3) between the Reguibat cratonic block and the northern margin of the Toudeni sedimentary basin.

In summary, the genesis of the Brazilian CAMP magmatism appears consistent with thermal anomalies leading to hot mantle incubation under thicker continental domains of the lithosphere and edge-driven mantle convection at the boundary between thick and comparatively thin subcontinental lithospheric plates. Therefore, the CAMP magmatism cannot be easily explained through deep "plume head" (active) models,

Figure 13. Histograms of $\varepsilon^t Sr$ and $\varepsilon^t Nd$ for the Brazilian CAMP tholeiites. Circles and solid circles = Cassiporé LTi and HTi dykes, respectively; grey field= all the other low-TiO_2 Brazilian CAMP tholeiites.

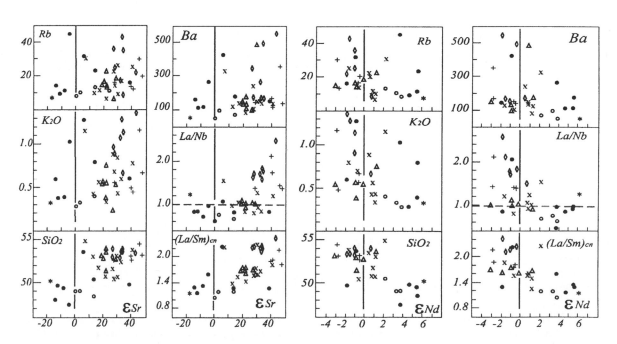

Figure 14. $\varepsilon^t Sr$ and $\varepsilon^t Nd$ vs. Rb, Ba (ppm), K_2O, SiO_2 (wt%), La/Nb and (La/Sm)$_{cn}$; symbols as in Fig. 6.

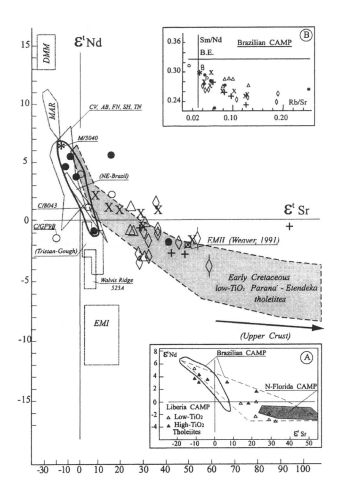

Figure 15. ε^tSr vs. ε^tNd for the Brazilian CAMP tholeiites; symbols as in Fig. 6. Bold solid field: Low- and high-TiO_2 Cassiporé dykes which span from Middle Atlantic Ridge [MAR; *Ito et al.*, 1987] to Tristan da Cunha-Gough island basaltic rocks. The other low-TiO_2 Brazilian CAMP tholeiites fit the distribution (grey dashed field) of the Early Cretaceous (ca. 132 Ma) low-TiO_2 analogues from the Paraná-Etendeka (and Angola) which trend to crustal components (see text for explanation). Inset A: Rb/Sr vs. Sm/Nd plot for the Brazilian CAMP tholeiites. Inset B: ε^tSr vs. ε^tNd plot for the Liberian (West Africa) Early Jurassic tholeiites compared with the Brazilian and the North Florida coeval CAMP tholeiites; B. E. = Bulk Earth. Source data: DMM = depleted MORB mantle [*Zindler and Hart*, 1986]; CV = Cape Verde islands, [*Davies et al.*, 1989], B = Abrolhos islands [*Fodor et al.*, 1989; *Montes Lauar*, 1993], FN = Fernando da Noroñha island [*Gerlach et al.*, 1987; *Lopes*, 1997, *Fodor et al.*, 1998], SH = S. Helena [*Chaffey et al.*, 1989]; TN = Trindade island [*Halliday et al.*, 1992 and *Marques et al.*, 1999]; NE-Brazil = Tertiary Naalkaline volcanics containing spinel-peridotite nodules from NE-Brazil [*Novello*, 1995; *Fodor et al.*, 1998]; Walvis Ridge 525A [*Richardson et al.*, 1982]; EMI = enriched mantle I [*Zindler and Hart*, 1986]; EMII = enriched mantle II [*Weaver*, 1991]; Paraná-Etendeka [e.g. *Petrini et al.*, 1987; *Piccirillo et al.*, 1989; *Peat*, 1997]; Liberia [*Dupuy et al.*, 1988] and North Florida [*Heatherington and Mueller*, 1999].

being instead consistent with "passive" models involving "upper" mantle geodynamics, where the unstable buoyancy of "supercontinents" [e.g. *Anderson, 1994b*] played an essential role to approach isostatic stabilization through the Pangea break-up.

CONCLUSIONS

a) The Early Jurassic Brazilian CAMP tholeiites outcrop in the central-northern part of the South American platform and span from the continental margin (i.e. Cassiporé

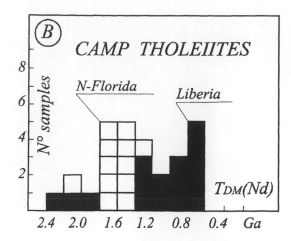

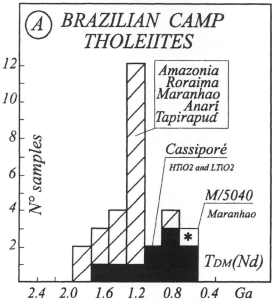

Figure 16. Histograms of $T_{DM}(Nd)$ model ages for the Brazilian (A), and the North Florida and Liberia (B) Early Jurassic CAMP tholeiites [*Heatherington et al.*, 1998 and *Dupuy et al.*, 1988]. $T_{DM}(Nd)$= Sm/Nd model ages relative to depleted mantle source.

dykes) to about 2,000 km inside the continent (Anarí and Tapirapuá flows). Permian/Triassic sediments of fluvial to aeolian environments preceded the emplacement of the Brazilian CAMP magmatism.

b) $^{40}Ar/^{39}Ar$ ages of the Brazilian CAMP tholeiites yielded a mean age of 199.0±2.4 Ma, with the youngest (e.g. 190 Ma) magmatism occurring near the continental margin.

c) The Brazilian CAMP tholeiites are somewhat evolved, the samples approaching primary melt compositions (mg# 0.64-0.66) being rare. The basalt compositional variability is related to fractional crystallization (low pressure and near-anhydrous and low fO_2 conditions) starting from different parental melts.

d) Most Brazilian CAMP tholeiites have low TiO_2 (<2wt%), while the Cassiporé dykes usually have high TiO_2 (>2wt%). Petrology and major-trace elements support the view that the HTi Cassiporé basalts cannot be derived from the LTi ones by fractional crystallization.

e) The Cassiporé dykes (LTi and HTi types) are characterized by positive Nb anomaly and Sr-Nd isotopes that conform to a "typical" mantle array. Relative to Bulk Earth, the Cassiporé tholeiites span from the depleted (e.g. Mid Atlantic Ridge) to the enriched quadrant (EMI-type: Tristan da Cunha, Walvis Ridge 525A). The other CAMP Brazilian tholeiites have εSr and εNd values trending towards crustal components which might be related to Proterozoic subduction processes and low-pressure crustal contamination during magma emplacement.

f) The Brazilian CAMP tholeiites have high Rb/Sr and low Sm/Nd values which conflict with the corresponding low $^{87}Sr/^{86}Sr$ and high $^{143}Nd/^{144}Nd$ values, respectively. Assuming that low-pressure crustal contamination was negligible, these data suggest that the Brazilian CAMP tholeiites may have experienced "mantle" metasomatism (i.e. "typical mantle" trend) and "subduction-related" crustal contamination (e.g. EMII-type trend) in Late and Middle Proterozoic times, respectively.

g) In general, all the data support the view that the CAMP tholeiites from different regions of Brazil are related to compositionally different mantle sources. We emphasize that these sources possibly reflect geochemically heterogeneous lower parts of the lithospheric thermal boundary layer.

h) In general, the paleomagnetic poles for CAMP rocks from South America, Africa and North America match an age of ca. 200 Ma, but also point to younger ages (e.g. 190 Ma). The latter ages mainly refer to South American poles of the northern Brazilian CAMP magmatism outcropping towards the continental edge.

i) The relationships between the nature and occurrence of the Brazilian CAMP tholeiites and their geo-tectonic environments suggest that this magmatism was probably related to hot "upper" mantle incubation under thick continental lithosphere and edge-driven convection at the boundary between subcontinental domains, characterized by different lithospheric thickness. Therefore, the genesis of the Brazilian CAMP magmatism, similar to that from

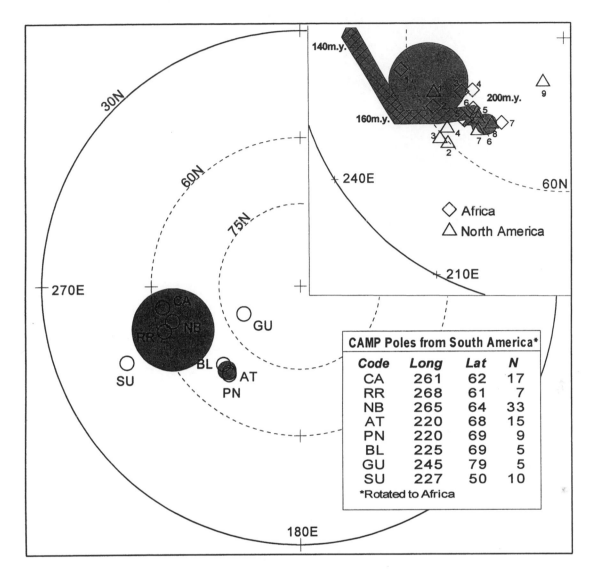

Figure 17. Paleomagnetic poles from the CAMP tholeiites of South America: CA – Cassiporé dykes; RR – Roraima dykes (this paper); NB – Northeastern Brazil; AT - Anari-Tapirapuã flows; PN – Penatecaua dykes, Amazonia; BL – Bolivar dykes, Venezuela; GU – Guiana dikes; SU – Suriname dykes [*Ernesto et al.*, this volume]. Shaded circles are mean poles as described in the text. The insert shows these mean poles compared to CAMP poles from North America (source data after *Van der Voo*, [1990]: 1 – Nova Scotia, 2 – Pennsylvannia diabases; 3 – New Jersey volcanics; 4 – Newark volcanis II; 5 – Piedmont dykes I; 6 - Connecticut Valley volcanics; 7 – Watchung basalts; 8 – Newark volcanics I; 9 – Piedmont dikes II), Africa (source: *Courtillot et al.*, [1994]: 1 – Foum-Zguidike, Morocco; 2 – Intrusives, Nigeria; 3 – dykes, Liberia; 4 – Hodh dolerite, Mauritania; 5 – Draa Valley sills, Morocco; 6 - Hank dolerite, Mauritania; 7 – Intrusives, Morocco), and the Jurassic APWP for North America according to *Besse and Courtillot* [1991]. Rotation poles: South America to Africa [*Martin et al.,* 1981]; North America to Africa [*Rabinowitz and La Brecque*, 1979].

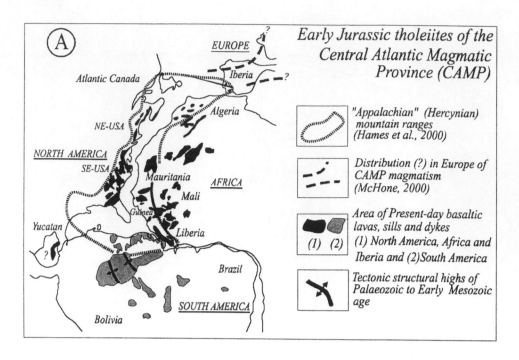

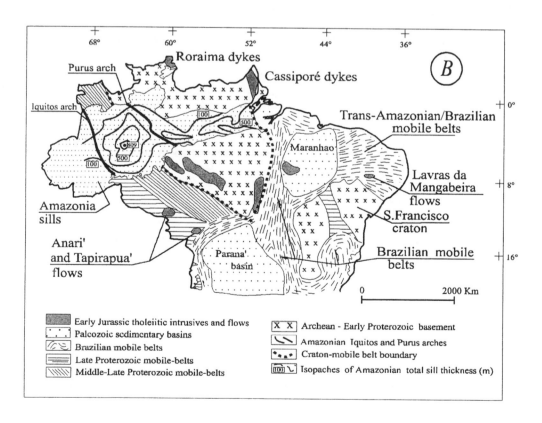

Figure 18. Distribution of the Early Jurassic magmatism in the Central Atlantic Pangea provinces (A) and in Brazil (B).

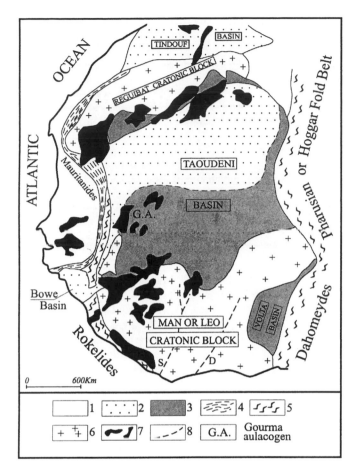

Figure 19. Geological sketch of Western Africa craton [modified after *Trompette*, 1994]: 1= Meso-Cainozoic sedimentary basins; 2= Palaeozoic sedimentary basins; Late-Proterozoic sedimentary basins (Pan-African cycle); Hercynian fold belts; 5= Pan-African fold belts; 6= cratonic blocks (mainly of Eburnean cratonization age, ≥ 2 Ga); 7= Early Jurassic CAMP magmatism [*McHone*, 2000]; 8= Sassandra (S) and Dimbokro (D) faults.

eastern USA and western Africa [*Hames et al.*, 2000; *McHone*, 2000], appears to be consistent with geodynamic passive models related to the isostatic stabilization of the "supercontinents" [*Anderson*, 1994b; *Tackely*, 2000].

Acknowledgements. This work was supported by MURST (60% and COFIN 1998) and CNR (Italian Agencies), FAPESP and CNPq (Brazilian Agencies) and by Gordon Getty Foundation (U.S.A.). This work benefitted from the helpful discussions with A. Cundari, P. Comin-Chiaramonti, M. Iacumin, P. Antonini, F. Princivalle and N. Ussami. F. Castorina, G. Cavazzini, R. Rizzieri, M. D'Antonio and R. Petrini kindly provided Sr-Nd isotope analyses. A.M.P. Mizusaki and Petrobras are acknowledged for the Amazonia core-samples. We thank L. Furlan, (Trieste University), and A. Giaretta and R. Carampin (Padova University) for technical and analytical assistance. Thanks are also due to Sandoval Pinheiro (CPRM-Brasil) for invaluable field assistance in Roraima, and to the Novo Astro and Serra do Navio mining companies for their kind logistical support in Amapá.

REFERENCES

Anderson, D. L., The sublithospheric mantle as the source of continental flood basalts: The case against the continental lithosphere and plume head reservoirs, *Earth and Planet. Sci. Lett.*, *123*, 269-280, 1994a.

Anderson, D. L., Superplumes or supercontinents?, *Geology*, *22*, 39-42, 1994b.

Antonini, P., E. M Piccirillo., R. Petrini, L. Civetta, M. D'Antonio, G. Orsi, Enriched mantle-Dupal signature in the genesis of the Jurassic Ferrar tholeiites from Prince Albert Mountains (Victoria Land, Antarctica), *Contrib. Mineral. Petrol.*, *136*, 1-19, 1999.

Arora, B. R., A. L. Padilha, I. Vitorello, N. B. Trivedi, S. L. Fontes, A. Rigoti, F. H. Chamalaun, Geoelectrical model for the Parnaiba Basin conductivity anomaly of Northeast Brazil and tectonic implications, *Tectonophysics*, *302*, 1-2, 57-69, 1999.

Almeida, F. F. M., Distribução regional e relações tectônicas do magmatismo pós-Paleozóico no Brasil, *Revista Brasileira de Geociências.*, *16*, 325-349, 1986.

Baksi, A. K. and D. A. Archibald, Mesozoic igneous activity in the Maranhão Province, northern Brazil; $^{40}Ar/^{39}Ar$ evidence for separate episodes of basaltic magmatism, *Earth and Planet. Sci. Lett.*, *151*, 139-153, 1997.

Bellieni, G., E. M. Piccirillo, B. Zanettin, Classification and nomenclature of basalts. I.U.G.S. subcommission on the systematics of igneous rocks. Circular 34, *Contribution No, 86*, 19 pp., 1981.

Bellieni, G., E. M. Piccirillo, G. Cavazzini, R. Petrini, P. Comin-Chiaramonti, A. J. R. Nardy, L. Civetta, A. J. Melfi and P. Zantedeschi, Low- and High TiO_2 Mesozoic tholeiitic magmatism of the Maranhão basin (NE-Brazil): K/Ar age, geochemistry, petrology, isotope characteristics and relationships with Mesozoic low- and high-TiO_2 flood basalts of the Paraná basin (SE-Brazil), *N. J. Miner. Abh.*, *162*, 1-33, 1990.

Berrangé, J. P., and R. Dearley, The Apoteri volcanic formation, tholeiitic flows in the North Savannah graben in Guyana and Brazil, *Geologische Rundschau 64*, Band 64, Heft 33, 1-82, 1975.

Besse, J. and V. Courtillot, Revised and synthetic apparent polar wander paths of the African, Eurasian, North American and Indian plates, and true polar wander since 200 Ma. *J. Geophys. Res.*, *96*, 4029-4050, 1991.

Boutelier, R. R., C. E. Keen, Small-scale convection and divergent plate boundaries, *J. Geophys. Res.*, *104*, 7389-7403, 1999.

Boynton, W. V., Cosmochemistry of the rare Earth elements: meteorite studies, in *Rare Earth Element Geochemistry*, edited by P. Henderson, pp. 63-114, Elsevier publ. Co., Amsterdam, 1984.

Buck, W. R., Small-scale convection induced by passive rifting: the cause for uplift or rift shoulders, *Earth and Planet. Sci. Lett.*, 77, 362-372, 1986.

Buddington, A. F., D. H. Lindsley, Iron-titanium oxide minerals and synthetic equivalents, *J. Petrol.*, 5, 310-57, 1964.

Campbell, I. H., R. W. Griffiths, Implications of mantle plume structure for the evolution of flood basalts, *Earth and Planet. Sci. Lett.*, 99, 79-93, 1990.

Carmichael, I. S. E, The iron-titanium oxides of salic volcanic rocks and their associated ferromagnesian silicates, *Contrib. Mineral. Petrol.*, 14, 36-64, 1967.

Chaffey, D. J., R.A. Cliff and B. M. Wilson, Characterization of the St Helena magma source, in *Magmatism in the Ocean Basin*, Spec. Publ., 42, edited by A.D. Saunders and M.J. Norry, pp. 257-276, The Geological Society, London, 1989.

Comin-Chiaramonti, P., A. Cundari, E. M. Piccirillo, C. B. Gomes, F. Castorina, P. Censi, A. De Min, A. Marzoli, S. Speziale, V. F. Velazquez, Potassic and sodic igneous rocks from eastern Paraguay: their origin from the lithospheric mantle and genetic relationships with the associated Paraná flood tholeiites, *J. Petrol.*, 38, 495-528, 1997.

Cordani, U. G., B. B. Brito Neves, R. A. Fuck, R. Porto, A. T. Filho, F. M. Bezzarra da Cunha, Estudo preliminar de integração do Pré-Cambriano com os eventos tectônicos das bacias sedimentarias Brasileiras, *Ciência Tecnica Petroleo, Seç. Expl. de Petrol.*, 15, 11-70, 1984.

Courtillot, V., J. Besse, and H. Théveniaut, North American apparent polar wander: the answer from other continents. *PEPI*, 82, 87-104, 1994.

Cox, K. G., The Karoo Province, in *Continental Flood Basalts*, Macdougall J.D. editor, Kluwer Academic, pp. 239-271, Dordrecht, 1988.

Davies, G. R., M. J. Norry, D. C. Gerlach and R. A. Cliff, A combined chemical and Pb-Nd isotope study of the Azores and Cape Verde hot-spots: the geodynamic implications, in *Magmatism in the Ocean Basin*, Spec. Publ., 42, edited by A.D. Saunders and M.J. Norry, pp. 231-255, The Geological Society, London, 1989.

Deckart, K., G. Féraud, H. Bertrand, Age of Jurassic continental tholeiites of French Guyana, Surinam and Guinea: Implications for the initial opening of the Central Atlantic Ocean, *Earth and Planet. Sci. Lett.*, 150, 205-220, 1997.

Deer, W. A, R. A. Howie, J. Zussman, Introdution to the *Rock Forming Minerals*, second edition, vol. 2A, 3-18, 1978.

De La Roche, H., P. Leterrier, P. Grandclaude, M. Marchal, A classification of volcanic and plutonic rocks using R1-R2 diagram and major-element analysis. Its relationships with current nomenclature, *Chem. Geol.*, 29, 183-210, 1980.

Dupuy, C., J. Marsh, J. Dostal, A. Michard and S. Testa, Asthenospheric and lithospheric sources for Mesozoic dolerites from Liberia (Africa): trace element and isotopic evidence, *Earth and Planet. Sci. Lett.*, 87, 100-110, 1988.

Encarnación, J., T. H. Flemming, D. H. Elliot, H. V. Eales, Synchronous emplacement of Ferrar and Karoo dolerites and the early breakup of Gondwana, *Geology*, 24, 535-538, 1996.

Erlank, A. J., J. S. Marsh, A. R. Dunkan, R. McGMiller, R. McG., C. J. Hawekesworth, P. J. Betton, D. C. Rex, Geochemistry and petrogenesis of the Etendeka volcanic rocks from SWA/Namibia, *Special Pubblication of Geological Society of South Africa*, 13, 195-246, 1984.

Ernst, R. E., J. W. Head, E. Perfitt, E. Grosfils, L. Wilson, Giant radiating dyke swarms on Earth and Venus, *Earth and Planet. Sci. Lett.*, 39, 1-58, 1995.

Ernesto, M., G. Bellieni, E. M. Piccirillo P. R. Renne, A. De Min, I. G. Pacca, G. Martins, J. W. P. Macedo, Paleomagnetic, geochemical and geochronological constraints in time and duration of the Mesozoic igneous activity in northeastern Brazil. *AGU, present volume*.

Farnetani, C. G. and M. A. Richards, Numerical investigation of the mantle plume initiation model for flood basalt events, *J. Geophys. Res.*, 99, 13,813-13,833, 1994.

Fleming, T. H., A. Heimann, K. A. Foland, D. H. Elliot, $^{40}Ar/^{39}Ar$ geochronology of Ferrar Dolerite sills from the Transantartic Mountains, Antarctica: Implications for the age and origin of the Ferrar magmatic province, *Geol. Soc. Am. Bull.*, 109, 533-546, 1997.

Fodor, R. V., A. N. Sial, S. B. Mukasa and E. H. McKee, Petrology, isotope characteristics, and K-Ar ages of the Maranhão, northern Brazil, Mesozoic basalt province, *Contrib. Mineral. Petrol.*, 104, 555-567, 1990.

Fodor, R. V., S. B. Mukasa, A. N. Sial, Isotopic and trace element indications of lithospheric and asthenospheric components in Tertiary alkalic basalts, northeastern Brazil, *Lithos*, 43, 197-217, 1998.

Gerlach, M. S., J. C. Stormer, P. A. Muller, Isotopic geochemistry of Fernando de Noronha, *Earth and Planet. Sci. Lett.*, 85, 129-144, 1987.

Govindarajau, K., G. Mevelle, Fully automated dissolution and separation methods for inductively coupled plasma atomic emission spectrometry rock analysis: application to the determination of rare earth elements, *J. Anal. Spectrom.*, 2, 615-621, 1987.

Guerreiro, S. D. C. and A. Schult, Paleomagnetism of Jurassic tholeiitic intrusions in the Amazon Basin, *Münchner Geophys. Mitteilungen*, 1, 37-48, 1986.

Gurnis, M., Large-scale mantle convection and the aggregation and dispersal of supercontinents, *Nature*, 332, 695-699, 1988.

Halliday, A. N., G. R. Davies, D. C. Lee, S. Tommasini, C. R. paslick, J. G. Fitton, D. E. James, Lead isotope evidence for young trace element enrichment in the oceanic upper mantle, Nature, 359, 623-627, 1992.

Hames, W. E., P. R. Renne, New evidence for geologically instantaneous emplacement of earliest Jurassic Central Atlantic magmatism province basalts on the North American margin, *Geology*, 28, 859-862, 2000.

Heatherington, A. L., P. A. Mueller, Lithospheric sources of North Florida, USA tholeiites and implications for the origin of the Suwannee terrane, *Lithos*, 46, 215-233, 1999.

Heimann, A., T. H. Flemming, D. H. Elliot, K. A. Foland, A short interval of Jurassic continental flood basalt volcanism in Antartica as demonstrated by $^{40}Ar/^{39}Ar$ geochronology, *Earth and Planet. Sci. Lett.*, 121, 19-41, 1994.

Hill, R. I., I. H. Campbell, G. F. Davies, and R. W. Griffiths, Mantle plumes and continental tectonics, *Science*, 256, 186-193, 1992.

Ito, E., W.M. White, and C. Göpel, The O, Sr, Nd and Pb isotope geochemistry of MORB, *Chemical Geology, 62,* 157-176.

Jaques, A. L. and D. H. Green, Anhydrous melting of peridotite at 0-15 Kb pressure and the genesis of tholeiitic basalts, *Contrib. Miner. Petrol.,73, 287-310,* 1980.

King, S. D., D. L. Anderson, Edge-driven convection, *Earth and Planet. Sci. Lett., 160,* 289-296, 1998.

Kretz, R., Transfer and exchange equilibria in a portion of the pyroxene quadrilater as deduced from natural and experimental data, *Geochim. Cosmochim. Acta, 46,* 411-412, 1982.

Kudo, A. M., D. F. Weill, An igneous plagioclase thermometer, *Contrib. Mineral. Petrol., 25,* 1, 52-65, 1970.

Le Bas, M. J., R. W. Le Maitre, A. Streckeisen, B. Zanettin, A chemical classification of volcanic rock based on the total alkali-silica diagram, *J. Petrol., 27,* 745-750, 1986.

Leitch, A. M., G. F. Davies, M. Wells, A plume head melting under a rifting margin, *Earth and Planet. Sci. Lett., 161,* 161-177, 1998.

Lima, M. I. C., R. M. G. Montalvão, R. S. Issler, A. S. Oliveira, M. A. S. Basei, J. V. F., Araujo and G. G. Silva, Geologia Folha Na/Nb.22 Macapa. Projeto RADAM, *Departamento Nacional da Producao Mineral (DNPM), 6,* 9-141, Rio de Janeiro (Brazil), 1974.

Lopes, R. P., Petrologia dos fonólitos do Arquipelago de Fernando de Noronha, PE. *Unpublished Ms. Thesis,* Instituto de Geociências, USP, Brasil, 125 pp., 1997.

Marques, L. S., B. Dupré, E. M. Piccirillo, Mantle source composition of the Paraná Magmatic Province (southern Brazil): evidence from trace element and Sr-Nd-Pb isotope geochemistry, *J. Geodynamics, 28,* 439-458, 1999.

Martin, A. K., C. J. H. Hartnady and S. N. Goodlad, A revised fit of South America and South Central Africa, *Earth and Planet. Sci. Lett., 54,* 293-305, 1981.

Marzoli, A., P. R. Renne, E. M. Piccirillo, M. Ernesto, G. Bellieni and A. De Min, Extensive 200-Million-Year-Old Continental Flood Basalts of the Central Atlantic Magmatic Province, *Science, 28a,* 616-618, 1999.

Mathez, E., Refinement of the Kudo-Weill plagioclase thermometer and its applications to basaltic rocks, *Contrib. Mineral. Petrol., 41,* 61-72, 1973.

McHone, J. G., Non-plume magmatism and rifting during the opening of the central Atlantic Ocean, *Tectonophysics, 316,* 287-296, 2000.

Mellini, M., S. Carbonin, A. Dal Negro and E. M. Piccirillo, Tholeiitic hypoabissal dykes: how many clinopyroxenes?, *Lithos, 22,* 127-134, 1988.

Menezez Leal, A. B., V. A. V. Girardi, L. R. Bastos Leal, Petrology and Geochemistry of the basic magmatism from the Apoteri basic suite, State of Roraima, *Geochimica Brasiliensis,* in press.

Menzies, M. A., C. J. Hawkesworth (eds). *Mantle metasomatism,* Academic Press, 472 pp., London, 1987.

Milner, S. C., A. P. Le Roex, Isotope characteristics of the Okenyanya igneous complex, northwestern Namibia: constrain on the composition of the early Tristan plume and the origin of the EM 1 mantle components, *Earth and Planet. Sci. Lett., 141,* 277-291, 1996.

Montes-Lauar, C.R., Paleomagnetismo de rochas magmáticas mesozóico-cenozóicas da Plataforma Sul-Americana: estudo dox complexos alcalino-carbonatíticos de Tapira (MG) e Salitre (MG) e das ilhas do Arquipélago de Abrolhos (BA), *PhD thesis,* IAG-USP, São Paulo, Brazil, 1993.

Montes-Lauar, C. R., I. G. Pacca, A. J. Melfi, E. M. Piccirillo, G. Bellieni, R. Petrini, R. Rizzieri, The Anarí and Tapirapuá Jurassic formations, western Brazil; paleomagnetism, geochemistry and geochronology, *Earth and Planet. Sci. Lett., 128,* 357-371, 1994.

Muir, I. D. and D. A. Mattey, Pyroxene fractionation in ferrobasalts from the Galápagos spreading centre, *Mineral. Mag., 45,* 193-200, 1982.

Muir, I. D. and C. E. Tilley, Iron enrichment and pyroxene fractionation in tholeiites, *J. Geol., 4,* 143-156, 1964.

Nimis, P., A clinopyroxene geobarometer for basaltic systems based on crystal-structure modeling, *Contrib. Miner. Petrol., 121,* 115-25, 1995.

Novello, A., Studio petrografico e geochimico di dicchi, sill e vulcaniti del Brasile nord orientale (Ceará, Piaui, Rio Grande do Norte), *Unpublished thesis,* University of Trieste, Trieste, Italy, 1995.

Oliveira, E. P. and J. Tarney, Geochemistry of the Mesozoic Amapá and Jari Dyke Swarm, northern Brazil: Plume-related magmatism during the opening of the central Atlantic, in *Mafic Dykes and Emplacement Mechanism,* A.J. Parker, P.C. Rickwood and D.H. Tuker eds., pp.173-183, Rotterdam, 1990.

Oyarzun, R., M. Doblas, J. Lopez-Ruiz, J. M. Cebria, Opening of the Central Atlantic and asymmetric mantle upwelling phenomena: implications for long-lived magmatism in western North Africa and Europe, *Geology, 25,* 727-730, 1997.

Papike, J. J., K. L. Cameron and K. Baldwin, Amphiboles and Pyroxenes: characterization of other than quadrilateral components and estimates of ferric iron from microprobe data, *Geol. Soc. Am. Abstr. Prog., 6,* 1053-1054, 1974.

Peate, D. W., C. J. Hawkesworth, Lithospheric to asthenospheric transition in Low-Ti flood basalts from southern Paraná, Brazil, *Chem. Geol., 127,* 1-24, 1996.

Peate, D. W., The Paraná-Etendeka Province, in *Large Igneous Provinces: Continental, Oceanic, and Planetary Flood Volcanism,* MaHoney J.J and Coffin M. F. (eds.), 217-245, 1997.

Petri, S. and J. W. Fulfaro, *Geologia do Brazil,* T.A. Queiroz e Universidade de São Paulo (Brazil), 1983.

Petrini, R., L. Civetta, E. M. Piccirillo, G. Bellieni, P. Comin-Chiaramonti, L. S. Marques, A. J. Melfi, Mantle heterogeneity and crustal contamination in the genesis of low-Ti continental flood basalts from the Paraná Plateau (Brazil); Sr-Nd isotope and geochemical evidence, *J. Petrol., 28,* 701-726, 1987.

Philips Procedure, *X40 Software for XRF analysis.* Software Operation Manual, 1994.

Piccirillo, E. M. and A. J. Melfi (eds.), *The Mesozoic Flood Volcanism of the Paraná Basin: Petrogenetic and Geophysical Aspects,* 600 pp., IAG-USP, São Paulo (Brazil), 1988.

Piccirillo, E. M., L. Civetta, R. Petrini, A. Longinelli, G. Bellieni, P. Comin-Chiaramonti, L. S. Marques, A. J. Mclfi, Regional variations within the Paraná flood basalts (Southern Brazil):

evidence for subcontinental mantle heterogeneity and crustal contamination, *Chem. Geol., 75,*103-122, 1989.

Rabinowitz, P. D. and J. La Brecque, The Mesozoic South Atlantic Ocean and evolution of its continental margins, *J. Geophys. Res., 84,* 5973-6002, 1979.

Rapalini, A. E., A. L. Abdeldayem and D. H. Tarling, Intracontinental movements in Western Gondwanaland: a palaeomagnetic test, *Tectonophysics, 220,* 127-139, 1993.

Renne, P. R., M. Ernesto, I. G. Pacca, R. S. Coe, J. M. Glen, M. Prevot, M. Perrin, Rapid eruption of the Paraná flood volcanics, rifting of southern Gondwanaland and the Jurassic-Cretaceous boundary, *Science, 258,* 975-979, 1992.

Renne, P. R., K. Deckart, M. Ernesto, G. Feraud, E. M. Piccirillo, Age of the Ponta Grossa dike searm (Brazil) and implications to Paraná flood volcanism. *Earth and Planet. Sci. Lett., 144,* 199-211, 1996.

Renne, P. R., C. C. Swisher, A. L. Deino, B. B. Karner, T. Owens, D. J. De Paolo, Intercalibration of Standards, Absolute ages and Uncertainties in ^{40}Ar/^{39}Ar Dating, *Chem. Geol., 145,* 117-152, 1998.

Richardson, S. H., A.J. Erlank, A. R. Duncan and D. L. Reid, Correlated Nd, Sr and Pb isotope variations in Walvis Ridge basalts and implication for the evolution of their mantle source, *Earth Planet. Sci. Lett., 59,* 327-342, 1982.

Schobbenhaus, C., D. A. Campos, G. R. Derze and H. E. Asmus, *Geologia do Brasil,* Departamento Nacional da Produção Mineral (DNPM), Brasilia (Brazil), 501 pp., 1984.

Stewart, K., S. Turner, S. Kelley, C. J. Hawkesworth, L. Kirstein and M. S. M. Mantovani , 3-D ^{40}Ar-^{39}Ar geochronology in the Paraná flood basalt province, *Earth and Planet. Sci. Lett., 143,* 95-110, 1996.

Sun, S.-s. and W. F. McDonough, Chemical and isotopic systematics of oceanic basalts: implications for mantle composition and processes, in *Magmatism in the Ocean Basin, Spec. Publ., 42,* edited by A.D. Saunders and M.J. Norry, pp. 313-345, The Geological Society, London, 1989.

Tackley, P. J., Mantle convection and plate tectonics: toward an integrated physical and chemical theory, *Science, 288,* 2002-2007, 2000.

Taylor, S. R., S. M. McLennan, *The continental crust; its composition and evolution,* Blackwell Scientific, 312 pp., Oxford, 1985.

Thompson, G. A., Deep mantle plumes and geoscience vision, *GSA Today (April),* 17-25, 1998.

Trompette, R. (eds), *Geology of Western Gondwana (2000-500 Ma) Pan African-Brasiliano aggregation of South America and Africa,* A.A. Balkema/ Rotterdam/ Brookfield, 1994.

Van der Voo, R., Phanerozoic paleomagnetic poles from Europe and North America and comparison with continental reconstructions, *Rev. Geophys., 28,* 167-206, 1990.

Veevers, J. J., Middle/Late Triassic (230±5Ma) singularity in the stratigraphic and magmatic history of the Pangean heat anomaly, *Geology, 17,* 784-787, 1989.

Wilson, M., Thermal evolution of the central Atlantic passive margins: continental break-up above a Mesozoic super-plume, *J. Geol. Soc. London, 154,* 491-495, 1997.

Weaver, B., The origin of the ocean island basalt end-member compositions: trace element and isotopic costrains, *Earth and Planet. Sci. Lett., 104,* 381-397, 1991.

[1]Angelo De Min (*corresponding author*) and Enzo M. Piccirillo, Dipartimento di Scienze della Terra, University of Trieste, Via E. Weiss 8, 34127 Trieste, Italy, demin@univ.trieste.it; picciril@univ.trieste.it

[2]Andrea Marzoli, Département de Minéralogie, Université de Genève, 13 Rue des Maraîchers, 1211 Genève 4, Switzerland, Andrea.Marzoli@terre.unige.ch

[3]Giuliano Bellieni, Dipartimento di Mineralogia e Petrologia, University of Padova, Corso Garibaldi 37, 35137 Padova, Italy, giuliano@dmp.unipd.it

[4]Paul Renne, Berkeley Geochronology Center, 2455 Ridge Road, Berkeley, CA 94720, USA, prenne@bgc.org

[5]Marcia Ernesto and Leila Marques, Departamento de Geofisica, Instituto Astronomico e Geofisico, University of São Paulo (USP), Cidade Universitaria, Rua de Matão 1226, CEP 05508-900 São Paulo, SP, Brazil, marcia@iag.usp.br; leila@iag.usp.br

Paleomagnetic and Geochemical Constraints on the Timing and Duration of the CAMP Activity in Northeastern Brazil

M. Ernesto[1], G. Bellieni[2], E.M. Piccirillo[3], L.S. Marques[1], A. de Min[3], I.G. Pacca[1], G. Martins[4], and J.W.P. Macedo[5]

The northeastern border of the Maranhão basin (NE Brazil) is dominated by low-Ti tholeiites which show strong geochemical affinity with the low-Ti Early Jurassic Mosquito Formation from the western side of the basin. The low-Ti tholeiitic flows (Early Jurassic) from Lavras basin, and the low-Ti dikes outcropping in the Ceará State are also comparable to the Mosquito tholeiites. Compositional similarity also exists between the Early Cretaceous Sardinha intrusives (eastern Maranhão basin), the high-Ti dikes from the Ceará State, and the high-Ti rocks of the Rio Ceará-Mirim dike swarm (mainly subswarms I and III) easterwards. The high-Ti and low-Ti rocks show paleomagnetic remanences that allow the discrimination of the two groups. However, the characteristic remanent magnetization from the Rio Ceará-Mirim subswarm II differs significantly from the other two groups; these rocks can be also distinguished from the other high-Ti rocks by slight different chemical characteristics. On paleomagnetic basis, three age groups were identified: Early Cretaceous (Sardinha Formation, high-Ti Ceará dikes, and Ceará-Mirim subswarms I, III, and V), Early Jurassic (Mosquito Formation, low-Ti Ceará dikes, Lavras basin flow, and Ceará-Mirim subswarm IV), and Late Triassic (Ceará-Mirim subswarm II), and corresponding paleomagnetic poles were calculated. The Early Jurassic pole fits well with other CAMP poles from South America, extending easterwards the area in northern South America affected by that magmatic event. The Late Triassic age might represent a magmatic manifestation preceding by about 20-30 My the CAMP activity in NE Brazil.

[1]Departamento de Geofísica, Universidade de São Paulo, São Paulo, Brazil.

[2]Dipartimento di Mineralogia e Petrologia, Universita di Padova, Padova, Italy.

[3]Dipartimento di Scienze della Terra, Universita di Trieste, Trieste, Italy.

[4]Departamento de Geologia, Univerisidade Federal do Ceará, Fortaleza, Brazil.

[5]Departamento de Física, Univerisidade Federal do Rio Grande do Norte, Natal, Brazil.

The Central Atlantic Magmatic Province:
Insights from Fragments of Pangea
Geophysical Monograph 136
Copyright 2003 by the American Geophysical Union
10.1029/136GM07

INTRODUCTION

Northeastern Brazil (Fig. 1) was affected by recurrent tholeiitic magmatic activity [e.g. *Sial*, 1976; *Almeida et al.*, 1988; *Bellieni et al.*, 1990, and 1992] since at least Early Jurassic to Late Cretaceous. Basalt flows of the Early Jurassic Mosquito Formation [*Bellieni et al.*, 1990; *Fodor et al.*, 1990; *Baksi and Archibald*, 1997; and *Marzoli et al.*, 1999] are found in the western side of the Maranhão basin whereas in the eastern side, sills, dikes and scarce basaltic flows of the Early Cretaceous Sardinha Formation are present. Eastwards the Ceará-Mirim dike swarm [*Sial*, 1976; *Gomes et al.*, 1981; *Martins*, 1991; *Bellieni et al.*, 1992; *Oliveira*, 1992] trends east-west for more than 300 km. *Sial* [1975] subdivided the Ceará-Mirim dike swarm into six subparallel lineaments or subswarms of which five

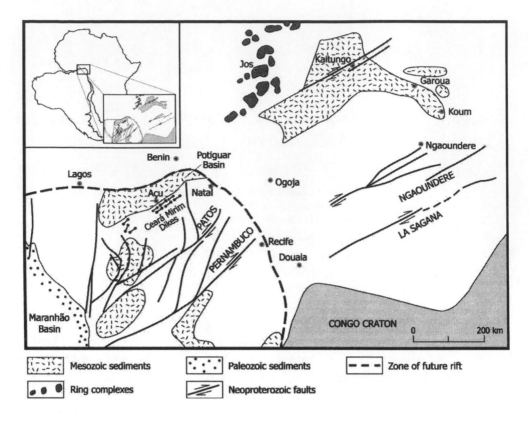

Figure 1. Schematic map of Northeastern Brazil showing the location of the Mosquito and Sardinha Formations (Maranhão basin), the Ceará-Mirim dike swarm, the Lavras da Mangabeira flows, and other magmatic occurrences in the Ceará and Piauí States. Full triangles represent the Tertiary volcanic necks and plugs. Sampling sites are also indicated. Dashed line represents the eastern limit of the Maranhão basin.

were also discriminated by paleomagnetic data [*Martins*, 1991; *Bellieni et al.*, 1992], and are shown in Fig. 2. Some isolated dikes trending E-W to NE-SW are also found in the Ceará State, between the Maranhão basin and the area occupied by the Ceará-Mirim dike swarm, forming a practically continuous magmatic area along Northeastern Brazil. The easternmost Early Jurassic rocks are represented by a few basalt flows [*Marzoli et al.*, 1999; and *De Min et al.*, this volume] located in the Lavras sedimentary basin (Fig 1).

In spite of the large amount of K-Ar, and some ^{40}Ar/^{39}Ar data for the Mesozoic magmatism in Northeastern Brazil, the age of the Ceará-Mirim dikes is still debatable. *Bellieni et al.* [1992] concluded that the distribution of the K-Ar and fission track ages delineated two maxima, 175-160 Ma and 145-125 Ma, indicating the existence of both Middle Jurassic and Early Cretaceous dikes in the Ceará-Mirim swarm. This was previously suggested by *Horn et al.* [1988] based on a minimum K-Ar age of 164±5 Ma [Middle Jurassic) calculated for two dikes belonging to subswarm II. However, K-Ar ages as old as 214±6 Ma and 228±3 Ma [*Bellieni et al.*, 1992] for dikes of subswarm II,

and an age of 245±13 Ma [*Gomes et al.*, 1981] were discarded as anomalous data.

In a pre-drift reconstruction (Fig. 3) the area occupied by the Ceará-Mirim swarm lies opposite to the Benue Trough (Nigeria) where two Mesozoic periods of magmatic activity, 147-106 Ma, and 97-81 Ma have been recognized by *Maluski et al.* [1995]. However, northwards in the Jos plateau the alkaline ring complexes vary in age from 213 to 141 Ma [*Maluski et al.*, 1995, and references therein].

Paleomagnetic results for the Ceará-Mirim dikes, and Maranhão Basin magmatism were first reported by *Guerreiro and Schult* [1983] and *Schult and Guerreiro* [1979], respectively. However, only the Ceará-Mirim dikes cropping out east of Augusto Severo (Fig. 2), corresponding to subswarms I, II and III, were investigated, and an unique paleomagnetic pole was calculated. *Bücker et al.* [1986] studied the magnetic mineralogy of these dikes and improved the existing paleomagnetic results, although the recalculated pole did not change significantly. *Horn et al.* [1988] obtained K-Ar ages for the Ceará-Mirim dikes in the range 167 to 130 Ma and concluded that two different poles should be calculated: an Early Cretaceous pole based

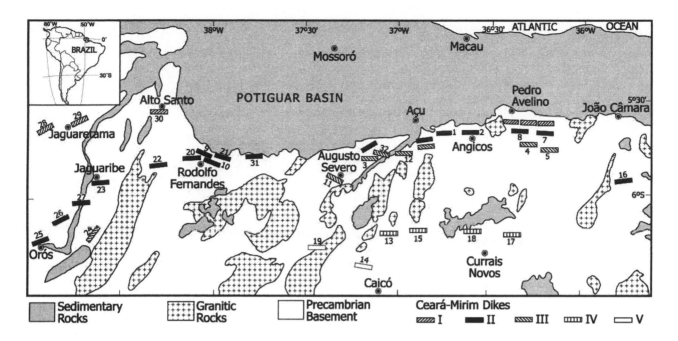

Figure 2. Sampling sites of the Ceará-Mirim dike swarm distinguishing dikes belonging to subswarms I to V, according to *Sial* [1975], and the magnetic characteristics (this paper).

on five sites from subswarms I and III, and a Jurassic pole based on five sites from subswarm II. The same subdivision was presented in *Martins* [1991] and *Bellieni et al.* [1992] for a higher number of sites.

Geochemical studies of both Early Jurassic and Early Cretaceous tholeiites of the Maranhão Basin [*Bellieni et al.*, 1990] evidenced that these two magmatic episodes can be distinguished by their low-Ti (Mosquito Formation) and high-Ti (Sardinha Formation) content, respectively. The same behavior is also observed for incompatible trace elements (e.g. Ba, La, Ce, Y, Zr, Sr), that also present relatively low and high concentrations in the Mosquito and Sardinha rocks, respectively. The Ceará-Mirim dike swarm has also been investigated by *Bellieni et al.* [1992]. Both Jurassic and Early Cretaceous dikes have high-Ti and incompatible trace element (ITE) concentrations, with the Jurassic ones showing slightly higher TiO_2 and ITE. Rare tholeiitic dikes low in TiO_2 and ITE (subswarm IV) are also present. *Bellieni et al.* [1992] concluded that the Ceará-Mirim tholeiitic dikes show close mineralogical, chemical and isotopic similarities with the Early Cretaceous high-Ti tholeiites from the Maranhão basin.

This paper presents new paleomagnetic data from the two age groups of the Maranhão basin, and extends the Ceará-Mirim sampling area including sites from subswarms IV and V (Fig. 2). Some isolated dikes outcropping to the west of the Ceará-Mirim swarm in the Ceará State, and two Jurassic flows from the Lavras basin were

also studied. New geochemical data from the Maranhão basin and Ceará dikes are also presented. The analyzed samples come from the same paleomagnetic sites (Fig. 1). Comparisons of this new data set with the preexisting results from the same areas, along with paleomagnetic information, will be used in the investigation of the age, and spatial distribution of the different magmatic episodes in Northeastern Brazil.

GEOLOGICAL ASPECTS

The Ceará-Mirim dikes subvertically cross-cut the Precambrian crystalline basement to the south of the Mesozoic Potiguar sedimentary basin, in a general east-west direction. However, to the west they progressively change towards NE-SW (Fig. 2). Thickness generally ranges from 5 to 30m although it may reach about 100m [*Rolff*, 1965 and *Sial*, 1975]. According to *Bellieni et al.* [1992] the Ceará-Mirim magmatism is essentially represented by high-Ti ($TiO_2 > 2$ wt%) evolved tholeiites (andesi-basalts; mg number: 0.48-0.30) excepting for subswarm V which is constituted by alkaline rocks (tephrite and hawaiite). The Ceará-Mirim tholeiitic dikes show fine- to medium-grained size textures and mineral assemblages commonly consisting of plagioclase, augite, pigeonite, Ti-magnetite, apatite and ilmenite. In the tholeiitic dikes olivine is scarce, and the small crystals are completely altered. It is worth noting that for similar MgO content, the dikes belonging to sub-

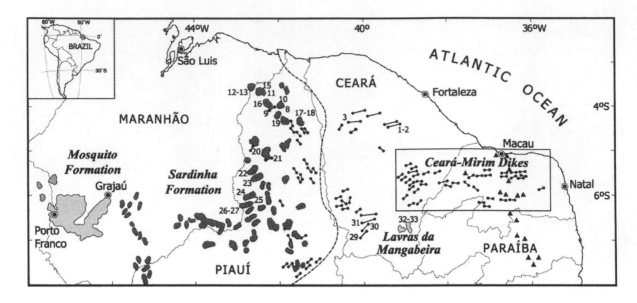

Figure 3. The Northeastern Brazil and the corresponding Northwestern African sector in a pre-drift reconstruction. Simplified from *Popoff* [1988].

swarm II show slightly higher TiO_2 and lower P_2O_5, K_2O, Y, Zr, La, Ce contents relative to the subswarms I and III.

The Paleozoic Maranhão sedimentary basin (Fig. 1) is located in Northern Brazil and covers about 600,000 km^2. During the Mesozoic it has been affected by a magmatic activity resulting in the emplacement of flows, dikes and sills. The Jurassic flows of Mosquito Formation appear restricted to the western side of the basin, whereas the Early Cretaceous intrusives of the Sardinha Formation (mainly dikes and sills) dominate towards the eastern border. The dikes mainly trend N50-70E and N20-40W [*Rezende*, 1964; this work] and thickness ranges from 2 to 60m. For the Sardinha Formation K-Ar ages [*Sial*, 1976] are close to 158 Ma, while $^{40}Ar/^{39}Ar$ data [*Baksi and Archibald*, 1997] vary from 129 to 124 Ma. However, *Nascimento et al.* [1981] reported a K-Ar age of 171±8 Ma in the northeastern area, which is in accordance with the 200-154 Ma [*Bellieni et al.*, 1990; *Fodor et al.*, 1990] interval defined by K-Ar ages for the Jurassic Mosquito Formation. A narrower interval of 200-190 Ma [*Baksi and Archibald*, 1997; *Marzoli et al.*, 1999] is defined by $^{40}Ar/^{39}Ar$ data for the Mosquito rocks.

The Maranhão magmatism is generally characterized by evolved basalts with mg number usually lower than 0.6. The Mosquito Formation is mostly represented by tholeiitic basalts with slightly porphyritic texture and mineral assemblages made up of augite, plagioclase, pigeonite and Ti-magnetite. Olivine, commonly confined in the groundmass, is scarce and completely altered. The Mosquito rocks

have low concentrations of TiO_2, P_2O_5, K_2O, FeO_t and ITE. The high-Ti Sardinha intrusive rocks have fine- to medium-grained size and mineral assemblages consisting of augite, plagioclase, pigeonite, Ti-magnetite and scarce ilmenite, completely altered olivine and apatite. They are more evolved than the Mosquito volcanics, and mainly correspond to andesi-basalts [*Bellieni et al.*, 1990].

The Lavras basin, near the locality of Lavras da Mangabeira, is a small Mesozoic sedimentary basin covering an area of about 70 km^2, and consisting of up to 80 m of sediments. Concordant vesicular basaltic flows occur at the base of the sedimentary column showing thickness of 1 to 8 m. These flows show slightly porphyritic to aphyric textures and mineral assemblages made up of augite, plagioclase, pigeonite, Ti-magnetite and scarce completely altered olivine. These basalts contain zeolite- and carbonate-bearing vesicles (L.O.I.=3.74 - 4.29 wt%) and are characterized by variable degree of alteration. Their tholeiitic nature is given by coexisting augite and pigeonite, while the elements less influenced by alteration (e.g. Ti, Zr, Y) indicate that they are of low-Ti type tholeiites, as the Mosquito ones. K-Ar data [*Priem*, 1978] for these basalts gave a mean age of 175±4 Ma, whereas one $^{40}Ar/^{39}Ar$ age [*Marzoli et al.*, 1999] is of 198.4±1.4 Ma.

During the mid-Tertiary (~29-36 Ma) alkaline magmatism occurred [*Sial et al.*, 1981; *Almeida et al.*, 1988] affecting the Ceará and Rio Grande do Norte States; necks and plugs (the Macau magmatism) cross-cut the Ceará-Mirim dike swarm (Fig. 1) in a nearly N-S direction.

PALEOMAGNETISM

Eighty-nine hand samples were taken from 32 sampling sites (dikes) of the Ceará-Mirim dike swarm, and 6 to 9 cylinders were cut in the field from 28 sites in the Maranhão basin and Ceará State. Sample orientation was performed by both magnetic and sun compasses whenever possible. One cylinder (2.5cm in diameter and 2.2cm long) of each sample was submitted to detailed AF cleaning up to 100mT (200mT in a few cases) in a Molspin - MSA2 tumbling demagnetizer and a second cylinder underwent detailed thermal cleaning (up to 600 °C) in a Schonstedt - TSD1 shielded furnace. A Molspin - MS1 spinner magnetometer was used for the remanence measurements. In general, the most stable component of magnetization was isolated by applying fields of moderate intensity but in some cases relatively high fields were necessary to erase secondary magnetizations as is shown by the vector diagrams in Fig. 4. Thermal cleaning was in general as efficient as AF cleaning but samples were previously submitted to a demagnetizing 10 to 30 mT peak field. During thermal cleaning samples became completely demagnetized at temperatures ranging from 530°C to 600°C. Characteristic components of magnetization for each sample were identified by means of vector analyses (Figs. 4 and 5) using the least square fit of *Kirshvink* [1980]. Site mean characteristic direction was calculated by giving unit weight to specimens, and uncertainties were evaluated by *Fisher*'s [1953] statistics.

Paleomagnetic results for the Ceará-Mirim sites are displayed in Table 1, along with those from *Bücker et al.* [1986]. Wulff plot of the site mean magnetization directions is on Fig. 6. Subswarms I to IV are clearly identified in this figure. Only one site from subswarm V (alkaline dikes) gave reliable result and plots close to subswarm I data. The other analyzed site of group V gave anomalous direction probably due to weathering. The magnetically defined subswarms correspond to the lineaments shown in Fig. 2. Secondary components were mainly identified in samples from site groups II, III and IV (Fig. 7), and will be discussed later.

Results for the Maranhão basin and Ceará State are listed in Table 2, and site mean remanence directions are on Fig. 6. Data from the Mosquito and Sardinha Formations [*Schult and Guerreiro*, 1979] are also included in Fig. 6. Site means are discriminated by their content of TiO$_2$ (see section 5) in the low- and high-Ti groups. The low-Ti rocks, including the Lavras basin samples, plot close to the existent Mosquito data, whereas the high-Ti group is undistinguishable from the Sardinha data. Therefore, the Jurassic sites can be distinguished from the Cretaceous sites by their geochemical characteristics as well as by their magnetic polarity, mainly normal and reversed for the Jurassic and Cretaceous, respectively (Fig. 6).

ROCKMAGNETISM

Rockmagnetic investigations by *Bücker et al.* [1986] in the Ceará-Mirim dikes revealed the presence of Ti-magnetites with low TiO$_2$ content and ilmenite exsolution lamellae as the result of high-temperature oxidation during slow cooling in the inner parts of the dikes. The origin of the magnetization was therefore partly attributed to a thermochemical rather than a thermal process owing to the low (500 to 600 C) temperatures calculated [*Buddington and Lindsley*, 1964] from Ti-magnetite and ilmenite compositions. Evidence of low-temperature oxidation, and hydrothermal alteration with the formation of maghemite was also found in the majority of samples.

The Fe-Ti oxides of samples belonging to the same dikes investigated here were analyzed by *Bellieni et al.* [1992]. They observed that Ti-magnetite is quite abundant (ca. 5-10 vol%) while ilmenite is comparatively scarce and is given by thin elongated laths. Ti-magnetite may be present in large crystals up to 4mm across. The compositions of homogeneous groundmass magnetite-ilmenite pairs yielded temperatures [*Buddington and Lindsley*, 1964] significantly higher than those reported previously: 1215 °C for a fine-grained sample from subswarm III, 740 °C for medium-grained samples from subswarm II and 721 °C for subswarm IV. These temperatures suggest a thermal origin for the remanence. Thermomagnetic analysis under fields of at most 400 mT (3184×10^{-2} Am^{-1}) gave J$_S$/T curves of the same types as those reported by *Bücker et al.* [1986]. They are all indicative of Ti-magnetites highly oxidized with low TiO$_2$ content. About 60% of the investigated samples gave maximum-final to initial magnetization ratios (J$_f$/J$_i$) of about 0.70. Only 20% showed significant contents of maghemite.

Most of the Ceará-Mirim samples showed Curie temperatures in the range 536-559 C, except for samples from subswarm V with Curie temperatures as high as 596 C. The oxidation state (Fe^{3+}/Fe^{2+}) of the Ti-magnetites [*Bellieni et al.*, 1992] in the andesi-basalt rocks has an average of 0.48 for the early crystallized Ti-magnetites and 0.50 for the late crystallized ones. The lati-andesite group has lower ratios of 0.27 and 0.23 for the early and late crystallized minerals, respectively.

The low- and high-Ti samples from the Maranhão basin, and from other localities investigated in this paper, have Ti-magnetites as the main remanence carrier. Unblocking temperatures are in the range of 536-572 C, clustering

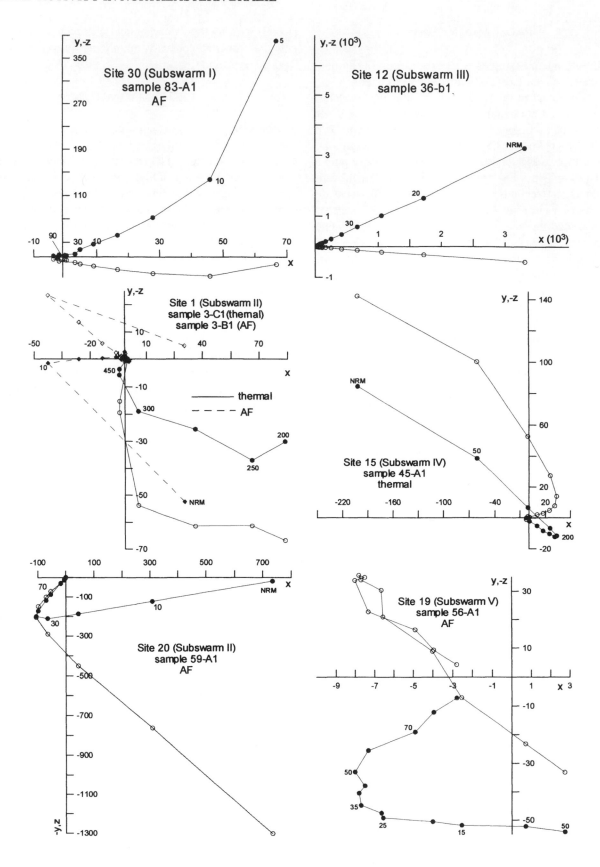

around 560 C, indicating that the Ti-magnetites are highly oxidized at high temperatures. The Lavras basin samples yielded the lowest unblocking temperatures. Coercivity varies from 15 to 23 mT.

GEOCHEMISTRY

Major, minor and trace elements of 23 samples from the Maranhão basin and dikes from the Ceará State (Fig. 1; Table 3) were determined by using X-ray fluorescence, at the University of Trieste (Italy). Procedures are those described by *Bellieni et al.* [1983], and analytical precision is better than 3% for major and minor elements, whereas it is generally between 5 and 10% for trace elements. Selected samples from the Maranhão basin already investigated by *Bellieni et al.* [1990] were reanalyzed [*De Min et al.*, this volume] to provide a larger set of trace elements than presented before (Table 3). According to *De La Roche et al.* [1980], the investigated samples correspond to tholeiitic basalts, andesi-basalts and lati-andesites. Only two samples plot in the field of dacites (Fig. 8).

The basic rocks ($SiO_2 < 54$ wt%) have fine- to medium-grain size textures and mineral assemblages consisting of augite (Wo_{44-32}), plagioclase (An_{69-26}), pigeonite (Wo_{9-12}), Ti-magnetite (av. ulvöspinel 49%), apatite and occasionally ilmenite. Olivine is scarce, occurring as small completely altered crystals. The dacites present medium-grained size textures and are made up of plagioclase (An_{41-57}), augite (Wo_{31-41}), K-feldspar, Ti-magnetite, abundant quartz (more than 20 vol%) and apatite as accessory.

The investigated basic rocks are somewhat evolved, presenting low atomic $Mg/(Mg+Fe^{2+})$ ratios (mg number < 0.61). For similar MgO, the basic rocks may be divided into two groups, with very different geochemical characteristics (Fig. 9). The low-Ti group is characterized by low concentrations of TiO_2 (< 2 wt%), FeO_t, K_2O, P_2O_5 and ITE (e.g. light REE, Y, Zr, Nb and Sr), whereas the high-Ti (TiO_2 > 2%) group has higher abundance of such elements. Note that the low-Ti rocks (tholeiitic basalts and andesi-basalts) are lesser evolved (mg number: 0.47-0.61) than the high-Ti ones (andesi-basalts and lati-andesites) (mg number: 0.35-0.48). The high-Ti tholeiites have lower $SiO_2/(Na_2O+K_2O)$ values, straddling the basaltic trachy-andesite field of the TAS diagram [*Le Bas et al.*, 1986; inset of Fig. 8].

The geochemical characteristics (Figs. 8 and 9), spatial distribution (Fig. 1) and paleomagnetic results (Fig. 6)

make apparent that the analyzed high-Ti magmatism is very similar to that of the Sardinha Formation (Early Cretaceous). The new data strengthen the compositional similarity between the Sardinha intrusives and the high-Ti tholeiitic dikes from the Ceará-Mirim (Fig. 9), as previously shown by *Bellieni et al.* [1992]. Analogously, the low-Ti tholeiites presented in this paper (Figs. 8 and 9) show geochemical and paleomagnetic signatures very similar to the Early Jurassic basalt flows (Mosquito Formation) of the Maranhão basin [*Bellieni et al.*, 1990; *De Min et al.*, this volume]. Besides that, these low-Ti basalts are also very similar in composition (e.g. Ti, Zr, La, Ce, Y; Fig. 9) to the Early Jurassic low-Ti tholeiites from the Lavras basin [*De Min et al*, this volume].

The two analyzed dacites have important geochemical differences (Tab. 3; Fig. 8). The dacite from site 31 is significantly enriched in SiO_2, Na_2O, K_2O and incompatible trace elements relative to the dacite of site 17. These felsic magmas are believed derived from low-Ti tholeiitic melts through complex petrogenetic processes [*Novello*, 1995]. The paleomagnetic data indicate that dacites are related to the Jurassic tholeiitic magmatism, supporting the hypothesis of the involvement of low-Ti parental magmas in their genesis.

DISCUSSION AND CONCLUSIONS

Virtual geomagnetic poles (VGP) calculated for each site mean are displayed in Tables 1 and 2, and plotted in Fig. 10. VGPs from subswarms I and III are very coherent with the results from the Sardinha Formation, and the other high-Ti intrusive rocks. This comparison is also supported by the chemical characteristics of the corresponding rocks (Figs. 8 and 9), as discussed in the previous section. Although of different geochemical composition (alkaline rocks), subswarm V may be also incuded in the Lower Cretaceous paleomagnetic group. However, the high- and low-Ti subswarm IV seems to be an independent data set, comparable only with the low-Ti Jurassic data from the Lavras basin flow, and the Mosquito Formation (including also the other low-Ti dikes from the Maranhão basin). Therefore, it is suggested that subswarm IV might be of Early Jurassic age. VGPs from the low- and high-Ti dikes from the Ceará State are also in accordance with the Mosquito and Sardinha data, respectively. Therefore, two well characterized groups can be considered: the Early Cretaceous (Sardinha, Ceará-Mirim I, III and IV, and high-Ti

Figure 4. (Opposite.) Orthogonal vector plots for the Ceará-Mirim samples. Full symbols, horizontal plane; open symbols, vertical plane. AF demagnetization in mT, and thermal demagnetization in degrees Celsius.

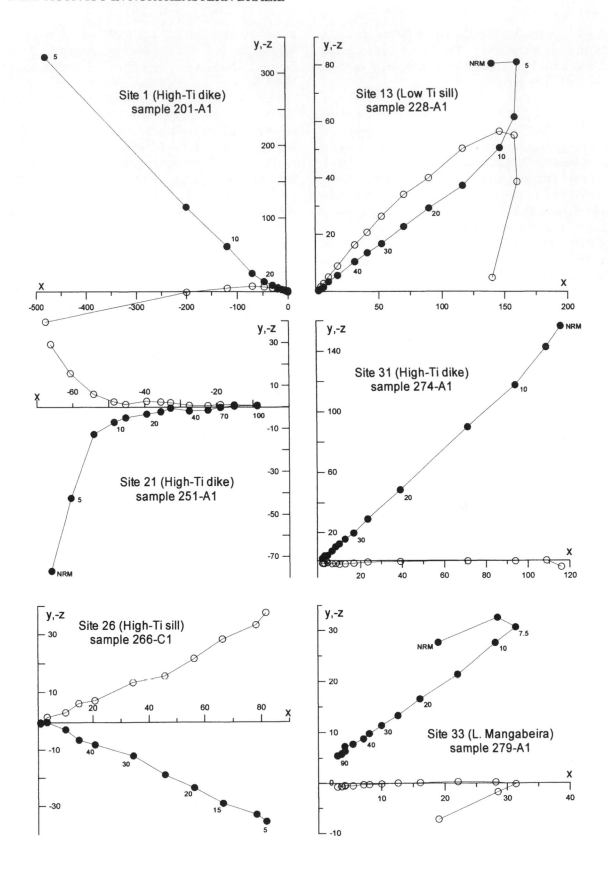

Table 1. Paleomagnetic results for the Ceará-Mirim dikes

	Site			Mean Magnetization Direction				Virtual Geomagnetic Pole			
Code	Long. (°W)	Lat. (°S)	N	Dec. (°)	Inc. (°)	α_{95} (°)	k	Long. (°)	Lat. (°)	dp (°)	dm (°)
Subswarm I											
28	38°57'	5°39'	6	175.2	-9.9	2.7	607	116.7	-78.3	5.4	2.7
29	38°47'	5°37'	6	169.3	-3.9	9.6	49	86.2	-76.9	19.2	9.6
30	38°20'	5°37'	3	216.6	2.8	6.1	408	226.7	-53.2	12.2	6.1
44[†]			9	177.2	-5.3	2.6	387	125.3	-81.1	5.2	2.6
49[†]			7	178.2	-2.3	6.1	97	129.2	-82.8	12.2	6.1
Subswarm II											
1	36°46'	5°38'	4	198.4	28.4	10.5	77	261.1	-69.9	21.0	11.9
2	36°36'	5°38'	9	185.2	45.8	5.7	83	311.1	-67.8	11.4	8.2
6				random							
7	36°12'	5°37'		random							
8	36°21'	5°37'	9	184.2	54.3	5.2	98	316.6	-60.5	10.4	8.9
9	38°03'	5°46'	9	179.6	60.5	5.1	104	322.5	-54.3	10.2	10.4
10	38°03'	5°48'	5	225.8	43.8	6.1	158	261.6	-41.9	12.2	8.4
16	35°46'	5°54'	9	183.2	22.8	5.4	90	296.5	-83.2	10.8	5.9
20	38°05'	5°47'	9	196.5	51.3	6.7	61	293.4	-59.6	13.4	10.7
21	37°57'	5°46'	3	207.4	48.5	3.1	1569	277.8	-54.9	6.2	4.7
22	38°27'	5°50'	9	185.6	53.4	4.9	108	311.3	-61.4	9.8	8.2
23	38°47'	5°57'	8	194.6	42.7	4.9	126	286.5	-66.3	9.8	6.6
25	38°57'	6°15'	6	159.0	48.1	6.3	112	359.8	59.7	12.6	9.4
26	38°52'	6°10'	3	184.5	61.5	8.7	202	315.5	-53.3	17.4	18.2
27	38°50'	6°03'	6	70.3	-57.2	10.7	40	89.2	19.1	21.4	19.7
31	37°50'	5°48'	9	187.3	51.1	2.2	527	308.3	-63.1	4.4	3.5
43[†]			9	190.1	51.6	3.9	174	305.2	-61.7	6.1	5.3
47[†]			8	210.1	33.5	4.2	169	259.0	-57.9	7.8	4.8
50[†]			5	183.5	38.6	15.8	25	312.0	-73.5	28.3	18.8
B2[†]			6	191.9	40.1	3.2	436	290.0	-69.4	6.4	4.2
B4[†]			4	196.1	34.5	8.3	125	275.0	-69.5	16.6	10.1
Subswarm III											
3	37°29'	5°50'		random							
4	36°16'	5°42'	6	355.6	0.5	6.9	93	287.1	82.6	13.8	6.9
5	36°10'	5°44'	5	355.1	0.6	10.6	53	284.6	82.2	21.2	10.6
11	37°19'	5°53'	9	8.0	2.8	7.2	52	10.6	79.2	14.4	7.2
12	36°58'	5°46'	6	1.5	-0.6	4.2	247	338.4	84.3	8.4	4.2
24	38°40'	6°13'	6	359.1	-26.5	9.2	53	147.7	82.2	18.4	10.3
32	36°54'	5°46'		random							
51[†]			8	1.9	-11.1	5.8	91	51.0	88.1	11.5	5.9
46[†]			14	4.2	-3.3	4.1	91	10.5	84.1	8.2	4.1
B5[†]			7	0.9	-0.9	5.1	139	153.4	84.7	10.2	5.1
Subswarm IV											
13	37°00'	6°11'		random							
15	36°52'	6°10'	3/6	209.0	0.7	2.4	739	222.9	-60.5	4.8	2.4
17	36°21'	6°11'	3/9	203.2	-15.2	5.6	83	202.9	-63.0	11.2	5.8
18	36°35'	6°10'	3/8	196.6	-18.2	2.6	471	190.4	-67.3	5.2	2.7
Subswarm V											
14	37°10'	6°21'	3/8	331.3	32.4	7.6	55	273.5	52.9	15.2	9.0
19	37°24'	6°16'	2/5	182.7	12.6	2.7	808	235.1	-87.3	5.4	2.7

N, number of specimens for mean calculations; Dec. and Inc., Declination and Inclination; α_{95} and k, *Fisher*'s [1953] statistical parameters; Pol., polarity (N=normal, R=reversed); Long. and Lat., Longitude and Latitude; dp and dm, confidence oval of 95%. [†]Data from *Bücker et al.* [1986].

Figure 5. (Opposite.) Orthogonal vector plots for the Maranhão basin samples and Ceará dikes. Full symbols, horizontal plane; open symbols, vertical plane. AF demagnetization in mT, and thermal demagnetization in degrees Celsius.

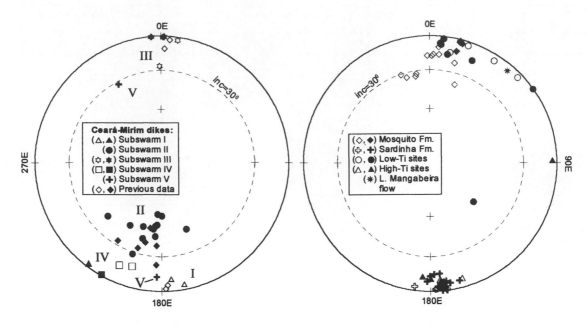

Figure 6. Site mean magnetization directions for a) Ceará-Mirim dikes, including data from *Bücker et al.* [1986], and b) Maranhão basin and Ceará dikes, and the Mosquito and Sardinha data from *Schult and Guerreiro* [1979]. Open and full symbols correspond to negative and positive inclinations, respectively.

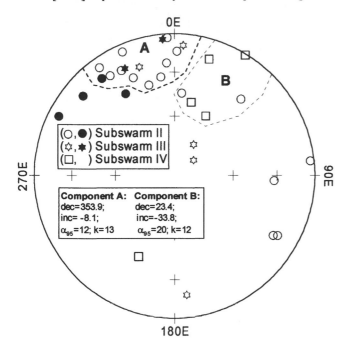

Figure 7. Secondary magnetization components identified in the Ceará-Mirim dikes. Open and full symbols correspond to negative and positive inclinations, respectively.

sites), and the Early Jurassic (Mosquito, Lavras basin, Ceará-Mirim IV, and low-Ti sites) groups. Corresponding paleomagnetic poles were calculated, and are referred as to Northeastern Brazilian Magmatism poles in Table 4.

Results from subswarm II are more striking in the sense that they do not match either the Early Cretaceous or the Early Jurassic data, although a Jurassic age is currently assigned in literature [*Bücker et al.*, 1986; *Horn et al.*, 1988; *Bellieni et al.*, 1992]. Therefore, an independent paleomagnetic pole was calculated based on this subswarm. Note that slight geochemical differences between subswarm II and the other Ceará-Mirim subswarms (Fig. 9) were already pointed out by *Bellieni et al.* [1992]. These differences are also valid in relation to the subswarm IV that fits well the Early Jurassic data.

Paleomagnetic poles calculated for the Early Cretaceous (NBM1), Early Jurassic (NBM2), and the Ceará Mirim subswarm II (CM2) are compared with the apparent polar wander path (APWP) for South America (Fig. 11). This APWP is well defined for the Early Cretaceous based on the robust Serra Geral [*Ernesto et al.*, 1999] and other time-equivalent poles from southeastern Brazil, Paraguay and Argentina (Table 4). The Early Cretaceous NBM1 pole plots in good agreement with the ~130 Ma segment of the polar path. The Jurassic pole calculated in this paper (NBM2) plots between the Late Jurassic poles from Argentina [*Rapalini et al.*, 1993, and references therein; *Vizán*, 1998], and the Early Jurassic (~200 Ma) from the Central Atlantic Magmatic Province - CAMP [*Marzoli et al.*, 1999]. Actually, NBM2 pole fits other Brazilian CAMP poles obtained from the Roraima and Cassiporé tholeiitic dikes [*Marzoli et al.*, 1999; *De Min et al.*, this volume; unpublished data] and the CAMP pole from the

Table 2. Paleomagnetic results from the Low- and High-Ti rocks, and Lavras da Mangabeira flows

	Sampling Sites					Mean Magnetization Direction					Virtual Geomagnetic Pole			
Code	Long. (°W)	Lat. (°S)	Trend	TiO₂	N	Dec. (°)	Inc. (°)	α95 (°)	k	Pol.	Long. (°)	Lat. (°)	dp (°)	dm (°)
High-Ti														
1	39°16'44"	4°21'57"	E-W	3.10	3	164.5	-7.9	6.7	339	R	78.9	-72.4	13.4	6.7
2	39°15'45"	4°21'41"	N80E	3.12	6	179.8	11.2	5.3	163	R	328.6	-88.5	10.5	5.4
21	42°20'06"	5°11'39"	N70E	3.41	12	188.9	- 4.1	4.0	121	R	187.3	-78.3	8.0	4.0
22	42°38'3"4	5°32'01"	sill	3.45	12	176.0	5.7	3.1	195	R	81.0	-85.2	6.2	3.1
23	42°35'26"	5°39'30"	sill	3.41	6	173.4	8.2	8.5	63	R	60.3	-83.2	16.9	8.6
24	42°45'11"	6°05'24"	sill	3.25	9	174.7	9.2	2.6	404	R	62.7	-84.5	5.2	2.6
25	42°43'28"	6°06'38"	sill	2.92	6	174.3	3.6	2.9	532	R	84.2	-82.8	5.8	2.9
26	42°49'39"	6°14'47"	sill	3.18	6	177.6	1.9	2.8	571	R	112.8	-84.2	5.6	2.8
27	42°50'34"	6°15'02"	sill	3.22	6	187.4	3.5	3.3	405	R	195.8	-81.1	6.6	3.3
Low-Ti														
3	40°08'53"	4°13'59"	N80E	1.19	3	9.3	-1.4	10.3	146	N	29.7	80.1	20.6	10.3
8	41°57'44"	4°03'30"	?	1.29	12	12.2	-16.8	2.4	338	N	69.4	77.1	4.7	2.5
9	42°08'02"	3°58'03"	N40W	1.63	3	5.2	6.3	5.9	431	N	353.9	81.2	11.8	5.9
10	42°10'56"	3°56'08"	flow	1.27	6	32.2	- 5.8	7.0	92	N	47.2	57.9	14.0	7.0
11	42°17'33"	3°45'52"	flow	1.08	6	9.5	15.6	7.0	94	N	356.6	74.9	13.8	7.2
12	42°37'44"	3°40'23"	sill?	1.58		ramdom								
13	42°35'12"	3°41'13"	sill?	1.85	9	18.3	- 3.7	6.9	57	N	42.3	71.6	13.8	6.9
14	42°32'24"	3°41'56"	sill?	1.19	3	10.6	-13.6	4.6	708	N	65.3	78.9	9.1	4.7
15*	42°26'56"	3°38'09"	sill?	0.95	2	22.6	14.8	10.7	544	N	20.9	64.8	21.2	11.0
16	42°13'58"	3°54'13"	sill?	1.35	6	6.7	2.2	6.4	109	N	9.1	81.3	12.8	6.4
17	41°28'32"	4°26'01"	sill	1.85	9	5.7	7.6	4.2	152	N	353.2	79.9	8.4	4.2
18*	41°28'29"	4°24'13"	sill?	1.44	6	132.4	56.7	5.8	133	R	4.3	-35.2	11.2	10.3
19	41°51'29"	4°23'39"	sill?	0.96	3	184.0	12.8	2.8	1885	R	255.9	-85.5	5.6	2.9
20	42°36'56"	4°52'17"	N30E	1.59	6	45.7	- 4.4	8.5	62	N	45.8	44.3	17.0	8.5
29	40°00'40"	6°48'21"	N40E	1.99	3	183.5	26.6	11.5	116	R	294.9	81.9	23.0	12.8
30	39°56'29"	6°40'18"	N40E	2.21	5	ramdom								
31	39°50'15"	6°34'44"	N55E	1.51	3	54.4	0.3	5.7	470	N	45.3	35.3	11.4	5.7
Lavras da Mangabeira														
32	38°58'38"	6°48'04"	flow	1.06	6	ramdom								
33	38°57'25"	6°47'30"	flow	1.09	6	39.9	6.1	9.8	47	N	41.8	49.1	19.4	9.8

N, number of specimens for mean calculations; Dec. and Inc., declination and inclination; α95 and k, *Fisher's* [1953] statistical parameters; Pol., polarity (N=normal, R=reversed); Long. and Lat, Longitude and Latitude; dp and dm, confidence oval of 95%; * Data not included in the means.

French Guyana dikes [*Nomade et al.*, 2000]. These poles tend to be slightly younger (possibly ~190 Ma) than the other CAMP group (Anari-Tapirapuã flows, and Amazonian sills; Table 4). This age interval (200-190) for the CAMP activity had already been predicted by *De Min et al.* [this volume].

The new CM2 pole obtained from the Ceará-Mirim subswarm II does not match the CAMP groups but is close to some of the available Triassic poles for South America, mainly those from the northern part of the continent (Table 4). Poles from the southern part (poles PV, IS, TE from

Argentina; Fig. 11) form a distinct group closer to the Jurassic poles. This could be an indication of imprecise age definition of the rocks or even the presence of secondary magnetizations. However, if these poles do indicate a very slow movement of the South American plate during the Triassic-Jurassic interval, then the CM2 pole could be even older. The existence of K/Ar Triassic ages like 214±6 Ma (dike 2), 228±3 Ma (dike 16) obtained by *Bellieni et al.* [1992] for subswarm II, and the ~213 Ma age reported for the Jos plateau [*Maluski et al.*, 1995] in the contiguous portion of Africa (Fig. 3), point to a Triassic age for CM2

Table 3. Bulk rock compositions of low- and high-Ti rocks from the Maranhão basin and Ceará State.

Site	1	2	3	8	9	10	11	13	14	15	16	17	18	19	20	21	22
Sample	8200	8201	8202	8207	8208	8209	8210	8212	8213	8214	8215	8216	8217	8218	8219	8220	8221
SiO_2 (wt%)	53.87	53.73	52.82	52.82	54.13	52.60	52.75	54.03	52.42	52.78	53.24	59.91	53.44	52.83	51.89	53.70	53.79
TiO_2	3.10	3.12	1.19	1.07	1.56	1.04	1.08	1.85	1.19	0.95	1.35	1.85	1.44	0.96	1.59	3.41	3.45
Al_2O_3	13.67	13.26	15.49	14.80	13.71	14.63	15.62	12.52	15.11	15.22	13.91	11.91	14.72	15.39	14.60	14.01	13.50
FeO_t	12.81	12.68	10.26	9.79	12.36	9.73	9.76	13.21	10.29	9.22	11.40	13.44	11.47	9.87	11.95	12.26	12.60
MnO	0.19	0.19	0.17	0.17	0.19	0.16	0.16	0.20	0.17	0.16	0.19	0.21	0.19	0.17	0.21	0.19	0.19
MgO	3.36	3.73	5.93	7.53	5.42	7.68	6.67	5.68	6.49	7.26	6.58	1.51	5.56	7.07	5.60	3.29	3.46
CaO	6.65	7.47	10.68	11.00	9.17	11.46	10.82	8.81	11.06	11.53	10.28	5.43	9.32	10.74	10.92	7.47	6.93
Na_2O	3.07	2.76	2.50	2.14	2.43	2.06	2.40	2.52	2.36	2.20	2.34	3.14	2.73	2.24	2.39	2.70	2.98
K_2O	2.44	2.20	0.80	0.57	0.84	0.54	0.62	0.99	0.78	0.57	0.55	2.04	0.97	0.62	0.72	2.20	2.22
P_2O_5	0.84	0.86	0.16	0.11	0.19	0.10	0.11	0.20	0.13	0.10	0.16	0.56	0.17	0.11	0.13	0.77	0.88
Fe_2O_3	3.21	2.29	1.67	1.89	3.09	1.78	1.14	3.24	2.10	1.97	1.98	4.30	2.91	2.14	3.09	3.16	5.38
FeO	9.92	10.62	8.76	8.09	9.58	8.13	8.73	10.29	8.40	7.45	9.62	9.57	8.85	7.94	9.17	9.42	7.76
L.O.I.	2.02	1.99	0.80	1.31	1.60	1.03	1.08	1.71	1.17	1.02	1.47	1.89	1.44	1.29	2.58	1.87	1.76
mg number	0.35	0.37	0.54	0.61	0.47	0.61	0.58	0.47	0.56	0.61	0.54	0.19	0.50	0.59	0.49	0.35	0.36
Rock Type	LA	AB	TB	TB	AB	TB	TB	AB	TB	TB	TB	DAC	AB	TB	TB	AB	LA
Cr (ppm)	38	37	219	217	31	234	197	20	184	169	67	1	67	104	100	46	10
Ni	37	25	60	104	43	113	104	47	107	100	66	7	64	85	82	34	26
Ba	623	601	196	155	186	115	124	234	316	106	163	440	194	129	212	963	674
Rb	72	53	22	14	24	15	15	29	19	14	14	57	26	18	19	51	49
Sr	606	564	241	185	197	186	185	194	203	215	199	201	238	214	221	692	615
La	42	47	14	8	12	7	7	13	9	8	11	37	12	7	11	39	49
Ce	101	96	30	23	34	22	24	40	20	24	28	85	40	19	31	98	113
Nd	55	52	16	13	20	11	13	22	14	12	17	46	16	9	17	52	59
Zr	330	332	119	93	145	91	94	171	106	85	121	363	129	88	135	308	359
Y	50	51	28	24	34	24	25	38	28	34	30	76	32	26	33	46	51
Nb	27	28	10	6	9	6	6	11	7	6	8	22	9	7	10	27	30

Table 3. Continued

Site	23	24	25	26	27	31	SF	SF	SF	SF	SF	SF	SF	SF	SF	SF	SF
Sample	8222	8223	8224	8225	8226	8230	5001	5002	5005	5006	5046	5047	5048	5049	5050	5051	5052
SiO_2 (wt%)	52.61	54.32	52.05	52.99	52.84	61.74	52.92	53.03	52.20	52.33	53.97	52.03	51.36	55.92	53.59	52.36	51.07
TiO_2	3.41	3.25	2.92	3.18	3.22	1.51	3.05	3.40	3.34	3.24	3.33	4.01	4.09	2.41	3.31	3.52	3.84
Al_2O_3	13.10	13.10	13.83	13.01	12.95	14.81	13.77	14.62	14.47	14.72	13.87	14.63	14.30	13.18	14.15	13.07	14.80
FeO_t	12.55	12.38	11.90	12.46	12.54	7.56	13.01	11.73	13.48	12.24	12.13	11.93	12.27	12.25	12.01	14.13	12.07
MnO	0.18	0.19	0.17	0.18	0.19	0.17	0.20	0.18	0.20	0.17	0.18	0.17	0.17	0.19	0.18	0.20	0.17
MgO	4.59	3.34	5.37	5.08	4.94	2.33	3.85	4.55	3.63	4.38	3.91	4.24	4.43	3.49	3.72	3.89	4.45
CaO	8.17	6.68	8.90	7.91	8.17	4.13	6.79	5.82	7.40	7.48	7.06	7.72	8.32	5.17	6.95	7.38	8.59
Na_2O	2.76	3.32	2.58	2.53	2.53	3.68	2.72	3.52	2.87	2.59	2.90	2.77	2.67	3.25	3.13	2.65	2.82
K_2O	1.84	2.39	1.70	1.93	1.88	3.38	2.27	2.34	1.81	1.97	1.94	1.74	1.69	2.84	2.06	2.12	1.60
P_2O_5	0.79	1.02	0.58	0.73	0.74	0.69	1.42	0.81	0.60	0.88	0.71	0.76	0.70	1.30	0.90	0.68	0.59
Fe_2O_3	5.21	3.72	2.66	3.56	4.09	4.68	4.50	3.33	5.20	3.31	4.38	3.16	3.42	3.74	3.26	5.39	3.14
FeO	7.86	9.03	9.51	9.26	8.86	3.35	8.69	8.53	8.49	9.01	7.97	8.87	8.92	8.60	8.87	8.90	9.07
L.O.I.	1.91	1.87	1.50	1.84	1.90	2.87	1.98	1.66	2.41	1.94	1.72	1.71	2.13	2.22	1.61	2.59	1.36
mg number	0.43	0.35	0.48	0.45	0.44	0.38	0.37	0.44	0.35	0.42	0.39	0.42	0.42	0.37	0.39	0.36	0.43
Rock Type	AB	LA	AB	AB	AB	DAC	LA	LT	AB	AB	AB	AB	AB	LA	AB	AB	AB
Cr (ppm)	85	14	170	120	119	1	16	19	34	90	37	19	17	8	12	19	73
Ni	48	10	71	52	50	79	8	26	37	42	31	41	42	6	22	13	65
Ba	570	721	464	556	533	2280	727	632	536	572	607	507	490	823	628	581	451
Rb	39	56	32	39	37	98	43	45	34	33	33	34	28	62	46	46	25
Sr	591	609	673	537	555	705	533	462	545	553	531	570	582	482	574	598	622
La	41	49	29	39	41	62	56	51	52	50	55	39	47	71	55	53	36
Ce	97	121	69	97	94	147	107	93	91	99	108	90	77	140	112	90	69
Nd	52	64	40	47	50	83	74	63	60	63	67	53	50	86	63	57	41
Zr	304	404	236	284	282	522	344	336	323	279	300	227	211	451	314	319	187
Y	47	55	38	43	45	62	60	51	51	46	49	41	41	66	49	51	36
Nb	26	34	19	24	24	34	27	26	28	23	25	20	20	33	27	25	17

Oxides are normalized to 100% on a volatile free basis, with total iron expressed as FeO_t. Original Fe_2O_3, FeO and loss on ignition (L.O.I.) are also reported; mg number, $Mg/(Mg+Fe^{2+})$, is calculated assuming $Fe_2O_3/FeO = 0.15$. SF sites are the reanalyzed samples from the Sardinha Formation [Bellieni et al., 1990]; sites 1 to 31 are indicated in Fig. 2. Rock type according to De La Roche et al. [1980]: TB = tholeiitic basalt, AB = andesi-basalt, LA = lati-andesite, LT = latite, DAC = dacite.

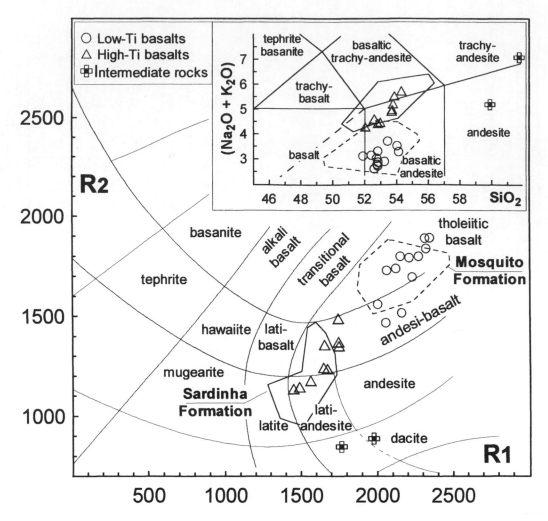

Figure 8. R1-R2 classification diagram [*De La Roche et al.*, 1980] of the low- and high-Ti rocks from the Maranhão basin and Ceará State. The fields of the Mosquito [*Bellieni et al.*, 1990; *De Min et al.*, 2000] and the Sardinha formations (*Bellieni et al.*, 1990; this paper] are also shown for comparison. The TAS classification [*Le Bas et al.*, 1986] is shown in the inset.

pole. In Fig. 11 the African Triassic pole [*Westphal et al.*, 1986] plots close to CM2, therefore strengthening this conclusion.

Based on the above evidence it comes out that the magmatic activity in northeastern Brazil started earlier than previously thought, probably during Late Triassic, extending through Early Jurassic, Early and Late Cretaceous, and ending in the Tertiary, as testified by various flows and plugs located towards the Northern border of the Rio Grande do Norte and Ceará States. This recurrent magmatic activity in the area could be responsible for a resetting of the Ar system in the older rocks without causing a reset of the magnetization in dikes II and IV. This hypothesis is supported by the indication given by the susceptibility and coercivity variation (Fig. 12) within the

Ceará-Mirim dike swarm. The fact that the susceptibility in subswarm II tends to be lower than that of the Jurassic and Early Cretaceous dikes, while coercivity tends to be higher (smaller magnetic grains), is consistent with a colder crust by the time the Triassic dikes were emplaced, which were re-heated by the younger magmatic pulses. Secondary magnetizations identified in some sites of subswarms II, III, and IV (Fig. 7) give another indication for a possible alteration of the radiometric ages. Most of the sites where the secondary component A was found belong to subswarm II; this component fits well the APWP for South America at ~130 Ma (pole A in Fig. 11), indicating that this magnetic imprint may be of thermal origin, which was probably accompanied by argon losses. Secondary component B was identified mostly in samples from subswarm IV

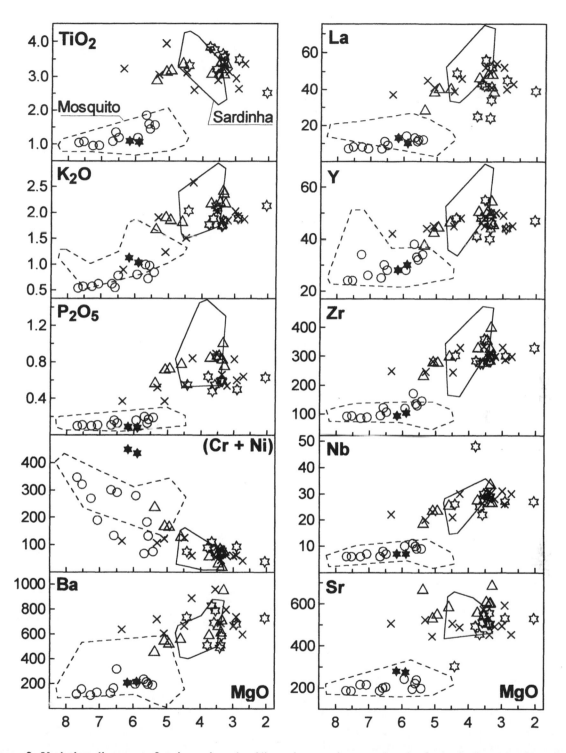

Figure 9. Variation diagrams of major, minor (wt.%), and trace elements (ppm) of the basic rocks from the Maranhão basin and Ceará State. The high-Ti rocks of the Ceará-Mirim dike swarm and the low-Ti tholeiites from the Lavras da Mangabeira, as well as the fields of the Mosquito and Sardinha formations are shown for comparison. Symbols: open circles and triangles = low- and high-Ti tholeiites from the Maranhão basin and Ceará State, respectively; solid stars = Lavras da Mangabeira flows; oblique crosses and open stars = Ceará-Mirim subswarms I, III and IV, and subswarm II, respectively.

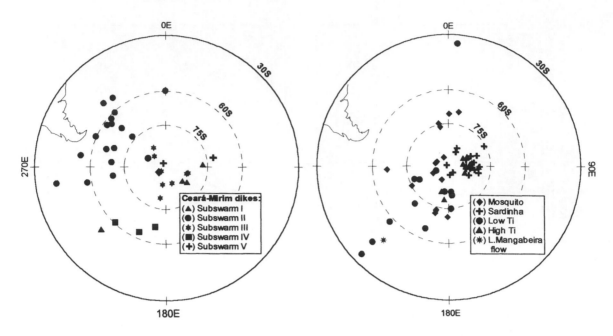

Figure 10. Virtual geomagnetic poles plotted in the southern hemisphere for a) the Ceará-Mirim dikes, and b) the Maranhão basin igneous rocks, Ceará dikes, and the Lavras da Mangabeira flow plotted in the southern hemisphere. Data from *Bücker et al.* [1986], and *Schult and Guerreiro* [1979] are also included.

(Jurassic), as well as in some from the subswarm II, and seems to be of Early Jurassic age (pole B in Fig. 11), in spite of the large uncertainty as indicated by the statistical parameters (Fig. 7).

One of the conclusions that can be addressed from the data presented in this paper is that the subdivision of age provinces in the Maranhão basin (Early Jurassic Mosquito Formation on the west, and the Early Cretaceous Sardinha Formation on the east) must be reconsidered. The northeastern part of the Maranhão basin in Piauí State is dominated by low-Ti tholeiites which present remanent magnetization and geochemical signature of Early Jurassic age, allowing to consider these rocks as correlating with the Mosquito Formation. Based on this new spatial distribution, it is possible to conclude that the Early Jurassic magmatism was more widespread and more voluminous than was supposed so far (confined to the Porto Franco-Grajaú area; Fig. 1), occurring also in the northeastern border of the basin. Eastwards, the Jurassic Ceará-Mirim subswarm IV and the low-Ti Ceará dikes testify that the CAMP-related (~200-190 Ma) activity in the extreme northeastern South America was more expressive than the limited area occupied by the remnants of the Lavras basin flows could indicate. Furthermore, it is possible to conclude that the CAMP activity in NE Brazil was preceded by Late Triassic magmatism, possibly at ca. 215-220 Ma based on the similar ages found in the magmatic activity in northwestern Africa [*Maluski et al.*, 1995]. The small sedimentary

basins in northeastern Brazil, mostly filled with Mesozoic sediments, may show conglomeratic sandstones of Paleozoic age (e.g. the Araripe and Iguatú basins; *Priem et al.*, 1978], indicating that extension tectonics occurred before the emplacement of the Early Jurassic CAMP magmatism. *Hames et al.* [2000] proposed a geodynamical model for the formation of the Central Atlantic Magmatic Province (Appalachian-Mauritanian belt) where the CAMP incubating melt (ca. 230 Ma) was responsible for the subsidence of onshore basins. The subsequent (ca. 230-200 Ma) extensional tectonics and crustal thinning of the Appalachian orogen would have later (ca. 200 Ma) promoted the emplacement of the CAMP magmatic activity from Roraima plateau northwards. Similarly, in NE-Brazil extensional tectonics, and crustal thinning [*Castro et al.*, 1998] allowed the local emplacement of small volumes of Late Triassic magmatism (dikes of the Ceará-Mirim subswarm II) about 20-30 My before the huge and widespread Early Jurassic CAMP magmatic activity.

CONCLUDING REMARKS

The Mesozoic magmatism in NE Brazil is represented by two contrasting rock types, i.e. the low-Ti and high-Ti tholeiites.

In general the low-Ti tholeiites from the Northeastern Maranhão basin, Ceará and the Lavras basin show strong geochemical affinity with the Mosquito tholeiites. Petro-

Table 4. Mesozoic paleomagnetic poles for South America

Formation	Age (Ma)	N	Paleomagnetic Pole Long. (°E)	Paleomagnetic Pole Lat. (°S)	α_α (°)	References
Mean Late Cretaceous	70-90*	5	349.7	80.3	4.1	Calculated from *Montes-Lauar et al.* [1995] and *Butler et al.* [1991]
Cabo de Santo Agostinho, Brazil	~100*	9	315.1	87.6	4.5	*Schult and Guerreiro* [1980]
Cañadón Asfalto Basin, Argentina	~116	16	159.0	87.0	3.8	*Geuna et al.* [2000]
Florianópolis Dikes, SE Brazil	~121†	65	3.3	89.1	2.6	*Raposo et al.* [1998]
Central Alkaline Province, Paraguay	127-130*	75	62.3	85.4	3.1	*Ernesto et al.* [1999], and reference therein
Ponta Grossa Dikes, SE Brazil	129-131†	115	58.5	84.5	2.0	*Ernesto et al.* [1999], and reference therein
Serra Geral Formation., Paraná Basin	133±1†	339	90.1	84.3	1.2	*Ernesto et al.* [1999]
Cordoba Province, Argentina	133-115*	55	75.9	86.0	3.3	*Geuna and Vizán* [1998]
Northeastern Brazil Magmatism-1	125-145	44	97.6	85.2	1.8	This paper
Northeastern Brazil Magmatism-2	175-198	33	223.9	78.1	5.2	This paper
Marifil Formation, Argentina	174-164*	8	133.4	77.7	19.0	*Vizán* [1998], and reference therein
Chon-Aike Formation, Argentina	168±2#	54	197.0	85.0	6.0	*Vizán* [1998], and reference therein
Lepá-Osta Arena Formations, Argentina	Je	13	129.4	75.5	6.8	*Vizán* [1998]
Guacamayas Group and Bolivar dikes, Venezuela	195-199*	10	265.8	76.1	11.3	*Rapalini et al.* [1993], and reference therein
French Guyana CAMP Dikes	198-192†	26	235.1	81.2	4.0	*Nomade et al.* [2000]
Guyana dykes	Tr-J	7	222.0	63.3	12.4	*Rapalini et al.* [1993], and reference therein
Caciporé Dikes, North Brazil§	~200-192†	17	208.6	79.8	5.2	*De Min et al.* [2000]; unpublished data
Roraima Dikes, North Brazil§	198.8±2.0†	7	235.1	80.1	6.6	*De Min et al.* [2000]; unpublished data
Penatecaua Dikes, North Brazil§	164±26*	9	249.5	65.6	8.9	*Guerreiro and Schult* [1986]
Anari-Tapirapuá Lavas, Brazil§	~197†	15	250.3	65.5	3.6	*Montes-Lauar et al.* [1994]
Ceará-Mirim Dike II, NE Brazil	245-160*	18	297.5	65.1	6.5	This paper
Suriname Dikes	242±12*	10	320.0	81.9	7.1	*Rapalini et al.* [1993], and reference therein
Puesto Viejo Formation, Argentina	232	6	236	-76	18.0	*Rapalini et al.* [1993], and reference therein
Ischigualasto Volcanics, Argentina	224*	9	239.2	79.2	15.0	*Rapalini et al.* [1993], and reference therein
Cacheuta Group, Argentina	Tr		266.0	74.0	14.0	*Rapalini et al.* [1993], and reference therein
C. Terreros Volcanics, Argentina	Tr		228.0	80.0	10.0	*Rapalini et al.* [1993], and reference therein
Amaná Formation, Argentina	Tr	21	317.0	83.0	8.0	*Rapalini et al.* [1993], and reference therein

N= number of sites; § Brazilian CAMP poles; * K-Ar ages; † $^{40}Ar/^{39}Ar$ ages; # Rb-Sr ages;

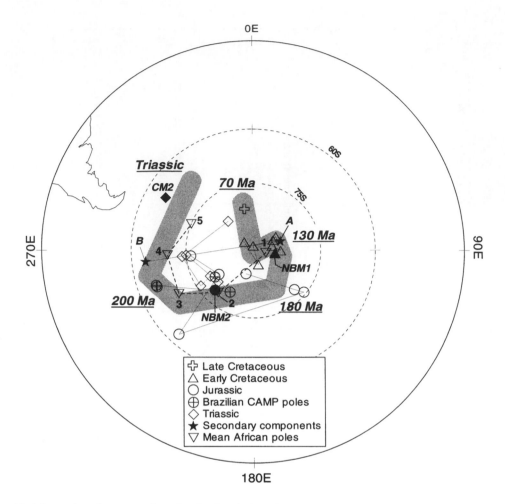

Figure 11. Mesozoic paleomagnetic poles for South America (Table 4). Full symbols represent the poles calculated in this paper. Full stars are the paleomagnetic poles corresponding to the secondary magnetization components A and B. Mean African poles rotated to South America; codes 1 to 5 refer to ages 130 Ma [*Gidskehaug et al.*, 1975], 154-180 Ma, 180-205 Ma, 205-230 Ma [*Kies et al.*, 1995], and 240-245 Ma [*Westphal et al.*, 1986], respectively.

logic similarities also exists between the dikes from the Ceará-Mirim (subswarms I and III), the Ceará high-Ti rocks, and the Sardinha intrusives.

Remanent magnetizations allow distinguishing distinct sets of Ceará-Mirim dikes as proposed by *Sial* [1975]. Remanences also permit to discriminate the low- and high-Ti groups from the Maranhão basin (Mosquito and Sardinha formations, respectively) and other intrusives from the Ceará State.

Paleomagnetic poles calculated for the Early Cretaceous (NBM1; Ceará-Mirim I, III and V, the Sardinha Formation, and other high-Ti intrusives), and the Early Jurassic (NBM2; Ceará-Mirim IV, Lavras basin flows, Mosquito Formation and other low-Ti intrusives) are in accordance with other South American and African paleomagnetic poles of same age.

A Late Triassic age is proposed for the paleomagnetic pole from the Ceará-Mirim subswarm II (CM2), which differs from the other poles of the studied area. Two K-Ar ages older than 200 Ma have already been reported for this intrusives, and the pole position agrees with coeval poles from Africa and South America.

The new paleomagnetic and geochemical data presented in this paper demonstrate that the previous age subdivision for the Mesozoic magmatic rocks in the Maranhão basin (Early Jurassic Mosquito Formation on the west, and the Early Cretaceous Sardinha Formation on the east) must be reconsidered. The northeastern part of the basin (in Piauí State) is dominated by low-Ti rocks of Early Jurassic age corresponding to the Mosquito Formation.

The Early Jurassic (CAMP) magmatism in the Maranhão basin was more widespread and voluminous than was

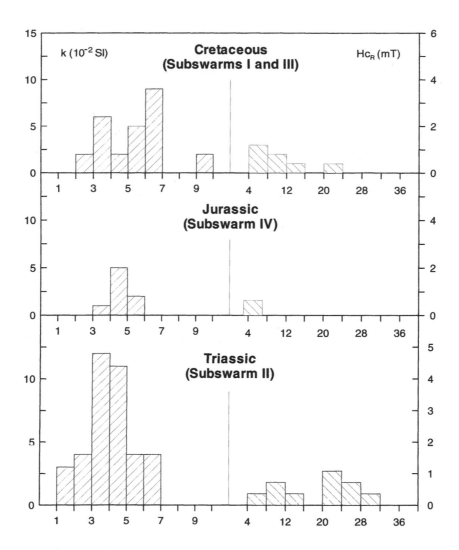

Figure 12. Histograms of susceptibilities in 10^{-2} SI units (on the left), and coercivities in mT (on the right) for the Ceará-Mirim dikes.

supposed so far, occurring also in the northeastern border of the basin. Eastwards, the Ceará-Mirim subswarm IV testifies that the CAMP-related activity in the extreme northeastern South America was more expressive than the limited area presently occupied by the remnants of the Lavras basin flows.

The Ceará-Mirim subswarm II represents a Late Triassic magmatism which preceded the CAMP activity in the Northeastern Brazil.

Acknowledgments. This project has been supported by the Brazilian agencies PADCT/FINEP (Project 0568/92), FAPESP (Project 93/2796-1) and CNPq (40.4085/91-1), and by the Italian agencies MURST and CNR. Thanks are due to J. Marins and R. Siqueira for the help in the various phases of the paleomagnetic work. The authors are grateful to N. Bardelli for his help in preparing figures and diagrams. Logistic facilities during field work were provided by University of Rio Grande do Norte and Petrobrás. Thanks are also due to W. MacDonald and A. Grunow by their thorough reviews of the manuscript.

REFERENCES

Almeida, F.F.M., Carneiro, C.D.R., Machado Jr., D.L., and Dehira, L.K., Magmatismo pós-Paleozóico no Nordeste Oriental do Brasil. *Rev. Brasil. Geoc.*, 18, 451-462, 1988.

Baksi, A.K., and Archibald, D.A., Mesozoic igneous activity in the Maranhão province, Northern Brazil: $^{40}Ar/^{39}Ar$ evidence for separate episodes of basaltic magmatism. *Earth Planet. Sci. Letters*, 151, 139-153, 1997.

Bellieni, G., Macedo, M.H.F., Petrini, R., Piccirillo, E.M., Cavazzini, G., Comin-Chiaramonti, P., Ernesto, M., Macedo, J.W.P., Martins, G., Melfi, A.J., Pacca, I.G., and De Min, A., Evidence of magmatic activity related to Middle Jurassic and

Lower Cretaceous rifting from NE-Brazil (Ceará-Mirim): K/Ar age, palaeomagnetism, petrology and Sr-Nd isotope characteristics, *Chem. Geology*, 97, 9-32, 1992.

Bellieni, G., Piccirillo, E.M., Cavazzini, G., Petrini, R., Comin-Chiaramonti, P., Nardy, A.J.R., Civetta, L., Melfi, A.J., and Zantedeschi, P., Low- and High-TiO₂ Mesozoic tholeiitic magmatism of the Maranhão basin (NE-Brazil): K/Ar age, geochemistry, petrology, isotope characteristics and relationships with Mesozoic low- and high-TiO2 flood basalts of the Paraná basin (SE-Brazil), *Neues Jhr. Miner. Abh.*, 162, 1-33, 1990.

Bellieni, G., Brotzu, P., Comin-Chiaramonti, P., Ernesto, M., Melfi, A. J., Pacca, I.G., Piccirillo, E.M., and Stolfa, D., Petrological and paleomagnetic data on the plateau basalts to rhyolite sequences of the southern Paraná Basin (Brazil). *An. Acad. Bras. Ciênc*, 55, 355-383, 1983.

Bücker, C., Schult, A., Bloch, W., and Guerreiro, S.D.C., Rock-magnetism and palaeomagnetism of an Early Cretaceous/Late Jurassic dike swarm in Rio Grande do Norte, Brazil, *J. Geophys.*, 60, 129-135, 1986.

Buddington, A.F., and Lindsley, D.H., Iron-titanium oxide minerals and synthetic equivalents, *J. Petrol.*, 5, 310-357, 1964.

Butler, R.F., Hervé, F., Munizaga, F., Beck, M.E.Jr., Burmester, R.F., and Oviedo, E.S., Paleomagnetism of the Patagonian plateau basalts, southern Chile and Argentina. *J. Geophys. Res.*, 96(B4), 6023-6034, 1991.

Castro, D.L., Medeiros, W.E., Jardim de Sá, E.F., and Moreira, J.A.M., Gravity map of part of Northeast Brazil and adjacent continental margin and its interpretation based on the hypothesis of isostasy. Brazil. *J. Geophys.* 16, 115-129, 1998.

De La Roche, H., Leterrier, P., Grandclaude, P., and Marchal, M., A classification of volcanic and plutonic rocks using R1-R2 diagram and major-element analyses. Its relationships with current nomenclature, *Chem. Geol.*, 29, 183-210, 1980.

De Min, A., Piccirillo, E.M., Marzoli, A., Bellieni, G., Renne, P.R., Ernesto, M., and Marques, L.S.. The Central Atlantic Magmatic Province (CAMP) in Brazil: Petrology, Geochemistry, ⁴⁰Ar/³⁹Ar ages, paleomagnetism and geodynamic implications, in *AGU*, this volume.

Ernesto, M., Raposo, M.I.B., Marques, L.S., Diogo, L.A., and De Min, A., Paleomagnetism and geochemistry of the magmatic rocks from Northeastern Paraná Magmatic Province: regional tectonic implications. *J. Geodynamics*, 28, 321-340, 1999.

Fisher, R.A., Dispersion on a sphere. *Proceedings of Royal Society of London*, Ser. A 217, 295-305, 1953.

Fodor, R.V., Sial, A.N., Mukasa, S.B., and McKee, E.H., Petrology, isotope characteristics and K-Ar ages of the Maranhão, northern Brazil, Mesozoic basalt province. *Contrib. Mineral. Petrol.*, 104, 555-567, 1990.

Geuna, S.E., and Vizán, H., New Early Cretaceous palaeomagnetic pole from Córdoba Province (Argentina): revision of previous studies and implications for the South American database. *Geophys. J. Int.*, 135, 1085-1100, 1998.

Gidskehaug, A., Creer, K.M., and Mitchell, J.G., Palaeomagnetism and K-Ar ages of the South-West African basalts and their bearing on the time of the initial rifting of South Atlantic Ocean. *Geophys. J.R. Astron. Soc.*, 42, 1-20, 1975.

Gomes, J.R.C., Gatto, C.M.P.P., de Souza, G.M.C., da Luz, D.S.,

Pires, J.L., and Teixeira, W., Geologia: mapeamento regional, Folhas SB.24/25 - Jaguaribe/Natal, *Projeto Radambrasil, Levantamento de Recursos Naturais*, 23, 29-300, 1981.

Guerreiro, S.D.C., and Schult, A., Palaeomagnetism of Jurassic tholeiitic intrusions in the Amazon Basin. *Münchner Geophysik. Mitteil.*, 1, 37-48, 1986.

Guerreiro, S.D.C., and Schult, A., Paleomagnetismo de um enxame de diques toleíticos de idade meso-cenozóica, localizados no Rio Grande do Norte, *Rev. Brasil. Geof.*, 1, 89-98, 1983.

Hames, W.E., Renne, P.R., and Ruppel, C., New evidence for geologically instantaneous emplacement of earliest Jurassic Central Atlantic magmatic province basalts on the North American margin. *Geology*, 28, 859-862, 2000.

Horn, P., Müller-Sohnius, D., and Schult, A., Potassium-Argon ages on a Mesozoic tholeiitic dike swarm in Rio Grande do Norte, Brazil, *Rev. Brasil. Geociênc.*, 18, 50-53, 1988.

Kies, B., Henry, B., Merabet, N., Derder, M.M., and Daly, L., A new Late Triassic-Liasic paleomagnetic pole from superimposed and juxtaposed magnetizations in the Saharan craton. *Geophys. J. Int.* 120, 433-444, 1995.

Kirshvink, J., The least-squares lines and planes and the analysis of paleomagnetic data. *Geophys. J.R. Astron. Soc.*, 62, 699-718, 1980.

Le Bas, M.J., Le Maitre, R.W., Streckeisen, A., and Zanettin, B., A chemical classification of volcanic rocks based on the total alkali-silica diagram. *J Petrol.*, 27, 745-750, 1986.

Maluski, H., Coulon, C., Popoff, M., and Baudin, P., ⁴⁰Ar/³⁹Ar chronology, petrology and geodynamic setting of Mesozoic to early Cenozoic magmatism from the Benue Trough, Nigeria. *J. Geol. Soc.*, 152, 311-326, 1995.

Martins, G., Caracterização petrológica e geoquímica do enxame de diques máficos Rio Ceará-Mirim. M.Sc. thesis, University of São Paulo, São Paulo, 1991.

Marzoli, A., Renne, P.R., Piccirillo, E.M., Ernesto, M., Bellieni, G., and De Min, A., Extensive 200-million-year-old continental flood basalts of the Central Atlantic Magmatic Province. *Science* 284, 616-618, 1999.

Montes-Lauar, C.R., Pacca, I.G., Melfi, A.J., and Kawashita, K., Late Cretaceous alkaline complexes, southeastern Brazil: Paleomagnetism and geochronology. *Earth Planet. Sci. Lett.*, 134, 425-440, 1995.

Montes-Lauar, C.R., Pacca, I.G., Melfi, A.J., Piccirillo, E.M., Bellieni, G., Petrini, R., and Rizzieri, R., The Anari and Tapirapuã Jurassic formations, western Brazil: paleomagnetism, geochemistry and geochronology. *Earth Planet. Sci. Lett.*, 128, 357-371, 1994.

Nascimento, D.A., Gava, A., Pire, J.L., and Teixeira, W., Folha AS-24, Fortaleza, *Projeto RADAMBRASIL, MME/SG*, 212pp., Rio de Janeiro, Brazil, 1981.

Nomade, S., Théveniaut, S., Chen, Y., Pouclet, A., and Rigollet, C., Paleomagnetic study of French Guyana Early Jurassic dolerites: hypothesis of a multistage magmatic event. *Earth Planet. Sci. Lett.*, 184, 155-168, 2000.

Novello, A., Studio petrografico-geochimico di dicchi, sill e vulcaniti del Brasile Nord Orientale (Ceará, Piaui, Rio Grande do Norte), Tesi di Laurea, University of Trieste, 1995.

Oliveira, D.C., O papel do enxame de diques Rio Ceará Mirim

na evolução tectônica do Nordeste Oriental (Brasil): implicações na formação do rifte Potiguar, M.S. thesis, Federal University of Ouro Preto, Brazil, 1992.

Popoff, M., Du Gondwana á l'Atlantique sud: les connexions du fossé de la Bénoué avec les bassins du Nord-Est brésilien jusqu'á l'ouverture du golfe de Guinée au Crétacé inférieur, *J. Afr. Earth Sci.*, 7, 409-431, 1988.

Priem, H.N.A., Boelrijk, N.A.I.M., Verschure, R.H., Hebeda, E.H., Verdurmen, E.A.T., and Bon, E.H., K-Ar dating of a basaltic layer in the sedimentary Lavras Basin, Northeastern Brazil. *Rev. Brasil. Geociênc.*, 8,262-269, 1978.

Rapalini, A.E., Abdeldayem, A.L., and Tarling, D.H., Intracontinental movements in Western Gondwanaland: a palaeomagnetic test. *Tectonophysics*, 220, 127-139, 1993.

Raposo, M.I.B., Ernesto, M., and Renne, P.R., Paleomagnetism and $^{40}Ar/^{39}Ar$ dating of the Early Cretaceous Florianópolis dike swarm (Santa Catarina Island), Southern Brazil. *Phys. Earth Planet. Int.*, 108, 275-290, 1998.

Rolff, P.A.A., O pico vulcânico do Cabugi, Rio Grande do Norte, Notas Prelim. Estudos, *DNPM*, 26 pp., Rio de Janeiro, 1965.

Rezende, W.M., Bacia do Maranhão - estudos dos processos de intrusões e extrusões de magmas básicos, 23pp., *Petrobrás/Depex*, Rio de Janeiro, 1964.

Schult A., and Guerreiro, S.D.C., Palaleomagnetism of Upper Cretaceous volcanic rocks from Cabo de Sto. Agostinho, Brazil. *Earth Planet. Sci. Lett.*, 50, 311-315, 1980.

Schult A., and Guerreiro, S.D.C., Palaeomagnetism of Mesozoic igneous rocks from the Maranhão Basin, Brazil, and the time of the opening of the South Atlantic. Earth Planet. Sci. Lett., 42, 427-436, 1979.

Sial, A.N., Long, L.E., Pessoa, D.A.R., and Kawashita, K., Potassium-argon ages and strontium isotope geochemistry of Mesozoic and Tertiary basaltic rocks, Northeastern Brazil, *An. Acad. Brasil. Ciênc.*, 53, 116-122, 1981.

Sial, A.N., The post-Paleozoic volcanism of Northeast Brazil and its tectonic significance, *An. Acad. Brasil. Ciênc.*, 48(supl.), 299-311, 1976.

Sial, A.N., Petrologia e significado tectônico dos diabásios mesozóicos do Rio Grande do Norte e Paraíba, in *Atas VII Simpósio de Geologia*, Fortaleza, 207-221, 1975.

Vizán, H., Paleomagnetism of the Lower Jurassic Lepá and Osta Arena formations, Argentine Patagonia. *J. South Am. Earth Sci.*, 11, 333-350, 1998.

Westphal, M., Bazhenov, M.L., Lauer, J.P., Pechersky, D.M., and Sibuet, J.C., Paleomagnetic implications on the evolution of the Tethys belt from the Atlantic ocean to the Pamirs since the Triassic. *Tectonophysics*, 123, 37-82, 1986.

Zijderveld, J.D.A., AC demagnetization of rocks, analysis of results, in *Methods in Paleomagnetism*, edited by D.W. Collinson, K.M. Creer and S.K. Runcorn, 254-286, Elsevier, Amsterdam, 1967.

A.de Min and E.M. Piccirillo, Dipartimento di Scienze della Terra; Universita di Trieste, Via E. Weiss 8, 34127 Trieste, Italy.

G. Bellieni, Dipartimento di Mineralogia e Petrologia, Universita di Padova; Corso Garibaldi, 37, 35137 Padova, Italy.

G. Martins, Departamento de Geologia, Universidade Federal do Ceará, Campus do Pici, 60455-760 Fortaleza, Brazil.

I.G. Pacca, L.S. Marques and M. Ernesto, Departamento de Geofisica, Universidade de São Paulo, Rua do Matão, 1226, Cidade Universitária; 05508-900 São Paulo, Brazil, marcia@iag.usp.br.

J.W.P. Macedo, DFTE/CCET, Universidade Federal do Rio Grande do Norte, 59072-970 Natal, Brazil

A Reactivated Back-arc Source for CAMP Magma

John H. Puffer

Department of Geophysical Sciences, Rutgers University, Newark, New Jersey

Although several Mesozoic magma types are found within the geographic boundaries of CAMP, magmatic activity peaked during the extrusion of basalt characterized by 1 wt. % TiO_2 and a uniform composition that is found on each of the circum-Atlantic continents. Other CAMP magmas pre-date or post-date this peak CAMP event or are confined to restricted portion of the province. Evidence is presented indicating that most CAMP basalt is the product of decompression induced eutectic melting of a back-arc source that had been dormant since Paleozoic magmatism associated with the assembly of Pangea. The composition of most CAMP basalt is unlike OIB, plume, hot –spot, N-MORB, or E-MORB magma types but resembles typical arc-related basalt and back-arc basin basalt in particular. Similarities to arc and back-arc basalts include a distinct negative Nb anomaly and overlapping lines plotted on silicate earth normalized spider diagrams. The uniform composition of CAMP basalt indicates that fractionation and contamination played only minor roles and that CAMP magma was quickly generated and extruded during the break-up of Pangea. The composition of CAMP basalt also closely resembles Paleozoic volcanics that stratigraphically underlie Eastern North American CAMP flows including the Ordovician Ammonoosuc and Partridge Formations and the Silurian Newbury and Lieghton Formations that were extruded onto arcs and back-arc basins during the assembly of Pangea. The enriched mantle involved in the Paleozoic magmatism remained trapped under the Laurentian – Gondwana suture until early Jurassic extentional tectonism forced renewed melting.

INTRODUCTION

Most published models of continental flood basalt petrogenesis depend heavily on a plume or hot spot to initiate volcanism that is sustained by highly varying amounts of contamination or mixing with subcontinental lithospheric material. For example, each of the continental flood basalts (CFBs) described in the AGU Monograph "Large Igneous Provinces: Continental, Oceanic, and Planetary Flood Volcanism" edited by *Mahoney* and *Coffin* [1997], with one possible exception, are interpreted as the product of deep mantle plume induced melting involving either 1) a low degree of interaction with subcontinental lithosphere such as the North Atlantic Igneous Province [*Saunders et al.* 1997], 2) or a high degree of interaction such as the Parana-Etendeka Province [*Peate*, 1997], the Siberian Traps [*Lassiter* and *DePaolo*, 1997], and the Columbia River Basalt [*Hooper*, 1997]. *Sharma* [1997] is noncommittal about the origin of the Siberian traps but

The Central Atlantic Magmatic Province:
Insights from Fragments of Pangea
Geophysical Monograph 136
Copyright 2003 by the American Geophysical Union
10.1029/136GM08

considers the non-plume or "perisphere" source proposed by *Anderson* [1994] for continental flood basalts. However, *Sharma* [1997] accurately indicates that the existence of the perisphere "...as a viable mantle source is not accepted by many geochemists".

Anderson [1994] argues that a low density and low viscosity region of the mantle (the perisphere) is a more realistic source for CFBs than subcontinental lithosphere or plumes. *Anderson* [1994] proposes that melting of the perisphere triggered by lithospheric pull-apart can account for at least some CFBs. This paper does not argue that CFBs are never genetically related to deep mantle plumes but strongly supports the application of a modified *Anderson* [1994] model to the petrogenesis of CAMP magmas. This paper also supports *Pegram's* [1990] proposal that the source of CAMP magma was enriched by previous subduction events. *Pegram* [1990] based his proposal on isotopic data that indicated a close resemblance of several eastern North American CAMP basalts to arc-related volcanism. This study supports a similar enriched source, a back-arc basin basalt (BABB) source for most CAMP basalt, largely on the basis of trace element data, particularly high field strength elements (HFSEs). This study also attempts to place the CAMP event in the context of a sequence of Paleozoic magmatic events that were active during the assembly of Pangea, and then reactivated during the break-up of Pangea.

GEOLOGIC SETTING

The lithologies and aerial distribution of CAMP rocks have been defined and described by *Marzoli et al.* [1999] and *McHone* [2000]. The tectonic setting was clearly extensional and the magmatism was very short-lived [*Olsen et al.*, 1996; *Hames et al.*, 2000; *Et-Touhami et al.*, 2001] which are important characteristics of A-type (arc-type) continental flood basalts as defined by *Puffer* [2001].

Most CAMP magmatism occurred at the initial stages of Pangean break-up along the plate boundaries that separated to form the Atlantic Ocean. The distribution of most CAMP magmatism, however, also approximates the boundaries of plate sutures that closed during the assembly of Pangea, perhaps even more closely than the break-up boundaries. For example, the Paleozoic sutures on both sides of the Piedmont terrain are closer to the major dikes and sills that correlate with most eastern North American CAMP volcanism than the current continental margin of North American. A large portion of eastern North American CAMP flood basalt extruded onto the Piedmont accreted terranes including the basalts of the Culpepper,

Newark, and Hartford basins. Even the eastern margin of Laurentia more closely approximates the center of most eastern North American CAMP magmatism than the current North American continental margin. The rifts that were the source of CAMP magmatism are commonly described as "failed rifts".

The Central Atlantic Magmatic Province may be just as accurately described as the Circum Iapitus Magmatic Province, although during the closure of the Iapitus, compressional tectonism was dominated by arc magmatism. However, the key to the understanding of the source of CAMP magma may have more to do with what was happening under the Iapitan sutures than what was happening under the initial Mid-Atlantic-Ridge.

EXTRUSIVE AND INTRUSIVE CAMP ROCKS

Although the size and scope of the CAMP event is best measured by including both extrusive rocks and related dike swarms [*McHone*, 2001] the chemical composition of the main phase of igneous activity is best represented by extrusives (Table 1). Extrusive rocks are less likely to be influenced by in-situ fractionation, wall-rock assimilation, hydrothermal alteration, and their stratigraphic position is more easily determined than intrusions.

The composition of some individual CAMP sills, such as the Palisades Sill, New Jersey, ranges from ultramafic to granodioritic [*Walker*, 1940] and includes compositions that are not represented by any extrusives. Wall rock assimilation and hydrothermal alteration effects on a large Palisades related intrusion have been shown by *Puffer and Benimoff* [1997] to be largely confined to within 10 m of the exterior intrusive contact, but similar effects may significantly influence the composition of thin dikes and sills. However, the diverse intrusive and extrusive rocks plotted within the boundaries of the CAMP province [*Marzoli et al.* 1999] generally fit into four chemistries: 1.) Initial CAMP, 2.) Secondary CAMP, 3.) Low Ti, olivine normative, and 4.) High Ti, REE enriched.

Initial CAMP

Only one basalt chemistry is common to the whole CAMP province. The high-Ti quartz normative (HTQ) basalt type as defined by *Weigand and Ragland* [1970] is exposed as thick extrusive flows on each of the circum-Atlantic continents and includes each of the basalt averages listed in Table 1a. This initial extrusive basalt composition together with related dikes represents the peak of CAMP magmatic activity. It is this HTQ magma that is the main focus of this paper.

Table 1. Average analyses of CAMP basalts.

	A. Initial HTQ-type Flows								B. Secondary LTQ-type Flows		
Name	Orange Mt	Talcott	Mt.Zion	North Mt.	H. Atlas	Algarve	Maranhão	CAMP	Holyoke	Preakness	Sander
Location	NJ	CN	VA	N. Scotia	Morocco	Portugal	Brazil	average	CN	NJ	VA
n	11	7	7	6	15	2	16	64	28	31	16
SiO2	52.68	51.79	52.65	53.81	52.80	51.90	52.98	52.41	53.65	52.09	52.69
TiO2	1.12	1.07	1.09	1.08	1.19	1.01	1.10	1.11	0.97	0.95	1.09
Al2O3	14.25	14.25	14.62	13.79	14.38	14.44	14.87	14.50	14.67	14.33	14.22
FeOt	10.15	10.84	9.97	9.73	9.57	10.36	9.77	10.00	11.74	12.66	12.43
MnO	0.18	0.16	0.17	0.16	0.16	0.16	0.17	0.17	0.20	0.21	0.22
MgO	7.81	7.97	7.68	8.09	8.13	7.91	7.26	7.80	5.97	6.55	5.76
CaO	10.80	11.23	10.69	10.38	10.98	11.23	10.43	10.84	9.40	9.49	9.40
Na2O	2.45	2.06	2.68	2.06	2.10	2.31	2.21	2.37	2.61	3.11	3.44
K2O	0.43	0.50	0.31	0.81	0.52	0.54	1.07	0.66	0.73	0.52	0.61
P2O5	0.13	0.13	0.14	0.09	0.17	0.14	0.15	0.14	0.05	0.12	0.14
Rb	15	22	11	24	18		30	20	18	16	
Ba	119	174	145	216	108		331	178	142	123	
Sr	192	186	191	169	173	235	275	207	153	153	
Th	2.00			2.88	2.04		0.00	1.55		2.29	
Zr	99	87	99	96	104	98	111	102	81	76	
Hf	2.40			2.90	2.64		0.00	1.97		2.02	
Nb	6.90			9.00	8.72		11.44	9.08	5.30	3.89	
Ni	99	86	79	64	89	87	76	88	36	52	
Cr	309	322	282	225	221	241	301	282	11	111	
La	10.2	11.1	10.8	13.0	10.4	14.6	13.3	11.2	18.0	8.2	
Ce	22.1	23.9	24.1	26.0	22.5	29.7	27.1	24.0	29.0	17.2	
Nd	11.8			12.8			15.8	13.1		9.0	
Sm	3.40	3.66	3.51	3.30	3.45		3.34	3.42		2.81	
Eu	1.10	1.15	1.13	1.03	1.11		1.11	1.12		0.93	
Tb	0.61	0.67	0.65	0.63	0.65		0.63	0.64		0.64	
Yb	2.10	2.35	2.30	2.13	2.41		1.91	2.22		2.81	
Lu	0.30	0.33	0.34	0.33	0.41		0.27	0.33		0.41	
Y	20			26	30		34	28	27	24	

Note all Fe is listed as FeO_t, major elements are in wt.% normalized to 100% anhydrous, trace elements are in ppm. Listed averages for Orange Mt. Basalt from *Tollo and Gottfried* [1992]; Talcott Basalt from *Puffer* [1992]; Mt Zion Church Basalt from *Puffer* [1992]; North Mt. Basalt from *Papezik et al.* [1988]; High Atlas Basalt from *Bertrand et al.* [1982]; Algarve basalt is new data; western Maranhão Basalt from *Fodor et al.* [1990]; Holyoke Basalt from *Philpotts et al.* [1996], Preakness Basalt from *Tollo and Gottfried* [1992]; and Sander Basalt from *Tollo* [1988].

Secondary CAMP

The HTQ flows of eastern North America are overlain by a thick sequence of basalt flows defined by *Weigand and Ragland* [1970] as low-Ti quartz normative (LTQ) and a thin uppermost sequence of high-Fe quartz normative (HFQ) flows. These secondary basalts extruded within 580 thousand years of the underlying HTQ flows [*Olsen et al.*, 1996]. The LTQ flows and related dikes occur throughout the central and southern portions of the eastern North American sub-province whereas the HTQ flows are restricted to the flows of New Jersey and Connecticut. There are also several iron enriched quartz normative dikes, such as Laurel Hill, New Jersey (*Puffer and Benimoff*, 1997), that are fractionation products of HTQ magma, but the LTQ and HFQ flows probably represent second and third magma batches that quickly followed initial HTQ magmatism [*Puffer and Philpotts*, 1988; *Tollo and Gottfried*, 1989].

All other Mesozoic chemical types within the boundaries of CAMP either proceed, follow, or are only marginally related to the brief main magmatic event. Wherever early Jurassic sections of volcanic activity are well exposed, only HTQ flows are present, with or without LTQ and/or HTQ flows. All other magma types are absent from this peak CAMP extrusive sequence.

Low Ti, olivine normative

Several examples of olivine normative basaltic rocks appear in the literature many of which can be dismissed as alteration products of HTQ basalt (chloritization) including those described by *Puffer and Benimoff* [1997]. However the eastern North American olivine normative dike swarms described by [*Ragland et al.*, 1992] are distinctive chemical types that are unlike any extrusive CAMP rocks. Their genetic relationship to the main CAMP extrusive event is unclear. One possibility is the depth of erosion throughout the southeastern US has reached a level where olivine enrichment has occurred. It is logical to expect the concentration of dense phases to increase with depth. The chemistry of most of the olivine normative dikes plots as a continuous series that includes the secondary LTQ type CAMP magmas [*Ragland et al.*, 1992]. Perhaps many of the olivine normative dikes fed LTQ flows of relatively fractionated composition. Alternatively some or most of the olivine normative dikes may represent a transition from secondary CAMP magma to subsequent MORB chemistry.

High Ti, REE enriched

Several Mesozoic and undated intrusions and extrusions characterized by TiO_2 contents approaching and exceeding 2 %, and REE contents resembling plume related ocean island basalt, occur within the boundaries of the CAMP province. However, it is unlikely that any of these rocks are genetically related to initial CAMP magmatism. Examples include portions of the Liberian dike swarm [*Dupuy et al.*, 1988], the dikes of the eastern Maranhão province, Brazil [*Fodor et al.*, 1990], the Mesozoic dikes of the coastal New England (CNE) igneous province [*McHone*, 1992], and some thin upper or "group 3" flows in Morocco including those described by *Manspeizer and Puffer* [2000].

Of the five groups of dikes included in the Liberian dolerite dike swarm described by *Dupuy et al.* [1988] only the group that intrudes the Paynesville Sandstone (Group IV) has been dated within a time range (173-193 My) approximating the CAMP event. Group IV is also the only group that is chemically the same as initial CAMP magma type. The other dikes that intrude the Precambrian basement should be interpreted with caution. Most of the late Proterozoic diabase dikes intruded into the New Jersey Highlands [*Volkert and Puffer*, 1995] were previously mapped as Triassic and interpreted erroneously as genetically related to Palisades magmatism.

The High-Ti basalts of the eastern Maranhão province, Brazil are much younger (115-122 Ma) than the CAMP flood basalts of the western Maranhão or Mosquito province [*Fodor et al.*, 1990]. *Fodor et al.* [1998] relate these dikes to the Fernando de Noronha hotspot about 250 km offshore northeastern Brazil. After the break-up of Pangea it is plausible that Brazil and other circum-Atlantic continents may have drifted over hotspots that have little to do with initial CAMP volcanism. A Cape Verde mantle plume has also been proposed as an important influence on CAMP magmatism [*Wilson*, 1997]. However, as pointed out by *McHone* [2000] a major plume responsible for all or most of CAMP is inconsistent with the contrasting chemistries of pre- and post-CAMP magmas and with the absence of a hotspot tract consistent with the northward drift of Pangea during the Mesozoic [*Olsen*, 1997].

The Mesozoic alkalic dikes of coastal New England (CNE) igneous province [*McHone*, 1992; *Pe-Piper et al.*, 1992] are also TiO_2 enriched and plot within the geographic boundaries of the CAMP. CNE magma was interpreted by *Pe-Piper et al.* [1992] as related to a hotspot together with the progression from alkalic to tholeiitic CAMP magmatism. However, *McHone et al.*, [1987] present compelling evidence that CNE magma as well as subsequent alkalic early Cretaceous dikes and seamounts are the product of parallel leaky transform fault zones oriented perpendicular to the eastern US coast line.

Some thin TiO_2 enriched flows that are locally ultrapotassic [*Manspeizer and Puffer*, 2000] overly the HTQ flows of Morocco. However these flows are separated from the HTQ flows by a thick stratigraphic section and, again, are probably unrelated to the massive extrusions of early Jurassic CAMP basalt

CHEMICAL CHARACTERISTICS OF CAMP BASALT

Table 1 includes several of the component parts of the initial and secondary CAMP events. Table 1a lists averages representing the initial HTQ outpourings of basalt including the North Mountain, Talcott, Orange Mountain, and Mt. Zion Church basalts of eastern North America, the Algarve basalt of Portugal, the High Atlas basalt of Morocco and the western Maranhão (Mosquito) basalt of Brazil. Table 1b lists averages representing secondary

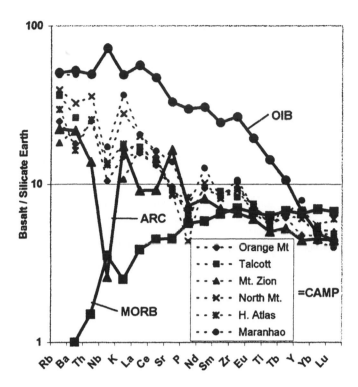

Figure 1. Basalt incompatible element compositions normalized to the Silicate Earth [*McDonough and Sun*, 1995] and arranged according to compatibility in oceanic basalt. Standard Ocean Island Basalt (OIB) and normal Mid-Ocean-Ridge Basalt (MORB) from *Sun and McDonough*, [1989], and standard ARC sample 272825 from Puyehue within the southern active volcanic zone of the Andes Mountains after *Hickey et al.* [1986] are compared to the average CAMP values of Table 1a.

LTQ outpourings that overly or follow the initial HTQ basalts including the Holyoke, Preakness, and Sander basalts of eastern North America.

Most of the chemical characteristics of the CAMP basalts of Table 1a are displayed on spider diagrams (Figure 1) that have been normalized to the silicate earth [*McDonough and Sun*, 1995] with elements arranged according to incompatibility in oceanic basalt. Probably the most obvious characteristic displayed by Figure 1 is the extreme degree of chemical uniformity among the initial CAMP basalts. The degree of uniformity is particularly close within the range of high field strength elements Nd through Ti (Figure 1).

The very high degree of chemical uniformity of CAMP basalts is also nicely displayed on a Zr/Ti variation diagram (Figure 2). The location of the CAMP cluster is completely outside the field of WPB or "within plate basalts" as defined by several of the earth's largest continental flood basalt provinces [*Pearce*, 1982]. Although CAMP is one of the earth's larger continental

flood basalt provinces it instead plots within the arc and MORB fields of *Pearce* [1982]. This arc characterization may be quite meaningful.

Another important observation from Figure 1 is the lack of any similarity of CAMP with the standard ocean island basalt (OIB) or normal mid-ocean-ridge basalt (MORB) of *Sun* and *McDonough* [1989]. The CAMP basalts plot in a field about mid-way between OIB and MORB. Unlike OIB the CAMP basalts are much less enriched in incompatible elements relative to the composition of the silicate earth, particularly LILs but also HFSEs.

Compared to N-MORB, CAMP basalts contain similar HFSE concentrations but much higher LIL values (Figure 1). E-MORB, a hybrid magma type, intermediate between N-MORB and OIB, (including the E-MORB standard of *Sun and McDonough*, 1989), is a somewhat better match for CAMP except for it's positive Nb anomaly with respect to Th and K. A positive Nb anomaly is a consistent characteristic of N-MORB, E-MORB and OIB magmas, however, CAMP basalts consistently display negative Nb anomalies.

By far the best fit for CAMP basalt among the three genetic choices of Figure 1 is ARC although any typical

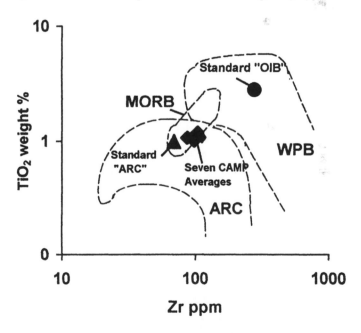

Figure 2. Ti/Zr variation diagram of *Pearce* [1982] with dashed field boundaries for arc basalt, MORB, and within-plate basalt (WPB) which includes several of the earth's largest continental flood basalt provinces. The standard arc sample [*Hickey et al.*, 1986] and the standard OIB sample [*Sun and McDonough*, 1989] are the same as used in Figure 1. Each of the CAMP basalt averages of Table 1a are plotted for comparison and fit into the arc field.

calc-alkaline basalt derived from any of the worlds arc systems would plot close to CAMP. The ARC basalt plotted onto Figure 1 is commonly used as a calc-alkaline basalt or ARC standard [for example, *Wilson*, 1989]. It is sample 272825 from Puyehue within the southern active volcanic zone of the Andes Mountains after *Hickey et al.* [1986].

CAMP BASALT COMPOSITIONS COMPARED TO ARC AND BACK-ARC BASALTS

Although the extensional tectonic setting of CAMP petrogenesis is the opposite of the compressional setting of most arc basalts, the spider diagrams of Figure 1 indicate a close geochemical resemblance.

Despite clear enrichment of alkalis and other LILs among both CAMP and ARC basalts relative to the silicate earth, the Nb contents are not as enriched. *Cann* [1970] has shown that Nb is a highly immobile element during hydrous metasomatic processes. It is not involved in the metasomatic processes that tend to enrich arc mantle sources in LILs above subducting oceanic lithospheric slabs. Nb is, therefore, a good magma source indicator.

One difference between CAMP and arc basalts is the absence of a positive Sr anomaly among CAMP basalts. The positive Sr anomaly of arc basalts (Figure 1) is consistent with the plagioclase enrichment of virtually all calc-alkaline rocks. Plagioclase phenocrysts are, however, rare to absent in the initial CAMP flows, and no Sr anomaly is observed. Contrasting plagioclase phenocryst content may be due in part to the contrasting tectonic setting of CAMP and arc basalts. The extensional tectonic setting of CAMP would allow for wide-open conduits through the crust in contrast to the compressional setting of arc volcanism that commonly results in ponding in the crust, cooling, plagioclase crystallization, and calc-alkaline fractionation.

CAMP basalts also chemically resemble some Paleozoic volcanics that stratigraphically underlie the New England portion of the CAMP. The chemical composition of the Ordovician Ammonoosuc and Partridge Formations (Figure 3) plot very close to the average of each of the seven CAMP basalts of Table 1a, particularly in regard to HFSE concentrations. *Robinson* and *Hall* [1980] proposed that the Ammonoosuc lavas were extruded during the middle Ordovician onto an oceanic plate that was accreting onto the eastern margin of Laurentia during the Taconic Orogeny. The Ammonoosuc Formation and associated ophiolites were then overlain by the volcanic rocks of the Partridge Formation during a continuation of compression and thrusting during the late middle Ordovician. These Paleozoic basalts are part of the Bronson Hills

anticlinorium that was metamorphosed during the Devonian Acadian Orogeny, at which time additional arc lavas were extruded; including the Lieghton [*Gates* and *Moench*, 1981], and the Newbury Formation [*McKenna et al., 1993*].

Compared with the initial CAMP basalts, the secondary CAMP basalts, (as represented by the average composition of the Preakness Basalt), are even more arc-like (Figure 4). The secondary CAMP magma source was probably depleted in incompatible elements, particularly HFSEs, during the melting event that produced the initial CAMP basalts. A similar scenario distinguishes arc magma from back-arc related magma. *Woodhead et al.* [1993] have shown that HFSE content can be reliably used to distinguish arc basalts from BABB. They show that arc basalts have lower average concentrations of incompatible

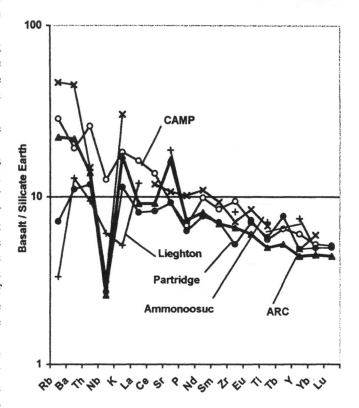

Figure 3. Basalt compositions normalized to the Silicate Earth [*McDonough and Sun*, 1995] including an average of 48 CAMP basalts from Table 1a, the ARC standard from [*Hickey et al.*, 1986], and Paleozoic calc-alkaline volcanics including an average of 5 meta-basalts for the Ammonoosuc Formation of New Hampshire, Vermont, and Massachusetts analyzed by *Leo* [1985], with an average of 24 analyzed Nb determinations after *Schumacher* [1988], an average of 23 basalt samples of Partridge Formation of the Bronson Hill Anticlinorium of Massachusetts analyzed by *Hollocher* [1993], and a meta-basalt sample from the Leighton Formation analyzed by *Gates and Moench* [1981] from the Machias-Eastport Area of Maine.

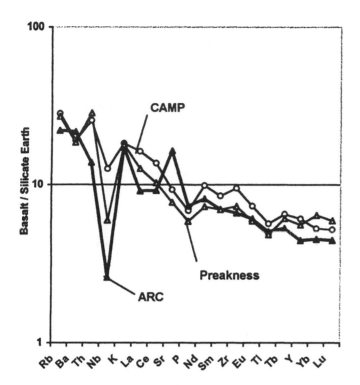

Figure 4. Basalt compositions normalized to the Silicate Earth [*McDonough and Sun*, 1995] illustrating the relative incompatible element depletion of secondary CAMP basalts (as represented by the average Preakness basalt) compared to the initial CAMP flows (represented by the average CAMP of Table 1a) with the arc standard of *Hickey et al.* [1986] for comparison.

HFSEs and Y, and consistently higher Ti/Zr ratios than BABB. They conclude that the relatively depleted arc basalts are more consistent with derivation from a source that has undergone previous melt extraction. The Preakness basalt (Table 1b) is an example of secondary melt extraction.

In addition, the secondary HFQ basalts of New Jersey and Connecticut are also are-like compared to initial HTQ basalt. The HFQ basalt of New Jersey contains only 4.1 ppm Nb, similar to the 3.9 ppm Nb content of the underlying Preakness flows [*Tollo and Gottfried*, 1992] but much less than the average 9.1 ppm Nb content of HTQ flows (Table 1).

The generation of a second and third batch of melt from relatively arc-like sources to derive LTQ and HFQ magmas is consistent with *Tollo and Gottfried's* [1989] proposal that eastern North American CAMP magmas were derived from heterogeneous source materials and remained largely independent during ascent and eventual eruption. *Tollo and Gottfried's* [1989] trace element data support conclusions by *Puffer and Philpotts* [1988] that HTQ, LTQ and HFQ magmas are not related by differentiation.

A wide range of samples are available to use as representative arc and back-arc basalts. The choices made by *Falloon et al.* [1999] are consistent with the observations of *Woodhead et al.* [1993]. They chose sample ODP Leg 135 site 839B-25R-1, 27-32, unit 3 (matrix) analyzed by *Ewart el al.* [1994] to represent the arc portion of the Lau-Tonga arc and sample M-2212-2, KTJ analyzed by *Falloon et al.* [1992] to represent the back-arc portion (Figure 5).

Figure 5 shows that both the Lau arc basalts and the secondary (Preakness) CAMP basalts are relatively depleted in HFSEs compared with both the Lau BABB and the initial CAMP basalt. The initial CAMP composition is, therefore relatively less depleted which is a BABB characteristic.

Ewart el al. [1994] also show that BABB magmatism can develop in an extensional tectonic setting. They describe the basin between the Lau Ridge and the Tonga Ridge as it occurred 3 million years ago as a "basin and range" style basin. The Newark Basin of Eastern North America is commonly described in similar terms.

THE SOURCE OF CAMP

Figure 6 is a slightly modified version of a model proposed by *Anderson* [1994] as a plausible way to generate at least some continental flood basalts. Figure 6 shows that the thermal boundary layer (TBL) under the Craton becomes enriched in H_2O and large ion lithophile elements (LIL) during subduction before it melts to generate arc magma at shallow depths. Relatively depleted and slightly more mafic BABB is then generated at greater depths where a higher degree of partial melting would occur.

Anderson [1994] uses the term thermal boundary layer in preference to lithosphere in Figure 6A partly because of geophysical evidence indicating that it is not a cool rigid layer with long term strength as is meant by the term lithosphere. *Anderson* [1994] argues that the TBL is too weak to be carried by a continental plate or to remain attached to an overlying continental crust in motion. It is, therefore unlikely that any old (1 Ga) Grenville or pre-Grenville subcontinental lithosphere such as proposed by *Pegram* [1990] would remain attached to Laurentia during its rapid Neoproterozoic to Cambrian displacement from the south pole to the equator following the break-up of Rodinia [*Dalziel*, 1997], or remain attached to Gondwana during complex Paleozoic drifting as Pangea was assembled. However, after the assembly of Pangea was largely completed by the end of the Silurian, the Laurentian-Gondwana or North American-African suture

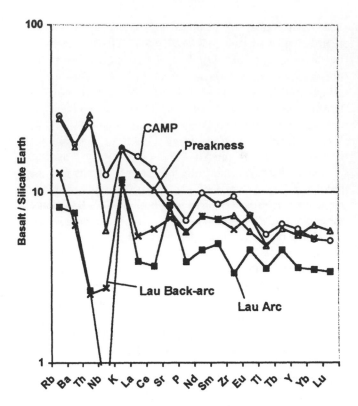

Figure 5. Basalt compositions normalized to the Silicate Earth [*McDonough and Sun*, 1995] comparing the incompatible element depletion of the average secondary CAMP (represented by the Preakness Basalt) [*Tollo and Gottfried*, 1992] with respect to the average initial CAMP basalt. Also shown is the relative depletion of the Lau ARC [*Ewart el al.*, 1994] with respect to the Lau back-arc [*Falloon et al.*, 1992].

zone located near the center of Pangea, underwent relatively minor displacement, largely confined to late Paleozoic transpressional motion [*Gates et al.*, 1988].

Once the TBL becomes enriched in LILs and fluxed with H_2O from subducted slabs it is converted to enriched mantle (Figure 6B) or "perisphere". This perisphere is described by *Anderson* [1994] as a mantle source that would be a suitable source for CFB magmas. Melting is triggered by decompression melting induced by lithospheric pull-apart at cratonic boundaries. Then as the enriched mantle is depleted, it is replaced by normal MORB mantle (Figure 6C).

A similar scenario may have occurred during the assembly of Pangea. If Laurentia was the craton in Figure 6 some of the Paleozoic volcanic rocks of the Piedmont may have been extruded as arc and BABB lavas during stage A. After Pangea was assembled during stage B it began to break up with a reversal from a compressional tectonic setting to an extensional setting. Lithospheric pull-apart would trigger melting under the craton (Pangea)

that would escape through the plate suture to produce CAMP lavas. The subsequent opening of the Atlantic (Figure 6C) would separate the Eastern North American portion of CAMP from the African and South American portions.

Figure 6 is too simple to be directly applied to what actually took place during the events that preceded the extrusion of CAMP lavas but resembles the tectonic models proposed by *Robinson* and *Hall* [1980], *Cook et al.* [1979], *Whitney et al.* [1978], *Hatcher* [1989], *Scotese* [1997], and *Torsvik et al.* [1996] for the assembly of Eastern North America during the Paleozoic.

Figure 7 is a model that includes several of the Paleozoic events described by these authors and adds the CAMP event at 200 Ma. It begins with Cambrian Arc volcanism over an eastward dipping subduction zone on what is described by *Cook et al.* [1979] as a continental fragment or Piedmont microcontinent (Figure 7A). This southern Appalachian volcanism coincided with northern Appalachian Ammonoosuc volcanism (Figure 3) in a similar tectonic setting as described by *Robinson* and *Hall* [1980] and *Hatcher* [1989]. This volcanism was followed by Partridge Formation volcanism that occurred during the Taconic accretion of the continental fragment [*Robinson and Hall*, 1980]. This first accreted arc terrane later became the Blue Ridge Mountains of the southern Appalachians and was thrust westward to form the Taconic allochthons of the New England Appalachians [*Robinson and Hall*, 1980].

As the Piedmont arc was being accreted during the Taconic Orogeny a second island arc was migrating toward Laurentia. *Scotese* [1997] and *Torsvik et al.* [1996] on the basis of paleontological and palaeomagnetic evidence, including the occurrence of *Paradoxides* trilobites, indicate that this second arc was located on the northwest edge of Gondwana. *Torsvik et al.* [1996] show that during the early Ordovician (490 Ma) the width of the Iapetus Ocean between Laurentia and the Avalonian portion of Gondwana was about 5000 km. Their modeling indicates that during early to mid-Ordovician time the arc rifted away from the northwest Gondwana margin and began to become very rapidly displaced across the Iapetus to a position close to Laurentia by Late Ordovician to Early Silurian time (about 440 Ma), (Figure 7B).

Then, during the Middle Silurian, (425 Ma) Avalonia was accreted onto Laurentia (Figure 7C). *Cook et al.* [1979] describe this accretion as the source of the eastern Piedmont or Carolina Slate Belt. In the New England Appalachians, Avalonian terranes represent this second arc. *Robinson* and *Hall* [1980] indicate that during Silurian accretion this second arc was undergoing Newbury volcanism in the Nashoba and Boston tectonic terranes of

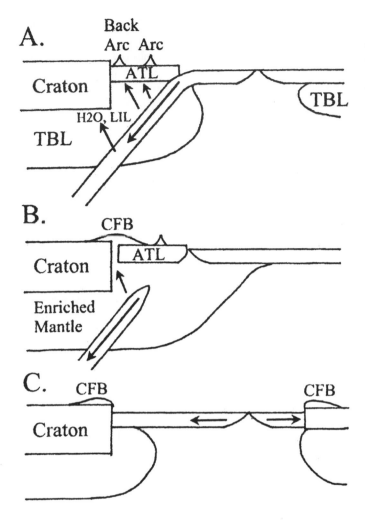

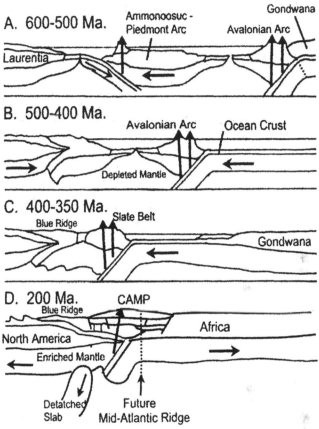

England. The calc-alkaline basalts of the Leighton Formation (Figure 3) are part of this Silurian volcanic event.

Although only one Avalonian arc is modeled on Figure 7 for the sake of simplicity and in agreement with *Scotese* [1997] and *Torsvik et al.* [1996], the actual tectonic

Figure 6. A slightly modified version of a model proposed by *Anderson* [1994]. A. The mantle wedge fluxed by H_2O and enriched in LILs produces typical ARC and BABB magmas but only a fraction of the available LIL enrichment ends up in the volcanics. B. As the tectonic setting changes from compressional to extensional, continental flood basalts (CFB form as a result of plate pull-apart of an accreted terrane layer (ATL) and cratonic lithosphere). The former mantle wedge becomes part of the shallow mantle (the perisphere), [*Anderson*, 1994]. C. During seafloor spreading the shallow enriched mantle is displaced and normal depleted basalts (MORB) replace enriched CFB basalts.

Figure 7. A model illustrating the Paleozoic and Mesozoic development of the source of CAMP basalt. A through C are modified versions of a tectonic synthesis by *Cook et al.* [1979]. A. the Ammonoosuc – Piedmont Arc is developed over a subducting plate during the early Paleozoic; B. the Ammonoosuc and Piedmont are accreted onto Laurentia while a second Gondwandaland arc (Avalonia) begins to migrate toward Laurentia. C. More arc-derived basalts, including the Newbury and Leighton are extruded during Silurian (Acadian) accretion and subduction of an oceanic plate below the Piedmont. D. Detachment and sinking of a dense oceanic slab displaces undepleted asthenosphere into a zone of enrichment above the undetached portion of the oceanic plate; a BABB source. Decompression melting of enriched asthenosphere quickly penetrates through Paleozoic allochthons and through the Triassic rift basin sediments during the early Jurassic to extrude as CAMP basalt. Secondary CAMP basalt is derived from asthenosphere that has been depleted by previous melting. Subsequent development of the Mid-Atlantic ridge occurs at the Alleghanian Suture (dashed vertical line).

Massachusetts. Palentological evidence indicates that Newbury volcanism continued post accretion with late Silurian to earliest Devonian time, depositing over 3000 m of subaerial rhyolite, basalt, and andesite (*Shride*, 1976). The basalt and andesite portions of the Newbury complex are calc-alkaline and display distinct negative Nb anomalies on spider diagrams (*McKenna et al.*, 1993). *Gates* and *Moench* [1981] describe similar Silurian volcanism in the Machias-Eastport area of Maine that is near the center of the Coastal Volcanic Belt of eastern New scenario may have been much more complex. *Hatcher* [1989] describes two Avalonian arcs, a western and eastern separated by a "Theic-Rheic Ocean" only one of which was accreted during the Acadian Orogeny. The Ammonoosuc – Piedmont arc was accreted during the Penobscottian Orogeny. The western Avalonian terranes accreted during the Taconic Orogeny and the eastern Avalonian terrane during the Acadian Orogeny according to [*Hatcher*, 1989]. However, the final complex along the eastern margin of North America may have included at least six large terranes and several additional smaller terranes accreted during the Paleozoic (*Williams and Hatcher*, 1982). The tectonic synthesis of eastern Laurentia and other plate margins that may have influenced the source of CAMP magma is, therefore, controversial and is the subject of ongoing revision.

Although most tectonic details are not depicted in Figure 7, subduction events such as the two that are modeled are important precursors to the CAMP event because they were each driving lithospheric plates into the mantle and, to the extent that mantle metasomatism was occurring, they were modifying and enriching the chemistry of the mantle under the Piedmont. The most enriched portion of the sub-Pangean mantle was relatively shallow compared to deep plumes that were not involved.

The Late Carboniferous to Permian Alleghanian Orogeny that followed Silurian accretion and subduction did not involve widespread arc volcanism. The absence of such volcanism supports arguments that Alleghanian tectonism was largely transpresional [*Gates* et al., 1988]. Newbury and Leighton arc volcanism was the last widespread volcanism until the extrusion of CAMP, however the magmatic source regime remained largely unchanged or dormant during the intervening time.

Figure 7D is a model illustrating the proposed early Jurassic tectonic setting of the CAMP magma source. It is consistent with the tectonic setting of continental flood basalts proposed by *Anderson* [1994], (Figure 7) and with the development of BABB proposed by *Ewart et al.* [1994]. In essence it proposes that the initial CAMP volcanic rocks were generated in a back-arc tectonic setting and from a source that was enriched during mid-Paleozoic volcanism.

Figure 7D shows CAMP lavas extruding onto rift basin sediments such as those of the Newark Basin of eastern North America, the Argana Basin of Morocco, the Algarve Basin of Portugal, and the Amazon Basin of South America. The basalt flows are fed through mid-basin conduits in agreement with the typical absence of intrusive rock along border-faults. The rift basin sediments of Eastern North America were deposited onto Piedmont terrains or onto complex stacks of Paleozoic allochthons

that are bounded by normal faults against early Paleozoic or Precambrian basement.

Melting was triggered by the decompression of enriched mantle (perisphere) below the continental lithosphere, near the base of the thermal boundary layer, that was fluxed by water and enriched in LILs as it migrated upward across the upper surface of a subducted slab. Figure 7D proposes that the oceanic lithospheric plate that was involved in Newbury and Leighton volcanism detached and forced the upward displacement of the asthenosphere. However, because of multiple Paleozoic arc accretion events there may have been several lithospheric plates to choose from. The resulting extensional tectonic setting would have allowed rapid transmission of magma through an attenuated mid-Pangean suture system; fractionation and contamination of the magma would be minimized. Eutectic melting of the enriched, back-arc source would account for the uniformity in melt compositions among CAMP basalts. However, the second batch of melt would be relatively depleted in incompatible elements, as is the case of the Preakness Basalt and its correlatives (Figures 4 and 5). Finally, after the source of CAMP magmas became drained of its limited amount of enriched and fluxed source rock, it was quickly superseded by the onset of N-MORB magmatism. This occurred along the Alleghanian Suture separating North America from Africa. The Alleghanian Suture is located distinctly to the east of the Piedmont [*Hatcher*, 1989] and was the site of the first MORB volcanism in a completely new tectonic setting.

Acknowledgments. I thank Alec Gates and Warren Manspeizer for sharing with me their expertise in the area of Paleozoic and Mesozoic Appalachian tectonics and thank Angelo DeMin and an anonymous reviewer for their careful reviews.

REFERENCES

Anderson, D. L., The sublithospheric mantle as the source of continental flood basalts; the case against the continental lithosphere and plume head reservoirs, *Earth Planet. Sci. Lett.*, 123, 269-280, 1994.

Bertrand, H., Dostal, J., and Dupuy, C., Geochemistry of early Mesozoic tholeiites from Morocco, *Earth Planet. Sci. Lett.*, 58, 225-239, 1982.

Cann, J. R., Rb, Sr, Y, Zr, and Nb in some ocean-floor basaltic rocks, *Earth Planet. Sci. Lett.*, 10, 7-11, 1970.

Cook, F. A., D. S. Albaugh, L. D. Brown, S. Kaufman, J. K. Oliver, and R. D. Hatcher, Jr., Thin-skinned tectonics in the crystalline southern Appalachians; COCORP seismic-reflection profiling of the Blue Ridge and Piedmont, *Geol.*, 7, 563-567, 1979.

Dalziel, I. W. D., Neoproterozoic-Paleozoic geography and tectonics: Review, hypothesis, environmental speculation, *Geol. Soc. Amer. Bull.*, 109, 16-42, 1997.

Dupuy, C., J. Marsh, J. Dostal, A. Michard, and S. Testa, Asthenospheric and lithospheric sources for Mesozoic dolorites from Liberia (Africa): trace element and isotopic evidence: *Earth Planet. Sci. Lett.*, 87, 100-110, 1988.

Et-Touhami, M., P. E. Olsen, and J. H. Puffer, Lithostratigraphic and biosyratigraphic evidence for brief and synchronous Early Mesozoic basalt eruption over the Maghreb (Northwest Africa). *Eos, Transactions, American Geophysical Union*, Supplement, 82, no. 20, S276, 2001.

Ewart, A., W. B. Byran, B. W. Chappell, and R. L. Rudnick, Regional geochemistry of the Lau-Tonga arc and backarc systems. in *Proceedings of the Ocean Drilling Program, Scientific Results*, edited by Hawkins, J., L. Parson, J. Allan, J. Resig, and P. Weaver, 135, College Station, TX: Ocean Drilling Program, 385-425, 1994.

Falloon, T. J., D. H. Green, A. L. Jacques, and J. W. Hawkins, Refractory magmas in back-arc basin settings – Experimental constraints on the petrogenesis of a Lau Basin Example, *J. Petrol.*, 40, 255-277, 1999.

Falloon, T. J., A. Malahoff, L. P. Zonenshain, and Y. Bogdanov, Petrology and geochemistry of back-arc basin basalts from Lau Basin spreading ridges at 15°, 18°, and 19°S., *Mineral. Petrol.* 47, 1-35, 1992.

Fodor, R. V., S. B. Mukasa, and A. N. Sial, Isotopic and trace element indications of lithospheric and asthenospheric components in Tertiary alkalic basalts, northeastern Brazil: *Lithos*, 43, 197-217, 1998.

Fodor, R., A. N. Sial, S. B. Mukasa, and E. H. McKee, Petrology, isotopic characteristics, and K-Ar ages of the Maranhão, northern Brazil, Mesozoic basalt province: *Contrib. Miner. Pet.*, 104, 555-567, 1990.

Gates, A. E., J. A. Speer, and T. L. Pratt, The Alleghanian southern Appalachian Piedmont: A transpressional model: *Tectonics*, 7, 1307-1324, 1988.

Gates. O. and R. H. Moench, Bimodal Silurian and Lower Devonian volcanic rock assemblages in the Machias-Eastport area, Maine, *U.S. Geological Survey Professional Paper* 1184, 32 pp 1981.

Hames W. E., P. R. Renne, and C. Ruppel, New evidence for geologically instantaneous emplacement of earliest Jurassic Central Atlantic magmatic province basalts on the North American margin, *Geol.*, 28, 859-862, 2000.

Hatcher. R. D., Tectonic synthesis of the U.S. Appalachians: in *The Geology of North America*, v. F-2, *The Appalachian-Ouachita orogen in the United States*, edited by Hatcher, R. D., Thomas W. A., and Viele, G. W., 511-536, 1989.

Hickey, R. L., F. A. Frey, and D. C. Gerlach.. Multiple sources for basaltic arc rocks from the southern zone of the Andes: trace element and isotopic evidence for contributions from subducted oceanic crust, mantle, and continental crust. *J. Geophys. Res.*, 91, 5963-5983, 1986.

Hollocher, K., Geochemistry and origin of volcanics in the Ordovician Partridge Formation, Bronson Hill Anticlinorium, West-central Massachusetts, *Amer. J. Sci.*, 293, 671-721, 1993.

Hooper, P. R., The Columbia River flood basalt province: Current status, in *Large Igneous Provinces: Continental, Oceanic, and Planetary Flood Volcanism*, edited by Mahoney, J. J. and M. F.

Coffin, Washington D. C. American Geophysical Union, 1-29, 1997.

Lassiter, J. C., and D. J. DePaolo, Plume/lithosphere interaction in the generation of continental and oceanic flood basalts: Chemical and isotopic constraints, in *Large igneous Provinces: Continental, Oceanic, and Planetary Flood Volcanism*, edited by Mahoney, J. J. and M. F. Coffin, Washington D. C. American Geophysical Union, 35-356, 1997.

Leo, R. W, Trondhjemite and metamorphosed quartz keratophyre tuff of the Ammonoosuc Volcanics (Ordovician), western New Hampshire and adjacent Vermont and Massachusetts, *Geol. Soc. Amer. Bull.*, 96, 1493-1507, 1985.

Mahoney, J. J. and M. F. Coffin, *Large igneous Provinces: Continental, Oceanic, and Planetary Flood Volcanism*, Washington D. C. American Geophysical Union, 438 p. 1997.

Manspeizer, W. and J. H. Puffer, Oldest Atlantic MORB extrusion records early ridge event on the passive margin of Morocco, *Geol. Soc. Amer. Abstracts with Program*, 31, A-57, 2000.

Marzoli, A., P. R. Renne, E. M. Piccirillo, M. Ernesto, G. Bellieni, and A. De Min, Extensive 200-million-year-old continental flood basalts of the Central Atlantic Magmatic Province, *Science*, 284, 616-618, 1999.

McDonough, W. F., and S. S. Sun, The composition of the Earth, *Chemical Geol.*, 120, 223-253, 1995.

McHone, J. G., Mafic dike suites within Mesozoic igneous provinces of New England and Atlantic Canada: in *Eastern North American Mesozoic Magmatism*, edited by Puffer, J. H., and P. C. Ragland, Geol. Soc. Amer. Special Paper, 268, 1-11, 1992.

McHone, J. G., Non-plume magmatism and rifting during the opening of the Central Atlantic Ocean: Tectonophysics, 316, 287-296, 2000

McHone, J. G., Mapping the Central Atlantic Magmatic Province: *Geol. Soc. Amer. Abstracts with Program*, 32, A-78, 2001.

McHone, J. G., M. E. Ross, and J. D. Greenough, Mesozoic dike swarms of Eastern North America: in *Mafic dike swarms* edited by H. C. Halls, Geological Association of Canada Special Paper 34, 279-288, 1987.

McKenna, J. M., J. C. Hepburn, and Rudolph Hon, Geochemistry of the Newbury volcanic complex, Northeastern Massachusetts, *Geol. Soc. Amer. Abstracts with Program*, 25, 63, 1993.

Olsen, P. E., Stratigraphic record of the early Mesozoic breakup of Pangea in the Laurasia – Gondwana rift system: Ann. Rev. Earth Planet Sci., 25, 337-401, 1997.

Olsen, P. E., R. W. Schlische and M. S. Fedosh, 580 Ky duration of the Early Jurassic flood basalt ement in Eastern North America estimated using Milankovitch cyclostratigraphy, in *The Continental Jurassic* edited by M. Morales, Museum of Northern Arizona Bull. 60, 11-22, 1996.

Papezik, V. S., J. D. Greenough, J. Colwell, and T. Mallinson. North Mountain basalt from Digby, Nova Scotia, *Can. J. Earth Sci.*, 25, 74-83, 1988.

Pearce, J. A., Trace element characteristics of lavas from destructive plate boundaries. in Andesites, edited by R. S. Thorp, Wiley, Chichester, p. 525-548, 1982.

Peate, D. W., The Parana-Etendeka Province in *Large igneous Provinces: Continental, Oceanic, and Planetary Flood Volcanism*, edited by J. J. Mahoney and M. F. Coffin, Washington D. C. American Geophysical Union, 217-246, 1997.

Pegram, W. J., Development of continental lithospheric mantle as reflected in the chemistry of the Mesozoic Appalachian Tholeiites, U. S. A., *Earth Planet. Sci. Lett.*, 97, 316-331, 1990.

Pe-Piper, G., L. F. Lubomir, and R. St. J. Lambert, Early Mesozoic magmatism on the eastern Canadian margin: petrogenetic and tectonic significance: in *Eastern North American Mesozoic Magmatism*, edited by Puffer, J. H., and P. C. Ragland, Geol. Soc. Amer. Special Paper, 268, 13-36, 1992.

Philpotts, A. R., Maureen Carroll, and J. M. Hill, Crystal-mush compaction and the origin of pegmatitic segregation sheets in a thick flood-basalt flow in the Mesozoic Hartford Basin, Connecticut, *J. Petrol.*, 37, 811-836, 1996.

Puffer, J. H. Eastern North American flood basalts in the context of the incipient breakup of Pangea, in *Eastern North American Mesozoic Magmatism*, edited by J. H. Puffer, and P. C. Ragland, Geol. Soc. Amer. Special Paper 268, 95-118, 1992.

Puffer, J.H., Contrasting HFSE contents of plume sourced and reactivated arc-sourced continental flood basalts: *Geology*, 29, 675-678, 2001.

Puffer, J.H. and A. I. Benimoff, Fractionation, hydrothermal alteration, and wall-rock contamination of an early Jurassic diabase intrusion: Laurel Hill, New Jersey: *J. of Geology*, 105, 99-110, 1997.

Puffer, J. H., and Philpotts, A. R., 1988, Eastern North American quartz tholeiites: geochemistry and petrology, in *Triassic-Jurassic Rifting, Continental Breakup and the Origin of the Atlantic Ocean*, edited by W. Manspeizer, Developments in Geotectonics, Elsevier Scientific Publishers, Amsterdam, 22, 579-606, 1988.

Ragland, P. C., L. E. Cummins, and J. D. Arthur, Compositional patterns for early Mesozoic diabases from South Carloina to central Virginia, in *Eastern North American Mesozoic Magmatism*, edited by Puffer, J. H., and P. C. Ragland, Geol. Soc. Amer. Special Paper, 268, 309-331, 1992.

Robinson, P. and L. M. Hall, Tectonic synthesis of southern New England: in *The Caledonides in the USA*, edited by D. R. Wones, Proceedings 1979 meeting, Blacksburg, Virginia, I.G.C.P. Project 27: Caledonide Orogen, 73-82, 1980.

Saunders, A. D., J. G. Fitton, A. C. Kerr, M. J. Norry, and R. W. Kent, The North Atlantic Igneous Province, in *Large igneous Provinces: Continental, Oceanic, and Planetary Flood Volcanism*, edited by J. J. Mahoney, and M. F. Coffin, Washington D. C. American Geophysical Union, 45-94, 1997.

Schumacher, J. C., Stratigraphy and geochemistry of the Ammonoosuc volcanics, Central Massachusetts and Southwestern New Hampshire, *Amer. J. Sci.*, 288, 619-663, 1988.

Scotese, C. R., Continental drift (seventh edition): Arlington, Texas, *PALEOMAP Project*, 79 p., 1997.

Sharma, Mulul, Siberian Traps, in *Large igneous Provinces: Continental, Oceanic, and Planetary Flood Volcanism*, edited by J. J. Mahoney and M. F. Coffin, Washington D. C. American Geophysical Union, 273-296, 1997.

Shride, A. F., Stratigraphy and correlation of Newbury volcanic complex, northern Massachusetts, in *Contributions to the stratigraphy of New England*, edited by L. R. Page, Geological Society of America Memoir 148, 147-177, 1976.

Sun, S., and W. F. McDonough, Chemical and isotopic systematic of oceanic basalts: Implications for mantle composition and processes, in *Magmatism in the ocean basins*, edited by A. D. Saunders, and M. J. Norry, Geological Society Special Publication, 42, 313-345, 1989.

Tollo, R. P., Petrographic and major-element characteristics of Mesozoic basalts, Culpeper basin Virginia, in *Studies of the early Mesozoic basins of the eastern United States*, edited by A. J. Froelich, and G. R. Robinson, U. S. Geological Survey Bulletin 1776, 105-113, 1988.

Tollo, R. P., and David Gottfried, Early Jurassic quartz-normative magmatism of the Eastern North American province: evidence for independent magmas and distinct sources: International Association of Volcanology and Chemistry of the Earth's Interior, Continental Magmatism Abstracts, New Mexico Bureau of Mines and Mineral Resources, 131, 1989.

Tollo, R. P., and David Gottfried, Petrochemistry of Jurassic Basalt from eight cores, Newark Basin, New Jersey: Implications for the volcanic petrogenisis of the Newark Supergroup, in *Eastern North American Mesozoic Magmatism*, edited by J. H. Puffer and P. C. Ragland, Geol. Soc. Amer. Special Paper 268, 233-259, 1992.

Torsvik, T. H., M.A.. Smethurst, J. G. Meert, R. Van der Voo, W. S. McKerrow, M. D. Brasier, B. A. Sturt, and H. J. Walderhaug, Continental break-up and collision in the Neoproterozoic and Palaeozoic – A tale of Baltica and Laurentia, *Earth Sci. Reviews*, 40, 229-258, 1996.

Volkert, R.A. and J. H. Puffer, J.H., Late Proterozoic Diabase Dikes of the New Jersey Highlands - A Remnant of IapetanRifting in the North-Central Appalachians: *U.S. Geol. Survey Prof. Paper*, 1565-A, 22, 1995.

Walker, F., The differentiation of the Palisades diabase, New Jersey: *Geol. Soc. Amer. Bull.* 51, 1059-1106, 1940.

Weigand, P. W., and P. C. Ragland, Geochemistry of Mesozoic dolerite dikes from eastern North America, *Contrib. Mineral. Petrol.*, 29, 195-214 , 1970.

Whitney, J. A., T. A. Paris, R. H. Carpenter, and M. E. Hartley, 111, Volcanic evolution of the Southern Slate Belt of Georgia and South Carolina: A primitive Oceanic Island Arc, *J. Geol.*, 86, 173-192, 1978.

Williams, H. and R. D. Hatcher, Suspect terrains and accretionary history of the Appalachian orogen, *Geol.*, 10, 530-536, 1982.

Wilson, M., *Igneous Petrogenesis*, Unwin Hyman, London, 466p., 1989.

Wilson, M., Thermal evolution of the Central Atlantic passive margins: continental break-up above a Mesozoic super-plume: *J. Geol. Soc., London*, 154, 491-495, 1997.

Woodhead, J., S. Eggins, and J. Gamble, High field strength and transition element systematics in island arc and back-arc basin basalts: evidence for multi-phase melt extraction and a depleted mantle wedge, *Earth Planet. Sci. Lett.*, 114, 491-504, 1993.

John D. Puffer, Department of Geological Sciences, Rutgers University, Newark, New Jersey 07102.

Temporal Chemical Variations Within Lowermost Jurassic Tholeiitic Magmas of the Central Atlantic Magmatic Province

Vincent J.M. Salters[1], P.C. Ragland[2], W.E. Hames[3], K. Milla[4] and C. Ruppel[5]

The Central Atlantic Magmatic Province (CAMP), of greater extent than any other large igneous province (LIP) yet identified surrounds the Central Atlantic in eastern North America, northeastern South America, western Africa, and southwestern Europe. It covers over 7×10^9 km^2 and was active for no more than 4 Ma. Virtually all CAMP rocks are mafic tholeiites, and include both intrusives and extrusives. The most extensive intrusives, diabase (dolerite) dikes, occur in three main swarms on Pangaea: NW-, NE-, and NS trending. These mafic tholeiites can be classified based on their Ti contents into low-Ti (LTi), intermediate-Ti (ITi), and high-Ti (HTi). The NE swarm contains primarily the ITi magma type, whereas the NW swarm is heterogeneous and contains all three types. The N-S swarm contains highly evolved (high-Fe) quartz tholeiites in North America and ITi rocks in South America. These dike swarms can be correlated across the Atlantic basin on the basis of composition and attitude.

The two principal magma types within the CAMP, LTi and ITi, were derived from mantle sources that were compositionally similar and contained both continental lithospheric and asthenospheric components. Compared with other large igneous provinces the CAMP basalts show depleted geochemical characteristics. Compositional differences between them are primarily due to differences in depth and degree of melting; LTi represents the deepest and greatest degrees of melting. The temporal progression of the chemical characteristics indicate deeper melting with time, which is consistent with a shallow (such as crustal thinning) and passive origin for the break-up of the Pangaean continent.

INTRODUCTION

The Central Atlantic Magmatic Province (CAMP; [*Marzoli et al.*, 1999], Earth's most extensive large igneous province (LIP), is coincident in time and space with the breakup of Earth's largest supercontinent, Pangaea, and one of the most pronounced climate change and mass extinction events [*Courtillot et al.*, 1996]. The CAMP, of Lower Jurassic age, surrounds the Central Atlantic and extends on Pangaea from present-day central Brazil to

France and from west Africa to eastern North America (Fig. 1 [*Sundeen*, 1989]. As such, it encompasses over 7×10^9 km^2 and was apparently active for no more than 4 Ma [*Marzoli et al.*, 1999]; within northeastern North America

The Central Atlantic Magmatic Province:
Insights from Fragments of Pangea
Geophysical Monograph 136
Copyright 2003 by the American Geophysical Union
10.1029/136GM09

[1] National High Magnetic Field Laboratory and Department of Geological Sciences, Florida State University, Tallahassee, Florida

[2] Department of Geological Sciences, Florida State University, Tallahassee, Florida

[3] Department of Geological Sciences, Auburn University, Auburn, Alabama

[4] Division of Agricultural Sciences, Florida A&M University, Tallahassee, Florida

[5] School of Earth and Atmospheric Sciences, Georgia Institute of Technology, Atlanta, Georgia

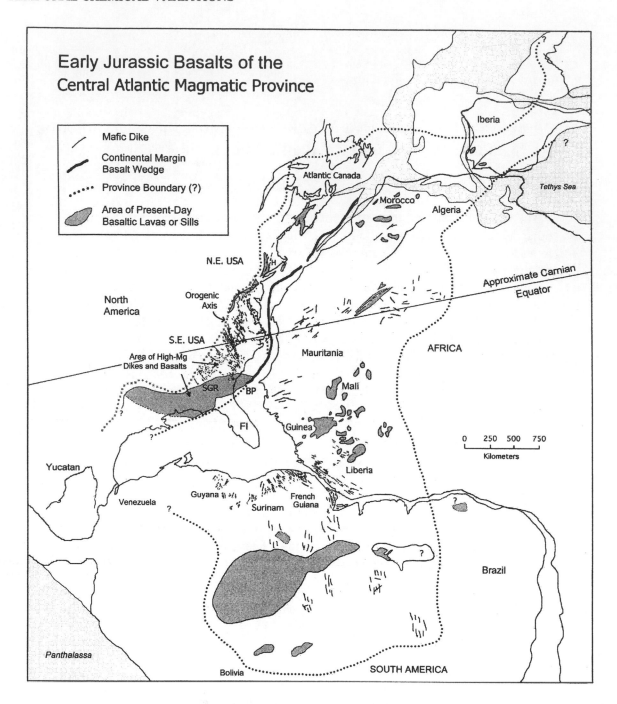

Figure 1. Map of CAMP on Pangaea (from McHone, http://jmchone.web.wesleyan.edu/CAMP.html).

its duration was perhaps less than 1 Ma [*Olsen et al.*, 1996]. CAMP rocks are mafic tholeiites in composition, and include basalt flows, as well as predominantly diabase (dolerite) sills, small lopoliths, dikes, and very probably crustal underplates.

In recent years, a number of developments have not only expanded estimates of the size of this now-rifted LIP, but also suggested a geologically instantaneous emplacement history [*Deckart et al.*, 1997; *Marzoli et al.*, 1999]. Results obtained during the past decade suggest that about 200 Ma ago some 6000 linear kilometers of the continental margin on four continental masses were affected by tholeiitic magmatism a few million years in duration. The majority of well constrained radiometric ages place the CAMP in the lowermost Jurassic, at about 198-202 Ma [*Marzoli et al.*, 1999]. Although CAMP basalts have not yet been found exactly at or below the Triassic-Jurassic stratigraphic boundary [*Olsen*, 1997; *Olsen et al.*, 1996], the available data indicate CAMP magmatism is linked to marine and terrestrial faunal changes at the boundary [*McElwain et al.*, 1999; *Pálfy et al.*, 2000].

The CAMP has been identified as a LIP, yet uncertainties in the age and geochemical characteristics of key components limit interpretation of the fundamental geodynamic processes that formed it. Its enormous size and apparently short time span are cause for speculation about what processes active in the mantle could have caused such an event. Like other ancient LIPs, the CAMP has been fragmented by tectonic activity and now consists of a deeply eroded and extensive ganglion of dominantly mafic dikes and variably exposed basalt flows (Fig. 1). Traditionally, workers have traced the origin of LIPs to the impingement of a mantle plume head at the base of the lithosphere, implying that large-scale mantle dynamics drives LIP formation. An alternate class of models has turned that causal relationship around and instead has explained LIP formation in terms of the role of the lithosphere and lithospheric processes in modifying the thermal and dynamic properties of the upper mantle (e.g., [*King and Anderson*, 1995; *Mutter et al.*, 1988; *Zehnder et al.*, 1990]. To some extent these endmember models for the origin of LIPs represent incarnations of the classic framework of active (plume-driven) and passive (driven by the mantle's response to far field plate stresses that affect the overlying lithosphere) rifting processes (in addition to the review below, see also [*Hames et al.*, 2000; *Ruppel*, 1995]).

In the plume (active) models, hot material originates at the core-mantle boundary and rises buoyantly through the mantle [*Hill et al.*, 1992]. The plume head eventually encounters the base of the rigid lithosphere, where it spreads laterally beneath an area of hundreds (thousands?) kilometers in diameter. Recent numerical experiments have shown that extremely fast mantle upwelling rates, up to 10m/yr, are possible [*Larsen and Yuen*, 1997]. These large-scale plumes may be associated with the rapid production of voluminous melt over length scales of hundreds of kilometers, while the plume tail, which may become detached from the plume head, could produce a distinct hotspot track. The normal stresses exerted on the base of the lithosphere by plume upwelling can lead to regional dynamic uplift on the order of several hundred meters [*Campbell and Griffiths*, 1990] or kilometers [*Farnetani and Richards*, 1994]. Crustal deformation caused by the thermal and dynamic effects of plume impingement and melt generation should generally postdate uplift. However, pre-existing extensional deformation of the overlying lithosphere may trigger release and localization of plume-derived magmas [*Hill*, 1991] and greatly increase the volume of melt [*Farnetani and Richards*, 1994]. Plume models predict that the volcanic rocks are derived mostly from the plume, that the initial degree of melting is low, and that the average depth of melting is relatively deep. As the plume impinges on the "cold" lithosphere the melting column is initially short. With time the lithosphere is displaced, and the degree of melting increases and the average depth of melting decreases [*Campbell and Griffiths*, 1990; *Hill et al.*, 1992; *Saunders et al.*, 1997].

Traditional passive models invoke thinning of the lithosphere above a pre-existing mantle thermal anomaly to drive decompression melting of the asthenosphere [*White*, 1989]. Since this mechanism is thought not to produce effusive, rapid volcanism, it has long been discredited as a possible explanation for the origin of LIPs. However, other models that rely on lithospheric processes to perturb the upper mantle are now considered viable for LIP formation. The plume incubation hypothesis [*Anderson*, 1994] postulates that upper mantle trapped beneath stable super continents with thick thermal ("perisphere") or chemical ("tectosphere") roots heats up over time. Eventually, this heating of the upper mantle may lead to thermally driven uplift, rifting, and magmatism within the supercontinental mass. A second type of passive model involves edge-driven convection -- the evolution of small-scale convective rolls in the upper mantle beneath an abrupt change in lithospheric thickness [*Boutilier and Keen*, 1999; *Buck*, 1986]. It has been proposed that this type of convective partial melting at passive margins will produce igneous crustal thickness that are observed at continental margins [*Mutter et al.*, 1988]. A temperature perturbation of only 0.1% of the ambient adiabatic temperature across the leading edge of a rifting continental margin may provoke edge-driven convection and produce volumetrically significant, short-lived magmatism [*King and Anderson*, 1998]. This edge-driven convection occurs at a zone of weakness and is preceded by basin formation that creates a "thin" in the crust and underlying lithosphere. During the initial stages the source of the magmatic products will have a significant lithospheric component, but with time the involvement of lithosphere decreases. Furthermore, there is no strong variation in depth of melting as is observed in the "active" (plume) case [*Mutter et al.*, 1988], making it difficult to distinguish between these two passive processes based solely on the geochemical variations of the magmas.

Although it has been argued that the fragmentation of Pangaea and opening of the central Atlantic was associated with flood basalts and plume type magmatism [*Courtillot et al.*, 1999; *Leitch et al.*, 1998; *Wilson*, 1997], several researchers have argued that the association with a plume is less clear e.g. [*Heatherington and Mueller*, 1999; *Kele-*

men and Holbrook, 1995; *McHone and Puffer,* 1997; *Ruppel and Hames,* 1999]. Wilson [1997] used the term "superplume" to explain the origin of the CAMP. Field and petrological work [*McHone,* 2000; *Ragland et al.,* 1983] has shown that there are multiple and overprinting dike swarms with distinct chemical signatures, and thus May's [1971] proposed radial pattern for the CAMP dikes is oversimplified. Furthermore, there is no clear plume track associated with the CAMP, although it has been suggested that CAMP is associated with the Fernando da Nohorona hot spot or the Caribbean hot spot [*Dalziel et al.,* 2000; *Hill,* 1991]. Consequently, several hybrid models that incorporate aspects of both passive and active geodynamic models have been proposed for the origin of the CAMP. This paper will examine these models using an extensive compilation of major element, trace element, and isotopic analyses of the tholeiites.

GEOLOGICAL SETTING

For the purpose of this paper we focus on CAMP dikes that can provide information on the pressure, temperature and composition during melting, with a minimum discussion of subdivision and classification of magma types. Many classifications have been used for intrusive and extrusive rocks of the CAMP, based primarily on orientation or structural setting, major- or trace-element geochemistry, or both. Exposed basaltic rocks of the CAMP can be related to three main tholeiitic dike swarms [*McHone,* 2000; *Ragland et al.,* 1997]: 1) a northeast-trending swarm of quartz tholeiites with limited compositional variability; 2) a northwest-trending swarm of tholeiites that includes a variety of compositions; ranging from relatively depleted picritic basalts to high-Ti tholeiites and, 3) a north-south trending swarm of tholeiites with compositions that are either highly evolved (high-Fe), or similar to the northeast swarm.

Dike swarms of CAMP

The NE-trending dike swarm occurs in the northeastern US and Canada (northern Virginia to southeastern Canada), in northeastern Africa (Morocco, Algeria, Mauritania) as well as Guyana and Surinam in South America, France and Iberia. The dikes are thought to have intruded during the initial rifting stage of the Pangaean breakup [*Withjack et al.,* 1997]. The NE-trending swarms are characterized mainly by quartz tholeiites, with limited chemical variation that can be explained by fractional crystallization. Crystallizing phases are orthopyroxene and plagioclase which is at later stages followed by augite crystallization. These rocks have been referred to as the Initial Pangaean Rifting (IPR) basaltic rocks, [*Puffer,* 1994] and high-Ti quartz (HTQ) tholeiites [*Weigand and Ragland,* 1970]. Relative to LIP basaltic rocks on a worldwide scale, their Ti contents of 1-1.5 wt. % are intermediate. Thus, we will refer to this group as the intermediate Ti (ITi) group on the suggestion of J.G. McHone (pers. comm., 2000).

A second, northwest-trending dike swarm, occurs in the southeastern US (Alabama to Central Virginia, partly in the subsurface), in Africa (Liberia, Guinea and Sierra Leone) as well as in South America (Surinam and French Guyana). This NW-trending swarm is thought to have intruded after the initial rifting stage, during the early drifting stage of the breakup of Pangaea [*Withjack et al.,* 1997]. The NW-trending dikes exhibit a large compositional range, although low-Ti olivine-tholeiite dikes (the most primitive of which are picritic basaltic rocks), and their low Ti-quartz tholeiites derivatives dominate in the southeastern US [*de Boer et al.,* 1988; *Philpotts and Reichenbach,* 1985; *Ragland et al.,* 1992]. A large part of the chemical variation in this group can be explained by low pressure crystal fractionation of olivine and plagioclase. It has previously been noted however that the large variation in FeO content (at constant MgO) indicates that processes other than low pressure fractionation (variations in depth of melting?) can be important. This group, again based on McHone's suggestion, will be referred to as the low-Ti (LTi) group. Truly high Ti-quartz tholeiites (4-5 wt% TiO₂, HTi) also occur in the NW-swarm; they are common in Africa and South America [*Bertrand et al.,* 1999; *Dupuy et al.,* 1988] but not in North America. They have, however, been reported in well cuttings under the Coastal Plain in Georgia [*Chowns and Williams,* 1983]. Furthermore, NW-trending dikes consisting of quartz tholeiites similar in composition to ITi also occur in Liberia and northeast South America. The NW-trending dikes are thus considerably more variable in composition than the NE-trending suite.

In addition to these two swarms a third swarm that trends north-south exists in the SE USA and also in Brazil and Surinam. In the SE USA these dikes consist of high-Fe quartz tholeiites and have been reported to be related by crystal fractionation to the LTi of the NW-swarm [*Ragland et al.,* 1992]. The possibility remains, however, that they could also be derived from an ITi parental magma. Dikes of comparable age and N-S orientation in Brazil have compositions similar to ITi [*Bellieni et al.,* 1992; *Bellieni et al.,* 1990]. In fact, N-S trending dikes in North and South America may not be related. In addition to their chemical differences, the North and South American dikes have different characteristics in the field. The dikes in the SE US occupy a narrow belt that converges southward toward an area near Charleston, SC [*Ragland et al.,* 1983], whereas N-S dikes in South America occupy a much broader belt [*Bellieni et al.*, 1992; *Bellieni et al.*, 1990].

Diabases of the NS-trending suite in the SE US contain several percent of magnetite and form magnetic anomalies that can be traced under the Coastal Plain and linked to a major magnetic anomaly near Charleston, SC [*Ragland et al.*, 1983]. Flows drilled at Clubhouse Crossroads, SC (near to Charleston) are similar in composition to these high-Fe quartz tholeiites. The top of these flows are also a strong seismic reflector and this reflector can be traced to the seaward dipping reflector of the coast of Georgia and South Carolina [*Austin Jr. et al.*, 1990]. Cross-cutting relationships indicate that the NW-swarm is older than the NS-swarm [*Ragland et al.*, 1983]. However, there are no known field relations that allow the determination of the relative age of the NW- or NS-swarm with respect to the NE-swarm.

GEOCHRONOLOGY

The diabase dikes have been notoriously difficult to date with standard geochronological techniques. Many of the rocks have undergone varying degrees of alteration and most previous studies relied on bulk rock K-Ar and ^{40}Ar/^{39}Ar techniques. Typically, these bulk rock K-Ar and ^{40}Ar/^{39}Ar ages are systematically younger than U/Pb age determinations and age estimates based on precise stratigraphy. The Newark Supergroup includes numerous, thick basalt flows that appear structurally related to the Palisades sill [*Ratcliffe*, 1987] and geochemically-related to NE-trending dike swarms. Milankovitch cyclicity of lacustrine sediments in the Newark Supergroup suggests an overall duration of ca. 580 Ka within the lowermost Jurassic for the eruptive history of the Newark Basin basalts [*Olsen et al.*, 1996] and correlation to U/Pb dating studies of accessory minerals in the Palisades sill [*Dunning and Hodych*, 1990] suggests this event occurred at ~201 Ma. Recently, a ^{40}Ar/^{39}Ar study of plagioclase separated from the lower and upper basalt flows in the Newark Supergroup (Watchung Flow I and III) yielded ages of 201±1.5 and 199±1.5 Ma, confirming the short duration and timing of the Newark Basin volcanism [*Hames et al.*, 2000]. The Newark Basin basalts and NE-trending dikes in eastern North America can be correlated with similar NE-trending swarms from Guinea (Africa) that have yielded a broad range of dates from 189-200 Ma [*Deckart et al.*, 1997].

Hames et al. [2000] dated three samples from a set of NW-trending dikes in the South Carolina Piedmont and obtained ^{40}Ar/^{39}Ar plateau ages of 199.5 to 198.4 Ma for plagioclase separates. This result indicates that the NW-trending olivine-tholeiites of the southeast US and NE-trending quartz tholeiites of the Newark basin formed at

approximately the same time. NW-trending dikes similar to the swarms of the southeastern USA (Figure 1) can be found in Liberia, Sierra Leone, and French Guyana. K/Ar and early ^{40}Ar/^{39}Ar ages for samples from Liberia range from 173-193 Ma [*Dalrymple et al.*, 1975]. Two single grains of amphibole picked from NW trending tholeiites of French Guyana yielded concordant laser incremental heating plateau ages of 196.0±5.7 and 196.1±7.5 Ma [*Deckart et al.*, 1997]. We note that the composition of whole rock samples in the study by [*Deckart et al.*, 1997] is unspecified.

Data were compiled on samples from all four present-day circum-Atlantic continents on which the province is known to exist. To date, 767 major-element analyses, 119 rare-earth patterns and trace-element analyses have been compiled from the literature and from unpublished theses and dissertations. Of far less abundance are Sr and Nd isotopic data; however, the available information on these isotopes was also included in the database. The database and attendant references are available on the CAMP-website at http://jmchone.web.wesleyan.edu/CAMP.html.

In fact, well over 1000 major-element analyses were available, but we applied several criteria in compiling the database. Of greatest concern are secondary alteration: the chemical effects of weathering, hydrothermal alteration, or burial metamorphism. Samples were excluded if there was petrographic or chemical evidence for alteration, such as high percentages of hydrous minerals or loss on ignition (LOI). Samples with greater than 2 % LOI were generally excluded. Samples with abundant large phenocrysts that were suspect of being accumulative were also omitted. In the absence of petrographic information, samples whose chemical compositions were strongly indicative of crystal accumulation were also omitted. Only representative samples were taken from studies in which a large number of samples were obtained from a single igneous body. Of particular importance was the identification of the least evolved, aphyric samples, thought to be representative of primary mantle melts.

GEOCHEMICAL CHARACTERISTICS OF THE CAMP DIKES

Mafic rocks of the CAMP are either olivine or quartz tholeiites, and a distinct bimodal distribution is evident in the entire dataset (Fig. 2). The modifiers *quartz* and *olivine* refer to both their normative and modal compositions. Most olivine tholeiites contain normative olivine and early-formed olivine crystals ("primocrysts," which are not necessarily phenocrysts). Quartz is present with potassium feldspar in late-crystallizing granophyre of the mesostasis in many quartz-normative tholeiites.

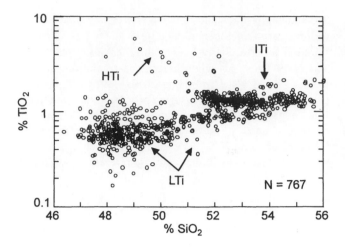

Figure 2. Plot of SiO$_2$ versus TiO$_2$ showing magma groups and the bimodal distribution of the data.

The geochemical data indicate that there are at least three types of parental magmas in the CAMP. The NW trending swarm in the SE US contains LTi and ITi only, while the NW trending swarms in Africa (Liberia, Sierra Leone, Guinea; [*Dupuy et al.*, 1988] and south America [*Bertrand et al.*, 1999] contains all three magma types. However, the NE-swarm contains the ITi-type only. As the LTi and ITi magma types are by far the most common magma types in the CAMP, the following discussion will concentrate on them.

Based on the trace element and isotopic characteristics of mostly ITi diabases from the US, Pegram [1986; 1990] concluded that crustal contamination was not a major factor in defining the geochemical characteristics of these basalts. The recognition of basalts with similar geochemical characteristics throughout the 6000km long province also makes it unlikely that these characteristics were acquired through uniform crustal assimilation in this vast province. LTi and ITi could be related by crystal fractionation based on major elements, but variations in incompatible trace elements (especially the REE) exclude this scenario. We therefore will attempt to explain these characteristics and differences between the LTi and ITi magma types as differences in mantle compositions and melting conditions.

Table 1 lists the composition of the most likely parental magmas for the LTi and ITi. These estimated parental melt compositions are based on regression of the fractionation trends in a similar fashion as originally described by Klein and Langmuir [1987] for Si$_8$, Na$_8$ etc. The goal of these calculations is to correct for low pressure fractional crystallization. After correction for fractionation differences in (standardized) major element chemistry indicate differences in the melting process or differences in

source. To correct for fractionation each sample is projected back along the fractionation trend to estimate the concentrations of the major elements at common point (composition) of reference which is close to a primary composition. Fractionation trends are determined using simple linear regression techniques. Normalization to a common MgO content like Klein and Langmuir [1987] is not practical as the LTi and ITi tholeiites have different low pressure fractionation assemblages. Consequently, the fractionation trends for the two series are not parallel and differences in the normalized values are highly dependent on the MgO value to which the data is standardized. We have chosen to standardize the LTi and ITi samples to Mg#=70 (molar Mg/(Mg+Fe)) and Mg#=64, respectively as these calculated compositions are similar to the high Mg# , high MgO aphyric endmembers of the groups. The highest Mg# of the aphyric samples is around 70 for the LTi suite, which is what is expected for melts in equilib-

Table 1. Regression-based estimates** of primary magma compositions, MgO-standardized compositions.

	ITi			LTi		
	mean	std dev	n	mean	std dev	n
	(Mg# = 64)			(Mg# = 70)		
SiO$_2$	51.84	0.81	277	48.33	1.01	317
TiO$_2$	1.01	0.16	277	0.45	0.13	317
Al$_2$O$_3$	14.92	0.60	277	14.47	0.90	317
FeO*	9.32	0.50	277	10.66	0.94	317
MgO	8.83	0.29	277	13.58	0.91	317
CaO	11.57	0.46	277	10.51	0.50	317
Na$_2$O	2.01	0.23	277	1.68	0.20	317
K$_2$O	0.51	0.19	277	0.33	0.14	317
P$_2$O$_5$	0.10	0.04	77	0.08	0.02	47
La	8.61	2.67	47	4.86	1.82	72
Ce	20.9	6.5	47	11.3	4.1	72
Sm	2.64	0.58	47	1.62	0.44	72
Eu	0.90	0.15	47	0.52	0.13	72
Tb	0.52	0.09	47	0.39	0.12	66
Yb	1.79	0.23	47	2.23	0.32	72
Lu	0.27	0.05	47	0.33	0.05	72
Ba	124	58	40	118	36	51
Hf	2.26	0.69	47	1.11	0.41	35
Nb	6.90	2.86	31	2.17	0.96	44
Rb	18.5	8.2	40	13.0	6.7	70
Sr	181	32	40	101	38	56
Ta	0.33	0.20	47	0.13	0.06	32
Th	1.38	0.76	30	0.76	0.40	59
V	269	36	33	193	19	72
Y	18.4	5.1	32	14.2	5.2	72
Zr	81.6	18.3	40	49.0	11.3	71

*Total Fe

**Estimates of primary compositions are for major oxides (in wt.%) and trace elements (in ppm). Ratios are calculated from non-normalized data.

rium with mantle olivine (Fo90-92). The LTi suite does include basalts with higher MgO content, but these show a straight line relationship on a MgO versus Ni plot indicating olivine accumulation [*Hart and Davis*, 1978]. In addition basalts with more than 14 wt% MgO have olivine phenocrysts. The most primitive aphyric basalts have MgO content between around 13 wt. % and do not show evidence for olivine accumulation. The aphyric samples of the most mafic members of the ITi-suite have Mg#=64, which is within the range of melts derived from a fertile peridotite source [*Kushiro*, 1996]. Thus although some of the differences between the high MgO basalts of the LTI suite can be explained by olivine accumulation, the differences between LTi and ITi suite have to lie in either the source or in the melting process.

The most striking difference between the normalized values of LTi and ITi is the higher SiO_2 and TiO_2 and lower normalized FeO for the ITi magmas. In particular the lower Fe content is surprising as the magma is derived from a source with a lower Mg#. The low sodium content of the parental basalts (≈ 2wt% Na_2O) indicates degrees of melting in excess of 10% if melts are derived from a moderately depleted mantle [*Langmuir et al.*, 1992]. A comparison of ITi and LTi indicates that they are also significantly different in trace element composition, especially the high field-strength elements and the REE. LTi magmas have lower concentrations of incompatible elements than ITi magmas and have lower values for trace element ratios like Ti/Zr and La/Yb, however ITi magmas have lower concentrations of heavy REE.

The lower Mg# and higher TiO_2 content for the ITi indicate that these basalts are derived from a more fertile/enriched source. One would expect that this enrichment is not restricted to TiO_2 only but would also be evident in the enrichment of other major elements, especially FeO and Na_2O and it is expected that FeO, Na_2O, and TiO_2 all increase with increasing fertility. FeO and Na_2O are also used as indicators for degree and depth of melting. The silica content of the estimated primary magmas is higher for the ITi than for the LTi, indicating that there is a pressure component to the major element variations whereby ITi magmas are derived from shallower depths than LTi magmas. Although ITi is thought to be derived from a more enriched source, the lower FeO content of the primary magmas confirms the differences in depth of melting derived from the difference in SiO_2 content as low FeO magmas are thought to be derived from shallower depths. The proposed enriched source for the ITi makes the lower FeO content even more significant. It is difficult to estimate the degree of melting as the Na_2O content of the ITi source is most surely higher than that of the LTi source. The similar Na_2O for the parental magmas of the

two suites indicates that ITi most likely represent a higher degree of melting than the LTi.

Estimates for the degree of melting within the LTi suite in the southeastern USA show systematic geographic variations [*Ragland et al.*, 2000]. Standardized major-element contents (i.e. Si_8, Na_8 etc.) shows that Na_8 and Ti_8 increases from north (Virginia) to south (Georgia and Alabama), suggesting decreasing degree of melting from north to south. Major element indicators of depth of melting (Si_8 and Fe_8) suggest deeper melts for tholeiites in the southern part of the province [*Ragland et al.*, 2000].

Although the distinction between the magma types is based on major elements, the distinction also bears out in the trace elements throughout CAMP (see table 1 and fig 3). These chemical characteristics persist over a large geographic area indicate that these magmas reflect a large scale feature in the Earth. Averages for La/Yb in Table 1 confirm the steeper negative slope of the ITi as compared to LTi. Although U-shaped REE pattern are generally rare, LTi tholeiites show U-shaped REE patterns throughout the whole province [*Dupuy et al.*, 1988; *Pegram*, 1990]. U-shaped REE patterns in peridotites have been shown to be indicative of a complex history in which a depletion event (resulting in LREE depletion) is overprinted by an enrichment event (i.e. addition of a component with high LREE/HREE ratio; [*Bodinier et al.*, 1990; *Takazawa et al.*, 1992]. There are no likely mineral phases that could fractionate the REE during melting such that a non-concave REE pattern of the source results in a U-shaped pattern in the ITi magmas. Therefore we have to assume that the U-shaped pattern reflect a U-shaped source pattern and a relatively complex history for the source of the LTi magmas.

A comparison of the trace element ratios in Table 1 and Fig. 3 indicates the following: 1) most of the averages

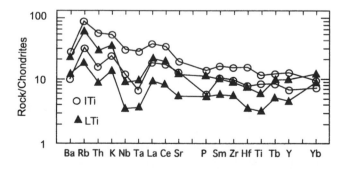

Figure. 3. Spidergrams (upper limit: mean + one standard deviation; lower limit: mean − one standard deviation) for estimates of primary LTi (open circles) and primary ITi (closed triangles). Plots are based on data from Table 1. Note the strong negative Nb-Ta anomaly

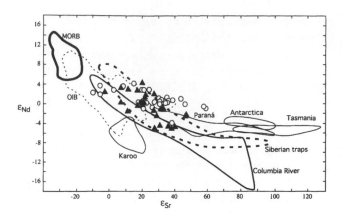

Figure 4. Plot of ε_{Nd} and ε_{Sr} for ITi (circles) and LTi (triangles). Also shown are fields for other Paraná-Etendeka low Ti basalts [*Peate and Hawkesworth*, 1996], Portal Peak, Antarctica [*Hergt et al.*, 1989], Siberian Traps [*Lightfoot et al.*, 1993], Karoo [*Ellam and Cox*, 1989; *Ellam and Cox*, 1991], Columbia River basalts [*Hooper and Hawkesworth*, 1993]. Fields for MORB and OIB taken from Zindler and Hart [1986].

suggest that all CAMP rocks are more enriched in light ion lithophile elements than MORB; 2) CAMP rocks show light REE enrichments that are less than OIB but similar to IAT; 3) both LTI and ITi show depletions in high field strength elements on spider diagrams which are indicative of an island arc-type component; 4) ITi and LTi magmas have La/Nb and Ba/Nb indicative of a continental component (either crust or lithosphere); LTi magmas have higher La/Nb indicating a larger continental component that in the ITi magmas; 5) ITi magmas have higher Ce/Yb, Ti/Zr and Hf/Lu than LTi magmas again indicating either a more enriched source and/or a smaller degree of melting.

Fig. 4 shows a plot of Nd- and Sr-isotopic compositions from eastern North America. No obvious differences in isotopic composition exist between the ITi and LTi samples. Although most of the data fall in the mantle array, their trend is not quite parallel to the mantle array. Compared to other provinces and compared to OIB the field for CAMP basalts seems to be offset to higher ε_{Sr} values for given ε_{Nd}. This offset can potentially be explained by alteration; none of the samples analyzed have been leached and the Sr-isotopic compositions could be affected by alteration. Results from our initial unpublished leaching studies shows that alteration can be a problem for these 200 Ma old tholeiites. Therefore the Sr-isotopic values should be considered maximum values for these samples. The range of isotopic compositions is quite large, even if alteration is a factor, and indicates the involvement of either continental lithosphere or crust.

REE patterns can provide an independent test for the differences in degrees of melting for ITi and LTi. The

source composition and the degree of melting can be constrained using the chondrite-normalized REE patterns in Fig. 3 for primary melt compositions in combination with the isotopic characteristics. The strongest constraint from the REE pattern is the slope of the pattern. The ITi magmas have a REE patterns with small LREE enrichments, which is consistent with their Nd-isotopic compositions which are close to chondritic. The U-shaped pattern for the LTi magmas indicate a complex history for the source which involves both depletion and enrichment [*Bodinier et al.*, 1988]. This complex history leaves the source composition for the LTi rather unconstrained. Pegram [1986] has shown that the isotopic compositions do not show any correlations with indicators of crustal contamination and that the isotopic and trace element characteristics of magmas suggested a source older than 1 Ga. Using this 1 Ga estimate for the age of the source, the Nd-isotopic characteristics can be used to estimate the time-integrated Sm/Nd of the source. The difference between the Sm/Nd ratio derived from the isotopic composition and the measured Sm/Nd of the basalts has to be explained by recent processes, i.e. melting. The average ε_{Nd} of the basalts (both LTi and ITi) is approximately zero (Figure 4). Assuming the simplest history for the source this implies a close to chondritic REE pattern. Time integrated non-chondritic Sm/Nd ratios would (in 1 Ga) lead to non-zero ε_{Nd}. ITi parental magma has $La_N/Lu_N = 2.7$ and $La_N = 25$ (N stands for chondrite normalized). Assuming a REE pattern and concentrations for the source similar to bulk silicate Earth, the ITi parental magmas can be generated by 10% melting; of such melting 70% would take place in the garnet stability field. The melting model employed is one similar to the model of Salters [1996] using partition coefficients (K_D's) from Hart [1992] for low pressure high-Ca clinopyroxene and partition coefficients from Salters and Longhi [1999] for all other phases. An upper limit on the degree of melting can be set by taking the isotopically most enriched magma for the ITi source ($\varepsilon_{Nd} = -4$) which allows a (Sm_N/Nd_N) of 0.93 for the source. In this case the degree of melting can be increased to a maximum of 15%, and still a significant amount of melting in the garnet stability field is required. The REE pattern of the ITi can also be reproduced by shallower melting (in the spinel stability field only) but this requires that the absolute concentrations in the source of the ITi basalts are lower. However, the major element characteristics indicate a fertile source for the ITi (see above) and it is therefore unlikely that the REE concentrations in the source are 3-4 times below those of bulk silicate earth. The lower LREE content of these LTi magmas as well as lower contents of large ion lithophile elements indicates a higher degree of melting for the LTi than for the ITi or a more depleted source for the ITi both consistent with the major element estimates.

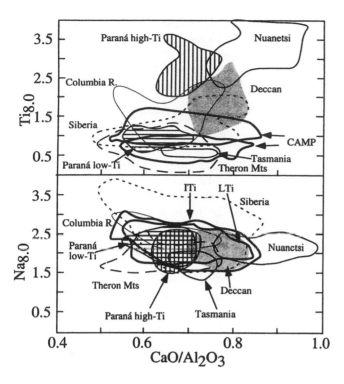

Figure 5. Ti8 and Na8 versus CaO/Al_2O_3. Figure taken from *Turner and Hawkesworth*, [1995] and the CAMP fields were added.

Fe_8 values are higher than values for these low-Ti-provinces. This indicates that the Ti-content in the CAMP tholeiites are lower than expected based on the other major element characteristics. Some similarities exist between the CAMP and the Columbia River basalts, although the compositional range of the Columbia River basalts is larger and extends to more enriched compositions.

In terms of isotopic characteristics CAMP basalts fall on the depleted end of the spectrum on a Sr-Nd isotope correlation diagram, again indicating that the amount of crustal contamination must be minimal. Although the range in composition is large, the relatively depleted character of the CAMP LTI and ITi (compared to other CFB) is also seen in the La/Sm, Ce/Yb and Ti/Zr ratios of these magmas, as some CAMP basalts always form the depleted end of the spectrum. Trace element ratios of the highly incompatible elements like Ba/La, Ba/Rb, Rb/Th show that the ITi and LTi magmas are similar to other CFB provinces and do not show this depletion.

The large depletion Nb compared to other CFB "fits" with the low Ti-characteristics of these magmas indicating that indeed the Ti content is anomalously low compared to the other major element characteristics. High La/Nb and

As there are differences in the source composition between LTi and ITi we have to evaluate the possibility of a more hydrous source for either magma type. For a given pressure and degree of melting, magmas that form under relative hydrous conditions should be higher in SiO_2, and lower in TiO_2, FeO* (total iron), MgO, CaO, and Na_2O [*Turner and Hawkesworth*, 1995]. Relative to LTi, ITi rocks are higher in SiO_2, but also higher in TiO_2, CaO, and Na_2O, so differences in the magma compositions cannot be caused by differences in the water content of the source alone and have to relate to variations in degree of melting and variations in average depth of melting.

COMPARISON WITH OTHER LIPS

Compared to other flood basalts provinces the LTi series fall to the low-Ti end (see fig 5) with intermediate Na-content, and fall to the high-Fe end of the spectrum (fig 6). In addition, CAMP magmas are average in terms of CaO/Al_2O_3 ratio and $Mg\#_8$. which are both an indicator for source fertility. Both LTi and ITi tholeiites from the CAMP are similar in Ti and Na content to the low-Ti basalts from Paraná (Brazil), Tasmania and Theron Mountains. However, $Mg\#_8$ values ($Mg\#$ = molar $100*Mg/(Mg+Fe^{2+})$) for CAMP tholeiites are lower and

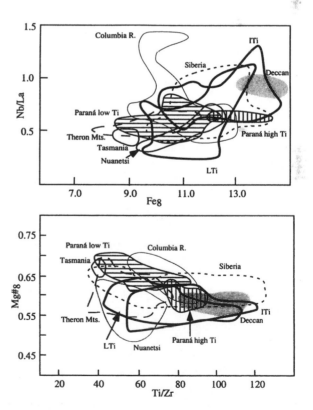

Figure 6. Nb/La vs Fe8 and Mg#8 vs Ti/Zr for a number of LIPs. Figure taken from *Turner and Hawkesworth*, [1995] with CAMP fields added.

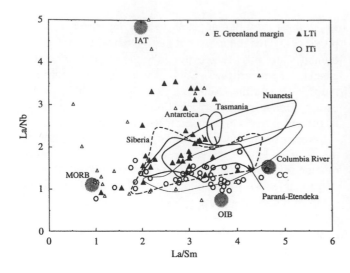

Figure 7. La/Nb vs La/Sm. Data sources and symbols same as in Fig 5. In addition NE Greenland data from Fitton et al. [1998]. Average for MORB and OIB are indicated [*Sun and McDonough*, 1989]. IAT is island arc tholeiites, average from Mariana Arc basalts [*Elliott et al.*, 1997]; CC is continental crust estimate [*Rudnick and Fountain*, 1995].

high field strength element depletions are characteristic of subduction related volcanism. Based on the high field strength element depletions in the ITi Pegram [1986; 1990] concluded that the lithosphere beneath CAMP had seen the addition of a subduction component. For subsurface samples from CAMP basalts in the South Georgia Rift Basin of northern Florida (LTi basalts) Heatherington and Mueller [1999] concluded that a lithospheric source is most likely. It thus appears that the LTi magmas contain the largest subduction component, which would mean that these lavas have the largest lithospheric component of all CAMP magmas. In addition, compared to other provinces (see fig. 7) the higher La/Nb in CAMP rocks seems to indicate that either the mantle source was more extreme in composition (larger subduction component to the lithosphere) or that the influence of an asthenospheric component was less (i.e. a smaller "plume" component than in other provinces). The relatively low ε_{Sr}, high ε_{Nd} and relatively large ion lithophile characteristics of the CAMP basalts is in agreement with a smaller plume component in CAMP than in other provinces.

CONSEQUENCES FOR CONTINENTAL BREAKUP

The opening of the Atlantic Ocean happened in three stages, each beginning with the formation of a LIP, as in the case of the CAMP at the opening of the central Atlantic approximately 200 Ma ago. The Paraná-Etendeka (P-E; 115-135 Ma) province is associated with the opening of the South Atlantic and the North Atlantic Tertiary province (NAT; initiated at 62 Ma) is associated with the opening of the North Atlantic. Thus, it is appropriate to make a brief comparison of the CAMP tholeiites to the extensive geochemical and isotopic literature on the P-E and NAT. Each of the younger provinces is associated with clearly defined hot spots; however, whether or not the plume initiated the break up is still debated for both P-E and NAT [*Harry and Sawyer*, 1992; *Mutter et al.*, 1988; *Mutter et al.*, 1982; *Zehnder et al.*, 1990] and both crustal and lithospheric controls on the breakup have been suggested. Translation of the geochemistry of the basalts in the paleo P-T conditions of the melting regime in combination with geochemical constrains on the basalts' sources tightly constrain mechanisms of continental break up. For NAT the major- and trace-element systematics show that with time the average depth of melting decreased while the degree of melting increased [*Fram and Lesher*, 1993; *Fram et al.*, 1998]. In addition, both the isotopes and trace elements argue against a large contribution of a continental lithosphere component to the magmatism [*Fitton et al.*, 1998; *Fram et al.*, 1998].

For Paraná the Open University group [*Peate*, 1997; *Peate and Hawkesworth*, 1996; *Peate et al.*, 1999] has clearly shown the important contribution the continental lithosphere makes to the volcanism. In particular, the high La/Nb ratios and Sr-Nd-isotopic compositions indicate a significant lithosphere component during the early stages and a larger astenospheric component at later stages [*Garland et al.*, 1996]. Again, there is a progression and depth of melting that is decreasing with time as the lithosphere is removed from the rift.

Further evidence for an extensive magmatic event has been found in the subsurface of the eastern US by seismic experiments which image the entire crust across the continent-ocean transition. Seismic stratigraphy of the US continental margin reveals seaward dipping reflectors forming a 15-20 km thick sequence of igneous rocks in the margin which thins to "normal" thickness igneous oceanic crust across less than 100 km [*Holbrook et al.*, 1994a; *Holbrook et al.*, 1994b]. It is unclear whether the igneous rocks "imaged" by the seaward-dipping reflectors are later or contemporaneous with the dike swarms, although, as discussed above, they seem to correlate with the flows of the South Georgia Rift basin and the NS-trending dike swarm. Cross sections at different locations up to 1000 km apart all show the same thickness for the igneous sequence indicating that these basalts do not find their origin in a local phenomenon, such as a hot spot, and that the amount of igneous rocks produced is voluminous. The igneous sequence in the margin has relatively high seismic velocities.

Table 2. Summary of characteristics for CAMP magmatism

Timing Stage	early rifting	middle earliest drifting	late drifting	later early spreading
Thermal event	waxing	maximum	waning	?
Swarm	NE	NW	N-S	Seaward dipping reflectors
Dikes parallel to	Atlantic rifts	Atlantic transforms	failed arm?	?
Magma types	ITi	LTi, ITi, HTi	LTi?, ITi?	?
Magma heterogeneity	homogeneous	heterogeneous	homogeneous	?
Components*	lith.>asthen.	Lith.>>asthen.	lith.>>asthen.	?
Depth of melting	2.0-2.5GPa	2.0-3.0GPa	2.0-3.0GPa	3.5 GPa
Degree of melting	10-15%	10-20%	10-20%	5%
Dike abundance	few	abundant	few	?
Dike size	long, wide	short, narrow	long, wide	?

lith. – lithospheric components in the source; asthen. - asthenospheric components in the source.

Kelemen and Holbrook [1995] parameterized seismic velocities as a function of pressure, temperature and composition. Compositional dependence of seismic velocity is controlled by the silica and magnesium content of the rock. Silica and magnesium are also good indicators of the depth and degree of melting. These relationships allowed Kelemen and Holbrook to translate seismic velocity into key major element parameters which allowed estimates for degree and depth of melting. They concluded that the high velocities of the igneous sequence (vp=7.3 km s-1) in the US continental margin reflect a high MgO content and a high percentage of olivine in the igneous sequence. From the inferred chemical characteristics an average pressure of melting in excess of 3.5 GPa was estimated and melt fractions well below 10% were estimated. Lizarralde and Holbrook [1997] modeled the subsidence history of the eastern US continental margin and combined with the estimates for the degree and depth of melting for the seaward dipping reflectors [Kelemen and Holbrook, 1995] placed the thermal lid at a minimum depth of 150 km. The estimate for depth of melting is significantly higher than estimates based on major elements for either LTi and ITi suites. Thus, data presently available are interpreted to indicate a progressive increase in the average depth of melting. The chronology of melting depth is thus the following (see also table 2): during the early rifting stage the NE-trending dikes (ITi) were feeders for relatively homogeneous volcanism. Melts were generated at average pressures of at most 2.5 GPa and degree of melting was 10-15%. Magmatism during the drifting stage (the NW-trending suite), which was contemporaneous or slightly later than the drifting stage, became more heterogeneous in composition, the pressure of melting was between 2.5 and 3.0 GPa, and the degree of melting was 15-20%. The geo-chronological data discussed above indicates perhaps only 1-2 Ma between the NW-and the NE-trending suites in North America [Hames et al., 2000]. During the initial spreading stage, which is expected to occur after the drifting stage, the seaward dipping reflectors were produced and the depth of melting increased (>3.5GPa) and degree of melting decreased to less than 10% [Kelemen and Holbrook, 1995]. During the mature spreading stage and the formation of normal oceanic crust depth of melting decreased again to 3.0-1.0 GPa and degree of melting increased again to 10-20%. Thus until the formation of the ocean crust the depth of melting was increasing with time. Furthermore, during the initial spreading stage the magma source was homogeneous while during later stages the magma source became more heterogeneous and contained a larger lithosphere component. The progression of degree and depth of melting is consistent with a shallow origin for the spreading in a crustal thinning regime. During the early stages of crustal thinning the melting is restricted to the upper part of the mantle, but as the thinning progresses melts can be derived from relatively deeper parts in the mantle. If there is limited lateral movement of mantle material the upper parts of the mantle will be depleted due to previous melting events and the melting can only progress at deeper levels in the mantle. The seaward dipping reflectors are the tail end of this process and the basalts derived from deep levels are thought to be MgO-rich [Kelemen and Holbrook, 1995] and thus might be derived from a more depleted source. In the mature spreading stage where lateral movement of the shallow mantle is established the depth of melting bounces back up to normal levels.

At first glance, the trace element characteristics (e.g., La/Nb-ratios) are inconsistent with this story as they indicate that the largest contribution from the lithosphere is expected to occur in the earliest stages of magmatism. Thus some additional explanation is required. In addition to lower La/Nb ratios, the ITi magmas are also thought to come from a more fertile source. One could argue that the ITi magmas are the most easily fusible parts of the lithosphere which are more enriched (perhaps containing a basaltic component) and are therefore less influenced by the "subduction overprint" of the continental lithosphere. If this were the case, our estimates for degree and depth of melting for the ITi suite would decrease and the progression to increasing depth of melting would become more obvious. This scenario would also fit with the more heterogeneous character of the earliest drifting stage in which all parts of the lithospheric mantle are melting with perhaps an addition of the asthenosphere. This addition would result in a large range of magma compositions as compared to the early stage where only the most enriched parts are able to melt. Recent work on the oldest sampled MORB (160-140 Ma in age) from the central Atlantic show an enriched isotope signature however these basalts do not show the negative Nb-Ta-anomalies seen in the CAMP basalts [*Janney and Castillo*, 2001]. These characteristics indicate the presence of an enriched mantle component in the oldest recovered basalts from the ocean basin and hint at the possibility of the existence of a plume.

Our proposed scenario points to a passive break-up of the continent. Testing of this scenario requires sampling and analysis of the seaward dipping reflectors, which are key in this scheme as well as combined analyses on critical samples. For example, more data on single samples are needed to determine whether trace element characteristics such as La/Nb are correlated with isotopic characteristics.

In conclusion, we have shown that the estimates of degree and depth of melting based on major and trace element characteristics can provide important constraints on the mechanism for continental break-up associated with the CAMP. The temporal progression with time of the chemical characteristics indicate deeper melting with time which is consistent with a shallow origin (such as crustal thinning) and passive origin for the break-up of the Pangaean continent.

Acknowledgements. J. De Boer, A. Marzoli, G. Mchone, P. Olsen, J. Puffer, and M. Withjack are gratefully acknowledged for many fruitful discussions, not only in connection this paper but also about the camp event in general. We also thank A. Stracke for helping to formulate some of the ideas developed herein.

REFERENCES

Anderson, D.L., The sublithospheric mantle as the source of continental flood basalts; the case against the continental lithosphere and plume head reservoirs, *Earth and Planetary Science Letters*, *123*, 268-280, 1994.

Austin Jr., J.A., P.L. Stoffa, J.D. Phillips, J. Oh, D.S. Sawyer, G.M. Purdy, E. Reiter, and J. Makris, Crustal structure of the southeast Georgia embayment-Carolina trough: Preliminary results of a composite seismic image of a continental suture(?) and a volcanic passive margin, *Geology*, *18*, 1023-1027, 1990.

Bellieni, G., M.H.F. Macedo, R. Petrini, E.M. Picrillo, G. Cavazzini, P. Comin-Chiaramonti, M. Ernesto, J.W.P. Macedo, G. Martins, A.J. Melfi, I.G. Pacca, and A. De Min, Evidence of magmatic activity related to Middle Jurassic and Lower Cretaceous rifting from northeastern Brazil (Ceara-Mirim): K/Ar age, paleomagnetiism, petrology, and Sr-Nd isotope characteristics, *Chemical Geology*, *97*, 9-32, 1992.

Bellieni, G., E.M. Picrillo, G. Cavazzini, R. Petrini, P. Comin-Chiaramonti, A.J.R. Nardi, and A.J. Melfi, Low and high-TiO2 Mesozoic tholeiitic magmatism of the Maranhao basin (NE Brazil): K/Ar age, geochemistry, petrology, isotope characteristics, and relationships with Mesozoic low and high-TiO2 basalts of the Paranha basin (SE Brazil), *Neues Jahrbuch Mineralogische Abhandlungen*, *162*, 1-33, 1990.

Bertrand, H., J.P. Liegeois, K. Deckart, and G. Feraud, High-Ti tholeiites in Guiana and their connection with the Central Atlantic CFB province: Elemental and Nd-Sr-Pb isotope evidence for a preferential zone of mantle upwelling in the course of rifting, *Transactions, American Geophysical Union*, *80*, 317, 1999.

Bodinier, J.L., C. Dupuy, and J. Dostal, Geochemistry and petrogenesis of Eastern Pyrenean perodotites, *Geochim. Cosmochim. Acta*, *52*, 2893-2907, 1988.

Bodinier, J.L., G. Vasseur, J. Vernieres, C. Dupuy, and J. Fabries, Mechanisms of mantle metasomatism: Geochemical evidence from the Lherz orogenic peridotite, *J. Petrol.*, *31*, 597-628, 1990.

Boutilier, R.R., and C.E. Keen, Small-scale convection and divergent plate boundaries, *Journal of Geophysical Research*, *104*, 7,389-7,403, 1999.

Buck, W.R., Small-scale convection induced by passive rifting: the cause for uplift of rift shoulders, *Earth and Planetary Science Letters*, *77*, 362-372, 1986.

Campbell, I.H., and R.W. Griffiths, Implications of mantle plume structure for the evolution of flood basalts, *Earth and Planetary Science Letters*, *99*, 79-93, 1990.

Chowns, T.M., and C.T. Williams, Pre-cretaceous rocks beneath the Georgia Coastal Plain-regional implications, in *Studies related to the Charleston, South Carolina, earthquake of 1886-tectonics and seismicity*, edited by G.S. Gohn, pp. L1-L42, US Geological Survey, 1983.

Courtillot, V., J. Jaeger, Z. Yang, G. Feraud, and C. Hofman, The influence of continental flood basalts on mass extinctions: Where do we stand?, *Geological Society of America Special Paper*, *307*, 513-525, 1996.

Courtillot, V., C. Jaupart, I. Manighetti, P. Tapponier, and J.

Besse, On causal links between flood basalts and continental breakup, *Earth and Planetary Science Letters*, *166*, 177-195, 1999.

Dalrymple, G.B., C.S. Gromme, and R.W. White, Potassium-argon and paleomagnetism of diabase dyke in Liberia: initiation of Central Atlantic rifting, *Geological Society of America Bulletin*, *86*, 399-411, 1975.

Dalziel, I.W.D., L.A. Lawvera, and J.B. Murphy, Plumes, orogenesis, and supercontinental fragmentation, *Earth and Planetary Science Letters*, *178*, 1-11, 2000.

de Boer, J., J.G. McHone, J.H. Puffer, P.C. Ragland, and D. Whittington, Mesozoic and Cenozoic magmatism, in *The geology of North America, v. I-2, the Atlantic continental margin, U.S.*, edited by R.E. Sheridan, and Grow, G. A., pp. 217-241, Geological Society of America, Boulder, 1988.

Deckart, K., G. Féraud, and H. Betrand, Age of Jurassic continental tholeiites of French Guyana, Surinam, and Guinea: Implications for the initial opening of the Central Atlantic Ocean, *Earth and Planetary Science Letters*, *150*, 205-220, 1997.

Dunning, G.R., and J.P. Hodych, U/Pb zircon and baddeleyite ages for the Palisades and Gettysburg sills of the northeastern United States: Implications for the age of the Triassic/Jurassic boundary, *Geology*, *18*, 795-798, 1990.

Dupuy, C., J. Marsh, J. Dostal, A. Michard, and S. Testa, Asthenospheric and lithospheric sources for Mesozoic dolerites from Liberia (Africa): trace element and isotopic evidence, *Earth Plan Sci Lett*, *87*, 100-110, 1988.

Ellam, R.M., and K.G. Cox, A Proterozoic lithospheric source for Karoo magmatism: evidence from the Nuanetsi picrites, *Earth Planet. Sci. Lett.*, *92*, 207—218, 1989.

Ellam, R.M., and K.G. Cox, An interpretation of Karo picrite basalts in terms of interaction between asthenospheric magmas and the mantle lithosphere, *Earth and Planetary Science Letters*, *105*, 330-342, 1991.

Elliott, T., T. Plank, A. Zindler, W. White, and B. Bourdon, Element transport from slab to volcanic front at the Mariana arc, *Journal of Geophysical Research*, *102*, 14,991-15,019, 1997.

Farnetani, C.G., and M.A. Richards, Numerical investigations of the mantle lume initiation model for flood basalt events, *Journal of Geophysical Research*, *99*, 13,813-13,833, 1994.

Fitton, J.G., B.S. Hardarson, R.M. Ellam, and G. Rogers, Sr-, Nd-, and Pb-isotopic composition of volcanic rocks from the southeast Greenland margin at 63°N: temporal variations in crustal contamination during continental breakup, *Proceedings of the Ocean drilling Program, Scientific results*, *152*, 351-357, 1998.

Fram, M.S., and C.E. Lesher, Geochemical constraints on mantle melting during creation of the North Atlantic Basin, *Nature*, *363*, 712-715, 1993.

Fram, M.S., C.E. Lesher, and A.M. Volpe, Mantle melting systematics; transition from continental to oceanic volcanism on the southeast Greenland margin, *Proceedings of the Ocean drilling Program, Scientific results*, *152*, 373-386, 1998.

Garland, F.E., S.P. Turner, and C.J. Hawkesworth, Shifts in the source of Parana basalts through time, *Lithos*, *37*, 223-243, 1996.

Hames, W.E., P.R. Renne, and C. Ruppel, New evidence for geologically-instantaneous emplacement of the earliest Jurassic Central Atlantic Magmatic Province basalts on the North American margin, *Geology, in press*, 2000.

Harry, D.L., and D.S. Sawyer, Basaltic volcanism, mantle plumes, and the mechanics of rifting; the Parana flood basalt province of South America, *Geology*, *20*, 207-210, 1992.

Hart, S.R., and K.E. Davis, Nickel partitioning between olivine and silicate melt, *Earth Planet. Sci. Lett.*, *40*, 203-219, 1978.

Hart, S.R., and T. Dunn, Experimental cpx/melt partitioning of 23 trace elements, *Contrib. Min. Petrol.*, *113*, 1-18, 1992.

Heatherington, A.L., and P.A. Mueller, Lithospheric sources of north Florida, USA tholeiites and implications for the origin of the Suwannee terrane, *Lithos*, *46*, 215-233, 1999.

Hergt, J.M., B.W. Chappel, G. Faure, and T.M. Mensing, The geochemistry of Jurassic dolerites from Portal peak, Antarctica, *Contributions to Mineralogy and Petrology*, *102*, 298-305, 1989.

Hill, R.I., Starting plumes and continental break-up, *Earth and Planetary Science Letters*, *104*, 398-416, 1991.

Hill, R.I., I.H. Campbell, G.F. Davies, and R.W. Griffiths, Mantle plumes and continental tectonics, *Science*, *256*, 186-193, 1992.

Holbrook, S.W., E.C. Reiter, G.M. Purdy, D. Sawyer, P.L. Stoffa, J.A. Austin Jr., J. Oh, and J. Makris, Deep structure of the U.S. Atlantic continental margin, offshore South Carolina, from coincident ocean bottom and multichannel seismic data, *Journal of Geophysical Research*, *99*, 9155-9178, 1994a.

Holbrook, W.S., G.M. Purdy, R.E. Sheridan, L. Glover III, M. Talwani, J. Ewing, and D. Hutchinson, Seismic structure of the U.S. Mid-Atlantic continental margin, *Journal of Geophysical Research*, *99*, 17,871-17,891, 1994b.

Hooper, P.R., and C.J. Hawkesworth, Isotopic and geochemical constraints on the origin and evoilution of Columbia River basalts, *Journal of Petrology*, *34*, 1203-1246, 1993.

Janney, P.E., and P.R. Castillo, Geochemsitry of the oldest Atlantic oceanic crust suggests mantle plume involvement in the early history of the central Atlantic Ocean, *Earth and Planetary Science Letters*, *192*, 291-302, 2001.

Kelemen, P.B., and W.S. Holbrook, Origin of thick, high velocity igneous crust along the US East Coast Margin, *Journal of Geophysical Research*, *100*, 10,077-10,094, 1995.

King, S.D., and D.L. Anderson, Edge-driven convection, *Earth and Planetary Science Letters*, *160*, 289-296, 1998.

King, S.D., and D.L. Anderson, 1995, An alternative mechanism of flood basalt formation: Earth and Planetary Science Letters, v. 136, p. 269-279., An alternative mechanism of flood basalt formation, *Earth and Planetary Science Letters*, *136*, 269-279, 1995.

Klein, E.M., and C.H. Langmuir, Global correlations of ocean ridge basalt chemistry with axial depth and crustal thickness, *J. Geophys. Res.*, *92*, 8089-8115, 1987.

Kushiro, I., Partial melting of fertile mantle peridotite at high pressures: an experimental study using aggregates of diamond., in *Earth processes: Reading the isotopic code*, edited by S.R. Hart, and A. Basu, pp. 109-122, AGU, Washington, DC, 1996.

Langmuir, C.H., E.M. Klein, and T. Plank, Petrological Systematics of Mid-Ocean Ridge Basalts: Constraints on Melt Generation beneath Ocean Ridges, in *Mantle flow and melt generation at Mid-Ocean Ridges*, edited by J.P. Morgan, D.K. Blackman, and J.M. Sinton, AGU, Washington, D C, 1992.

Larsen, T.B., and D.A. Yuen, Ultrafast upwelling bursting through the upper mantle, *Earth and Planetary Science Letters*, *146*, 393-399, 1997.

Leitch, A.M., G.F. Davies, and M. Wells, A plume head melting under a rifting margin, *Earth and Planetary Science Letters*, *161*, 161-177, 1998.

Lightfoot, P.C., C.J. Hawkesworth, J.M. Hergt, A.J. Naldrett, N.S. Gorbachev, V.A. Fedorenko, and W. Doherty, Remobilisation of the continental lithosphere by a mantle plume: major-, trace-element, and Sr-, Nd-, and Pb-isotope evidence from picritic and tholeiitic lavas fo the Noril'sk District, Siberian trap, Russia, *Contributions to Mineralogy and Petrology, 114*, 171-188, 1993.

Lizarralde, D., and W.S. Holbrook, U.S. mid-Atlantic margin structure and early thermal evolution, *Journal of Geophysical Research, 102*, 22,855-22,875, 1997.

Marzoli, A., P.R. Renne, E.M. Piccirillo, M. Ernesto, G. Bellieni, and A. De Min, Extensive 200-million-year-old continental flood basalts of the Central Atlantic Magmatic Province, *Science, 284*, 616-618, 1999.

May, P.O., Pattern of Triassic-Jurassic diabase dikes around the North Atlantic in the context of pre-drift positions of the continents, *Geological Society of America Bulletin, 82*, 1285-1292, 1971.

McElwain, J.C., D. Beerling, and F.I. Woodward, Fossil plants and global warming at the Triassic-Jurassic boundary, *Science, 285*, 1386-1390, 1999.

McHone, J.G., Non-plume magmatism and rifting during the opening of the central north Atlantic ocean, *Tectonophysics, in press*, 2000.

McHone, J.G., and J.H. Puffer, Flood basalt provinces of the Pangaean Atlantic rift: regional extent and environmental significance, in *Proceedings on aspects of Triassic-Jurassic rift basin geoscience*, edited by P.M. LeTourneau, and Olsen, P.E., pp. submitted, Columbia University Press, New York, 1997.

Mutter, J.C., W.R. Buck, and C.M. Zehnder, Convective partial melting I: A model for the formation of thick basaltic sequemces during initiation of spreading, *J Geophys Res, 93*, 1031-1048, 1988.

Mutter, J.C., M. Talwani, and P.L. Stoffa, Origin of seaward-dipping reflectors in oceanic crust off the Norwegian margin by "subaerial sea-floor spreading", *Geology, 10*, 353-357, 1982.

Olsen, P.E., Stratrigraphic record of the early Mesozoic breakup of Pangea in the Laurasia-Gondwana Rift System, *Annual Reviews in Earth and Planetary Sciences, 25*, 337-401, 1997.

Olsen, P.E., R.W. Schlische, and M.S. Fedosh, 580 Ky duration of the early Jurassic flood basalt event in eastern North America estimated using Milankovitch cyclostratigraphy, *Museum of Northern Arizona Bulletin, 60*, 11-22, 1996.

Pálfy, J., J.K. Mortensen, E.S. Carter, P.L. Smith, R.M. Friedman, and H.W. Tipper, Timing the end-Triassic mass extinction: First on land, then in the sea?, *Geology, 28*, 39-42, 2000.

Peate, D.W., The Paraná-Etendeka Province, in *Large Igneous Provinces; Continental, oceanic and planetary volcanism*, edited by J.J.M.a.M.F. Coffin, pp. 217-245, American Geophysical Union, Washington DC, 1997.

Peate, D.W., and C.J. Hawkesworth, Lithospheric to asthenospheric transition in low-Ti flood basalts from southern Parana, Brazil, *Chemical Geology, 127*, 1-24, 1996.

Peate, D.W., C.J. Hawkesworth, M.M.S. Mantovani, N.W. Roger, and S.P. Turner, Petrogenesis and stratigraphy of the high-Ti/Y Urubici magma type in the Parana flood basalt province and implications for the nature of "Dupal"-type mantle in the South Atlantic region, *Journal of Petrology, 40*, 451-476, 1999.

Pegram, W.J., Geochemical processes in the sub-continental mantle and the nature of crust mantle interaction: Evidence from the Mesozoic Appalachian Tholeiite Province, PhD thesis, Massachusetts Institute of Technology, Cambridge, Massachusetts, 1986.

Pegram, W.J., Development of continental lithospheric mantle as reflected in the chemistry of the Mesozoic Appalachian Tholeiites, U.S.A., *Earth Plan. Sci. Lett., 97*, 316-331, 1990.

Philpotts, A.R., and I. Reichenbach, Differentiation of Mesozoic basalts of the Hartford Basin, Connecticut, *Geological Society of America Bulletin, 96*, 1131-1139, 1985.

Puffer, J.H., Initial and secondary Pangean basalts, in *Pangea: Global Environments and Resources*, pp. 85-95, 1994.

Ragland, P.C., L.E. Cummins, and J.D. Arthur, Compositional patterns for early Mesozoic diabases from South Carolina to Central Virginia, in *Eastern North America Mesozoic magmatism*, edited by J.H. Puffer, and P.C. Ragland, pp. 309-331, Geological Society of America, 1992.

Ragland, P.C., R.D.J. Hatcher, and D. Whittington, Juxtaposed Mesozoic diabase dike sets from the Carolinas, *Geology, 11*, 394-399, 1983.

Ragland, P.C., S.A. Kish, and W.C. Parker, Compositional pattens for lower Mesozoic olivine tholeiitic diabase dikes in the Deep River basin, North Carolina, in *Proceedings from TRIBI Workshop on Triassic Basins*, 1997.

Ragland, P.C., V.J.M. Salters, and W.C. Parker, A geographic trend for MgO-Standardized major oxides in lower Mesozoic olivine tholeiites of the southeastern US, in *Proceedings on aspects of Triassic-Jurassic rift basin geoscience*, edited by P.M. LeTourneau, and P.E. Olsen, Columbia University Press, New York, 2000.

Ratcliffe, N.M., High TIO_2 metadiabase dikes of the Hudson Highlands, New York and New Jersey: Possible late proterozoic rift rocks in the New York recess, *Am. J. Sci., 287*, 817—850, 1987.

Rudnick, R.L., and D.M. Fountain, Nature and composition of the continental crust: a lower crustal perspective, *Review of Geophysics, 33*, 267-309, 1995.

Ruppel, C., Extensional processes in continental lithosphere, *J. Geophys. Res., 100*, 24,187-24,215, 1995.

Ruppel, C., and W. Hames, Tectonic framework of the Circum-Atlantic Large Igneous Province: Evaluation of active and passive mechanisms, *Transactions, American Geophysical Union*, *80*, 317, 1999.

Salters, V.J.M., The generation of mid-ocean ridge basalts from the Hf and Nd isotope perspective, *Earth and Planetary Science Letters*, *141*, 109-123, 1996.

Salters, V.J.M., and J.E. Longhi, Trace element partitioning during the initial stages of melting beneath ocean ridges, *Earth and Planetary Science Letters*, *166*, 15-30, 1999.

Saunders, A.D., J.G. Fitton, A.C. Kerr, M.J. Norry, and R.W. Kent, The North Atlantic Igneous Province, in *Large Igneous Provinces; Continental, oceanic and planetary volcanism*, edited by J.J.M.a.M.F. Coffin, pp. 45-93, American Geophysical Union, Washington DC, 1997.

Sun, S.-S., and W.F. McDonough, Chemical and isotopic systematics of oceanic basalts: implications for mantle composition and processes, in *Magmatism in the Ocean Basins*, edited by A.D. Saunders, and M.J. Norry, pp. 313-345, Geological Society, London, 1989.

Sundeen, D.A., Note concerning the petrography and K-Ar age of a Cr-spinel-bearing olivine tholeiite in the subsurface of Choctaw County, north-central Mississippi, *Southeastern Geology*, *30*, 137-146, 1989.

Takazawa, E., F.A. Frey, N. Shimizu, M. Obata, and J.L. Bodinier, Geochemical evidence for melt migration and reaction in the upper mantle, *Nature*, *359*, 55-57, 1992.

Turner, S., and C.J. Hawkesworth, The nature of the sub-continental mantle: constraints from the major-element composition of continental flood basalts, *Chemical Geology*, *120*, 295-314, 1995.

Weigand, P.W., and P.C. Ragland, Geochemistry of Mesozoic dolerite dikes from eastern North America, *Contributions to Mineralogy and Petrology*, *29*, 195-214, 1970.

White, R.S., Asthenospheric control on magmatism in the ocean basins, in *Magmatism in the Ocean Basins*, edited by A.D.a.N. Saunders, M.J. (eds.), pp. 17-27, Geological Society, 1989.

Wilson, M., Thermal evolution of the central Atlantic passive margin: continental break-up above a Mesozoic super-plume, *Journal of the Geological Society of London*, *154*, 491-495, 1997.

Withjack, M.O., R.W. Schlische, and P.E. Olsen, Diachronous rifting, drifting, and inversion on the passive margin of eastern North America: an analog for other passive margins, *American Association of Petoleum Geologists Bulletin*, *82*, 817-835, 1997.

Zehnder, C.M., J.C. Mutter, and P. Buhl, Deep seismic and geochemical constraints on the nature of rift-induced mamgmatism during breakup of the North Atlantic, *Tectonophysics*, *173*, 545-565, 1990.

Zindler, A., and S.R. Hart, Chemical Geodynamics, *Ann. Rev. Earth Plan. Sci.*, *14*, 493-571, 1986.

Vincent J.M. Salters, National High Magnetic Field Laboratory and Department of Geological Sciences, Florida State University, 1800 E. Paul Dirac Drive, Tallahassee, Florida, 32306, USA (email: salters@magnet.fsu.edu).

Paul C. Ragland, Department of Geological Sciences, Florida Syate University, Tallahassee, Florida; Professor Emiritus, present address: 52 arshall Road, Troy, Virginia, 22974, USA.

W.E. Hames, Department of Geology, 210 Petrie Hall, Auburn University, Auburn, Alabama, 36830, USA.

K. Milla, Division of Agricultural Sciences, Florida A&M University, Tallahassee, Florida, 32305, USA.

C. Ruppel, School of Earth and Atmospheric Sciences, 221 Bobby Dodd Way, Georgia Institute of Technology, Atlanta, Georgia, 30332-0340, USA.

The Late Triassic-Early Jurassic Volcanism of Morocco and Portugal in the Framework of the Central Atlantic Magmatic Province: an Overview

Nasrrddine Youbi[1], Línia Tavares Martins[2], José Manuel Munhá[2], Hassan Ibouh[3], José Madeira[4], El Houssaine Aït Chayeb[1] and Abdelmajid El Boukhari[1]

An overview on the Late Triassic-Early Jurassic Magmatic Province of Morocco and Portugal (TJMPMP) is presented. It comprises extrusive basalts, interbedded with clastic rocks sequences preserved in elongated rift basins, and their feeder dikes and sills.

Paleontologic ages range from Upper Ladinian–Lower Carnian to the Sinemurian for the sediments, while available $^{40}Ar/^{39}Ar$ analysis yield a mean age of 200±1,6 Ma for the volcanics.

The volcanologic characteristics of the TJMPMP are those of continental basaltic successions. It comprises subaerial lava flows and pyroclastic deposits, sometimes deposited in lacustrine environments, and feeder dikes, constituting an interesting volcanic sub-province of the Central Atlantic Magmatic Province (CAMP) as most preserved outcrops are extrusive volcanics.

These rocks correspond to Low-Ti (TiO_2<2wt%), quartz normative, tholeiites displaying uniform chemical and isotopic characteristics, and showing variable upper crustal contamination. The most primitive non-contaminated rocks display $^{87}Sr / ^{86}Sr$ ~ 0,70553 and Ba / Nb >11 suggesting that their source may be within the continental lithospheric mantle.

The nucleation of the rifting process may have started at two different triple junctions, an RRR junction near Florida and a RRT between Africa-Iberia-America.

[1] Department of Geology, Faculty of Sciences-Semlalia, Cadi Ayyad University, Marrakech, Morocco.
[2] Centro de Geologia, Departamento de Geologia, Faculdade de Ciencias, Universidade de Lisboa, Portugal.
[3] Department of Geology, Faculty of Sciences and Technics-Gueliz, Cadi Ayyad University, Marrakech, Morocco.
[4] LATTEX, Departamento de Geologia, Faculdade de Ciencias, Universidade de Lisboa, Portugal.

The Central Atlantic Magmatic Province:
Insights from Fragments of Pangea
Geophysical Monograph 136
Copyright 2003 by the American Geophysical Union
10.1029/136GM010

INTRODUCTION

The magmatic rocks of the Late Triassic-Early Jurassic Magmatic Province of Morocco and Portugal (TJMPMP) have always been studied in a regional perspective. They were seldom the objects of synthetic studies as a Magmatic Province [e.g., *Manspeizer et al.*, 1978; *Martins*, 1991; *Aït Chayeb*, 1997]. TJMPMP constitutes an excellent marker for the study of the geodynamic evolution of Central Atlantic. Continental rifting of Central Atlantic started during Late Triassic or even at the end of Late Permian progressing from the South to the North along the trend of the Late-Paleozoic Alleghanian-Hercynian orogenic belt [*Manspeizer*, 1988; *Piqué and Laville*, 1996; *Withjack et al.*, 1998]. Mapping and dating pairs of magnetic anomalies allowed precise reconstitutions of the different opening stages of Central Atlantic Ocean [*Olivet et al.*, 1984] and to date at about 170-175 Ma (Middle Jurassic) the beginning of the oceanic accretion [*Klitgord and Schouten*, 1986]; this event seems to be more or less simultaneous from Florida to the Azores-Gibraltar transform, along 2700 Km [*Klitgord and Schouten*, 1986]. Identical ages (178-180 Ma) were obtained from xenoliths of metagabbros and metabasalts of MORB type affinity from the Neogene-Quaternary volcanics of Canaries, which are interpreted as fragments of underlying Mesozoic oceanic crust [*Schmincke et al.*, 1998; *Hoernle*, 1998].

The TJMPMP comprises an important NE-SW trending dyke and sill system (dykes of Messejana and Foum Zguid), lava flows and pyroclastic-sedimentary sequences. Similar rocks are present in the western Armorican Massif (France), in Algeria, further south along the African continental margin [e.g., *Dalrymple et al.*, 1975; *Dupuy et al.*, 1988; *Mauche et al.*, 1989; *Bertrand*, 1991; *Sebai et al.*, 1991; *Caroff et al.*, 1995], and in the conjugated margins of North and South America [e.g., *Whittington*, 1988a, b; *de Boer et al.*, 1988; *Oliveira et al.*, 1990; *Cummins et al.*, 1992; *Pe-Piper et al.*, 1992; *Puffer*, 1992; *McHone*, 1996; *Deckart et al.*, 1997; *Marzoli et al.*, 1999; *McHone and Puffer*, 2000]. This set of tholeiitic magmatic rocks present around the Atlantic (basic lava flows, sills and dykes, and layered massifs spreading over an area of more than 7000 Km from SW to NE, and 3000 Km from E to W, with an estimated total volume of erupted materials of 2,3 to 4 million Km^3) have a similar chemistry to that of reference continental tholeiites of the world and an emplacement age around 200 Ma [e.g., *Sutter*, 1988; *Dunnig and Hodych*, 1990; *Sebai et al.*, 1991; *Hodych and Dunnig*, 1992; *Deckart et al.*, 1997; *Dunn et al.*, 1998; *Marzoli et al.*, 1999; *Hames et al.*, 2000]. They are linked in time and space to the fragmentation of Pangea and to the initial rifting stages of Central Atlantic defining one of the largest Phanerozoic continental volcanic provinces (Continental Flood Basalts - CFBs, or Large Igneous Provinces - LIPs) [e.g., *Coffin and Eldholm*, 1994] on Earth. This LIP has been recently designated as Central Atlantic Magmatic Province (CAMP) by *Marzoli et al.*, [1999] (Figure 1). It seems to coincide in time with the major mass extinction at the Triassic-Jurassic boundary [*Olsen*, 1999; *Palfy et al.*, 2000] and to be genetically related to the activity of a thermally and chemically anomalous mantle plume (super-plume) [e.g., *Oyarzun et al.*, 1997; *Wilson*, 1997; *Thompson*, 1998] or to the initiation and production of new oceanic crust during Middle Jurassic time [e.g., *Withjack et al.*, 1998; *Medina*, 2000].

The aims of this work are (1) to present a description of the different modes of occurrence in the TJMPMP, (2) to define the facies model prevailing during the emplacement of this volcanism, (3) to better constrain the emplacement age of the volcanic activity by compiling the most recent biostratigraphical and geochronological data, and thus constrain the timing of the rifting and magmatism affecting this part of the Eastern Central Atlantic margin, (4) to describe the petrography and mineralogy of the studied igneous rocks, (5) to interpret previously published and new chemical analyses of whole rock, their petrogenetic histories and to explain the observed isotopic and geochemical composition variations in time and space, (6) and to define the place occupied by the studied magmatism in the proposed geodynamic models for the opening of the Central Atlantic.

1. GEOLOGICAL SETTING

1. 1. Sedimentary Characteristics of the Late Triassic-Early Jurassic of Morocco and Portugal

The Early Mesozoic series (Late Triassic) of Morocco outcrop in numerous exposures that are the visible part of wider basins recognized by geophysical studies and boreholes. Fifteen individual basins, showing ill-defined limits, were recognized for the Triassic-Jurassic boundary (Figure 2). The Triassic-Liassic series of Morocco can be described as a mega-sequence constituted by the superposition of two lithologic sequences overlying the Hercynian basement (Figure 3); the lower sequence, composed of sandstones and polygenic conglomerates, is Carnian or even, locally, Permian in age. It is known in the High Atlas and in the Moroccan Atlantic margin. Overlying these deposits is a sequence of evaporites and pelites [e.g., *Piqué and Laville*, 1993a; *Le Roy et al.*, 1997]. This sequence also includes a succession of lava flows dated 196-200 Ma by $^{40}Ar/^{39}Ar$ [*Fiechtner*, 1990; *Sebai et al.*, 1991; *Fiechtner et al.*, 1992] and, thus approximately corresponds to the Triassic-Jurassic boundary. The upper sequence corresponds to the Carnian-Hettangian time interval, which is the admitted age of the post-rift unconformity or the Break-up Unconformity of *Falvey* [1974] across the Moroccan margin [e.g., *Van Houten*, 1977; *Manspeizer*, 1988; *Medina*, 1995; *Le Roy et al.*, 1997; *Piqué et al.*, 1998].

The Late Triassic-Early Jurassic sequences of Portugal occupy the base of the Lusitanian, Santiago do Cacém and Algarve sedimentary basins, but volcanic sequences are only present in the later two. (Figure 4b, c).

In the Algarve, Triassic-Liassic series may be described as a mega-sequence formed by the superposition of four lithologic formations (Figure 4d) overlying the Hercynian basement and separated by unconformities [*Manuppella*, 1988; unpublished data, 2000]. (1) The "Grés de Silves" (Silves Sandstones - unit AA of *Palain*, 1979; unit A of *Manuppella*, 1988; or "Grés de Silves" s. s. in the Geological Map of Portugal) comprises conglomerates and sandstones organized in cyclical sequences. This member, which shows cross bedding in the upper part, is always present in the basin. In Central Algarve, between São

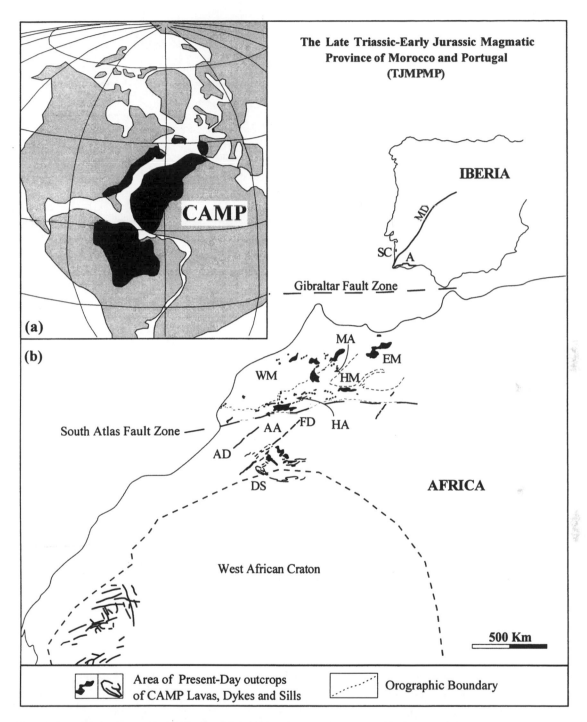

Figure 1. (a) Location map of the Central Atlantic Magmatic Province (CAMP) and (b) the Late Triassic-Early Jurassic Magmatic Province of Morocco and Portugal (TJMPMP) in a Pangea reconstruction at 200 Ma [after *Sebai et al.*, 1991; *Marzoli et al.*, 1999]. Abbreviations: Morocco, AA= Anti-Atlas; AD= Asdrem dike; DS= Draa sills; EM= Eastern Meseta; FD= Foum Zguid dike; HA= High Atlas; HM= High Moulouya; MA= Middle Atlas; WM= Western Meseta. Portugal, A= Algarve; MD= Messejana dike; SC= Santiago do Cacém.

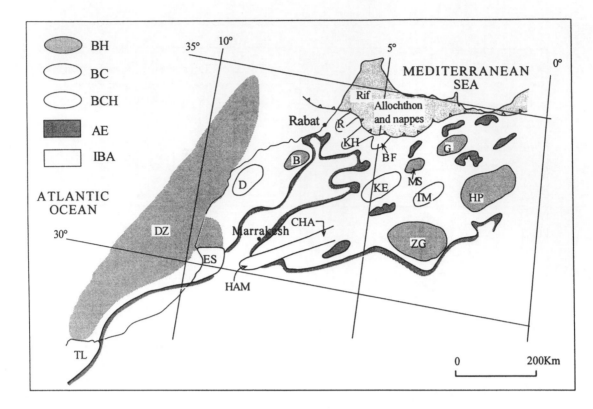

Figure 2. Generalized locations of rift basins and extent of Triassic deposits in Morocco [after *Salvan*, 1984; *Van Houten*, 1977; *Lorenz*, 1988; and *Laville and Piqué*, 1991]. Abbreviations: DZ= Diapiric zone of the Atlantic margin; TL= Tarfaya-Laayoun; D= Doukkala; ES= Essaouira; HAM= High Atlas of Marrakesh; CHA= Central High Atlas; B=Berrechid; R= Rharb; KH= Khemisset; BF=Bou Fekrane; KE= Kerrouchen; MS= Moussa ou Salah; G= Guercif; TM= Tamdafelt-Moulouya; ZG= Ziz-Guir; HP= High Plateaux. BH= Basins containing predominantly halites; BC = Basins containing predominantly clastics; BCH= Basins containing basal clastics and thick overlying halites; AE= Emergent areas during Triassic time; IBA= Interbasin areas covered mostly by evaporitic mudstones.

Bartolomeu de Messines and Silves, an essentially pelitic member ("Argilas de S. Bartolomeu de Messines"), containing siltitic intercalations, underlies the sandstones. (2) The "Complexo Pelítico, Carbonatado e Evaporítico" (term AB1 of unit AB of *Palain*, 1979; or unit B of *Manuppella*, 1988) comprises red pelites with fine to very fine sandstone intercalations, and dolomite layers with an irregular spatial distribution. This formation also contains evaporitic deposits (halite and gypsum) indicating a lagoonal or epicontinental sedimentary environment. (3) The "Complexo Vulcano-Sedimentar" is composed of alternating lava flows (up to 6 flows), pyroclastic layers (ash and breccia tuffs), fine detrital sediments and two discontinuous dolomitic levels located at the base or in variable positions in the sequence. Sometimes, the volcano-sedimentary sequence ends with a thin layer of red pelites. (4) The overlying sequence is formed by dolomites with a level of dolomitic breccia that corresponds to the post-rift unconformity, marking its base.

In the Santiago do Cacém region, the Triassic-Liassic sequence is similar to that of Algarve, with two small differences: (1) the sequence starts with a base conglomerate and does not include the "Argilas de S. Bartolomeu de Messines", and (2) a continuous dolomitic level is always present at the base of the "Complexo Vulcano-sedimentar".

1. 2. Volcanic Occurrences

Important dyke and sill swarms, lava flows, and pyroclastic-sedimentary sequences represent the magmatism of the TJMPMP.

Dykes and sills. The NE-SW Foum Zguid dyke is one of the major dykes of the Anti-Atlas domain. It extends for more than 200 Km, from the village of Foum Zguid in the South to Jbel Saghro in the North (Figure 5), with a thickness of 100-150 m. It cuts trough Precambrian and Paleozoic terranes as well as Hercynian structures, and is locally covered by Cretaceous formations [*Leblanc*, 1973; *Hollard*, 1973]. At the Bou Azzer El Graara inlier, a system of small parallel dykes (N30-50) with some perpendicular segments (N130-140), a few meters thick, locally form dense swarms accompanying the main dyke. Contact metamorphism originates the development of hornfels.

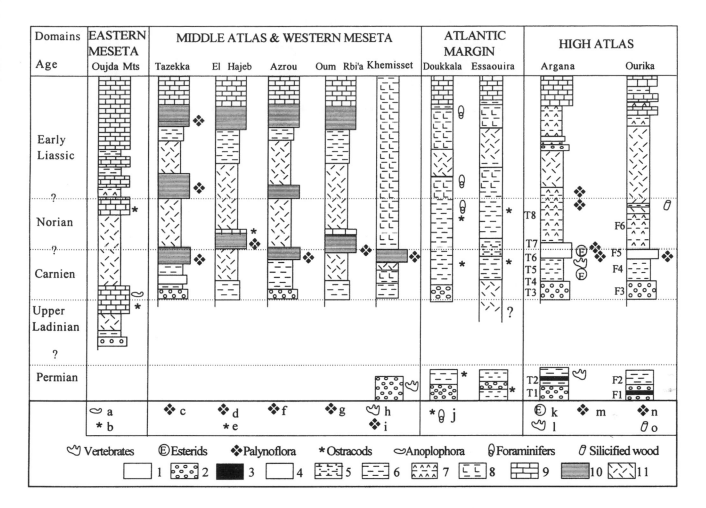

Figure 3. Generalized stratigraphic correlations of Permian-Triassic and Liassic series of Morocco [after *Salvan*, 1984; *Van Houten*, 1977; *Manspeizer et al.*, 1978; completed by *Aït Chayeb*, 1997; and *Oujidi et al.*, 2000]. Legend: 1= gap; 2= conglomerate; 3= magmatism of the second Permien eruptive cycle (?) of transitional affinity (necks and dykes of Argana and "andesites and spilites of Ourika"); 4= sandstones; 5= clayey sandstones; 6= silty claystones; 7= gypsiferous claystones; 8= halite and claystones; 9= carbonates; 10= grey claystones; 11= tholeiitic basaltic lavas. References: a= *Owodenco*, 1946; *Monition and Monition*, 1959; *Rakus*, 1979; and *Manspeizer et al.*, 1978; b= *Crasquin-Soleau et al.*, 1997; c= *Beaudelot and Charièrre*, 1983; *Sabaoui*, 1986; and *Beaudelot et al.*, 1990; d= *Ouarhache*, 1987; and *Beaudelot et al.*, 1986, 1990; e= *Ouarhache*, 1987; and *Beaudelot et al.*, 1986; f and g= *Ouarhache et al.*, 1999; h= *El Wartiti*, 1990; i= *Taugourdeau-Lanz*, 1978; j= *Slimane and El Mostaïne*, 1997; k= *Defretin and Fauvelet*, 1951; l= *Dutuit*, 1976; *Jalil and Dutuit*, 1996; *Lucas*, 1998a, b; m= *Lund*, 1996; *Fowell et al.*, 1996; *Tourani et al.*, 2000; and *Olsen et al.*, 2000; n= *Cousminer and Manspeizer*, 1976; *Biron and Courtinat*, 1982; *Le Marrec and Taugourdeau-Lantz*, 1983; *Doubinger and Beauchamp*, 1985; and *El Youssi*, 1986; o= *Desplats et al.*, 1983. The Triassic-Jurassic boundary was recently identified in the Argana Basin by *Olsen et al.* [2000].

The NE-SW Asdrem dyke (Figure 5), seldom represented in CAMP maps, is another of the main intrusive structures in the Anti-Atlas domain. It extends for 200 Km, from East of the Kerdous inlier in the South to the Siroua Massif in the North, crossing Precambrian and Paleozoic formations. The dyke was intruded along vertical tardi-Hercynian fractures. Its thickness varies from 60 to 100 meters, sometimes reaching 150 meters, and is locally covered by Neogene and Quaternary deposits. At its vicinity, the intruded formations were slightly metamorphosed.

In the Draâ plain, the sills are intruded in the folded sedimentary sequence closely following all structures [*Hollard*, 1973]. They present variable extent and thickness. Intrusion occurred preferentially along disharmonic folding between layers of contrasting competence. This led to the interpretation that they were older than Hercynian folding. However, neither the sill dolerites nor the dykes that fed them and link sills at different stratigraphic levels show any deformation.

The NE-SW Messejana dyke is one of the major geologic features of Iberia [*Schermerhorn et al.*, 1978]. With a thickness

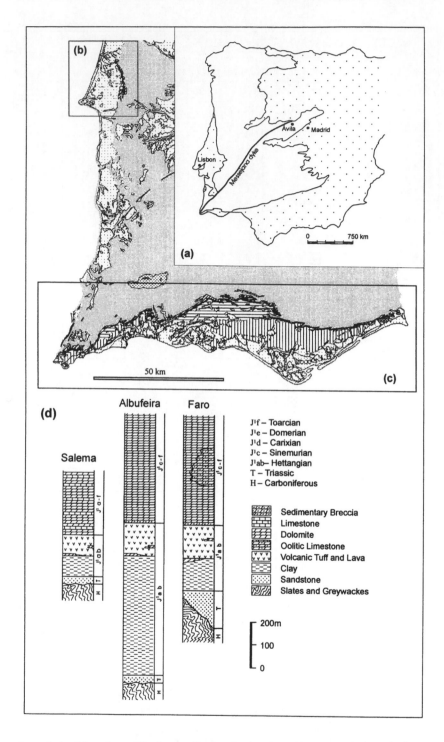

Figure 4. (a) Location of the Messejana dike in the Iberian Peninsula: white area - Paleozoic basement; dotted areas – Mesocenozoic cover (b) Late Triassic-Early Jurassic outcrops in Santiago do Cacém basin: gray – Paleozoic basement; oblique pattern – Triassic-Jurassic sandstones and clays; black – volcano-sedimentary sequences; horizontal pattern – Early Jurassic carbonated sediments; vertical pattern – Middle and Upper Jurassic sediments; dotted pattern - post-Jurassic cover (modified from Carta Geológica de Portugal at 1:500,000 scale, 1992). (c) Late Triassic-Early Jurassic outcrops in Algarve basin: conventions as in b. (d) Simplified stratigraphic columns of Late Triassic-Early Jurassic series of Algarve, Southern Portugal [after *Palain*, 1979; *Manuppella*, 1988; Manuppella, unpublished data, 2000].

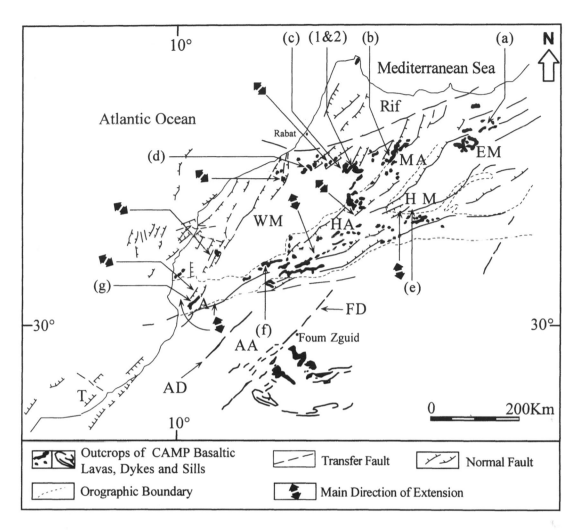

Figure 5. Main magmatic outcrops and Late Triassic-Early Jurassic basins in Northern Morocco and state of stress determined from fault striations [after *Laville and Piqué*, 1991; *Medina*, 2000]. Abbreviations: EM= Eastern Meseta; MD= Middle Atlas; HM= High Moulouya; HA= High Atlas; WM= Western Meseta; A= Argana Basin; AD= Asdrem dike; FD= Foum Zguid dike; AA= Anti-Atlas; T= Tarfaya Basin.

of 20 to 300 meters, it spreads over more than 530 Km along the tardi-Hercynian Messejana Fault. It crosses diagonally the Iberian Peninsula from the Portuguese Atlantic coast in the SW (Praia da Murração) to the Spanish Central Massif at Ávila (Figure 4a). It cuts vertically trough Precambrian and Paleozoic terranes as well as the Hercynian structures, and is locally covered by Tertiary deposits. The dyke is often divided into two or three parallel branches. Contact metamorphism originates the development of hornfels that are particularly evident where it crosses the Culm facies.

Lava flows. Lava flows can be found in all structural domains in Morocco except in the Anti-Atlas [e.g., *Salvan*, 1984; *Van Houten*, 1977]. These are also present in Algarve and in Santiago do Cacém region [e.g., *Manuppella*, 1988; *Martins*, 1991]. In most basins, the total thickness of lava flows is 100 to 200m. However, it may be as thick as 350 to 500m (in the Southern flank of the Central High Atlas), or just 8 to 50m in the inter-basin areas. The lava flows are usually inter-stratified with red clastic, evaporitic or carbonated sediments. The basaltic sequence is composed of one to several lava flows (up to 14 flows have been described by *Van Houten*, 1977) with or without inter-bedded clastic or carbonated sedimentary layers (Figure 6). Important variations of thickness from one section to another (from 8m to 500m) can be explained either by differential subsidence of a pre-volcanic basement during the emplacement of the lava flows (syn-rift series), or by the emplacement of the flows on a basement presenting an irregular paleotopography or a horst and graben structure.

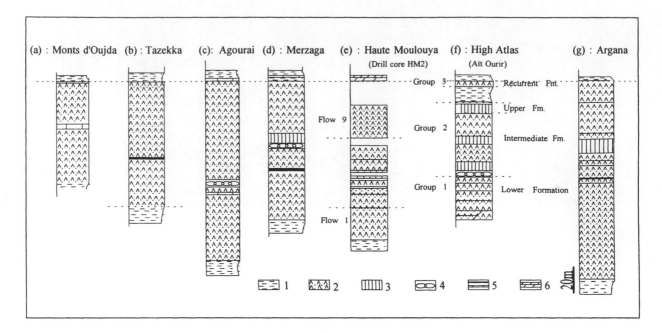

Figure 6. Stratigraphic columns of Late Triassic-Early Jurassic volcanic series of Northern Morocco. Legend: 1= red siltstones; 2= basalts; 3= basalts with prismatic jointing; 4= basaltic pillows lavas; 5= sedimentary layers (clastic and carbonated); 6= dolomitic limestones; Fm= Formation. [After *Aït Chayeb et al.*, 1998]. For the location of the lithostratigraphic columns (a) to (g): see figure 5. References: (a) = *Crasquin-Soleau et al.*, 1997; (b)= *Sabaoui*, 1986; (c) and (d) = *Cogney and Faugères*, 1975; (e)= *Fiechtner*, 1990; *Fiechtner et al.*, 1992; (f)= *Bertrand et al.*, 1982, modified by *De Pachtère*, 1983; (g)= *Aït Chayeb*, 1997; *Aït Chayeb et al.*, 1998.

In the Eastern Meseta (Oujda Mountains), the lava flows are sometimes associated with a system of dolerite dykes trending N50 to N90, and N135 in the Zekkara Massif and in Jbel Ez Zidour near Jerrada [*Oujidi*, 1996]. The occurrence, in the lava flows succession, of a limestone intercalation (3 to 13m thick) dated as Upper Ladinian-Lower Carnian [*Crasquin-Soleau et al.*, 1997] witnesses a shallow epicontinental sea open to the Tethys and reflects the existence of two eruptive periods separated by an important time interval (Figure 6).

In the Central High Atlas, the basaltic lava sequence ("High-Atlas dolerites") covers the arenitic ("Oukaimeden Sandstones") and siltitic formations ("Ramuncho Siltstones") (respectively, Formations F5 and F6 according to *Biron*, 1982) and is overlain by the first limestone layers of the Early Liassic-Sinemurian.

It is usually 15 to 170m thick, but may reach as much as 350 to 500m at the Southern flank of the Central High Atlas [*Van Houten*, 1977]. The volcanic pile consists of a succession of lava flows inter-stratified with sedimentary layers of varied lithologies (mudstones, sandstones, and carbonated sediments), or paleosols. These sometimes contain Late Triassic silicified or incarbonized wood [*Biron and Courtinat*, 1982; *Biron*, 1982; *De Pachtère*, 1983; *Le Marrec and Taugourdeau-Lantz*, 1983; *Desplats et al.*, 1983]. These layers represent short periods of volcanic rest. Usually, vesicular, sometimes clinkery, tops covered by iron pan horizons (paleosols?), and massive central

and lower parts characterize the lava flows. *De Pachtère* [1983] subdivided the volcanic pile in four units (the Lower, Intermediate, Upper and Recurrent Formations) separated by sedimentary levels or paleosols that represent minor periods of volcanic inactivity (Figure. 6). Basaltic pillow lavas, showing radial jointing and vitreous rinds (always of short lateral extent: 10 to 100m), identical to those found in the Western Meseta [*Cogney et al.*, 1971; *Cogney and Faugères*, 1975; *Cogney et al.*, 1974], are occasionally found. These are located in the base of the Intermediate Formation, immediately above clastic sediments, or in the Upper Formation. The pillow lavas represent subaerial flows that entered small ponds occupying depressions on the volcanic topography of the Lower and Intermediate Formations. Their occurrence does not imply a generalized subaqueous environment at the time of the lava emission [*Mattis*, 1977; *Lorenz*, 1988]. The short lateral extent of the pillow lavas and their constant stratigraphic position, the existence of lava flows with unequivocal subaerial characteristics associated to sediments containing fossilized wood, clearly indicate subaerial emplacement. Evidence for the presence of volcanic cones has never been found. However, several volcanic centers in the High Atlas of Marrakech allowed *De Pachtère*, [1983] and *De Pachtère et al.* [1985] to identify an eruptive fissure trending N110. This direction suggests the reactivation of tardi-Hercynian faults during rifting phases. The chemical composition of the lavas of those eruptive centers indicates that they were probably the source of the Upper Formation flows. At places, N60-80 normal faults displace lava flows and are fossilized by subsequent emissions, indicating an extensional tectonic event contemporary of the volcanism.

In Algarve (Figure. 4d), the succession of up to 6 flows is generally situated at the base of the "Complexo Vulcano-Sedimentar". A continuous or discontinuous dolomite layer marks the base of the lava flow succession [Manuppella, 1988]. The scarcity of pillow lavas and the presence of pyroclastic deposits containing highly vesicular juvenile fragments, clearly indicate a subaerial environment during the emplacement of this unit [Manuppella, 1988; Martins, 1991]. Nevertheless, evidence of phreato-magmatic deposits and pillow lavas has recently been found, showing the interference of superficial and groundwater in the volcanic activity.

Pyroclastic deposits. Pyroclastic deposits in volcano-sedimentary sequences are systematically found in Algarve and Santiago do Cacém [Manuppella, 1988; Martins, 1991]. These complexes are also found in the Middle Atlas (El Hajeb region) and in the junction of the Middle Atlas-High Moulouya (at Oued Kiss) [Ouarhache, 1987, 1993; Chalot-Prat et al., 1985]. Both present great similarities, however, some fundamental differences can be found: (1) the stratigraphic position of the volcano-sedimentary sequences of Algarve and its relation to the lava flows, is not always clear due to the effects of the Alpine deformation which complicates the geometry of the volcanic pile. Sometimes, it seems to belong to the "Complexo Pelítico, Carbonatado e Evaporítico" [Palain, 1979]; (2) the pyroclastic facies found in the Portuguese volcano-sedimentary sequences are usually coarser than those of Morocco, and contain lithics different in nature (dolomitic) of those found in Morocco (siltitic).

The volcano-sedimentary complex of the El Hajeb region was subject of a detailed study [Ouarhache, 1987; Chalot-Prat et al., 1985]. The volcano-sedimentary pile, presenting kilometric length and thickness of 10 to 50m, is called "Oued Defali Formation". It overlays the latest basalt flows of Carnian-Norian age [Ouarhache, 1987; Baudelot et al., 1990], and is covered by marly sediments of lagoonal character resembling the Hettangian-Sinemurian "Harira Formation" [Beaudelot and Charrière, 1983; Sabaoui, 1986, 1996; Baudelot et al., 1990]. The stratigraphic series continues with Liassic dolomites. The volcano-sedimentary sequence comprises red siltstones, tuffaceous siltstones, and lapilli or lapilli-breccia tuffites (Figure 7).

1. 3. Facies Model for the Late Triassic-Early Jurassic Volcanics of Morocco and Portugal

The TJMPMP volcanism has the following volcanological characteristics: (1) gradual vertical and lateral facies variations, except for the pyroclastic sequences, are not observed; (2) most eruptions occurred in subaerial continental environment; (3) the volcanism is characterized by dominantly basaltic magmas over basaltic andesites; real andesites, dacites, hybrid lavas and acid rocks are absent except for the Foum Zguid and Messejana dykes where magmatic differentiation by fractional crystallization can be seen in cross section, from olivine dolerites in the margin to granular rocks in the center [Rahimi, 1988; Aarab et al., 1994]; (4) the eruptive products are lava flows, sometimes with clinkery tops, followed systematically (Portugal) or occasionally (Morocco) by pyroclastic sequences. The later show volcanological characteristics (kilometric extents, average low thickness, permanence of low energy sedimentation during volcanism, variable vesiculation of pyroclastic material, bad sorting, presence of accretionary lapilli) indicating subaqueous basaltic eruptions in shallow sedimentary basins; (5) the volcanic succession contains clastic or carbonated sedimentary deposits and paleosols, indicating long intervals of sedimentation, non-deposition and/or erosive surfaces between short lived volcanic events. These sediments, associated to the basalts, represent the existence of continental (lacustrine, fluvial), lagoonal or even very shallow (intertidal to supratidal) marine environments. Sometimes the fluvial sequences are very thick (as in the Argana Basin) and of large lateral extent, and often precede the lava flows; (6) often the lava flows overlie discontinuities (angular and/or erosive); (7) volcanism occurred on the margin of a continental block (the West-African craton), and was followed by regional uplift, compensated by subsidence after the end of volcanism. This subsidence probably announces the Early Jurassic transgression; (8) no volcanic cones are preserved but the observed conduits define N40 to N110 trending alignments [De Pachtère et al., 1985; Chalot-Prat et al., 1985] that correspond to the reactivation of Hercynian fractures indicating fissure volcanism; (9) the basaltic rocks have an anorogenic character and a tholeiitic affinity of the continental tholeiite type (see below). The dykes present the same chemical signature of the flows. This suggests that they derive from the same mantle source [e.g., Bertrand, 1991; Bertrand et al., 1982; Martins, 1991] and can be considered as feeders of the flows, although a geometric link between them has never been observed. A similar proposal, with more evidences, has been recently presented for the Eastern North America Province [McHone, 1996].

The continental environment, basaltic composition and geometry of the TJMPMP volcanic succession indicate that the most adequate facies model is that of continental basaltic successions [Cas and Wright, 1987].

1. 4. Age of Late Triassic-Early Jurassic Magmatism

A compilation of paleontologic (Figure 3) and geochronologic data (Tables 1 and 2) is presented in order to constrain the emplacement age of the studied volcanism.

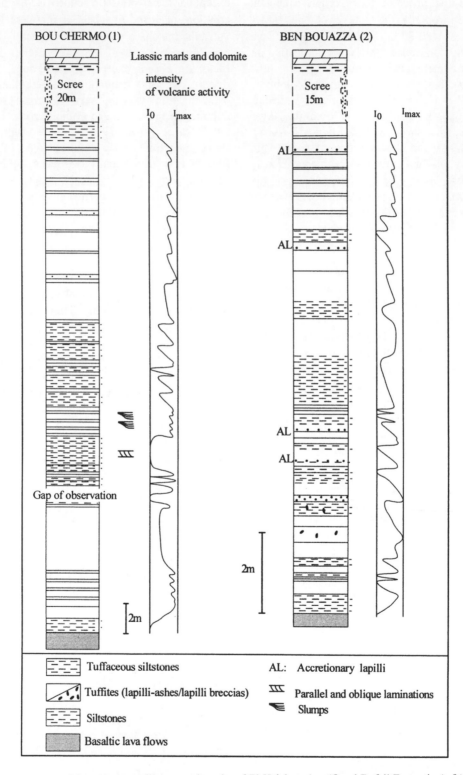

Figure 7. Type sequences of the volcano-sedimentary deposits of El Hajeb region (Oued Defali Formation). [After *Ouarhache*, 1987; *Chalot-Prat et al.*, 1985]. For the location of the lithostratigraphic columns (1) and (2): see figure 5.

Table 1. Compilation of available geochronologic data for the lava flows and dikes of Morocco and comparison with stratigraphic ages. European stages are taken from the Geological Time scale of *Odin* [1994]. Abbreviations: AD= Asdrem dike; BB= Berrechid basin; DB= Doukkala basin; EB= Essaouira basin; EM= Eastern Meseta; FD= Foum Zguid dyke; HM= High Moulouya; MD= Middle Atlas; P= Period; WM= Western Meseta. References: a) *Manspeizer et al.* [1975]; b) *Manspeizer et al.* [1978]; c) *Fiechtner et al.* [1992]; d) *Westphal et al.* [1979]; e) In *Medina* [2000]; f) In *Oujidi et al.* [2000]; g) *Bertrand and Prioton* [1975]; h) *Sebai et al.* [1991]; i) *Hailwood and Mitchell* [1971]; j) *Leblanc* [1973]; k) *Huch* [1988].

P	European stages	EM	MA			HM	WM				High Atlas					Anti-Atlas			

(Graphical chart of geochronologic age ranges plotted against European stages — Jurassic, Triassic, Permian — with data columns EM, MA, HM, WM, High Atlas, and Anti-Atlas. Ages shown include: EM 211±12; MA 182±13, 190±13, 199±15, 208±15, 196.3±1.2; HM 200.2±1.1, 210.4±2.1; WM 167±12, 205±12, 207±8, BB 205, DB 207, EB 210; High Atlas 153±1, 178±3, 196±17, 205±17, 188±2, 197±13, 206±13, 199±6, 203.3±2.6, 205.9±7.9, 197.1±1.8, 201.3±0.7; Anti-Atlas FD 152±5, 174±5, 180±4, 182±4, 168±5, 196.9±1.8, 187±4, 186±3, AD, 210±10, 235±10. Shaded band labeled "Stratigraphic age".)

References	a	b	b	c	c	d	b e e f				g	b	b	d	c	h	i	j	k	h	i
Methods	K/Ar		Ar/Ar				K/Ar				K/Ar				Ar/Ar		K/Ar		Ar/Ar	K/Ar	

Paleontologic data. In the Eastern Meseta (Oujda Mountains) (Figure 3), the ostracodes *Limnocythere keuperea Will,* 1969, and *Speluncella n. sp. Crasquin et al., 1997,* date the base of the "Upper Red Beds" overlying the second basaltic unit as Upper Norian [*Crasquin-Soleau et al.,* 1997], while *Lutkevichinella kristanae Crasquin et al., 1997* (similar to *Lutkevichinella lata Kozur, 1968*) found in association with *Anoplophora lettica Quenstedt,* allow the attribution of the interbedded carbonate

level in the basalts to the Upper Ladinian-Lower Carnian boundary. These data confirm and complete those of *Owodenco* [1946], *Monition and Monition* [1959], *Rakus* [1979], and *Manspeizer et al.* [1978].

In the Middle Atlas, the sedimentary layers underlying the basalts yielded an association of spores and pollen indicating a Carnian to Upper Carnian age, respectively in the Tazekka Massif and the Azrou region, while the sedimentary layers

Table 2. Compilation of available geochronologic data for the Messejana dyke and a lava flow from southern Portugal and comparison with stratigraphic ages. European stages are taken from the Geological Time scale of *Odin* [1994]. Abbreviations: P= Period. References: a) *Sebai et al.* [1991]; b) *Dunn et al.* [1998]; c*)* *Schott et al.* [1981]; d) *Teixeira and Torquato* [1975]; f) *Soares and Andrade* [1972]; g) *Ferreira et al.* [1977].

P	European stages		Messejena Dyke			Algarve
Jurassic	Oxfordian 154±5					
	Callovian 162±2		137±4	148±8	156±2	
	Bathonian 164±2					
	Bajocian 170±4					
	Aalenian 175					188±4
	Toarcian 184					
	Pliensbachian 191		198±0.4			
	Sinemurian 200±4				Stratigraphic age	
	Hettangian 203±3					
Triassic	Rhetian Norian 220±10		204.7±2.5	203±3	186±6	
	Carnian 230±6			209±6		
	Ladinian 233±6					
	Anisian 240±5					
	Olenkian Indusian 250±5				220±13	
Permian	Tatarian 255±5					
	Kazanian 258±9					271±5
	Kungurian 265±8					
	Artinskian 275±8					278±5
	Sakmarian					
	Asselian 285±8					
References		a	b	c	d	e	f	g
Methods		Ar/Ar		K/Ar			

interstratified in the basalts contain Upper Carnian-Lower Norian (Oum Rbi'a), Lower to Middle Norian (El Hajeb) and Early Liassic (Tazekka Massif) spores and pollen assemblages [*Sabaoui*, 1986; *Ouarhache*, 1987; *Beaudelot et al.*, 1986, 1990; *Ouarhache et al.*, 1999]. Additionally the carbonated layers

intercalated in the basalts of the El Hajeb region furnished ostracodes (*Sulcocythere hajbensis Colin*, associated with *Darwinula*) of Lower to Middle Norian age [*Ourahache*, 1987; *Beaudelot et al.*, 1986]. Finally, the sedimentary layers located above the upper siltstones and mudstones which overly the youngest basalts, contain spores and pollen associations of Hettangian-Sinemurian age [*Sabaoui*, 1986; *Beaudelot and Charièrre*, 1983; *Beaudelot et al.*, 1990].

In the Khemissat basin, *Taugourdeau-Lantz* [1978] has proposed a Carnian age for the palynological assemblage found in a silicified level located within the basalts. However, this assemblage, dominated by Cirumpolloid pollen, is rather similar to the Rhaetian-Hettangian assemblages of the nearby Middle Atlas [*Baudelot et al.*, 1990].

In the High Atlas of Marrakech, the "Oukaimeden Sandstones" (Formation F5 according to *Biron*, 1982) contain pollen assemblages from Middle [*Cousminer and Manspeizer*, 1976; *Biron and Courtinat*, 1982; *Le Marrec and Taugourdeau-Lantz*, 1983; *Doubinger and Beauchamp*, 1985] to Upper Carnian age [*El Youssi*, 1986]. Moreover, the sedimentary layers intercalated in the basalts often contain fossilized wood of the new species *Dadoxylon (Araucarioxylon) ourikense* similar to species described in the Late Triassic of Europe [*Desplats et al.*, 1983].

In the Argana Basin, the "Tourbihine Sandstones Member" (T2 according to *Tixeront*, 1973) has been recently attributed to Late Permian (Kazanian) based on invertebrate fauna (*Diplocaulidae* and *Captorhinidae*) [*Dutuit*, 1976; *Jalil and Dutuit*, 1996]. This member also contains specimens of *Voltzia heterophylla* and reptile tracks (*Rhynchosauroides sp.*) also known from the Spanish Buntsandstein [*Brown*, 1980]. The "Aglegal Sandstones Member" (T4) and the "Irohalene Mudstones Member" (T5) are assigned to Late Triassic (Middle to Upper Carnian) based on the presence of big *Metoposauridae* in the T5 member and of a Cyclotosaure in the T4 member. Recent work on global Triassic tetrapod biostratigraphy, allowed *Lucas* [1998a, b] to attribute the "Irohalene Mudstones Member" to Upper Carnian based on the presence of the genders *Palaeorhinus* and *Metoposaurus*. The "Bigoudine Formation" (members T6, T7 and T8) delivered a microflora that indicates Upper Carnian [*Lund*, 1996], Lower Norian [*Fowell et al.*, 1996] or Upper Carnian to Lower Norian age [*Tourani et al.*, 2000], or yet Norian-Hettangian [*Olsen et al.* 2000]. *Olsen* [1997] and *Olsen et al.* [2000] suggest that the Triassic-Jurassic boundary is located just a few meters below the Argana basalts based on palynology and similarities between Morocco and North America.

In Essaouira Basin, the red silty-mudstone sequence, underlying the evaporitic deposits, furnished *Darwinula sp.* and *Limnocythere sp.* of Carnian-Norian age. The red formations that overly the basaltic flows contain abundant ostracodes of the gender *Bairdiacypris sp.* that indicate a Liassic age [*Slimane and El Mostaïne*, 1997].

In Doukkala Basin, the base of the same sequence contains the ostracodes *Lutkevichinella sp.*, cf. *Pulviella sp.* and *Gemmanella sp.*, which indicate a Carnian-Norian age. The mudstones

underlying the basaltic flows furnished an assemblage of ostracodes (*Ogmoconcha sp.*, *Klinglerella sp.* and *Ljubimovella*) and foraminifers (*Haplophragmoïdes*) of Early Liassic age (probably Sinemurian) [*Slimane and El Mostaïne*, 1997].

In Algarve (Figure 4), the sediments underlying the basaltic flows contain faunal (bones) and floral associations (macroflora and palynoflora) indicating an age from Late Triassic (Keuper) to Early Liassic (Hettangian) [*Palain*, 1979, and references therein]. The upper part of the "Argilas de S. Bartolomeu de Messines" of the "Grés de Silves" furnished stegocephalus bones (*Estegocefalus Labyrinthodonta*) attributed to the Triassic. The upper part of the "Grés de Silves" contains *Euestheria minuta* and *Pseudoamussia destombesi* and a macroflora of Late Triassic age. The "Complexo Pelítico, Carbonatado e Evaporítico" furnished pollens of Triassic affinity and Hettangian mollusks, from the middle and upper parts, respectively. The dolomites of the "Complexo Vulcano-Sedimentar" contain bivalves and gastropods of Early Liassic age (Hettangian). In Santiago do Cacém region, the dolomites underlying the basaltic flows contain bivalves and gastropods of the same age.

Apart from this compilation of paleontologic data, other several important facts must be stressed: (1) the period between the Kazanian and Middle Ladinian may correspond to a major sedimentary gap (22-25 Ma). During this period the territories of Morocco and Portugal seem to have been subject to environmental conditions favoring alteration and erosion that, according to *Beauchamp* [1988], may be represented by the conglomerates with calcretes of the Triassic series of High Atlas; (2) the Triassic-Jurassic boundary is marked by a cartographic unconformity just below the "Complexo Vulcano-Sedimentar" in Algarve and Santiago do Cacém regions. This unconformity, marked by an abrupt change of floral and faunal associations, is probably a few meters below the "Complexo Vulcano-Sedimentar" [*Palain*, 1979]. In the Argana Basin, this boundary is marked by an abrupt floral change within a cyclical lacustrine sequence. It is lithologically marked by black shales that are also a few meters below the basalts [*Olsen*, oral communication, 1999; *Olsen et al.*, 2000]; (3) the first basaltic effusions of TJMPMP occurred during Upper Ladinian-Lower Carnian (or even earlier) in NE Morocco (Oujda Mountains), possibly in Essaouira Basin [*Broughton and Trepanier*, 1993], and in Ourika [*Van Houten*, 1977; *Manspeizer et al.*, 1978; *Manspeizer*, 1988; *Beraaouz*, 1995]. While the last basalt flows were emplaced during Hettangian-Sinemurian times in Algarve and Santiago do Cacém, in Morocco they occurred during Sinemurian in the Atlantic margin (Doukkala Basin) and in the High Atlas (particularly in Argana Basin). The scheduling of these eruptions indicates two age gradients: decreasing age from NE to SW (from Oujda Mountains to the Western High Atlas) and from SW to NE (from Oujda Mountains to Iberia) if we account for the paleogeographic reconstruction of the continents at the Triassic-Jurassic boundary.

Geochronology. In Tables 1 and 2 a compilation of the available geochronologic data on the Late Triassic-Early Jurassic magmatism of Morocco and Portugal is presented. A simple reading of these tables allows an immediate verification of a great scattering of radiometric ages due to the utilization of the K/Ar method on whole rock, to the nature of the dated samples (low K_2O content dykes or tholeiitic flows), and to different hydrothermal, tectonic, thermal and weathering events that affected these rocks. However, $^{40}Ar/^{39}Ar$ ages, and particularly plateau ages obtained on mineral separates yield substantially less scattered results [*Sebai et al.*, 1991; *Fiechtner*, 1990; *Fiechtner et al.*, 1992; *Dunn et al.*, 1998].

In High Moulouya (Morocco), the plagioclase separates of two samples located at the bottom (flow 1) and at the top (flow 9) of a drill core yielded plateau ages of 208.2 ± 2.3 Ma and 203.6 ± 2.6 Ma [1 σ errors; *Fiechtner et al.*, 1992].

In the High Atlas, plagioclase separates of lava flows from the bottom and the top of the volcanic pile displayed $^{40}Ar/^{39}Ar$ plateau ages of 201.3 ± 0.7 Ma and of 197.1 ± 1.8 Ma, respectively [*Sebai et al.*, 1991]. Other $^{40}Ar/^{39}Ar$ non-plateau dates range from 205.9 ± 7.9 Ma to 203.3 ± 2.6 Ma (southern flank of the Central High Atlas) to 196.9 ±1.8 Ma (Foum Zguid dyke), and to 196.3 ±1.2 Ma (Middle Atlas flows) [*Sebai et al.*, 1991; *Fiechtner*, 1990; *Fiechtner et al.*, 1992].

Available $^{40}Ar/^{39}Ar$ non-plateau ages for the Messejana dyke indicate a rough age interval between 204.7 ± 2.5 and 198.0 ± 0.4 Ma [*Sebai. et al.*, 1991; *Dunn et al.*, 1998].

As a synthesis, we can say that: (1) $^{40}Ar/^{39}Ar$ ages obtained for the dykes and flows of Morocco and Iberian Peninsula are similar to those obtained by the same method and by U/Pb in other regions of CAMP like North and South America [e.g., *Sutter*, 1988; *Dunnig and Hodych*, 1990; *Sebai et al.*, 1991; *Hodych and Dunnig*, 1992; *Deckart et al.*, 1997; *Marzoli et al.*, 1999; *Hames et al.*, 2000]. (2) Overlapping $^{40}Ar/^{39}Ar$ age intervals for basalts from Morocco and Portugal indicate that they are part of the same eruptive cycle. (3) The apparent duration of the volcanic activity was of less than 5 Ma at High Moulouya and High Atlas. Such a short duration of the igneous activity is characteristic of most Continental Flood Basalts [e.g., *White and McKenzie*, 1995]. Paleontological data suggests that the tholeiitic effusive and intrusive activity in various basins of the TJMPMP occurred over long time (Ladinian-Sinemurian) while the available $^{40}Ar/^{39}Ar$ plateau ages indicate a much shorter period for the volcanic activity (208-197 Ma for extrusive volcanics). This difference can be rationalized as result of: (a) an over-reliance on comparisons with northern European palynology (e.g., Argana basin), (b) over-interpretation of poorly preserved fossils (e.g., Khemissat basin), and (c) rarity of Early Jurassic non-marine ostracode assemblages [*Et-Touhami et al.*,

2001]. According to the latter authors contemporaneous and synchronous basaltic eruptions occurred province-wide over a time span of less than 200 Ka, based on Milankovitch calibration, and within about a 20 Ka interval after the Triassic-Jurassic Boundary. (4) Apparently the oldest flows of the High Moulouya (208.2 ± 2.3 Ma) precede those of the High Atlas (201.3 ± 0.7 Ma; age of biotite LP-6 standard of 128.5 and 128.8 for *Fiechtner et al.*, 1992 and *Sebai et al.*, 1991). This propagation of volcanic activity, that can be assumed to represent the propagation of syn-rift faulting (rifting diachronism) from the NE to the SW, recalls the known diachronism of the Central Atlantic opening.

2. PETROGRAPHY, MINERALOGY AND GEOCHEMISTRY

A summary of the most pertinent petrographic, mineralogical and geochemical characteristics of the TJMPMP will be presented in this chapter; detailed descriptions and analytical methods used by the authors from whom database was collected may be found in references cited in the text and Table 3.

2. 1. Petrography and Mineralogy

The Late Triassic-Early Jurassic magmatic activity in Morocco and Portugal generated tholeiitic dykes/sills, lava flows and pyroclastic deposits. Basaltic textures (glomeroporphyritic to porphyritic, at times vesicular) are typical of lava flows (e.g., Argana and Algarve; see Figures 4 and 5), but they also occur in chilled margins of shallow level intrusions. Intrusive rocks range from fine-grained doleritic dykes/sills (with ophitic to subophitic, intersertal or intergranular textures) to coarse-grained gabbroic bodies (e.g. Foum Zguid and Messejana dykes; Draâ sills; see Figures 4 and 5).

Most dykes/sills and lava flows from different locations display identical crystallization trends and mineralogical compositions [*Bertrand*, 1991; *Martins*, 1991]. Early olivine (Fo78-70, inclusions in clinopyroxene) show partial resorption during cotectic precipitation of clinopyroxene (endiopside - augite + pigeonite) and plagioclase (An 82-70 = microphenocrysts). This is followed by precipitation of titanomagnetite (USP 86-40) + ilmenite (ILM 98-94) indicating late iron-titanium oxides crystallization (1000°C to 700°C) under fO_2 conditions close to the QFM buffer. Variable amounts of late-stage quartz-alkali feldspar (Or 72-53) intergrowths (+ apatite, biotite and hornblende) are commonly observed filling the interstitial spaces, reaching well developed micropegmatitic textures in coarse-grained rocks [*De Pachtère*, 1983; *Bertrand*, 1991; *Martins*, 1991; *Aït Chayeb*, 1997].

Pyroxene chemistry was used as diagnostic of physical and chemical changes of host TJMPMP magmas as discussed in detail by *Bertrand* [1991] and *Martins* [1991]. Dominant clinopyroxene chemical features are congruent with the evolution of the host tholeiitic liquids. Two characteristic chemical trends were identified: one for calcic clinopyroxenes, from endiopside-augite to ferrohedenbergite (Wo 41-28 : En 51-32 : Fs 8-40), and the other for coexisting subcalcic pigeonite (Wo 7-13 : En 69-34 : Fs 24-53). The low-level of non-QUAD substitutions in calcic pyroxenes, as well as the occurrence of an extended period of pigeonite crystallization, reflect a relatively high level of silica activity in the tholeiitic magmas, inhibiting the recurrence of fayalitic olivine [Carmichael et al., 1970] and promoting the development of late-stage, granophyric, quartz-feldspar intergrowths. Significant exceptions to these general features correspond to highly contaminated magmas occurring in the Algarve basin [Martins, 1991; Martins and Kerrich, 1998]. Here, "in situ", assimilation of Hettangian-Sinemurian dolomitic limestones by the uprising, regional, tholeiitic liquids led to late crystallization of Ca-Tschermakite rich clinopyroxenes (Al_2O_3=9.11 wt%, Ca53Mg27Fe21 rimming early endiopside-augite) and fayalitic olivine (Fo 48), reflecting the generation of low SiO_2, high Ca, high Sr, leucite-nefeline normative rocks at the contacts with the host carbonates.

The degree of low temperature alteration of the TJMPMP rocks is variable. In Morocco both hydrothermal and supergenic alteration occur; hydrothermal recrystallization gave rise to the appearance of epidote, pumpellyite, zeolites, chlorite and carbonate filling fissures and vesicles; supergenic alteration is expressed by substitution of the primary mineralogy by serpentine, calcite and saponite. In the Portuguese outcrops significant alteration products are serpentine, sericite and celadonite and only in thicker sections (200m) of the Messejana dyke uralite replaces (total or partially) the pyroxenes.

2. 2. Geochemistry

Geochemical characterization of TJMPMP is supported by a large database including more than 300 samples analyzed for major and trace elements; however, the available isotopic data is still very scarce (see Table 3).

The great majority of samples from the TJMPMP have a uniform subalkaline character. This is clearly shown both in the Total-Alkalis/SiO_2 classification diagram (Figure 8) and by their CIPW quartz normative tholeiitic compositions [Bertrand, 1991; Martins, 1991]. Exceptions are the olivine normative (2-10%) tholeiites in Argana Basin (Morocco) and the leucite-nefeline normative contaminated basaltic rocks of the Algarve Basin (Portugal).

Most samples have a differentiation index Mg # [=$100*Mg/(Mg+Fe^{2+})$; Fe^{3+}/Fe^{2+}=0,15] ranging from 0,76 to 0,26, with MgO<10wt%, Ni≤150ppm and Cr≤300ppm [Bertrand, 1991; Martins, 1991] indicating that they represent magmas that have undergone variable degrees of predominantly fractionation. These characteristics are recognized in tholeiitic Continental Flood Basalts elsewhere [Thompson et al., 1983; Cox, 1988]. Only few samples from High Moulouya and Eastern Meseta (see Figure 10) with #Mg>68, Ni: 200-500ppm and Cr:

250ppm represent compositions that are closer to primary basaltic liquids [Fiechtner, 1990; Fiechtner et al., 1992; Robillard, 1978; Manspeizer et al., 1978].

The TiO_2 values ranging from 0.18-1.82wt% allow us, according to Petrini et al. [1987] and Cox [1988], to consider the TJMPMP as a Low TiO_2 (<2wt%) province (Figure 9).

Differentiation trends displayed by major element variations are controlled by the dominant crystallizing mineralogy (ol ± plag ± cpx) (Figure 9) and correspond to typical tholeiitic trends in AFM diagrams [Bertrand, 1991; Martins, 1991].

Trace element contents (Figure 10) support the rarity of primitive liquids and extensive crystal fractionation: this is indicated by both the general low concentrations of compatible elements (Ni, Cr and Sc) and the strong negative correlation of Zr, Y, Rb, Sr, Ba, Th, La and Ce with Mg# (70-26).

REE chondrite-normalized patterns are very similar for both the Moroccan and Portuguese samples (Figure 11). All are enriched in light rare earth elements relatively to heavy rare earth elements (Table 4) with an average $(La/Yb)_n$ = 3,23-5,33; fractionation among HREE is poor and only slight Eu anomalies (positive or negative) are observed. But in Moroccan High Atlas region the REE patterns of continental tholeiites show a vertical zoning characterized by progressive depletion in LREE towards the top of the sequence. This is attributed by Bertrand et al. [1982] to increasing partial melting degrees (from 7% to 17%) of a source in the continental upper mantle, which is enriched in most incompatible elements (Ba, U, Th) comparatively to the oceanic tholeiites mantle source. In TJMPMP, chondritic normalized $(La/Ce)_n$ values (1.02 to 1.27) being close to those of N-MORB [e.g., Le Roex, 1987] are distinct from MORB in LILE/HFSE incompatible element ratios contents (Table 4) and their isotopic composition (Table 3).

Generalized (primitive mantle normalized) incompatible element patterns for Triassic-Jurassic magmatic rocks from Morocco and Portugal are very similar (Figure 12), and all enriched in LILE (Rb, Ba, K, Th) and LREE relatively to HSFE (Nb, Ta, Zr, Hf, Ti, Y) and HREE, displaying negative anomalies in Nb (Ta), Sr, P and Ti that become more pronounced with progressing differentiation. Sr (and minor Eu) negative anomalies may be related to plagioclase crystallization; however, P, Nb and Ti negative anomalies are not easily related to any liquidus phase, because both apatite and Fe-Ti oxides are late-stage crystallization minerals, corresponding to the granophyric paragenesis. Therefore, the observed P, Ti and Nb (Ta) negative anomalies, which are characteristic of most CFB patterns, should result from other processes beyond shallow level crystallization [Thompson et al., 1983]. In accordance, it is suggested that these geochemical features reflect mantle source region characteristics and/or the effects of magma/continental crust interaction [Alibert, 1985; Fiechtner et al., 1992 and Martins, 1991]. Recognized the significance of crustal contamination on the geochemistry of the TJMPMP magmas,

Table 3a. Representative analysis of TJMPMP, selected from the cited references. Morocco: EM = Eastern Meseta [Manspeizer et al., 1978; Robillard, 1978]; MA = Middle Atlas [Manspeizer et al., 1978; Fiechtner, 1990; Fiechtner et al., 1992]); HM = High Moulouya [Manspeizer et al., 1978; Fiechtner et al., 1992]; WM = Western Meseta [Manspeizer et al., 1978; Girard, 1987; Beraaouz, 1995]; HA = High Atlas [Bertrand and Prioton, 1975; Manspeizer et al., 1978; Bertrand et al., 1982; De Pachtère, 1983; Beraaouz, 1995; Aït Chayeb, 1997; Aït Chayeb et al., 1998; Youbi, Ibouh and Aït Chayeb, unpublished data]; AA = Anti-Atlas [Leblanc, 1973; Bertrand and Prioton, 1975; Bertrand et al., 1982; Huch, 1988; Rahimi, 1988; Beraaouz, 1995].

Samples	Fil/88	Fii100 88(1)	Fii16 88(1)	Fi58 88(9)	Fi88 88(9)	Fi85 88(9)	TH78	M23	YAR 172	HA 1137	HA200	HA 1209	L3	L1	BV21	BV23	L6	L5
Morocco	MA	HM	HM	HM	HM	HM	WM	WM	HA	HA	HA	HA	HA	HA	HA	HA	AA	AA
SiO$_2$ (wt%)	52.88	51.78	52.13	52.59	51.71	50.90	52.71	50.30	50.80	49.79	50.80	52.45	49.80	53.40	48.89	46.29	49.52	54.31
TiO$_2$	1.12	1.44	1.33	1.17	1.06	1.07	1.09	1.16	1.18	1.19	1.03	1.31	0.98	1.34	1.42	1.34	0.98	1.11
Al$_2$O$_3$	14.26	14.03	13.59	14.46	13.24	13.26	13.86	14.03	13.80	12.60	14.55	13.76	14.80	14.10	15.00	15.48	14.14	13.66
Fe$_2$O$_3$t	10.81	9.48	11.33	9.79	10.34	9.13	10.81	11.41	10.82	10.99	9.94	9.96	10.88	9.45	10.84	15.25	10.85	9.58
MnO	0.15	0.12	0.12	0.11	0.15	0.16	0.15	0.15	0.19	0.17	0.18	0.15	0.25	0.21	0.06	0.11	0.17	0.13
MgO	7.42	8.95	7.61	7.69	9.40	8.63	7.96	8.90	6.76	10.38	9.85	7.29	7.80	5.89	14.20	15.39	7.96	6.03
CaO	10.20	8.24	8.89	9.65	10.03	11.03	10.13	9.71	7.25	9.44	10.05	9.75	11.00	7.80	6.80	4.64	9.22	6.99
Na$_2$O	2.07	2.08	1.78	1.67	1.77	2.18	1.95	1.76	3.80	1.73	2.22	2.52	2.07	2.69	2.15	1.31	2.55	3.10
K$_2$O	0.34	1.05	1.11	0.71	0.67	0.66	0.67	0.61	1.64	0.72	0.40	1.50	0.37	2.17	0.49	0.09	0.62	1.38
P$_2$O$_5$	0.17	0.18	0.19	0.11	0.12	0.11	0.12	0.50	0.15	0.15	0.13	0.06	0.18	0.20	0.15	0.12	0.16	0.18
L.O.I.	0.95	1.93	0.95	1.06	1.19	2.37	0.48	2.76	3.51	0.97	1.51	1.72	1.75	2.50			3.69	3.31
Mg# [a]	61.00	68.00	61.00	64.00	68.00	65.19	59.34	60.71	55.31	68.00	67.00	59.17	63.00	59.00		66.66	63.00	60.00
Cr (ppm)	285	395	407	289	310	315	293	448	174	514	206	447	219	286	287.6	261.5	263	209
Ni	94	129	126	100	168	147	230	118	59	200	85	108	85.8	83.6	90	90.9	80.2	55.8
Rb	34	31	38	27	27	22	24	15	23	23	7	34	8.64	47.3	18	11	15.3	30.8
Sr	174	269	241	199	180	191	172	129	198	203	177	203	156	206	82	59	169	217
Y	21	23	26	22	19	20	20		23	23	25		19.3	25	27	23	18.8	19.6
Zr	107	144	160	105	98	95	104		92	102	80		80.1	141	133	104	83.6	104
Nb	8	14	16	7	6	4	10.5		7.2	10.5	5.7		4.56	11.9	14	11	4.91	6.32
Ba	140	272	255	186	150	162	157	105	300	179	70	180	76.8	451	78	24	174	431
Th									3.0	2.5	1.3		1.4	4.1	3.0	2.4	1.8	2.2
La (ppm)	11.99	14.81	17.64	10.23	8.91	11.24	10.96		11.3	13.8	7.7		7.2	16.83	17.51	9.14	7.87	11.04
Ce	28.18	39.03	43.09	26.35	25.1	29.1	28.85		25.5	27.5	16.2		17.18	37.04	31.26	21.18	18.73	24.18
Nd	12.5	17.1	20.27	11.84	10.13	13.27	12.26		15.5				10.24	19.31	15.52	12.37	10.78	13.11
Sm	3.7	4.82	5.24	3.21	2.83	3.77	3.37			3.71	2.78		2.82	4.64	3.55	3.08	2.89	3.19
Eu	0.99	1.39	1.27	0.89	0.79	1.05	0.91		1.18	1.12	0.98		0.93	1.48	1.29	0.79	1.07	1.28
Gd	3.61	4.39	4.81	3.25	3.15	3.87	3.38						2.8	4.29	3.99	3.43	2.84	3.16
Tb													0.5	0.74	0.68	0.56	0.52	0.57
Dy	3.74	4.69	4.69	3.49	3.09	3.83	3.49		4.2				3.16	4.56	4.19	3.52	3.28	3.65
Er	2.2	2.27	2.56	1.96	1.78	2.06	1.98		2.3				1.83	2.26	2.29	1.99	1.79	1.97
Yb	1.93	2.01	2.16	1.81	1.60	1.89	1.82		2.15	1.98	1.85		1.96	2.39	2.13	1.91	1.81	2.08
Lu	0.29	0.34	0.33	0.29	0.35	0.31	0.32			0.32	0.32		0.28	0.33	0.31	0.29	0.27	0.31
(^{87}Sr/^{86}Sr)$_0$ [b]	0.70684	0.70637	0.70666	0.70648	0.70672													

[a] Mg# = 100 Mg / (Mg+Fe^{2+}); Fe^{3+}/Fe^{2+} = 0.15. [b] Initial ratio (^{87}Sr / ^{86}Sr)$_0$ computed at 200Ma.

Table 3b. Representative analysis of TJMPMP, selected from the cited references. Portugal: ALG = Algarve and A-HCa = Algarve high calcium [Martins, 1991; Martins et Kerrich, 1998]; MD = Messejana dyke [Martins, 1991; Alibert, 1985; Bertrand, 1987], SC = Santigo do Cacém [Martins, 1991].

Samples Portugal	505-3	587-6	587-9	587-14	587-16	597-21	587-2	587-12	597-16	MUR-1	40C-4	472-1	37C-6	51C-7	51C-9	51C-10	44AV-2	44AV-3
	SC	ALG	ALG	ALG	ALG	ALG	A-HCa	A-HCa	A-HCa	MD	MD	MD	MD	MD	MD	MD	MD	MD
SiO_2 (wt%)	52.12	51.78	51.64	50.48	49.76	51.08	45.19	49.22	48.47	50.68	52.12	50.63	52.02	54.02	51.39	51.64	50.93	51.09
TiO_2	1.39	1.20	1.18	0.87	0.76	1.05	0.79	0.94	0.87	0.97	1.11	0.97	0.97	1.51	1.40	1.25	1.66	1.68
Al_2O_3	13.87	13.25	13.35	13.55	13.82	14.05	13.18	13.53	13.72	14.16	14.75	14.73	14.84	16.24	16.25	17.57	15.93	15.98
Fe_2O_3t	11.93	11.78	11.23	10.37	10.22	11.44	10.37	11.73	10.29	11.37	11.87	10.73	10.58	11.37	13.08	11.87	13.40	13.73
MnO	0.16	0.14	0.14	0.13	0.13	0.18	0.16	0.17	0.17	0.16	0.11	0.16	0.12	0.11	0.16	0.14	0.11	0.14
MgO	6.27	8.29	7.90	9.94	10.11	7.29	7.45	8.29	6.87	8.45	6.21	8.45	6.79	2.25	3.89	3.65	3.31	4.14
CaO	9.45	9.58	9.59	10.49	11.82	11.33	20.15	12.73	16.51	12.03	10.56	12.31	11.96	7.83	10.07	9.79	9.51	9.05
Na_2O	2.34	2.42	2.35	2.22	1.75	2.35	1.28	2.42	2.44	2.29	2.56	2.29	2.69	3.50	3.23	3.02	2.75	3.29
K_2O	0.76	0.77	0.69	0.62	0.42	0.64	0.40	0.50	0.49	0.45	0.80	0.37	0.60	1.66	0.97	0.90	0.84	0.94
P_2O_5	0.16	0.19	0.20	0.13	0.11	0.16	0.15	0.13	0.14	0.12	0.18	0.10	0.14	0.31	0.23	0.22	0.27	0.27
L.O.I.																		
Mg# [a]	54.73	61.76	61.80	68.79	69.47	59.44	62.30	61.91	60.56	62.51	54.42	65.03	59.43	31.11	40.43	41.24	36.05	40.77
Cr (ppm)	68	63	67	237		197	160	209	202	204	69	268	110	10	36	28	19	21
Ni	80	56	73	101	116	74	85	88	78	104	45	140	73	21	31	28	28	39
Rb	40	26	23	13	14	19	11	12	4	11	23	10	17	77	31	32	34	42
Sr	170	177	179	183	153	167	1267	407	733	152	169	150	185	208	222	201	193	198
Y	27	24	23	21	20	23	20	25	19	24	27	25	26	40	32	29	36	38
Zr	120	115	105	77	60	93	90	92	87	89	111	87	99	195	141	133	155	160
Nb	8	11	11	10	6	11	14	7	9	8	11	9	9	15	13	11	12	14
Ba	180	194	173	135	110	134	137	120	112	122	195	95	160	351	217	217	244	284
Th	2.4	2.5	2.1	1.6	1.2	1.4	1.9	1.5	1.9		2.5	1.1	1.8	4.9	3.5	3.3	3.5	3.7
La (ppm)	15.6	14.3	12.1	9.2	8.6	10.2	11.2	9.4			12.5	7.4	10.7	27.8	19.2	18.6	21.6	22.1
Ce	35	33	27	21	17	21	24	21			29	17	23	56	42	38	45	49
Nd	18	16	14	11	11	10	11	10			14	10	11	30	24	17	23	24
Sm	4.27	3.87	3.25	2.74	2.34	3.06	2.73	2.75			3.57	2.41	3.06	7.39	5.33	5.05	6.2	6.22
Eu	1.28	1.28	1.09	1.15	0.86	1.02	1.03	0.97			1.17	0.85	0.96	1.44	1.26	1.39	1.86	1.49
Gd																		
Tb	0.8	0.8	0.6	0.6	0.5	0.6	0.5	0.6			0.7	0.4	0.7	1.2	1	0.9	1.2	1.1
Dy						3.4	3.7	4.1										
Er																		
Yb	3.19	2.81	2.36	2.04	2.09	2.13	1.90	1.93			2.45	1.83	2.14	4.50	3.40	3.20	3.84	3.97
Lu	0.47	0.41	0.33	0.29	0.33	0.35	0.29	0.34			0.40	0.29	0.32	0.68	0.54	0.47	0.61	0.63
$(^{87}Sr/^{86}Sr)_0$ [b]	0.70619	0.70544			0.70728	0.70538	0.70746	0.7068	0.70728	0.70556		0.70553	0.70607					

[a] Mg# = 100 Mg / (Mg+Fe^{2+}); Fe^{3+}/Fe^{2+} = 0.15.   [b] Initial ratio (^{87}Sr / ^{86}Sr)$_0$ computed at 200Ma.

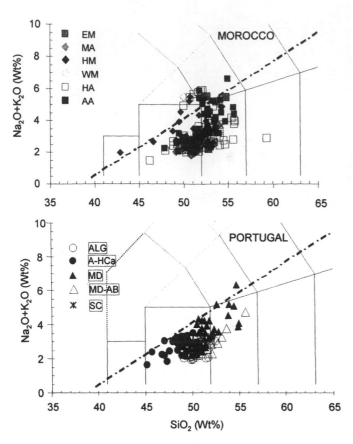

Figure 8. Total Alkalis-Silica diagram for chemical classification and nomenclature [*Le Bas et al.*, 1986] of Late Triassic-Early Jurassic magmatic rocks of Morocco and Portugal. Subalkaline versus Alkaline series boundary from *McDonald*, [1968, in *Rickwood*, 1989]. Data references in Table 3. Morocco: EM=Eastern Meseta; MA=Middle Atlas; HM=High Moulouya; WM=Western Meseta; HA=High Atlas; AA=Anti Atlas. Portugal: ALG=Algarve; A-HCa=Algarve high calcium; MD=Messejana Dyke; MD-AB=Messejana Dyke [data from *Alibert*, 1985 and *Bertrand*, 1991]; SD=Santiago do Cacém.

this become more obvious (Figure 13) for the Portuguese samples that display clear correlations between isotopic (^{87}Sr / $^{86}Sr)_0$ ratios, SiO_2 contents and TiO_2/K_2O values. Notwithstanding this, modeling simultaneous assimilation and crystal fractionation processes (ACF) of primitive TJMPMP magmas [*Bertrand*, 1991; *Martins*, 1991; *Fiechtner et al.*, 1992] do require further involvement of an enriched mantle source, besides continental crust contamination. Indeed, if it is accepted that TJMPMP magmas represent variable amounts (10-20%) of partial melting on their sources [*Bertrand et al.*, 1982; *Martins*, 1991; *Fiechtner et al.*, 1992], the depleted asthenosphere (N-MORB source) could not have produced the incompatible element enriched compositions of the most primitive magmas (with any or negligible contamination; see Table 4). So, as the convective asthenosphere does not preserve long-term high Sr_0

and low εNd_i isotope characteristics, a possible source providing the geochemical signature necessary to the production of the TJMPMP enriched magmas may be located in the continental lithospheric mantle reservoir, since it is characterized by the presence of ancient and metassomatically enriched domains [e.g. *Bertrand*, 1991; *Wilson*, 1993; *Turner and Hawkesworth*, 1995].

The TJMPMP constitutes a good example of the above statement as it extends over a vast region (more than 2500 km²), sampling a large mantle source located in the preceding Hercynian continental lithospheric mantle [*Bertrand*, 1991; *Alibert*, 1985; *Martins*, 1991]. How the mantle source was remobilized during the Central Atlantic rifting and oceanization events is still a point of enthusiastic and interesting discussion [see, *Manspeizer*, 1988, 1994; *Bertrand*, 1991; *Piqué and Laville*, 1993b, 1995, 1996; *Medina*, 2000; *Beraâouz*, 1995; *White and McKenzie*, 1989; *Oliveira et al.*, 1990; *Greenough and Hodych*, 1990; *Hill*, 1991; *Thompson*, 1998; *Wilson*, 1997; *Oyarzun et al.*, 1997; *McHone*, 2000).

3. THE LATE TRIASSIC–EARLY JURASSIC MAGMATISM FROM MOROCCO AND PORTUGAL IN THE GEODYNAMIC CONTEXT OF CENTRAL ATLANTIC OPENING

Integrating models that provide explanations for mechanisms responsible for the opening of the Central Atlantic are based on two hypotheses: the passive (lithospheric extensional hypothesis) or the active (plume related hypothesis) triggering processes [e.g., *McHone*, 2000; *Courtillot et al.*, 1999; *Hawkesworth et al.*, 1999; *Sheth*, 1999a, b; *Wilson* 1993, 1997; *King and Anderson*, 1995; and *Anderson*, 1994].

In the TJMPMP, the onset of the extensional process is poorly constrained, due to incertitude on the ages of Late Triassic formations both in Morocco and Portugal. Nevertheless, paleontological, stratigraphical and structural information suggest that rifting took place from Ladinian to Hettangian times. However, the older rifting events are difficult to separate from the Hercynian chain collapse extensional structures (*Medina*, 2000).

Paleontological data indicates that the earliest rifting related magmatism begun in NE Morocco (Oujda Mounts), in Essaouire basin [*Broughton and Trepanier*, 1993], and in Ourika [*Van Houten*, 1977; *Manspeizer et al.*, 1978; *Manspeizer*, 1988] during a time interval from Upper Ladinian to Lower Carnian. The youngest basaltic flows were extruded during Hettangian–Sinemurian times in Algarve and Santiago do Cacém and during Sinemurian times in the Morocco Atlantic margin (Doukkala Basin), High Atlas (Argana basin) and Anti-Atlas. This suggests that magmatic activity had a 30-39 Ma duration preceding by 55-75 Ma the onset of oceanic crustal accretion. However, available $^{40}Ar/^{39}Ar$ ages indicate a much shorter duration for the volcanic activity (comparing to paleontological data), which is in closer agreement with the values indicated by *Hames et al.* [2000] for North America.

The identification and interpretation of the inducing rifting mechanisms, involved in the production of this magmatism, allows the establishment of a geodynamic model.

Saunders et al. [1992] support that most Large Igneous Provinces (LIPs), and consequently CFB, may be related to mantle plume mechanisms, proposing topographic, thermal and compositional criteria for their identification.

The first criterion is the topographic uplift recognized in some LIP regions, which is related to the presence of magmatic underplating above a thermal mantle anomaly. In the Central Atlantic region, where TJMPMP is included, there is no evidence for uplift preceding the onset of magmatism [*Van Houten*, 1977; *Manspeizer*, 1988; *Medina*, 2000].

The second criterion correlates the extrusion of huge volumes of magma, in a short time interval, to the presence of an important thermal mantle anomaly [*White and McKenzie*, 1995]. The preserved volume of extruded basalts is estimated to be about 0,7-0,8 km^3 in TJMPMP, and $2,3 \times 10^6$ km^3 in CAMP, which may correspond to an initial volume of about 4×10^6 km^3 [*Olsen*, 1999; *McHone*, 2000]. These volumes are much higher than those predicted by mantle plume experimental models [*Campbell and Griffiths*, 1990].

Considering the topographic and thermal criteria and the ambiguity of the chemical and isotopic signatures, we favour the continental lithospheric mantle source hypotesis [*Seth*, 1999; *King and Anderson*; 1995; *Turner and Hawksworth*, 1995], although a mantle plume origin [*Wilson*, 1997; *Oyarzun et al.*, 1997; *Oliveira et al.*, 1990; *Greenough and Hodych*, 1990; *Thompson*, 1998] cannot be ruled out.

A geodynamic model for TJMPMP should take into account that:

- Geochemical and isotopic characteristics of the erupted magmas indicate that their sources were probably located in continental lithospheric mantle. These mantle sources acquired the enriched characteristics through previous subduction and magma extraction, preceding the production of asthenospheric liquids that gave rise to Atlantic Ocean floor.

- Partial melting occurred westwards, beneath the rift axis, where the lithosphere is more strongly stretched and weak, and melts migrate upwards along lithospheric discontinuities (particularly transform faults) as suggested by numerical modeling of the Atlantic rifting [*Sawyer and Harry*, 1991], field and geochemical data [*Bertrand*, 1991; *De Pachtère et al.*, 1985; and *Medina*, 2000];

- Structural data showing that basaltic volcanism is mainly controlled by Tardi-Hercynian faults, some of them corresponding to transform faults. These NE-SW faults (Messejana fault) or ENE-WSW (South Atlas fault) were active during Triassic and perhaps since Permian times; the nucleation of the rifting process may have started at two different triple junctions, an RRR junction near Florida and a RRT between Africa-Iberia-America (Terrinha, 1998);

- Remobilization of the mantle source seems to have taken place under the thicker and fragile Hercynian suture, and may have been related to extensional processes controlled by lithospheric discontinuities induced by isostatic compensation due to collapse and ablation of the Hercynian orogen.

These facts suggest that lithospheric extensional model is probably the most adequate to explain Late Triassic-Early Jurassic volcanism in the Magmatic Province of Morocco and Portugal.

CONCLUSIONS

The magmatic rocks and associated sediments of TJMPMP are preserved in graben and half-graben structures, separated by major transform fault zones. These major accidents, the South Atlas and the Azores-Gibraltar faults, commonly used as structural markers in the paleogeographic reconstitutions of Pangea prior to the opening of Central Atlantic, delimit three main blocks in the study area: the Iberian Peninsula, the Moroccan Meseta and the Anti Atlas-Sahara area. Despite the specific sedimentary, stratigraphic and magmatic characteristics of each block, they present common regional features that allow considering them as a sub-province of CAMP.

In the northern block, limited to the south by the Azores-Gibraltar Fault, sedimentary basins trending N-S (Lusitanian Basin) and NE-SW (Santiago do Cacém) and E-W (Algarve) preserve clastic, evaporitic and carbonated deposits associated with tholeiitic lava flows, pyroclasts and dykes of Hettangian age.

In the central block, limited by the Azores-Gibraltar and South Atlas faults, the sedimentary basins trend NNE-SSW (Berrchid Basin) or NE-SW (Khemissat Basin) and contain carbonated, clastic and evaporitic sediments with inter-stratified lava flows and pyroclasts of Upper Ladinian-Norian to Sinemurian ages.

The southern block, between the South Atlas Fault and the Anti Atlas craton, is characterized by an almost total lack of either sedimentary deposits (except for the Laayoum-Tarfaya sedimentary Basin) or extrusive volcanics of Triassic-Jurassic age, and by the occurrence of large dyke and sill swarms of Sinemurian age.

The timing of the onset of the extensional process is uncertain in Morocco as well as in Portugal. A time interval spreading from Ladinian (perhaps Kazanian?) to Hettangian is proposed.

These extensional processes are mainly controlled by inherited Hercynian structures due to over-thickening and weakening of the Variscan orogenic belt.

In the TJMPMP, the earliest basaltic successions are Upper Ladinian to Lower Carnian in the Eastern Meseta of Morocco, while the final extrusive events occurred during Hettangian-Sinemurian in Algarve and Santiago do Cacém, and as late as the Sinemurian in the Moroccan Atlantic margin (Doukkala

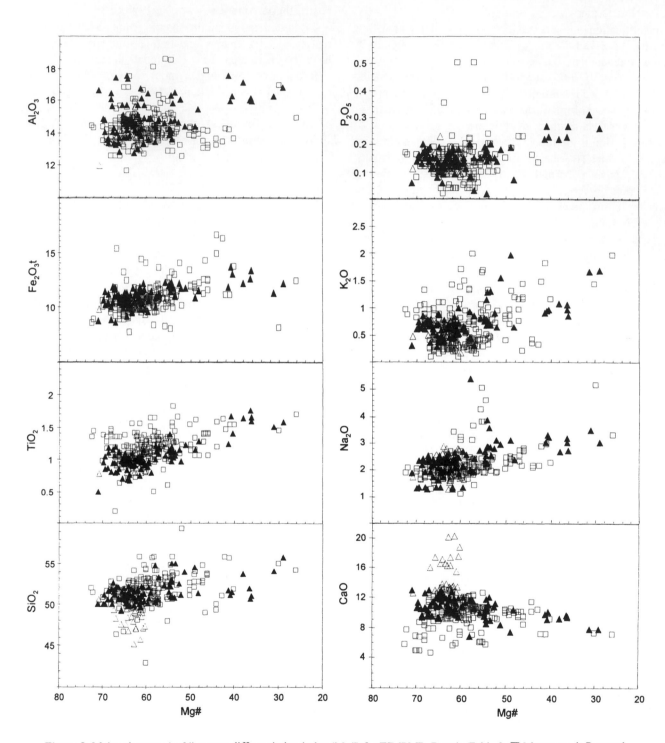

Figure 9. Major elements (wt%) versus differentiation index (Mg#) for TJMPMP. Data in Table.3. □ Morocco; ▲ Portugal (△ A-HCa).

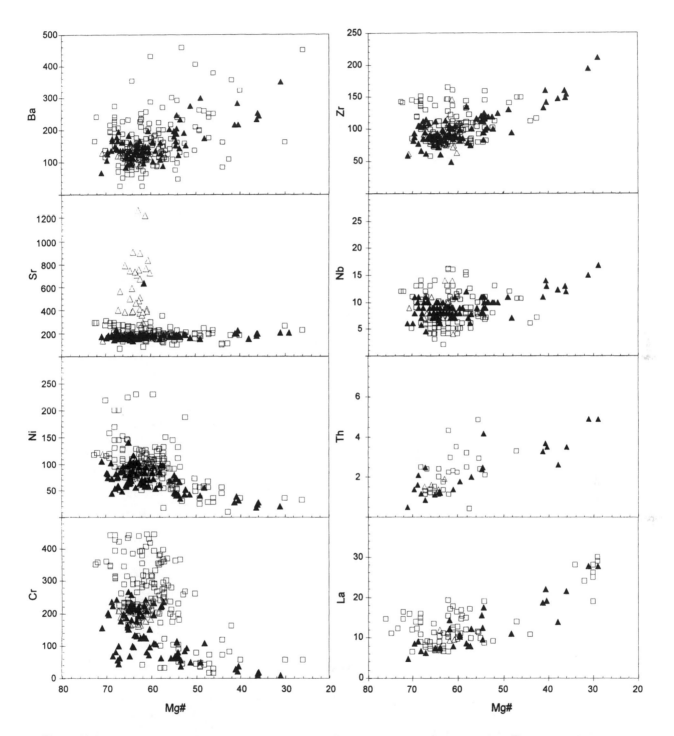

Figure 10. Trace elements (ppm) versus differentiation index (Mg#) for TJMPMP. Data in Table.3. □ Morocco; ▲ Portugal (△ A-HCa).

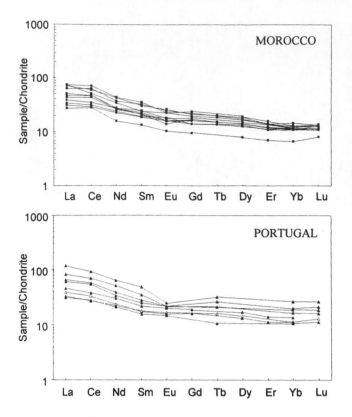

Figure 11. Chondrite normalized Rare Earth Elements patterns for representative samples of the TJMPMP. Normalizing values from *McDonough and Sun* [1995].

and SW to NE (Oujda Mountains to Iberia) according to the paleogeographic position of plates at the Late Triassic-Early Jurassic time (Iberia being at about 1000 km, or more, East of Morocco). However, all available $^{40}Ar/^{39}Ar$ ages range between 204-196 Ma for dikes and 210-196 Ma for flows.

The volcanological characteristics recognized in this magmatic province are very similar to those reported for continental basaltic successions model facies.

The TJMPMP presents very homogenous petrographical, mineralogical and geochemical characteristics, resembling those of world wide Low-TiO$_2$, quartz normative CFB.

Major and trace elements are in general correlated with the degree of evolution.

The enrichment in LREE and LILE with respect to N-MORB (and E-MORB) is attributed to magma/continental upper crust interaction modeled with ACF processes.

The correlations between isotopic (^{87}Sr / $^{86}Sr)_0$ ratios and major elements (SiO$_2$; TiO$_2$/K$_2$O) support the contamination process. However, the composition of the most primitive non-contaminated (or negligibly contaminated) samples reflect mantle source properties.

Although the TJMPMP mantle source presents (La/Ce)$_n$ ≤ 1, it cannot be considered representative of the asthenospheric (N-MORB like) mantle because it is enriched in high incompatible trace elements and isotopic (^{87}Sr / $^{86}Sr)_0$ ratios. Thus, we consider that these tholeiites are related to continental lithosphere mantle. This non-convective mantle layer preserves a vast enriched domain sampled by the TJMPMP from Ladinian to Sinemurian expressing the onset of rifting related magmatism in Central Atlantic. The mantle remobilization is envisaged as resulting from extensional processes controlled by lithospheric discontinuities coupled with isostatic compensation related to the collapse of the Variscan orogen.

Basin), in the High Atlas (particularly in Argana Basin), and in the Anti Atlas. These basaltic emissions show a time/space distribution determining an asymmetric zoning with two age gradients: NE to SW (Oujda Mountains to Western High Atlas)

Table 4. Mean values of incompatible element ratios (calculated from values of Table 3) for the different units, samples with Mg# > 60, in Morocco (MA, HM, WM, HA, AA) and Portugal (ALG, A-HCa, MD, SC). N-MORB and E-MORB values for comparison.

	La/Ba	La/Nb	Ba/Nb	Ti/Zr	Zr/Nb	Rb/Sr	(La/Yb)$_n$[b]	(La/Ce)$_n$[b]
MA	0.08	2.02	29.82	76.66	19.60	0.06	3.65	1.03
HM	0.06	1.37	22.57	62.09	14.19	0.16	4.80	1.02
WM	0.06	1.14	19.14	61.30	12.60	0.17	5.28	1.03
HA	0.12	1.30	15.90	70.54	11.54	0.14	4.17	1.21
AA	0.05	1.26	24.50	78.53	13.03	0.10	3.05	1.18
SC	0.09	1.95	22.50	69.50	15.00	0.24	3.51	1.15
ALG	0.07	1.19	16.56	64.95	10.82	0.10	3.38	1.18
A-HCa	0.09	1.08	14.60	63.07	10.49	0.03	4.03	1.27
MD	0.07	0.82	11.33	50.85	9.83	0.04	2.68	1.15
N-MORB[a]	0.39	1.07	2.70	102.00	31.76	0.01	0.59	1.03
E-MORB[a]	0.11	0.76	6.87	82.00	8.79	0.03	1.91	0.09

[a] Values from *Sun and McDonough* [1989]. [b] Chondritic normalizing values from *McDonough and Sun* [1995].

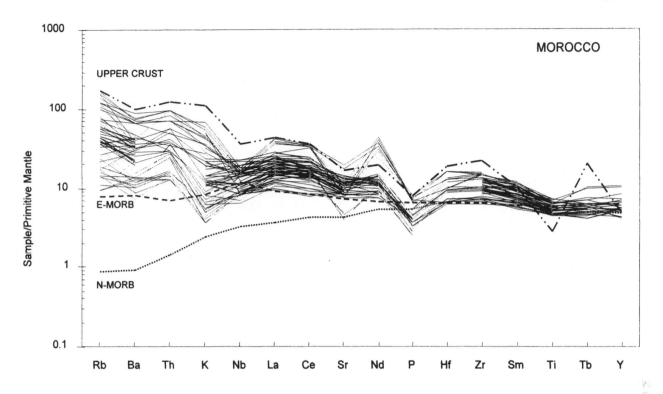

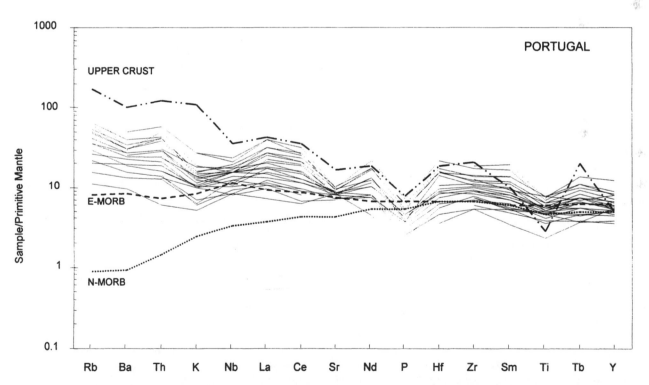

Figure 12. TJMPMP geochemical patterns. Trace element concentrations normalized to the composition of the primordial mantle and plotted from left to right in order of compatibility. The normalizing values are those of *McDonough and Sun* [1995]. N-MORB and E-MORB values from *Sun and McDonough* [1989], and upper crust from *Taylor and McLennan* [1981]. Data in Table 3.

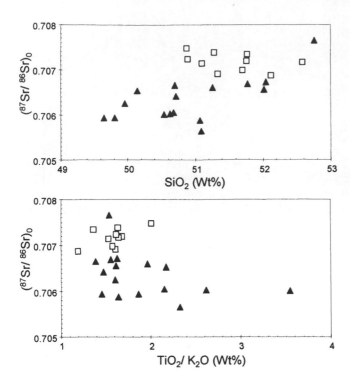

Figure 13. Initial ratios $(^{87}Sr/\ ^{86}Sr)_0$ versus SiO_2 and TiO_2/K_2O. Data in Table 3. Samples A-HCa not plotted. ☐ Morocco; ▲ Portugal.

Acknowledgements

Financial support for this work was provided by a bilateral co-operation project between the CNCPRST (Morocco), and ICCTI (Portugal).

We are grateful to the reviewers Andrea Marzoli and Greg McHone, and to the colleague Isabel Moitinho de Almeida, whose suggestions contributed to greatly improve this paper, and to the support and encouragement of the volume editors Bill Hames, Greg McHone, Paul Renne and Carolyn Ruppel.

REFERENCES

Aarab, E.M., A. Rahimi, and G. Rocci, Un exemple de différenciation transverse: le grand dyke de Foum Zguid (Anti-Atlas, Maroc), *C. R. Acad. Sci. Paris, t. 319, sér. II*, 209-215, 1994.

Aït Chayeb, E. H., Le volcanisme post-orogénique hercynien du bassin d'Argana (Haut-Atlas occidental, Maroc): lithostratigraphie, pétrographie, géochimie et implications géodynamiques, Thèse de Doctorat de 3ème cycle, Univ. Cadi Ayyad, Marrakech, 1997.

Aït Chayeb, E. H., N. Youbi, A. El Boukhari, M. Bouabdelli, and M. Amrhar, Le volcanisme Permien et Mésozoïque inférieur du bassin d'Argana (Haut-Atlas occidental, Maroc): un magmatisme intraplaque associé à l'ouverture de l'Atlantique central, *J. African Earth Sci., 26(4)*, 499-519, 1998.

Alibert, C., A Sr-Nd isotope and REE study of late Triasic dolerites from Pyrenees (France) and the Messejana (Spain and Portugal). *Earth and Planet. Sci. Lett., 73*, 81-90, 1985.

Anderson, D. L., The sublithospheric mantle as the source of continental flood basalts; the case against the continental lithosphere and plume head reservoirs, *Earth and Planet. Sci. Lett.*, 123, 269-280, 1994.

Beauchamp, J., Triassic sedimentation and rifting in the High Atlas (Morocco), in *Triassic-Jurassic rifting: Continental breakup and the origin of the Atlantic Ocean and passive margins*, edited by W. Manspeizer, pp. 477-497, *Developments in Geotectonics 22*, Elsevier, Amsterdam, 1988.

Beaudelot, S., and A. Charièrre, Définition et datation palynologique (Hettangien inférieur) de la formation de Harira, niveau de décollement sous les formations dolomitiques du Causse moyen-atlasique (Maroc), *C. R. Acad. Sci. Paris, t. 296, sér. II*, 1806-1811, 1983.

Beaudelot, S., A. Charrière, D. Ouarhache, and A. Sabaoui, Données palynologiques nouvelles concernant l'Ordovicien et le Trias-Lias du Moyen Atlas (Maroc), *Géologie Méditerranéenne, XVII(3-4)*, 263-277, 1990.

Beaudelot, S., J. P. Colin, and D. Ouarhache, Le niveau sédimentaire associé aux basaltes triasique de la bordure septentrionale du Causse d'El Hajeb (Maroc): données palynologiques et micropaléontologiques (Ostracodes), *Revue de Paléobiologie, 5(2)*, 281-286, 1986.

Beraâouz, E. H., Episodes magmatiques associées au rift atlasique et ouverture de l'Atlantique central. Thèse d'Etat Sci., Univ. Hassan II-Mohammadia, Casablanca, 1995.

Bertrand H., Le magmatisme tholéiitique continental de la marge ibérique, précurseur de l'ouverture de l'Atlantique Central: les dolérites du dyke de Messejana-Plasencia (Portugal, Espagne), *Comptes Rendus Acad. Sci. de Paris*, tome 304, Série II, pp. 215-220, Paris, 1987.

Bertrand, H., The Mesozoïc province of Northwest Africa: a volcano-tectonic record of the early opening of Central Atlantic, in *Magmatism in extentional structural setting*, edited by A. Kampunzu and R. T. Lubala, pp. 147-188, Springer-Verlag, Berlin-Heidelberg, 1991.

Bertrand H. and J. M. Prioton, Les dolérites marocaines et l'ouverture de l'Atlantique: étude pétrologique et géochimique, Thèse de 3ème cycle, Université Claude Bernard, 321p., Lyon, France, 1975.

Bertrand H., J. Dostal, and C. Dupuy, Geochemistry of Mesozoic tholeiites from Morocco. *Earth and Planet. Sci. Lett., 58*, 225-239, 1982.

Biron, P., Le permo-trias de la région de l'Ourika (Haut Atlas de Marrakech, Maroc), Thèse 3ème cycle, Univ. Scientifique et Médicale de Grenoble, Grenoble, 1982.

Biron, P., and F. Courtinat, Contribution palynologique à la connaissance du Trias du Haut Atlas de Marrakech, Maroc, *Geobios, 15*, 231-235, 1982.

Broughton, P., and A. Trepanier, Hydrocarbon generation in the Essaouira basin of western Morocco, *A. A. P. G. Bull., 77*, 999-1015, 1993.

Brown, R. H., Triassic rocks of Argana Valley, Southern Morocco, and their regional structural implication, *A. A. P. G. Bull., 64*, 988-1003, 1980.

Campbell, I. H., and R. W. Griffiths, Implications of mantle plume structures for the evolution of flood basalts, *Earth and Planet. Sci. Lett., 99*, 79-93, 1990.

Carmichael, I. S. E., J. Nicholls, and A. L. Smith, Silica activity in igneous rocks, *The Amer. Mineral., 55*, 246-263, 1970.

Caroff, M., H. Bellon, L. Chauris, J. P. Carron, S. Chevrier, A. Gardinier, J. Cotten, Y. Le Moan, and Y. Neidhart, Magmatisme fissural triasico-liasique dans l'ouest du Massif armoricain (France): pétrologie, géochimie, âge, et modalités de la mise en place, *Can. J. Earth Sci., 32*, 1921-1936, 1995.

Cas, R. A. F., and J. V. Wright, *Volcanic Successions: Modern and Ancient*, 1st ed., 528 pp., Chapman and Hall, London, 1987.

Chalot-Prat, F., A. Charrière, and D. Ouarhache, Découverte d'un volcanisme explosif fini-triastique sur la bordure occidentale du Moyen-Atlas (Maroc), *Bull. Fac. Sci. Marrakech, 3*, 127-141, 1985.

Coffin, M. F., and O. Eldholm, Large igneous provinces: crustal structure, dimensions and external consequences, *American Geophysical Union, 32(1)*, 1-36, 1994.

Cogney, G., and J. C. Faugères, Précisions sur la mise en place des épanchements basaltiques des formations triasiques de la bordure septentrionale du Maroc Central, *Bull. Soc. Géol. France, 7, XVII(5)*, 721-733, Paris, 1975.

Cogney, G., M. Normand, H. Termier, and G. Termier, Observations sur le basalte triasique de Rommani-Maaziz (Maroc occidental), *Notes et Mém. Serv. Géol. Maroc, 36(264)*, 153-173, 1974.

Cogney, G., H. Termier, and G. Termier, Sur la présence de "pilow-lavas" dans le basalte du Permo-Trias au Maroc central, *C. R. Acad. Sci. Paris, t. 273*, 446-449, 1971.

Courtillot V., C. Jaupart, I. Manighetti, P. Tapponnier, and J. Besse, On causal links between flood basalts and continental breakup, *Earth and Planet. Sci. Lett., 166*, 177-195, 1999.

Cousminer, H., and W. Manspeizer, Triassic pollen date Moroccan High Atlas and the incipient rifting of Pangea as Middle Carnien, *Science, 191*, 943-945, 1976.

Cox, K. G., The Karoo Province, in *Continental Flood Basalts*, edited by J. D. Macdougall, Kluwer Academic Publishers, 239-271, 1988.

Crasquin-Soleau, S., M. Rakus, M. Oujidi, L. Courel, M. Et Touhami, and N. Benaouiss, Découverte d'une faune d'Ostracodes dans le Trias des Monts d'Oujda (Maroc): relations paléogéographiques entre les plates-formes nord et sud de la Téthys. *C. R. Acad. Sci. Paris, t 324, Sér. II*, 111-118, 1997.

Cummins, L. E., J. D. Arthur, and P. C. Ragland, Classification and tectonic implications for early Mesozoic magma types of the Circum-Atlantic, in *Eastern North American Mesozoïc Magmatism*, edited by J. H. Puffer and P. C. Ragland, pp. 119-136, *Geological Society of America, Special Paper 268*, Boulder, 1992.

Dalrymple, G. B., C. S. Grommé, and R. W. White, Potassium-Argon Age and Paleomagnetism of Diabase Dikes in Liberia: initiation of Central Atlantic Rifting, *Geol. Soc. Amer. Bull., 86*, 399-411, 1975.

De Pachtère, P., Le volcanisme permien et fini-triasique dans le Haut Atlas de Marrakech (Maroc): Approche pétrologique et géochimique, Thèse 3ème cycle, Univ. Scientifique et Médicale de Grenoble, Grenoble, 1983.

De Pachtère P., H. Bertrand, and J. L. Tane, Mise en évidence de centres d'émissions dans la série volcanique fini-triasique du Haut Atlas de Marrakech (Maroc), *C. R. Acad. Sci. Paris, t. 300, II, 20*, 1029-1032, 1985.

de Boer, J., J. G. McHone, J. H. Puffer, P. C.Ragland, and D. W. Whittington, Mesozoic and Cenozoic magmatism, in *The Atlantic continental margin, U.S.*, edited by R. E. Sheridan and G. A. Grow,

pp. 217-241, Geological Society of America, The Geology of North America, v. I2, Boulder, Colorado, 1988.

Deckart, K., G. Féraud, and H. Bertrand, Age of Jurassic continental tholeiites of French Guyana, Surinam and Guinea: implications for the initial opening of the Central Ocean. *Earth and Planet. Sci. Lett., 150*, 205-220, 1997.

Defretin, S., and E. Fauvelet, Présence de phyllopodes triasiques dans la région d'Argana-Bigoudine (Haut Atlas occidental), *Notes et Mém. Serv. Géol. Maroc, 5(85)*, 129-135, 1951.

Desplats, D., C. Roy-Dias, and P. Biron, Etude d'un bois fossile: Dadoxylon (Araucarioxylon) Ourikense provenant des basaltes du Trias supérieur du Haut Atlas (Ourika-Maroc), *Geobios, 16(6)*, 717-725, 1983.

Doubinger, J., and J. Beauchamp, Description d'une palynoflore carnienne à la base des grès de l'Oukaïmeden de la région d'Asni (Haut Atlas): implication paléogéographiques (abstract), 5th Meeting of IGCP 183, Marrakech, 28, 1985.

Dunn, A. M., P. H. Reynols, D. B. Clarke, and J. M. Ugidos, A comparison of the age and composition of the Shelburne dyke, Nova Scotia, and Messejana dyke, Spain, *Can. J. Earth Sci., 35*, 1110-1115, 1998.

Dunnig, G. R., and J. P. Hodych, U/Pb zircon and baddeleyite ages for the Palisades and Gettysburg sills of the northeastern United States: Implications for the age of the Triassic/Jurassic boundary, *Geology, 18*, 795-798, 1990.

Dupuy, C., J. March, J. Dostal, A. Michard, and S. Testa, Asthenospheric and lithospheric sources for Mesozoïc dolerites from Liberia (Africa): trace element and isotopic evidence, *Earth and Planet. Sci. Lett., 87*, 100-110,1988.

Dutuit, J. M., Introduction à l'étude paléontologique du Trias continental marocain: Description des premiers Stégocéphales recueillis dans le couloir d'Argana. *Mémoires du Muséum national d'Histoire naturelle de Paris, 36, C*, 253 pp., 1976.

El Wartiti, M., Le Permien du Maroc Mésétien: étude géologique et implications paléogéographiques, Thèse d'Etat Sci., Univ. de Mohamed V, Rabat, 1990.

El Youssi M., Sédimentologie et paléogéographie du Permo-Trias du Haut Atlas central (Maroc), Thèse de Doctorat d'Université, Univ. Scientifique et Médicale de Grenoble, Grenoble, 1986.

Et-Touhami, M., P. E.Olsen, and J. Puffer, Lithostratigraphic and biostratigraphic evidence for brief and synchronous Early Mesozoic basalt eruption over the Maghreb (Northwest Africa). *Eos*, Supplement, v. 82, no. 20, p. S276, 2001.

Falvey, D. A., The development of continental margins in plate tectonic theory, *Australian Petroleum Exploration J., 14*, 95-106, 1974.

Fiechtner, L., Geochemie und Geochronologie frühmesozoischer Tholeiite aus Zentral-Marokko, *Berliner Geowissenschafliche Abhandlungen, 118*, 1-76, 1990.

Fiechtner, L., H. Friedrichsen, and K. Hammerschmidt, Geochemistry and geochronology of Early Mesozoic Tholeiites from Central Morocco. *Geol. Rundschau, 81*, 45-62, 1992.

Fowell, S. J., A. Traverse, P. E. Olsen, and D. V. Kent, Carnian and Norian palynofloras from the Newark Supergroup. Eastern United States, Canada, and the Argana Basin of Morocco: relationship to Triassic climate zones (abstract), 9th IPPC, Houston, *CIPM Newsletter, 51*, 4,1996.

Girard D., Géochimie et minéralogie des laves triasiques de la Meseta cotière (Maroc), *Bull. Institut Sci. Rabat*, N° 11, pp. 37-46, Rabat, Maroc, 1987.

Greenough, J. D., and J. P. Hodych, Evidence for lateral magma injection in the Early Mesozoic dykes of eastern North America, in *Mafic Dykes and Emplacement Mechanisms*, edited by A. J. Parker, P. C. Rickwood and D. H. Tucker, pp. 35-46, Proceedings of the Second International Dyke Conference, Rotterdam, Netherlands, Balkema, 1990.

Hames, W. E., P. R. Renne, and C. Ruppel, New evidences for geologically instantaneous emplacement of earliest Jurassic Central Atlantic magmatic province basalts on the North American margin, *Geology, 28(9)*, 859-862, 2000.

Hawkesworth C, S. Kelley, S. Turner, A. Le Roex, and B. Storey, Mantle processes during Gondwana break-up and dispersal, *J. African Earth Sci., 28(1)*, 239-261, 1999.

Hill, R., Strating plumes and continental break-up, *Earth and Planet. Sci. Lett., 104*, 398-416, 1991.

Hodych, J. P., and G. R. Dunnig, Did the Manicouagan impact trigger end-of-Triassic mass extinction? *Geology, 20*, 51-51, 1992.

Hoernle, K., Geochmistry of Jurassic Oceanic Crust beneath Gran Canaria (Canary Islands): Implications for Crustal Recycling and Assimilation, *J. Petrology, 39(5)*, 859-880, 1998.

Hollard, H., La mise en place au Lias des dolérites dans le Paléozoïque moyen du NE des plaines du Drâa et du bassin de Tindouf (sud de l'Anti-Atlas central, Maroc), *C. R. Acad. Sci. Paris, t. 277*, 553-556, 1973.

Huch K. M., Die Panafrikanische Khzama-Geosuture in Zentralen Anti-Atlas Morokko. Petrographie Geochemie und geochronologie des subduktions-Komplexes der Tourtit-ophiolithe und der Tachankacht-gneise Sowie einiger Kollisionsgesteine om Nordosten des Sirwa-Kristallindoms, PhD Dissertation, Freie Univesität Berlin, 122p., Berlin, Germany, 1988.

Jalil, N., and J. M. Dutuit, Permian Capthorinid reptiles from the Argana Formation, Morocco, *Palaeontology, 39*, 907-918, 1996.

King, D. C., and D. L. Anderson, An alternative mehanism of flood basalt formation, *Earth and Planet. Sci. Lett., 136*, 269-279, 1995.

Klitgord, K. D., and H. Schouten, Plate Kinematics of the Central Atlantic, in *The Geology of North America*, edited by P. R. Vogt and B. E.Tucholke, pp. 351-378, Bulletin of Geological Society of America, The Western North Atlantic region (M), Boulder, Colorado, 1986.

Laville, E., Rôle des décrochements dans les mécanismes de formation des bassins d'effondrement du Haut Atlas marocain au cours des temps triasique et liasique, *Bull. Soc. Géol. France,7, 23(3)*, 303-312, 1981.

Laville, E., and A. Piqué, La distension crustale atlantique et atlasique au Maroc au début du Mésozoïque: le rejeu des structures hercyniennes. *Bull. Soc. Géol. France, 162(6)*, 1161-1171, 1991.

Le Bas, M. J., R. W. Le Maitre, A. Streckeisen, and B. Zanetti, A chemical classification of volcanic rocks based on the total alkali-silica diagram, *J. Petrology, 27(3)*, 745-750, 1986.

Leblanc, M., Le grand dyke de dolérite de l'Anti-Atlas et le magmatisme jurassique du Sud-Marocain, *C. R. Acad. Sci. Paris, 276*, 2943-2946, 1973.

Le Marrec, A., and J. Taugourdeau-Lantz, Description

lihostratigraphique du Permien (?) et Trias du Haut Atlas de Demnat, Maroc, et nouvelles datations palynologiques, *Bull. Fac. Sci. Marrakech, Sect. Sci. Terre, n° sp. 1*, 52-60, 1983.

Le Roex, A. P., Source of mid-ocean ridge basalts: evidence for enrichment processes, in *Mantle Metasomatism*, edited by M. A. Menzies and C. J. Hawksworth, 389-422, 1987.

Le Roy, P., A. Piqué, B Le Gall, L. Aït Brahim, A. M. Morabet, and A. Demnati, Les bassins côtiers triasico-liasiques du Maroc occidental et la diachronie du rifting intra-continental de l'atlantique central. *Bull. Soc. Géol. France, 168(5)*, 637-648, 1997.

Lorenz, J. C., Synthesis on late paleozoïc and triassic redbed sedimentation in Morocco, in *The Atlas System of Morocco*, edited by V. H. Jacobshagen, pp. 139-168, Lecture Notes in Earth Sciences, 15, Springer-Verlag, Berlin, 1988.

Lucas, S. G., Global Triassic tetrapod biostratigraphy and biochronology, *Paleogeography, Paleoclimatology, Paleoecology, 143*, 347-384, 1998a.

Lucas, S. G., The aetosaur Longosuchus from the Triassic of Morocco and its biochronological significance, *C. R. Acad. Sci. Paris, 326, sér. II*, 589-594, 1998b.

Lund, J. J., Palynologie der tieferen Bigoudine Fm, Ober-Trias, Marokko (Abstract), AAP Tagung, AAP Rundbrief, 14-15, 1996.

Manspeizer, W., Triassic-Jurassic rifting and opening of the Atlantic: An overview, in *Triassic-Jurassic rifting: Continental breakup and the origin of the Atlantic Ocean and passive margins*, edited by W. Manspeizer, pp. 41-80, Developments in Geotectonics, 22, Elsevier, Amsterdam, 1988.

Manspeizer, W., The breakup of Pangea and its impact on climate: Consequences of Variscan-Alleghanide orogenic collapse, in *Pangea: Paleoclimate, Tectonics, and Sedimentation during Accretion, Zenith and Breakup of a Supercontinent*, edited by G. D. Klein, pp. 169-185, Geological Society of America, Special Paper 288, Boulder, 1994.

Manspeizer, W., J. H. Puffer, and H. L. Cousminer, Separation of Morocco and Eastern North America: A Triassic-Liassic stratigraphic record, *Geol. Soc. Amer. Bull., 89*, 901-920, 1978.

Manuppella, G., Litostratigrafia e Tectónica da Bacia Algarvia, *Geonovas (Lisboa), 10*, 67-71, 1988.

Martins, L. T., Actividade Ígnea Mesozóica em Portugal (Contribuição Petrológia e Geoquímica). PhD thesis, Universidade de Lisboa, Faculdade de Ciências, Lisboa, 1991.

Martins, L. T., and R. Kerrich, Magmatismo toleítico continental no Algarve (Sul de Portugal): Um exemplo de contaminação crustal "in situ", *Com. Inst. Geol. Mineiro, 85*, 99-116, 1998.

Marzoli, A., P. E. Renne, E. M. Piccirillo, M. Ernesto, G. Bellieni, and A. De Min, Extensive 200-Million-Year-Old Continental Flood Basalts of Central Atlantic Magmatic Province, *Science, 284*, 616-618, 1999.

Mattis, A. F., Nonmarine Triassic sedimentation, Central High Atlas mountains, Morocco, *J. Sedim. Petrol., 47*, 107-119, 1977.

Mauche, R., G. Faure, L. M. Jones, and J. Hoefs, Anomalous isotopic composition of Sr, Ar and O in the Mesozoic diabase dikes of Liberia, West Africa. *Contrib. Miner. Petrol., 101*, 12-18, 1989.

McDonough, W. F. and S. S. Sun, The composition of the Earth, *Chemical Geology*, 120, 223-253, 1995.

McHone, J. C., Broad-terrane Jurassic flood basalts across northeastern North America, *Geology, 4*, 319-322, 1996.

McHone, J.G., Non-plume magmatism and rifting during the opening of the central Atlantic Ocean, *Tectonopysics, 316*, 287-296, 2000.

McHone, J. G., and J.H. Puffer, Flood basalt province of the Pangean Atlantic rift: regional extent and environmental significance, in *Aspects of Triassic-Jurassic Rift Basin Geoscience*, edited by P. E. Olsen, and P. M. LeTourneau, Columbia University Press, New York, 2000 (in press). http://jmchone.web.wesleyan.edu/CAMP.html

Medina, F., Syn- and postrift evolution of the El Jadida-Agadir basin (Morocco): constraints for the rifting models of the Central Atlantic, *Can. J. Earth Sci., 32*, 1273-1291, 1995.

Medina, F., Structural styles of the Moroccan Triassic basins, *Zbl. Geol. Paläont. Teil I, 9-10*, 1167-1192, Epicontinental Triassic International Symposium, 2000.

Monition, A., and L. Monition, Quelques observations sur les formations "permo-triasiques" du Massif de Béni-Snassene (Maroc oriental), paper presented at 20 th International Geologic Congress, Mexico, Publications Association des Serv. Géol. Afric., 187-202, 1959.

Odin, S. G., Geological time scale, *C. R. Acad. Sci. Paris, t. 318, II*, 59-71, 1994.

Oliveira, E. P., J. Tarney, and X. J. João, Geochemistry of the Mesozoic Amapa and Jari Dykes Swarms, northern Brasil: Plume-related magmatism during the opening of the Central Atlantic, in *Mafic dykes and emplacement mecanisms*, edited by A. J. Parker, P. C. Rickwood, and D. H. Tucker, pp. 173-183, Proceedings of the Second International Dyke Conference, Rotterdam, Netherlands, Balkema, 1990.

Olivet, J. L., J. Bonnin, P. Benzart, and J. M. Auzende, Cinématique de l'Atlantique nord et central. *Rep., Rapports scientifiques et techniques du CNEXO, 54*, 108 pp., Publications Centre National pour l'Exploitation des Océans, 1984.

Olsen, P. E., Stratigraphic record of the early Mesozoic break up of Pangea in Laurasia, *Annual Reviews of Earth and Planet. Sci. Lett., 25*, 366-377, 1997.

Olsen, P. E., Giant Lava Flows, Mass Extinctions, and Mantles Plumes, *Science, 284*, 604-605, 1999.

Olsen, P. E., D. V. Kent, S. J. Fowell, R. W. Schilische, M. O. Withjack, and P.M. Le Tourneau, Implications of a comparison of the stratigraphy and depositional environments of the Argana (Morocco) and Fundy (Nova Scotia, Canada) Permian-Jurassic basins, in *Le Permien et le Trias du Maroc*, Oujidi, M. and Et-Touhami, M., eds., Actes de la Première et la Deuxième Réunion du Groupe Marocain du Permien et du Trias: Oujda, Hilal, 28 p, 2000. http://www.ldeo.columbia.edu/~polsen/ nbcp/peo.cv2.html

Ouarhache, D., Etude géologique dans le Paléozoique et le Trias de la bordure NW du Causse moyen Atlasique (S et SW de Fès, Maroc), Thèse de Doctorat de 3ème Cycle, Univ. Paul Sabatier, Toulouse, 1987.

Ouarhache, D., A. Charrière, G. Lachkar, and M. El Wartiti, Nouvelles datations palynologiques et précision de l'âge des premiers épanchements basaltiques triasiques dans le Moyen Atlas Tabulaire (abstract), *Deuxième Réunion du Groupe Marocain du Permien et du Trias, 33*, 1999.

Oujidi, M., Evolution tectono-sédimentaire des Monts d'Oujda (Maroc oriental) au cours du Trias et du Lias basal, In *Le Permien et le Trias du Maroc: Etat des connaissances*, edited by F. Medina, pp. 201-212, Editions PUMAG, Marrakech,.1996.

Oujidi, M. and S. Elmi, Evolution de l'architecture des monts d'Oujda (Maroc oriental) pendant le Trias et au début du Jurassique, *Bull. Soc. Géol. France, 171(2)*, 169-179, 2000.

Owodenco, B., Mémoire explicatif de la carte géologique du bassin houiller de Djerada et de la région du Sud d'Oujda, *Mém. Soc. Géol. Belgique, 70*, 162 p., 1946.

Oyarzun R., M. Doblas, J. López-Ruiz, and J. M. Cebriá, Opening of the central Atlantic and asymetric mantle upwelling phenomena: Implications for long-lived magmatism in western North Africa and Europe. *Geology, 25*, 727-730, 1997.

Palain, C., Connaissances stratigraphiques sur la base du mésozoïque portugais, *Ciências da Terra (UNL), Lisboa, 5*, 11-28, 1979.

Palfy, J., J. K. Mortensen, E. S. Carter, P. L. Smith, R. M. Friedman, and H. W. Tipper, Timing the end-Triassic mass extinction: First on land, then in the sea? *Geology:* Vol. 28, No. 1, pp. 39-42, 2000.

Pe-Piper G, L. F. Jansa, and R. S. J. Lambert, Early Mesozoic magmatism on the eastern Canadian margin: Petrogenetic and tectonic significance, in *Eastern North American Mesozoic Magmatism*, edited by J. H. Puffer and P. C. Ragland, pp.13-36, Geological Society of America, Special Paper 268, 1992.

Petrini, R., L. Civetta, E. M. Piccirillo, G. Bellieni, P. Comin-Chiaramonti, L. S. Marques, and A. J. Melfi, Mantle heterogeneity and crustal contaminations in the genesis of Low-Ti continental flood basalts from Paraná plateau (Brazil): Sr-Nd isotope and geochemical evidence, *J. Petrol., 28(4)*, 701-706, 1987.

Piqué, A., and E. Laville, Les séries triasiques du Maroc, marqueurs du "rifting" atlantique, *C. R. Acad. Sci. Paris, t. 317,II*, 1215-1220, 1993a.

Piqué, A., and E. Laville, L'ouverture de l'Atlantique Central: un rejeu en extension des structures paleozoïques?,. *C. R. Acad. Sci. Paris, t. 317,II*, 1325-1332, 1993b.

Piqué, A., and Laville E., L'ouverture initiale de l'Atlantique central, *Bull. Soc. Géol. France, 166*, 725-738, 1995.

Piqué, A., and Laville E., The Central Atlantic rifting: reactivation of Paleozoic structures? *J. Geodynamics, 21*, 235-255, 1996.

Piqué A., P. Le Roy, and M. Amrhar, Transtensive synsedimentary tectonics associated with ocean opening: the Essaouira-Agadir segment of the Moroccan Atlantic margin. *J. Geol. Soc. London, 155*, 913-928, 1998.

Puffer, J. H., Eastern North American flood basalts in the context of the incipient breakup of Pangea, in *Eastern North American Mesozoic Magmatism*, edited by J. H. Puffer and P. C. Ragland, pp. 95-118, Geological Society of America, Special Paper 268, 1992.

Rahimi, A., Le grand dyke jurassique de Foum Zguid (Anti-Atlas, Maroc): Un exemple de différenciation magmatique, Thèse de Doctorat de 3ème cycle, Univ. Cadi Ayyad de Marrakech, 1988.

Rakus, M., Evolution et position paléogéographique des Monts d'Oujda au cours du Mésozoïque, *Mines, Géologie et Energie, Rabat, 46*, 75-78, 1979.

Rickwood, P. C., Boundary lines within petrologic diagrams which use oxides and major and minor elements, *Lithos, 22*, 247-263, 1989.

Robillard, D., Etude pétrographique du complexe basaltique permo-triasique du Moyen Atlas septentrional (région de Taza - Maroc), *Ann. Soc. Géol. Nord*, 98, 135-144, 1978.

Sabaoui, A., Structure et évolution alpine de Moyen Atlas septentrional sur la transversale Tleta des Zerarda-Marhraoua (sud-ouest de Taza, Maroc), Thèse de Doctorat de 3ème cycle, Univ. Paul-Sabatier, Toulouse, 1986.

Salvan, H. M., Les formations évaporitiques du Trias marocain problèmes stratigraphiques, paléogéographiques et paléoclimatiques: Quelques réflexions, *Rev. Géol. Dyn. Géogr. Phys.*, 25, 187-203, 1984.

Saunders, A. D., M. Storey, R. W. Kent, and M. J. Norry, Consequences of plume-lithosphere interactions, in *Magmatism and Causes of Continental Break-Up*, edited by B. C. Storey, T. Alabaster, and R. J. Pankhurst, pp. 41-60, Geological Society of London, Special Publication 68, London, 1992.

Sawyer, D. S., and D. L. Harry, Dynamic modelling of divergent margin formation: application to the U.S. Atlantic margin, *Marine Geology, 102*, 29-42, 1991.

Schermerhorn L. J. G., H. N. A. Priem, N. A. I. M. Boelrijk, E. H. Hebeda, E. A. T Verdurmen, and R. H. Verschure, Age and origin of the Messejana dolerite fault-dike system (Portugal and Spain) in the light of the opening of the North Atlantic Ocean, *Journal of Geology*, 86, 299-309, 1978.

Schmincke H.-U., A. Klügel, T. H. Hansteen, K. Hoernle, and P. A. Bogaard, Samples from the Jurassic ocean crust beneath Gran Canaria, La Palma and Lazarote (Canary Islands), *Earth and Planet. Sci. Lett.*, 163, 343-360, 1998.

Sebai A., G. Feraud, H. Bertrand and J. Hanes, $^{40}Ar/^{39}Ar$ dating and geochemistry of tholeiitic magmatism related to the early opening of the Central Atlantic rift, *Earth and Planet. Sci. Lett.*, 104, 455-472, 1991.

Sheth, H. C., A historical approach to continental flood basalts volcanism: insights into pre-volcanic rifting, sedimentation, and early alkaline magmatism, *Earth and Planet. Sci. Lett.*, 168, 19-26, 1999a.

Sheth, H. C., Flood basalts and large igneous provinces from deep mantle plumes: fact, fiction and fallacy, *Tectonophysics, 311*, 1-29, 1999b.

Slimane A., and M. El Mostaïne, Observations biostratigraphiques au niveau des formations rouges de la séquence synrift dans les bassins de Doukkala et Essaouira (abstract), Première Réunion du Groupe Marocain du Permien et du Trias, 54, 1997.

Sun, S. S., and W. F. McDonough, Chemical and isotopic systematics of oceanic basalts: implication for mantle composition and processes, in *Magmatism in the Ocean Basins*, edited by A. D. Saunders, and M. Norry, pp 313-345, Journal of Geological. Society of London, 42, London, 1989.

Sutter, J. F., Innovative approaches to the dating of igneous events in the early Mesozoic basins of the eastern United States, *U. S. Geological Survey Bull., 1776*, 194-200, 1988.

Taugourdeau-Lanz, J., Pollens des niveaux sedimentaires associés aux basaltes du Trias sur la bordure septentrionale du Maroc Central: Précisions stratigraphiques, *Notes et Mém. Serv. Géol. Maroc, 40*, 275, 135-146, 1978.

Taylor, S. R. and S. M. McLennan, The composition and evolution of continental crust: REE evidence from sedimentary rocks, *Phil. Trans. R. Soc.*, A301, 381-399, 1981.

Terrinha, P. A. G., Structural Geology and Tectonic evolution of the Algarve Basin, South Portugal, PhD Thesis, Imperial College, University of London, London, 1998.

Thompson, G. A., Deep Mantle Plumes and Geoscience Vision, *GSA Today, 8(4)*, 17-25, 1998.

Thompson, R. N., M. A. Morrison, A. P. Dickin, and G. L. Hendry, Continental Flood Basalts... Arachnids Rule OK? in *Continental Basalts and Mantle Xenoliths*, edited by C. J. Hawkesworth and M. J. Norry, pp. 158-185, Shiva, Nanthwich, 1983.

Tixeront, M., Lithostratigraphie et minéralisation cuprifères et uranifères syngénétiques et familières des formations permo-triasiques du couloir d'Argana (Haut Atlas Occidental, Maroc), *Notes et Mém. Serv. Géol. Maroc, 33*, 147-177, 1973.

Tourani, A., Lund J. J., N. Benaouiss, and R. Gaupp, Stratigraphy of Triassic syn-rift deposition in Western Morocco, *Zbl. Geol. Paläont. Teil I, 9-10*, 1193-1215, 2000.

Turner, S., and C. Hawkesworth, The nature of the subcontinental mantle: constraints from the major-element composition of Continental Flood Basalts, *Chemical Geology, 120*, 295-314, 1995.

Van Houten, F. B., Triassic-Liassic deposits of Morocco and Eastern North America: Comparison, *A. A. P. G. Bull., 61(1)*, 79-99, 1977.

White, R. S., and D. McKenzie, Magmatism at rift zones: The generation of volcanic continental margins and flood basalts, *J. Geophys. Res., 94, B6*, pp. 7685-7729, 1989.

White R. S., and D. McKenzie, Mantle plumes and flood basalts, *J. Geophys. Res., 100, B9*, pp. 17,543-17,585, 1995.

Whittington, D., Chemical and physical constraints on petrogenesis and emplacement of ENA olivine diabase magma type, in *Triassic-Jurassic rifting: Continental breakup and the origin of the Atlantic Ocean and passive margins*, edited by W. Manspeizer, pp. 557-577, Developments in Geotectonics, 22, Elsevier, Amsterdam, 1988a.

Whittington, D., Mesozoic Diabase Dikes of North Carolina, Ph.D. thesis. The Florida State University, College of Arts and Sciences, Florida, 1988b.

Wilson, M., Magmatism and geodynamics of basin formation, *Sedimentary Geology, 86*, 5-29, 1993.

Wilson, M., Thermal evolution of the Central Atlantic passive margins: continental break-up above a Mesozoic superplume, *J. Geol. Soc. London, 154*. pp. 491-495, 1997.

Withjack, M. O., R. W. Schlische, and P. E. Olsen, Diachronous Rifting, Drifting, and Inversion on the Passive Margin of Central Eastern North America: An Analog for other Passive Margins, *A. A. P. G. Bull., 82, 5A*, 817-835, 1998.

Adelmajid El Boukhari - Department of Geology, Faculty of Sciences-Semlalia, Cadi Ayyad University, P.O. Box. 2390, Prince Moulay Abdellah Avenue, Marrakech 40000, Morocco. E-mail: elboukhari@ucam.ac.ma

El Houssaine Aït Chayeb - Department of Geology, Faculty of Sciences-Semlalia, Cadi Ayyad University, P.O. Box. 2390, Prince Moulay Abdellah Avenue, Marrakech 40000, Morocco.

Hassan Ibouh - Department of Geology, Faculty of Sciences and Technics-Guéliz, Cadi Ayyad University, P.O. Box.618, Marrakech 40000, Morocco. E-mail: ibouh@fstg-marrakech.ac.ma

José Madeira - Departamento de Geologia, Faculdade de Ciências, Universidade de Lisboa, Edif. C2 5° piso, Campo Grande, 1749-016 Lisboa, Portugal. E-mail: jose.madeira@fc.ul.pt

José Manuel Munhá - Departamento de Geologia, Faculdade de Ciências, Universidade de Lisboa, Edif. C2 5° piso, Campo Grande, 1749-016 Lisboa, Portugal. E-mail: jmunha@fc.ul.pt

Línia Tavares Martins - Departamento de Geologia, Faculdade de Ciências, Universidade de Lisboa, Edif. C2 5° piso, Campo Grande, 1749-016 Lisboa, Portugal. E-mail: ltm@fc.ul.pt

Nasrrdine Youbi - Department of Geology, Faculty of Sciences-Semlalia, Cadi Ayyad University, P.O. Box. 2390, Prince Moulay Abdellah Avenue, Marrakech 40000, Morocco. E-mail: youbi@ucam.ac.ma

The Northernmost CAMP: ^{40}Ar/^{39}Ar age, Petrology and Sr-Nd-Pb Isotope Geochemistry of the Kerforne Dike, Brittany, France

Fred Jourdan[1], Andrea Marzoli[1], Herve Bertrand[2], Michael Cosca[3], Denis Fontignie[1]

The Central Atlantic Magmatic Province (CAMP) is defined by tholeiitic basaltic flows and dikes associated with the initial break-up of Pangea at 200 Ma, preceding the opening of the Atlantic Ocean. These tholeiites occur in once-contiguous parts of North America, Africa, South America and Europe over a total area of about 7 million square kilometers. The Kerforne dike, located in Brittany (NW France), represents the northernmost outcrop of this province. Due to its orientation and location, more than 1500 Km from the Early Jurassic Atlantic rift, like other CAMP dikes, it can not be considered as the magmatic expression of the Central Atlantic rifting. Despite its distal position, this dike has an Ar/Ar age (193 ± 3 Ma, obtained on plagioclase separates) similar to the CAMP tholeiites. Kerforne dolerites are characterized by augite and minor plagioclase phenocrysts. According to petrographic observation and microprobe analyses, some of these plagioclases contain resorbed high-An (An$_{85}$) possibly xenocrystic cores which may be evidence of interaction with a mafic lower crust. The low-TiO$_2$ (1.0 wt%) tholeiitic Kerforne basalts are characterized by negative Nb anomalies, by a positive correlation between $\bullet_{Sr}$ and $\bullet_{Nd}$, by high radiogenic ^{207}Pb/^{204}Pb in comparison to relatively unradiogenic ^{206}Pb/^{204}Pb, and by an enrichment in LREE relative to HREE. These chemical features, along with the mineralogic observations, are indicative of a minor contamination with mafic lower crust, like that represented by granulitic xenoliths of the Massif Central, France. By contrast, contamination with the silicic upper crust (e.g., with the granitic basement) was negligible. The isotopic compositions of the little contaminated Kerforne basalts are similar to those of most other CAMP low-TiO$_2$ basalts, and are different from those of most oceanic basalts. It is suggested that this high ^{87}Sr/^{86}Sr and ^{207}Pb/^{204}Pb isotopic signature was inherited by interaction of primitive mantle with metasomatized portions of the continental lithospheric mantle, similar to the sources of Variscan lamproites of Brittany. Moreover, The contribution of an OIB mantle component may be ruled out as the Kerforne mantle source was isotopically different from those of the oceanic islands which have been suggested to represent the present-day expression of the hypothetic CAMP mantle plume.

The Central Atlantic Magmatic Province:
Insights from Fragments of Pangea
Geophysical Monograph 136
Copyright 2003 by the American Geophysical Union
10.1029/136GM011

[1] Section de Sciences de la Terre; Université de Genève, Switzerland

[2] Ecole Normale Supérieure et UCBL, Laboratoire des Sciences de la Terre,

[3] Institute of Mineralogy and Geochemistry, University of Lausanne, Switzerland

1. INTRODUCTION

At about 200 Ma, before the opening of the Central Atlantic and the breakup of Pangea vast volumes of continental tholeiitic basaltic magmas were emplaced over an area of at least 7 million square kilometers in South and North America, Western Africa and Southwestern Europe [e.g., *Bertrand*, 1991; *Deckart et al.*, 1997; *Marzoli et al.*, 1999; *McHone*, 2000; *Hames et al.*, 2000]. The Central Atlantic Magmatic Province (CAMP) is mainly represented by intrusions (dikes and sills) and minor erosional remnants of lava flows. Like most basalts of other large continental igneous provinces, including continental flood basalts (CFB), the CAMP basalts are subdivided into two geochemical subgroups, low- and high-Ti, respectively. The vast majority of CAMP basalts are low-TiO_2 (< 2.0 wt%) tholeiites characterized by a lithospheric (crust and/or mantle) isotopic signature. The only known high-TiO_2 CAMP tholeiites occur in the once contiguous Equatorial regions of Africa (Liberia) and South America [N-E Brazil and Guyana; *Dupuy et al.*, 1988; *Bertrand et al.*, 1999; *De Min et al.*, present volume].

The northernmost occurrence of the CAMP is represented by the Kerforne dike of Western Brittany, France. It is part of the European subprovince of the CAMP, which includes the lava flows, sills and dikes of the Pyrenees and of the Iberian peninsula [*Alibert*, 1985; *Bertrand et al.*, 1987; *Beziat et al.*, 1991; *Sebai et al.*, 1991; *Demant et al*, 1996; Fig. 1]. We present here the first integrated petrographic, mineralogic, geochemical, isotopic and Ar/Ar geochronological study of a CAMP dike. The present study is also an attempt to understand the petrogenesis of basalts of the CAMP, within the context of a relatively well studied region such as Brittany and France in general.

2. GEOLOGICAL SETTING AND SAMPLING

The Kerforne dike is located in Brittany in northwestern France (Fig. 1). The tholeiitic magma intruded a rejuvenated SE-NW oriented Hercynian fault system. By the time of magmatic intrusion, during the earliest Jurassic (see below), the Kerforne dike was situated at the western end of an extensional basin. This ranged from Catalonia to the Basque Countries, before the opening of the Gulf of Biscay, which occurred in the Cretaceous [*Moreau et al.*, 1997]. The orientation of this basin was orthogonal to the principal Central Atlantic and Alpine Tethys rifts [Fig. 1a; *Stampfli et al*, 1998], and roughly parallel to the Kerforne dike. Located at about 1500 km from the main Central Atlantic and Tethys rifts, the dike was aligned with the tholeiitic magmatic rocks of the Pyrenees [*Beziat et al.*, 1991], located 600 km to the south-west (Fig. 1a).

The comparable paleopole for Kerforne [*Sichler and Perrin*, 1993] and other CAMP basalts is indicative of a similar age. This is consistent with K/Ar dates obtained on whole rock, biotite and plagioclase of the Kerforne dike, despite the wide range displayed by these ages, from 170 to 210 Ma [*Caroff et al.*, 1995].

The dike intrudes igneous and metamorphic rocks of the Armorican massif, which was part of the Variscan arc [e.g., *Matte*, 1986]. Two major roughly E-W oriented faults, the North and South Armorican Shear Zones, separate three major units in western Brittany. The dike intrudes the central and southern structural units, and terminates near the intersection between the North Armorican Shear Zone and the Atlantic ocean. The Central Armorican domain consists mainly of low grade metasedimentary rocks, whose protoliths vary in age from Cambrian to Carboniferous. Minor volcanic rocks (lamproites, tholeiitic basalts and silicic flows) and granites intruding along the South Armorican Shear Zone were emplaced during the Variscan orogeny [*Turpin et al.*, 1988; *Caroff et al.*, 1996; *Bernard-Griffiths et al.*, 1985]. The South Armorican domain is a high grade metamorphic belt, built up of gneiss, schist and granite of Ordovician to Carboniferous age [*Bernard-Griffiths et al.*, 1985].

The 100 km long Kerforne dike has a thickness of several meters, up to a maximum of 30 m at the Brenterc'h North and Pors Milin outcrops. In all outcrops, the subvertical (75°) to vertical dike follows a 110-130°N orientation and generally crosscuts the foliation of the basement rocks. A complete description of the outcrops of Kerforne basalts is provided by *Caroff et al.* [1995]. For this study, only the most representative and best preserved sites within the Central Armorican domain have been sampled (Fig. 1b), which, from SE to NW are:

1) The 10 m thick dike close to Douarnenez (samples B1 and B2; Table 1) which intrudes a slightly metamorphosed trondhjemitic rock of Early Permian age (sample B3, 281 ± 5 Ma, $^{40}Ar/^{39}Ar$ plateau age; see below; Table1). Leucogranites cropping out a few kilometer to the South, still within the Central Armorican domain, have been dated (Rb/Sr) at 344 ± 7 Ma [*Bernard-Griffiths et al.*, 1985].

2) The slightly altered dike at Camaret, on the Crozon peninsula (B4 and B5), which is about 10 m thick. Its country rocks are Carboniferous meta-sedimentary units, weakly folded during the Hercynian orogeny.

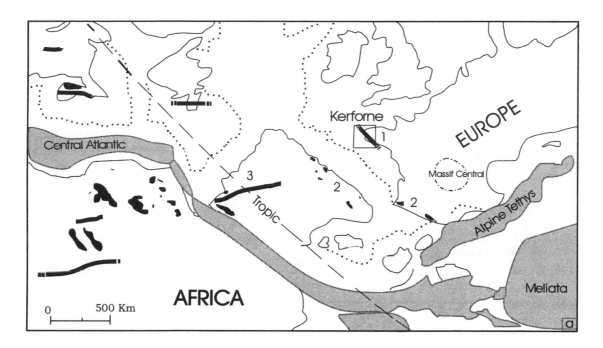

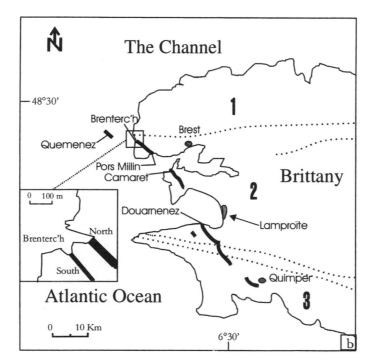

Figure 1. a: Location map of the European and North African CAMP *[Bertrand*, 1991; *Sebai et al.*, 1991; *McHone*, 2000] in an Early Jurassic Pangea reconstruction adapted from *Stampfli et al.* [1998]. Dikes are represented by bold line, magmatic flows by dark areas and continental margins by dotted lines; solid lines represent the present continental shape. 1: Kerforne dike; 2: Pyrenean sills; 3: Messejana dike.
(b): Sketch map of western Brittany showing sample locations of the Kerforne dike. The dike is represented by a bold discontinuous line and the North and South Armorican shear zones by dotted lines, separating the North (1), Central (2) and South (3) Armorican zones. Note location of lamproitic rocks *[Turpin et al.*, 1988]. Inset represents sketch map of Brenterc'h outcrops.

Table 1. Major (wt%), trace and REE (ppm) compositions, and CIPW-normative quartz and olivine of selected Kerforne dike samples plus one wall-rock sample. [Mg# = 100*Mg/(Mg+Fe^{2+}); Fe$_2$O$_3$/FeO= 0.17], as calculated for QFM buffered fO$_2$, at 1 kbar, by using MELTS; *Ghiorso and Sack* [1995].

| Sample | Kerforne basalts | | | | | | | | | | granite |
	B1	B 2	B 4	B 5	B 7	B 8	B 9	B 11	B 12	B 13	B3
SiO$_2$ (wt.%)	50.08	50.25	50.89	50.42	49.90	50.29	49.50	50.71	50.47	51.23	72.21
TiO$_2$	0.96	0.98	0.97	0.87	0.94	0.96	1.02	0.97	0.95	0.94	0.25
Al$_2$O$_3$	14.22	14.80	14.51	14.72	14.52	14.60	15.35	14.27	14.25	13.97	13.95
Fe$_2$O$_3$	11.03	11.02	10.69	10.39	11.03	11.19	11.59	11.04	10.87	10.77	2.83
MnO	0.19	0.18	0.16	0.16	0.17	0.18	0.17	0.17	0.17	0.17	0.04
MgO	7.93	7.86	7.47	7.81	8.15	8.04	7.31	7.62	7.71	7.71	0.80
CaO	11.66	11.72	11.55	11.52	11.34	11.75	10.85	11.52	11.71	11.67	1.88
Na$_2$O	1.90	2.02	1.94	1.90	2.00	1.95	2.04	2.00	1.92	1.96	3.30
K$_2$O	0.41	0.43	0.56	0.79	0.81	0.52	0.38	0.65	0.34	0.57	2.75
P$_2$O$_5$	0.10	0.11	0.11	0.09	0.10	0.10	0.12	0.11	0.10	0.10	0.10
LOI	0.99	0.35	0.79	0.95	0.61	0.23	1.82	0.56	0.77	0.45	1.19
Sum	99.51	99.77	99.70	99.68	99.62	99.86	100.21	99.67	99.30	99.57	99.29
Mg#	60.70	60.51	60.01	61.77	61.34	60.68	57.56	59.72	60.38	60.58	37.82
Nb (ppm)	5	5.5	5	5	5	5.3	5	5.4	5	5.3	3
Zr	74	81	75	75	69	78	78	80	74	78	191
Y	20	20	21	17.3	19	21	22	22	19	21	34
Sr	162	163	179	184	177	167	168	167	161	186	94
Rb	16	12.7	18	18.7	25	15.4	16	17.7	15	15.5	62
Ni	107	120	106	108	107	110	121	107	110	120	2
Cr	185	241	146	256	182	254	182	250	171	262	10
V	275	272	278	252	271	259	304	271	273	267	24
Ba	93	124	132	149	102	133	63	108	82	118	258
Sc	26	25	23	23	27	34	34	26	26	22	5
U	-	0.4	0.3	-	-	0.3	0.4	0.4	-	0.3	3
Th	-	1.2	1.2	-	-	1.2	1.3	1.2	-	1.2	-
Pb	-	2	-	3	-	4	3	2	-	2	-
La	-	7.4	7.6	-	-	7.6	8.4	7.9	-	7.5	-
Ce	-	17.6	17.7	-	-	17.9	19.8	18.8	-	17.7	-
Pr	-	1.9	1.9	-	-	1.9	2.1	2.0	-	1.9	-
Nd	-	10.3	10.7	-	-	10.4	11.5	10.9	-	10.4	-
Sm	-	2.9	2.9	-	-	2.9	3.2	3.0	-	2.8	-
Eu	-	0.92	0.94	-	-	0.96	1.06	0.94	-	0.9	-
Gd	-	3.0	3.3	-	-	3.2	3.7	3.4	-	3.3	-
Tb	-	0.6	0.6	-	-	0.6	0.6	0.6	-	0.6	-
Dy	-	3.5	3.5	-	-	3.5	3.7	3.6	-	3.4	-
Ho	-	0.76	0.75	-	-	0.74	0.81	0.78	-	0.73	-
Er	-	2.2	2.2	-	-	2.2	2.4	2.3	-	2.3	-
Tm	-	0.3	0.3	-	-	0.3	0.3	0.3	-	0.3	-
Yb	-	2.1	2.1	-	-	2.0	2.2	2.2	-	2.1	-
Lu	-	0.31	0.31	-	-	0.32	0.32	0.33	-	0.32	-
CIPW norm											
Quartz	1.04	0.29	0.87	0.51	-	0.06	0.75	1.03	1.94	2.02	-
Olivine	-	-	-	-	2.98	-	-	-	-	-	-

3) The dike close to Pors Milin (B12 and B13), about 30 m thick, which is weakly altered on its northern side, and intrudes the Brest orthogneiss.

4) The northern (B7, B8, B9) and southern (B10 and B11) branches of the dike at Brenterc'h, which reach a thickness of 30 m and 15 m, respectively. These two branches of the dike are separated by 100 m and both intrude the Ploumoguer basement gneiss.

3. ANALYTICAL METHODS

A set of 10 basalts and 2 wall-rock granites have been selected for major and trace element analyses which have been performed at the University of Lausanne (Switzerland), by X-ray fluorescence (XRF), following methods described in *Rhodes* [1988], and with average analytical uncertainties of 1% for major elements, and 5% for trace elements. REE, Th and U (uncertainties < 3%) were measured on 6 samples at XRAL laboratories, Toronto, Canada, by VG Plasma-Quad inductively coupled plasma-mass spectrometer (ICP-MS), using Na_2O_2 fusion. Mineral analyses were made at the University of Lausanne, with a CAMECA SX50 electron microprobe, using 15 kV accelerating voltage, and 15 to 30 nA current during analyses of plagioclase and mafic minerals, respectively.

For Pb-Sr-Nd isotope analyses at the University of Geneva, Switzerland, 500 mg of hand-picked grains of six basalts and one granite sample were leached (HCl) before dissolution with $HF+HNO_3$ and HCl, and followed by standard element separation with chromatographic ion-exchange columns. Measurements were performed with a 7-collector Finnigan MAT 262 thermal ionization mass spectrometer. $^{87}Sr/^{86}Sr$ ratios were measured in semi-dynamic mode and are mass-fractionation corrected to a $^{88}Sr/^{86}Sr$ ratio of 8.375209. Measurement of standard E & A yielded an average value of 0.708028 ± 5 (2σ), n = 52. $^{143}Nd/^{144}Nd$ analyses were performed in semi-dynamic mode with $^{146}Nd/^{144}Nd$ mass fraction corrected to 0.721903. La Jolla Nd standard gave a value of 0.511832 ± 6 (2σ), n = 28. Pb isotopic compositions were measured in dynamic mode. SRM NBS 981 standard yielded values of 0.92 ± 0.05 (2σ), n = 132 and Pb concentration were determined by XRF.

Three hand picked plagioclase separates (about 30 mg of 100-200 μm crystals) of the dike (B2, B8 and B11) and one biotite separate of the granitic basement (B3) were selected for $^{40}Ar/^{39}Ar$ dating. Samples were irradiated for 12 hours in the Triga reactor at Oregon State University, USA, along with Fish Canyon sanidine (FCs)

neutron fluence monitors, for which an age of 28.02 Ma is adopted [*Renne et al.*, 1998]. Incremental heating from 550 to 1600 °C (15 steps for the plagioclase, 20 for the biotite) in a resistance furnace and analyses on a MAP-215.5 mass spectrometer were performed at the University of Lausanne, Switzerland. Ages were calculated considering the ^{40}K decay constants of *Steiger and Jäger* [1977]

4. PETROGRAPHY AND MINERALOGY

The tholeiitic dolerites display a fine grained sub-ophitic to ophitic texture which becomes sub-hyaline at the dike's margins at Brenterc'h. The main minerals are augitic clinopyroxene (cpx), plagioclase, pigeonite, olivine, and Fe-Ti oxides (magnetite and minor ilmenite). A few samples contain secondary minerals (biotite, sericite, calcite and amphibole).

Cpx and plagioclase are the most abundant phenocryst phases in the Kerforne dolerites. Cpx consists of large optically homogeneous augite (30-40 vol. %; Wo_{31-42} En_{46-57} Fs_{6-11}) plus minor pigeonite (0-5 vol. %; Wo_{13-14} En_{61-62} Fs_{24}). Pigeonite is usually confined to the rim of augite and never occurs as isolated crystals. Augite is Cr_2O_3-rich (up to 0.8 wt%) and relatively Al_2O_3 poor (up to 3 wt%). Ca, Cr and Mg# [$100*Mg/(Mg+Fe^{2+})$, considering $Fe_2O_3/FeO= 0.17$] decrease toward the crystal rims, consistently with cpx evolution trends in tholeiitic magmas. The augite-pigeonite equilibrium crystallization temperature [*Lindsley*, 1983] is estimated at between 1150-1050 °C. For a temperature of 1100 °C the calculated augite crystallization pressure is about 3.5 ± 2 kbars [*Nimis*, 1995], suggesting a shallow crystallization.

Olivine represents 5 to 10 % of the rock and it occurs mainly as relatively large (up to 3 mm) resorbed crystals or as inclusions within augite microphenocrysts. Olivine is rarely fresh and is frequently replaced by bowlingite and iddingsite. Fresh portions of olivine phenocrysts are generally homogeneous and forsterite (Fo) poor (Fo_{70-71}). This suggests that olivine composition is out of equilibrium with the relatively little evolved host rock (Mg# = 57.6–61.8).

Plagioclase represents about 40-50 vol. % of the bulk rock and consists of two main textural and compositional types: (a) microlites and (b) sparse relatively large, optically zoned plagioclase phenocrysts, sometimes occurring as crystal aggregates (Table 2). The microlites have labradorite-andesine composition (An_{74-42}), which is consistent with the moderately evolved whole-rock compositions [Ca# = $100*Ca/(Ca+Na)$ = 73-76]. By contrast,

Table 2. Representative compositions of plagioclase phenocrysts and groundmass (gm) microlites. Molar % albite (Ab), anorthite (An) and orthoclase (Or) are shown.

Sample Location	B2 Plag3 core	B7 Plag1 core	B9 Plag1 core	B9 Plag1 rim	B9 Plag2 core	B12 Plag2 core	B2 Plgm microlite	B8 Plgm microlite	B11 Plgm microlite
SiO_2 (wt%)	47.28	46.41	46.20	56.02	46.37	46.82	49.10	53.48	49.92
Al_2O_3	32.81	33.58	33.30	27.68	33.66	33.63	31.35	28.12	31.37
Fe_2O_3	0.57	0.69	0.58	0.85	0.69	0.60	0.67	0.94	0.59
MgO	0.24	0.13	0.24	0.04	0.25	0.23	0.12	0.14	0.15
CaO	17.32	17.47	17.41	9.96	17.44	17.21	15.13	11.88	14.98
BaO	0.00	0.03	0.01	0.02	0.04	0.01	0.00	0.02	0.05
Na_2O	1.62	1.63	1.53	5.67	1.52	1.63	2.81	4.56	3.01
K_2O	0.05	0.04	0.03	0.39	0.05	0.05	0.09	0.20	0.08
Sum	99.89	99.99	99.31	100.62	100.01	100.17	99.28	99.35	100.16
Ab (Mol%)	14.45	14.39	13.71	49.59	13.55	14.61	25.02	40.52	26.55
An	85.27	85.30	86.09	48.13	86.08	85.07	74.43	58.30	72.97
Or	0.28	0.26	0.18	2.24	0.31	0.31	0.55	1.18	0.48

plagioclase phenocrysts consist of rounded and resorbed high-An cores (An_{80-86}), mantled by labradorite rims with compositions similar to those of the groundmass microlites.

Compositional profiles across ten plagioclase phenocrysts with An_{80-86} core have been obtained (cf., Fig. 2). According to experimental data [*Panjasawatwong et al.*, 1995] at anhydrous conditions, calcic plagioclase (An_{85}) crystallizes from extremely calcic magmas (e.g., Ca# = 85), i.e. substantially more calcic than Kerforne basalts and even more calcic than primary basalts obtained in peridotite melting experiments [e.g., *Hirose and Kushiro*, 1993]. Alternatively, high-An plagioclase crystallizes from hydrous magma ($Kd_{(Ca/Na)} > 1$), but there is no evidence that the tholeiitic Kerforne basalts were particularly rich in H_2O. We suggest therefore that the rounded-resorbed high-An cores are in fact xenocrysts crystallized from a relatively H_2O rich, little evolved magma and subsequently inherited by the Kerforne basalts.

Further constraints on the crystallization of plagioclase in Kerforne tholeiites has been obtained from minor element profiles (Fig. 2). In plagioclase phenocrysts with high-An cores, MgO decreases from about 0.25-0.30 wt% in the cores to less than 0.10 wt% at the rims. By contrast, K_2O increases from less than 0.10 wt% to 0.40-0.60 wt%. The composition of the magma from which the plagioclase crystallized can be calculated assuming equilibrium crystallization at a temperature of 1200 °C and adopting the partition coefficients of *Bindeman et al.* [1998]. High-An plagioclase cores may have crystallized from magmas with high MgO (> 10 wt%) and relatively high K_2O (ca. 0.5-1.0 wt%). Such compositions are different from those of Kerforne basalts. Minor element

compositions support the conclusion that the high-An plagioclase cores are not likely crystallized from the Kerforne basalts at equilibrium conditions.

The K_2O content in the outermost 20% of the plagioclase volume increases up to 0.60 wt%, while MgO decreases to low values. Such an important increase of K_2O and decrease of MgO requires that the plagioclase phenocrysts rims crystallized from an evolved residual liquid (with about 2 wt% K_2O and 2 wt% MgO).

5. AR/AR GEOCHRONOLOGY

For three basaltic samples only (B2, B8 and B11) we could obtain quantities of plagioclase separate (30-36 mg) sufficient for Ar/Ar dating. In general, a thin alteration zone affects plagioclase rims and fractures, as confirmed by an XRD analysis of a plagioclase separate of sample B8, which contains a little amount of sericite (about 5 wt%). Most of the alteration could be eliminated for sample B2, which contains plagioclase phenocrysts (cf., Fig. 2), but not for the fine grained plagioclase of B8 and B11 (about 100 µm diameter).

The three dated basalts yielded plateau ages of 189.0 ± 3.0 Ma (B2), 177.0 ± 3.0 Ma (B11) and 173.0 ± 3.0 Ma (B8; all errors are given at the 2σ confidence level; Fig. 3; Table 3). Apparent age and Ca/K spectra of B2 and of the Brenterc'h basalts (B8 and B11) are different. B2 displays plateau steps associated with high apparent Ca/K (generally >40), while Ca/K for plateau steps of B8 and B11 is generally low (<10). For B8 and B11, the oldest apparent ages are obtained for steps 5-6 (800-850 °C) which also have the highest apparent Ca/K (about 20), whereas subsequent plateau steps decrease in apparent

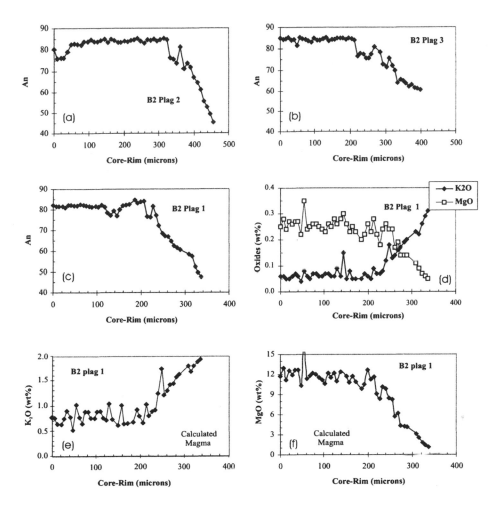

Figure 2. a-d: Representative plagioclase anorthite of B2 (An, molar %), K_2O and MgO (wt%) compositional profiles (Plagioclase 1, 2, and 3 of). Similar profiles were obtained for the analyzed plagioclase phenocrysts of B7, B9 and B12. **e** and **f**: calculated MgO and K_2O concentrations of magma in equilibrium with the analyzed plagioclase. Partition coefficients after *Bindeman et al.* [1998], assuming a temperature of 1200 °C.

age and Ca/K. In general, the absolute amount of released ^{40}Ar was substantially (up to 5 times) higher for B8 and B11 than for B2, suggesting that the two analyzed mineral separates of the Brenterc'h basalts were richer in potassium.

Isochron ages (MSWD = 0.48-1.33, 9-11 steps) of the three samples are slightly older than the plateau ages (193.4 ± 3.7 Ma, 178.0 ± 3.5 Ma, and 174.3 ± 3.3 Ma for B2, B11 and B8, respectively). The initial $^{40}Ar/^{36}Ar$ of the three samples is always lower than the atmospheric ratio (233-284), suggesting a depletion of $^{40}Ar*$ (i.e., an excess of $^{39}Ar/^{40}Ar$).

In general, the analyzed mineral separates, particularly those of B8 and B11, seem to consist of partially altered (sericitized) plagioclase, as suggested by the apparent Ca/K which is substantially lower than that of plagioclase (>40; cf., Table 2). Even if the XRD analyses suggest

that sericite doesn't exceed a few wt% of the dated mineral separate, the different K_2O concentration of plagioclase (<0.50 wt%) and sericite (probably about 10 wt%) suggests that the obtained age refers to a hydrothermal event (sericite crystallization) and not to the magmatic age (plagioclase crystallization). Moreover, the fine-grained mineral separate of the Brenterc'h samples may have been affected by some ^{39}Ar recoil, as suggested by the decreasing apparent ages associated with decreasing Ca/K [cf. *Onstott et al.*, 1995]. Altogether, (post-magmatic) alteration and recoil result in apparent ages younger than the magmatic age.

Notably however, the dated sample (B2) displaying the highest measured Ca/K, yields the highest plateau and isochron ages. These ages of B2, and particularly its isochron age (193.4 ± 3.7 Ma) are probably the closest estimate of the magmatic age of the Kerforne dike, and fall

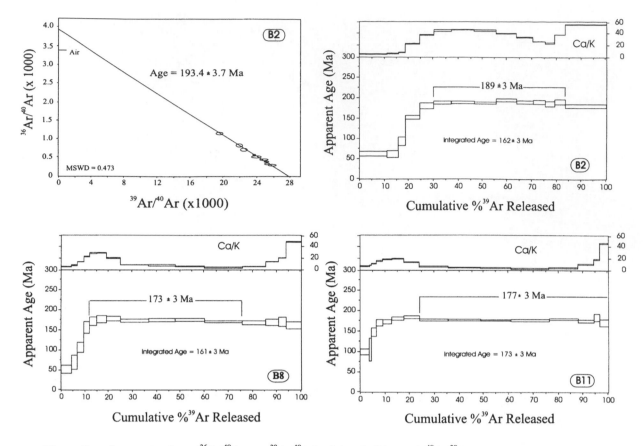

Figure 3. Inverse isochron ($^{36}Ar/^{40}Ar$ vs. $^{39}Ar/^{40}Ar$) of basalt B2 , and $^{40}Ar/^{39}Ar$ apparent age spectra for plagioclase separates of samples B2, B8 and B11. Ca/K is calculated from $^{37}Ar/^{39}Ar$. MSWD (mean square weighted deviation) for B2 inverse isochron is 0.473. Analytical uncertainties are quoted at the 2σ confidence level.

within the range of Ar/Ar ages of other CAMP basalts [191-205 Ma; *Sebai et al.*, 1991; *Deckart et al.*, 1997; *Marzoli et al.*, 1999; *Hames et al.*, 2000].

Note that the biotite separate of the basement trondhjemite B3 yielded a well defined plateau age of 281 ± 5 Ma, consistent with a late-Variscan crystallization.

6. WHOLE ROCK GEOCHEMISTRY

6.1. Major and Trace Elements

The analyzed rocks are low TiO_2 (0.94-1.02 wt%) CIPW quartz normative tholeiites (except olivine-normative sample B7; Table 1). According to the TAS diagram [*Le Bas et al.*, 1986] the Kerforne samples are classified as basalts (Fig. 4). The Mg# varies between 57.6 and 61.7, showing that the basalts are moderately evolved. Although major and trace element variations are moderate and mostly within analytical uncertainty within the whole dike, a relatively good correlation is observed

between decreasing MgO (8.2–7.3 wt%) versus increasing SiO_2 and Zr (with the exception of the dike margin B9). The correlation between MgO and other major and trace elements is not well defined (Fig. 5). Some incompatible elements, notably K_2O and Rb, display a relatively large variability (0.3-0.8 wt% for K_2O, and 15-25 ppm for Rb, c.f. Table 1).

On the multi-element mantle normalized diagram [*Sun and McDonough*, 1989] the Brittany tholeiites show an enrichment in the most incompatible trace elements (ITE; Fig. 6). The LILE (large ion lithophile elements, Rb, Ba, K) are the elements displaying the strongest variability. There is a significant negative Nb anomaly with respect to LILE and light rare earth elements (LREE). Similarly, Ti is slightly depleted compared to Zr and Y. Samples B1 and B5 show a rather marked positive Sr anomaly (mantle normalized Sr/Sr*= 1.65 and 1.90, respectively), whereas B9 has a negative Sr anomaly (0.80).

The chondrite-normalized [cn; *Boynton*, 1984] REE display only minor variability along the whole dike (Fig. 7). The pattern is slightly light REE enriched (La/Yb$_{cn}$ =

Table 3. Ar/Ar data for step heating analysis of sample B2. $^{40}Ar^*$ = radiogenic ^{40}Ar. Temperature is given in °C. 2σ errors are shown and don't include uncertainties in J value. J = 0.003152

Temperature°C	$^{40}Ar/^{39}Ar$	$^{37}Ar/^{39}Ar$	$^{36}Ar/^{39}Ar$	$^{40}Ar^*/^{39}Ar$	$\%^{40}Ar^*$	^{40}Ar Moles	Age (Ma)	+/-
550	33.557	3.67275	7.75E-02	10.973	32.6	1.17E-14	61.344	5.568
600	24.769	4.136846	4.76E-02	11.075	44.6	1.47E-15	61.903	8.000
650	30.233	5.092685	4.81E-02	16.505	54.4	2.40E-15	91.495	10.377
700	43.491	15.7496	5.81E-02	27.936	63.5	1.19E-14	152.248	4.1844
750	50.211	25.8863	6.64E-02	33.323	65.1	1.32E-14	180.184	6.526
800	43.877	30.93122	4.19E-02	34.809	77.5	1.58E-14	187.811	3.519
850	41.329	31.78116	3.31E-02	34.965	82.6	1.93E-14	188.617	2.602
900	38.682	30.58637	2.39E-02	34.890	88.1	1.76E-14	188.232	2.415
950	38.883	27.41187	2.12E-02	35.564	89.6	1.65E-14	191.682	3.069
1000	38.365	23.16169	1.91E-02	35.218	90.2	1.23E-14	189.908	3.345
1050	38.107	18.14978	1.65E-02	35.168	91	8.53E-15	189.653	4.193
1100	39.477	16.13693	2.27E-02	34.488	86.3	6.79E-15	186.170	5.517
1200	35.687	20.09528	2.20E-02	31.272	86.3	1.01E-14	169.599	5.348
1300	41.023	25.81309	2.89E-02	35.254	84.3	6.89E-15	190.094	5.760
1601	43.986	37.77531	4.94E-02	33.382	73.7	3.89E-14	180.487	3.170

2.37-2.57) which is typical of CAMP tholeiites [*Alibert*, 1985; *Dupuy et al.*, 1988; *Bertrand*, 1991; *De Min et al.*, present volume]. A small Eu anomaly characterizes the analyzed samples (Eu/Eu*=0.90-0.96).

The major, trace and REE compositions of Kerforne basalts are similar to those of other European CAMP basalts, e.g., Pyrenees, and Messejana dike [*Alibert*, 1985; *Beziat et al.*, 1991; *Demant et al*, 1996]. The mantle-normalized ITE patterns of Kerforne basalts are similar to those of most CAMP tholeiites (and CFB's in general), which also display a marked negative Nb anomaly. The REE patterns resemble those of South American and African low-TiO$_2$ CAMP basalts [*De Min et al.*, present volume; *Bertrand et al.*, 1982; *Bertrand*, 1991]. Compared to other CAMP tholeiites, the Kerforne basalts have among the lowest incompatible LILE and LREE enrichments and relatively low absolute ITE (including REE) concentrations.

6.2. Sr, Nd, Pb Isotopic Compositions

Initial $^{87}Sr/^{86}Sr$ and $^{143}Nd/^{144}Nd$ (recalculated to 200 Ma) display a moderate variation from 0.7053 to 0.7064 (ε_{Sr} between 15 and 30) and from 0.51232 to 0.51241 (ε_{Nd} from −1.22 to 0.55; Fig. 8; Table 4). The most radiogenic Sr isotopic composition belongs to sample B5, which appears to be slightly altered. However, its $^{143}Nd/^{144}Nd$ value (0.51239) is indistinguishable from those of the other samples. In general, the Sr-Nd isotopic composition of the dike suggests that the Kerforne basalts derived from a source enriched in radiogenic Sr

(i.e., with high time integrated Rb/Sr) and similar to the Bulk Silicate Earth (BSE) in terms of Nd isotopic composition (i.e., with chondrite-like REE pattern). In more detail, we observe a positive correlation between ε_{Sr} and ε_{Nd} (Fig. 9), i.e., the Kerforne samples define a trend which is different from that of the mantle array. No correlation is evident with MgO or Mg#.

The Sr-Nd isotopic compositions of Kerforne basalts fall within the field of other CAMP basalts [*Dupuy et*

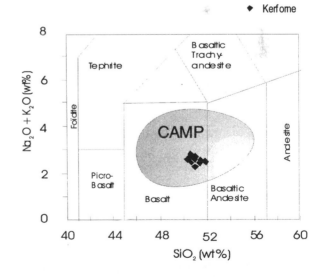

Figure 4. Total Alkalies-SiO$_2$ classification diagram [*Le Bas et al.*, 1986] of Kerforne samples (black diamonds). The circled field represents compositions of published CAMP rocks [e.g., *Alibert*, 1985; *Bertrand*, 1991; *McHone*, 2000].

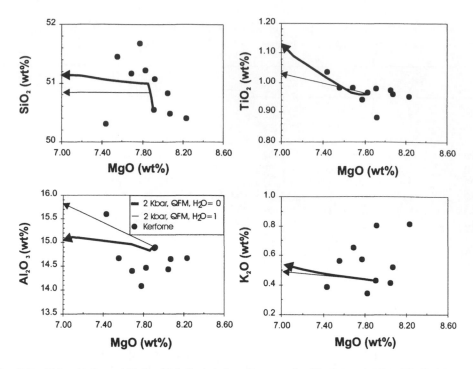

Figure 5. SiO_2, TiO_2, Al_2O_3 and K_2O vs. MgO variation diagrams for Kerforne samples. The bold and thin lines represent anhydrous and hydrous MELTS [*Ghiorso and Sack*, 1995]) fractional crystallization modeling at low pressure (2 kbar) and fO_2 (QFM buffer) and with H_2O concentrations of 0 and 1.0 wt% in the starting magma (B2). Note that for major elements an analytical uncertainty of 1 wt% has to be considered.

al., 1988; *Pegram*, 1990; *De Min et al.*, present volume], and most closely resemble those of the Pyrenees and of the Iberian Messejana dike [*Alibert*, 1985; Fig. 8]. In general, initial (200 Ma) Sr-Nd isotopic compositions of CAMP basalts display a wide variation, from E-MORB to high ε_{Sr} and relatively low ε_{Nd}. The previously described trend of decreasing ε_{Sr} and ε_{Nd} for Kerforne basalts seems to depart from the composition of the few analyzed Pyrenean basalts and to be different from the main trend of CAMP basalts. In terms of Sr-Nd isotopic compositions, Kerforne basalts plot close to some lower crustal basic (meta-) igneous xenoliths from the French Massif Central [Fig. 8; *Downes et al.*, 1990, 1991], whereas granitoid and meta-sedimentary basement rocks from Brittany have a substantially more radiogenic Sr and unradiogenic Nd composition [*Bernard-Griffiths et al.*, 1985; cf. B3, Table 4; Fig. 8].

Initial (200 Ma) $^{206}Pb/^{204}Pb$ (18.20-18.37), $^{207}Pb/^{204}Pb$ (15.59-15.64) and $^{208}Pb/^{204}Pb$ (38.18-38.34) are roughly constant throughout the dike (Fig. 8). $^{206}Pb/^{204}Pb$ isotopic compositions are relatively unradiogenic, while $^{207}Pb/^{204}Pb$ and $^{208}Pb/^{204}Pb$ are slightly higher than those of MORBs or OIBs, i.e., higher than the Northern Hemisphere Reference Line [NHRL, *Hart*, 1984]. No correlation is evident between ε_{Sr} and ε_{Nd} and Pb isotopic com-

positions. In general, Pb isotopic compositions of Kerforne basalts plot within the field of CAMP basalts from Eastern North America [*Pegram*, 1990]. Such compositions resemble those of lower crustal basic (meta-) igneous xenoliths from the Massif Central [*Downes et al.*, 1991; Fig. 8].

Finally, we note the difference between the Sr-Nd-Pb isotopic compositions of Kerforne basalts (and of the CAMP in general) compared to those of Ocean Island Basalts (OIB) from the Central Atlantic islands of Cape Verde, Fernando de Noronha and Ascension [e.g, *Gerlach et al.*, 1987, 1988; *Davies et al.*, 1985; *Halliday et al.*, 1992; Fig. 8] which have been suggested to be the present day expression of the hot spot source which generated CAMP basalts [*Wilson*, 1997; *Leitch et al.*, 1998].

7. PETROGENESIS OF KERFORNE BASALTS.

7.1. Fractional Crystallization

Kerforne basalts are moderately evolved, as they have MgO, Cr and Ni which are too low to be considered primary or near-primary mantle melts. According to MELTS [*Ghiorso and Sack*, 1995] calculations, Kerforne basalts may result from 30-40% fractional crystallization

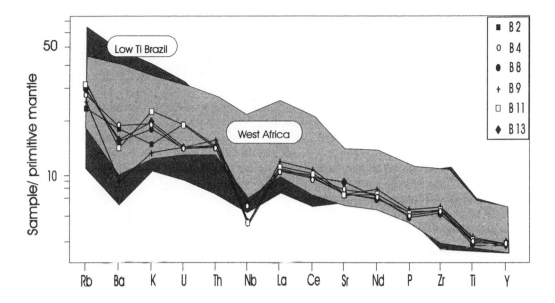

Figure 6. Primitive mantle normalized [*Sun and McDonough*, 1989] incompatible trace elements patterns for representative selected Kerforne basalts. Low TiO₂ CAMP rocks from Brazil [*De Min et al.*, present volume] and West Africa [*Bertrand*, 1991] are represented by the dark and light gray fields, respectively.

of olivine, minor cpx and spinel, starting from a primitive magma [e.g., *Hirose and Kushiro*, 1993]. Depending on the selected run conditions (pressure between 0.5 and 8 kbar, H₂O between 0 and 1 wt%; fO₂ either at the QFM or NNO buffer), the differentiation from less to more evolved Kerforne basalts (Mg# 62-58) would require 10-16 wt% fractionation of a gabbroic assemblage (augite, plagioclase and magnetite ± pigeonite ± olivine). However, the observed MgO vs. Al₂O₃ and K₂O trends of Kerforne basalts do not match the calculated trends (Fig. 5), which may be due to slightly lower crustal contamination and/or weak hydrothermal alteration of the samples.

7.2. Alteration and Crustal Assimilation

The sericitization that affected the plagioclase rims is evidence of a post-magmatic hydrothermal alteration which may have increased the K₂O (and Rb and Ba?) concentration of the basalts. There is some evidence that few samples (B1 and B5) have a positive Sr anomaly, and the Sr isotopic composition of B5 falls off the field of other Kerforne basalts (Table 4). It may be argued that the Sr concentration and isotopic composition of B5 is altered (i.e., enriched) by post-magmatic processes, even if samples were acid-leached before isotopic analyses. In general, however, no correlation is observed between Sr concentrations and Sr-Nd isotopic compositions; i.e., samples with similar εSr (e.g., B8 and B13) have different

Sr concentration (167 and 194 ppm, respectively; Table 1 and 4). In any case, it is unlikely that alteration (and weathering) affected the composition of the generally immobile REE, and therefore εNd (cf., Foland et al., 2000). In fact, εNd of B5 falls within the field of the other analyzed basalts.

The positive εNd-εSr trend of Kerforne basalts is different from mantle-generated oceanic and continental ba-

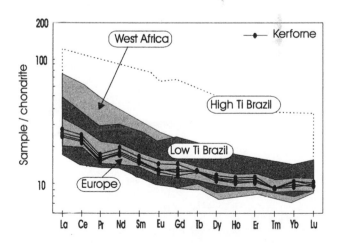

Figure 7. Chondrite-normalized [*Boynton*, 1984] REE compositions of Kerforne and CAMP basalts from Europe [*Alibert*, 1985; *Demant et al.*, 1996] West Africa [*Bertrand*, 1991], and Brazil [low and high TiO₂ samples; *De Min et al.*, present volume].

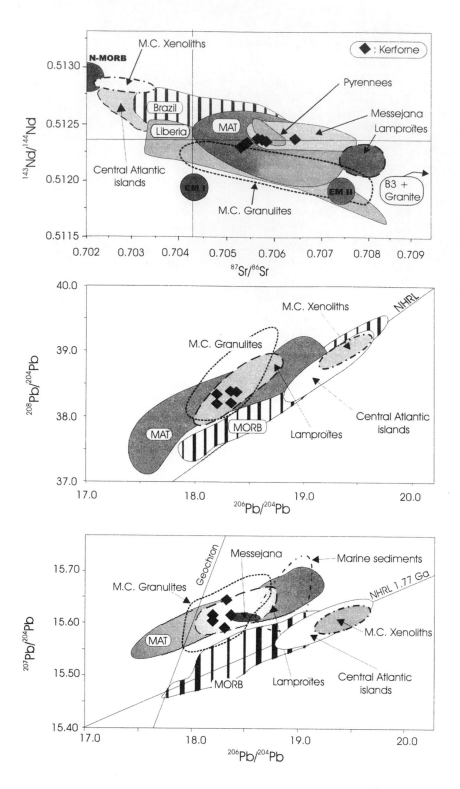

Figure 8. Initial (200 Ma) Sr, Nd and Pb isotopic compositions of Kerforne basalts (black diamonds) and of other CAMP basalts from Europe [Pyrenees and Messejana dike; *Alibert*, 1985], West Africa [*Dupuy et al.*, 1988], Brazil [*De Min et al.*, present volume] and Eastern North America [Mesozoic Appalachian Tholeiites, MAT; *Pegram*, 1990]. Isotopic compositions (recalculated to 200 Ma) of granulites from the Massif Central [*Downes et al.*, 1990, 1991], mantle xenoliths from the Massif Central [*Zangana et al.*, 1997] and lamproites from western Brittany [*Turpin et al.*, 1988] are contoured by a dotted, a dotted-dashed and a dashed line, respectively. Mantle poles after *Zindler and Hart* [1986] and *Sun and McDonough* [1989]. Marine sediments after *Pegram* [1990], Atlantic oceanic islands (Cape Verde, Fernando da Noronha and Ascension) after *Gerlach et al.* [1987, 1988], *Davies et al.* [1989], *Halliday et al.* [1992]. Sr-Nd compositions of granitic rocks from Brittany [B3, and *Bernard Griffiths et al.*, 1985] are located out of the figure and are indicated by an arrow.

Table 4. Measured and initial (200 Ma) Pb-Sr-Nd isotopic compositions of selected Kerforne basalts and of one wall-rock sample (B3). Initial ε_{Sr} and ε_{Nd} are indicated. Analytical uncertainty is quoted at the 2σ level, which, for Pb isotopic analyses is less than 0.01.

sample	B2	B5	B8	B9	B11	B13	granite B3
$^{206}Pb/^{204}Pb_m$	18.52	18.39	18.52	18.40	18.51	-	19.32
$^{206}Pb/^{204}Pb_i$	18.32	18.31	18.37	18.20	18.20	-	18.89
$^{208}Pb/^{204}Pb_m$	38.55	38.39	38.54	38.40	38.51	-	37.98
$^{208}Pb/^{204}Pb_i$	38.36	38.18	38.34	38.19	38.31	-	37.89
$^{207}Pb/^{204}Pb_m$	15.65	15.60	15.63	15.62	15.62	-	15.62
$^{207}Pb/^{204}Pb_i$	15.64	15.59	15.62	15.61	15.61	-	15.59
$^{87}Sr/^{86}Sr_m$	0.705946 (7)	0.707225 (11)	0.706521 (7)	0.706172 (6)	0.706612 (7)	0.706544 (7)	0.711821 (9)
$^{87}Sr/^{86}Sr_i$	0.705305	0.706389	0.705762	0.705388	0.705740	0.705858	0.709234
εSr	14.77	30.16	21.26	15.95	20.95	22.63	70.56
$^{143}Nd/^{144}Nd_m$	0.512541 (21)	0.512612 (12)	0.512591 (12)	0.512569 (8)	0.512554 (4)	0.512564 (8)	-
$^{143}Nd/^{144}Nd_i$	0.512318	0.512389	0.512370	0.512349	0.512409	0.512351	-
εNd	-1.22	0.17	-0.20	-0.62	0.55	-0.58	-

salts, including other CAMP basalts. Therefore, the Kerforne Sr-Nd isotopic compositions may have been modified during interaction with the continental crust. A possible assimilant would be the silicic upper crust, including the dike wall-rocks. However, the silicic basement rocks in western Brittany (including the here analyzed trondhjemite B3) are at least of Variscan age (i.e., Early Permian to Cambrian) and are characterized by negative ε_{Nd} and by high ε_{Sr} (Fig. 8) and therefore can not explain the ε_{Sr}-ε_{Nd} correlation of Kerforne basalts. Moreover, contamination with the silicic crust would substantially increase, for example, the SiO_2 content, which is not observed for Kerforne basalts that are, among CAMP basalts, relatively SiO_2-poor.

Alternatively, it could be suggested that the basaltic magma interacted with a mafic (lower) crust having negative ε_{Nd} and low ε_{Sr}. Outcrops of the lower crust are absent in Brittany, but basic (meta-) igneous xenoliths are present in Cenozoic alkaline volcanic rocks of the Massif Central, central France [*Downes et al.*, 1990, 1991]. These (meta-) igneous rocks were formed by deep crustal basaltic intrusions or metamorphism, probably during the Variscan orogeny. Some of these mafic rocks are characterized by moderately negative ε_{Nd}, moderately positive ε_{Sr} and high $^{207}Pb/^{204}Pb$ (Fig. 8).

In general, isotopic compositions do not correlate with major (e.g., SiO_2 and MgO) and trace elements. However, the basalts having the lowest ε_{Nd} (e.g., B2 and B9) contain a few plagioclase aggregates which are characterized by rounded high-An (bytownite) cores, quite different from the other microphenocrysts and microlites (An$_{75-42}$). The apparent textural and chemical disequilibrium of the high-An plagioclase cores suggests that they are xenocrysts in the Kerforne basalts. Modeling of incompatible minor elements, suggests that the high-An plagioclase cores may have crystallized at equilibrium from an MgO-rich and K_2O-poor, mafic magma.

Assimilation of mafic lower crustal rocks by CFBs has been proposed for example for some Deccan trap forma-

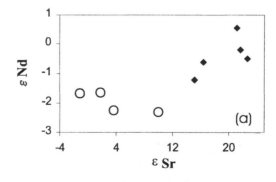

Figure 9. Detail of Sr-Nd isotopic composition of Kerforne basalts (diamonds) and of some Massif Central mafic xenoliths [white circles; *Downes et al.*, 1990].

tions [*Peng et al.*, 1994], and evidence of interaction between basaltic magma and the lower crust have been described for example for the deep crust section in the Ivrea zone [e.g., *Sinigoi et al.*, 1995]. *Reiners et al.* [1995] have shown that during initial stages of crystallization, a basaltic magma is capable of assimilating large amounts of granite. Performing isenthalpic MELTS runs, it can be shown that at lower crustal depths a primitive basaltic magma is capable of assimilating 10-15 wt% of a relatively hot (e.g., 800 °C) mafic granulitic wall rock (60% plagioclase + cpx + opx ± garnet), without substantially changing his basaltic composition.

In summary, Kerforne basalts were not substantially contaminated by Variscan or pre-Variscan upper silicic crust, but may have interacted with basic lower crust and may even have entrained plagioclase xenocrysts from it. Consequently, it may be concluded that the highest ε_{Sr} and ε_{Nd} (+23 and +0.55, respectively) of unaltered Kerforne basalts are close to the Sr-Nd isotopic composition of their mantle source. This mantle source may have been slightly enriched in Rb/Sr, compared to a primitive mantle, and may have had a near-chondritic REE pattern.

8.   THE MANTLE SOURCE OF KERFORNE BASALTS

8.1. REE Constraints

The moderately evolved Kerforne basalts display a slightly LREE enriched pattern with no significant Eu anomaly (Fig. 7). Such a pattern is typical of most CAMP rocks and may result from a differentiation process which involved: (a) melting of a mantle peridotite; (b) fractional crystallization dominated by olivine and (c) possible assimilation of mafic lower crust. We assume that the role of assimilation of mafic crustal rocks, which have low REE concentrations and LREE/HREE values similar to the basalts [*Downes et al.*, 1990], was negligible. Equilibrium melting and fractional crystallization have been modeled by using standard equations [*Shaw*, 1967; and *Rayleigh* fractionation], and considering the parameters (modal compositions, melting and partition coefficients) reported in the caption of Fig. 10. The inferred initial REE concentrations of the mantle peridotite are either chondritic or slightly enriched, as is suggested by the ε_{Nd} compositions of Kerforne basalts.

The moderately LREE-enriched pattern of Kerforne basalts cannot be due to equilibrium melting of a spinel peridotite or of a garnet peridotite for initial source compositions being either chondritic or slightly enriched (Fig. 10). This is striking for La/Yb vs. Gd/Yb of the basalts (Fig. 10b). The discrepancy between calculated and ob-

served REE ratios can not be explained by the fact that Kerforne basalts are not primitive, since fractional crystallization (± assimilation of a mafic rock) does not substantially change the La/Yb and Gd/Yb.

The REE compositions and ratios of Kerforne basalts are better reproduced by 11-13% equilibrium melting of a slightly enriched garnet + spinel peridotite and subsequent 20-30% fractional crystallization of olivine + cpx (Fig. 10). The slightly enriched nature of the modeled source is consistent with the close to chondritic Nd isotopic composition of the basalts.

8.2. Isotopic Constraints

In terms of Sr-Nd-Pb isotopic composition, Kerforne basalts are substantially different (i.e., enriched in ε_{Sr}) compared to those of oceanic basalts, but similar to those of most basalts from the CAMP and other CFB provinces. Generally, it has been suggested that the high ε_{Sr} of CFBs is due to assimilation of crustal material and/or to melting of an enriched continental lithospheric mantle. For Kerforne basalts, we suggest that crustal contamination with high ε_{Sr} material (upper crust) and hydrothermal alteration was negligible (except for outlier sample B5). The high Sr isotopic composition of Kerforne basalts (i.e., B13) may therefore be the closest estimate of their mantle source composition.

The composition of the continental lithosphere is known either from mantle xenoliths or from basaltic rocks of clear lithospheric origin. Samples of nearby lithospheric mantle are represented by spinel peridotite xenoliths entrained by Cenozoic alkali basalts of the Massif Central. Although these xenoliths may not be representative of the most fertile lithosphere. Their REE pattern are either depleted or substantially enriched in LREE [*Zangana et al.*, 1997; *Lenoir et al.*, 2000]. These peridotites have negative ε_{Sr} and positive ε_{Nd} and Pb isotopic compositions plotting on the NHRL [*Zangana et al.*, 1997]. Therefore, Massif Central mantle xenoliths are not a suitable mantle source for Kerforne basalts.

Late Variscan lamproites spread out over France, including western Brittany (Fig. 1) are witness of a strongly enriched and fertile lithosphere. These uncontaminated magmas were probably generated by melting of portions of the continental lithospheric mantle enriched by recycling of crustal material during the Variscan orogeny [*Turpin et al.*, 1988]. These lamproites have an incompatible element pattern which is similar in shape to those of Kerforne basalts, although substantially more enriched, with about 100 times chondritic LREE and LILE and relatively depleted HFSE (Ta) and Ti [*Turpin et al.*, 1988]. Their isotopic compositions are similar to those of

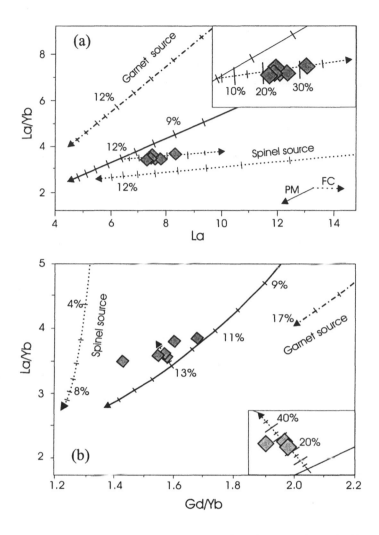

Figure 10. La/Yb vs. La and La/Yb vs. Gd/Yb compositions of Kerforne basalts and various peridotite melting models. Spinel + garnet peridotite melting: solid line; garnet peridotite melting: dashed line; spinel peridotite melting: dotted line. Small numbers close to melting curves represent melting percent. Modal composition of the peridotite: olivine = 55%, orthopyroxene = 25%, clinopyroxene = 15%, spinel and/or garnet = 5%. Initial composition (ppm) of the spinel and garnet peridotite: La = 0.78, Gd = 0.4; Eu = 0.11 and Yb = 0.34. Partition coefficients are from *McKenzie and O'Nions* [1991].
Fractional crystallization (dashed line, and inset of Fig. 10b) has been calculated considering the fractionating minerals calculated by MELTS for hydrous (1 wt% H_2O), low pressure (2 kbars) and QFM conditions.

Kerforne basalts for Pb, but different for Sr and Nd (Fig. 8). Although the similarity of the ITE pattern and of the Pb isotopic composition is striking, it is obvious that Kerforne basalts were not entirely generated by a mantle source similar to that of the lamproites. We suggest however, that a mantle component, similar to the "lamproitic source" may have contributed to the enriched isotopic signature of Kerforne basalts.

Contamination of magmas originating in the asthenosphere with enriched lithospheric mantle components has been proposed also for other CFB provinces, notably the Karoo [e.g., *Ellam and Cox*, 1991; *Luttinen and Furnes*,

2000]. Similarly, it has been suggested that CAMP basalts from Eastern North America were generated from a mantle source enriched by subducted oceanic sediments [*Pegram*, 1990].

Assuming that the "lamproitic" source represented the enriched mantle signature, a mantle with low ε_{Sr}, close to primitive ε_{Nd} and high $^{207}Pb/^{204}Pb$ is required to contribute to the origin of Kerforne basalts. Such a component may correspond to the bulk silicate earth (BSE), as calculated by *Allegre et al.* [1988]. We note that neither the lithospheric mantle xenoliths from the Massif Central [*Zangana et al.*, 1997] nor OIBs of the Central Atlantic (e.g., Cape Verde, Fernando de Noronha or Ascension) fit the required Sr-Nd and particularly Pb isotopic compositions (Fig. 8). Note that these ocean islands have been suggested to be the present-day expression of the mantle-plume which may have generated CAMP [*Hill*, 1991; *Wilson*, 1997; *Leitch et al*, 1998].

9. CONCLUSIONS

The Ar/Ar, geochemical and isotopic data presented here confirm that Kerforne basalts are similar in age and composition to the low-TiO_2 tholeiites of the CAMP. The Kerforne dike is the northernmost known outcrop of the CAMP, even if its emplacement is not demonstrably related to the Central Atlantic rifting (Fig. 1).

Based on petrographic, mineralogic and isotopic data, we argue that the basaltic Kerforne magmas were slightly contaminated with mafic lower crust, but were probably not substantially affected by upper crustal contamination. The "uncontaminated" endmember of Kerforne basalts points to relatively radiogenic Sr and $^{207}Pb/^{204}Pb$ and close to chondritic Nd isotopic compositions, shared also by the other basalts of the European sub-province of the CAMP. This suggests the presence of an isotopically enriched lithospheric mantle source component different from typical OIB or MORB mantle. This component may have been similar to that from which the Carboniferous lamproites of western Brittany originated. By contrast, it is unlikely that Kerforne magmas were generated only from a mantle source similar to the isotopically depleted continental lithosphere, as represented by mantle xenoliths from the Massif Central, nor from a mantle source similar to OIBs of the Central Atlantic (Cape Verde, Fernando de Noronha or Ascension). We speculate that Kerforne magmas may have been generated from a sublithospheric BSE-like mantle and were enriched en route to the surface first by "lamproitic" portions of the continental lithosphere and subsequently contaminated by the mafic lower crust.

Acknowledgments. J.L. Malfère (isotope analyses), S. Beuchat (XRD), D. Giorgis (Ar/Ar dating), and F. Bussy (electron microprobe), are thanked for analytical assistance. Discussions with M. A. Dungan, L. Melluso, E. M. Piccirillo, P. R. Renne, and C. Rapaille helped to improve the manuscript as well as critical reviews by A.L. Heatherington, and an anonymous reviewer. Editorial assistance by J.G. McHone was particularly appreciated. Financial support from the Swiss Science Foundation (grant FNRS 2100-057218.99 to A. M.) is acknowledged.

REFERENCES

Alibert, C., A Sr-Nd isotope and REE study of the late Triassic dolerites from the Pyrenees (France) and the Messejana dyke (Spain and Portugal), *Earth Planet. Sci. Lett., 73*, 81-90, 1985.

Allègre, C.J., E. Lewin, and B. Dupré, A coherent crust-mantle model for the uranium-thorium-lead isotopic system, *Chem. Geol., 70*, 211-234, 1988.

Bernard-Griffiths, J., J.J. Peucat, S. Sheppard, and P. Vidal, Petrogenesis of Hercynian leucogranites from the Southern Armorican Massif: contribution of REE and isotopes (Sr, Nd, Pb and O) geochemical data to the study of source rock characteristics and ages, *Earth Planet. Sci. Lett.*, 235-250, 1985.

Bertrand, H., Le magmatisme tholéiitique continental de la marge ibérique, précurseur de l'ouverture de l'Atlantique central: les dolerites du dyke de Messajana-Plasencia (Portugal-Espagne). *C. R. Acad. Sc. Paris, 304*, 215-220, 1987.

Bertrand, H., The Mesozoic tholeiitic province of northwest Africa: a volcano-tectonic record of the early opening of the Central Atlantic, in *Magmatism in extensional structural setting (the Phanerozoic African plate)*, edited by A.B. Kampunzu and R.T. Lubala, pp. 147-191, Springer, Heidelberg, New York, 1991.

Bertrand, H., J. Dostal, and C. Dupuy, Geochemistry of Early Mesozoic tholeiites from Morocco, *Earth Planet. Sci. Lett., 58*, 225-239, 1982.

Bertrand, H., J. Liegeois, K. Deckart, and G. Feraud, High-Ti tholeiites in Guiana and their Connection with the Central Atlantic CFB Province: Elemental and Sr-Nd-Pb Isotopic Evidence for a Preferential Zone of Mantle Upwelling in the Course of Rifting, in *Spring Meeting*, edited by AGU, 1999.

Béziat, D., J.L. Joron, P. Monchoux, M. Treuil, and F. Walgenwiz, Geodynamic implication of geochemical data for the Pyrenean ophites (Spain-France), *Chem. Geol.*, 243-262, 1991.

Bindeman, I.N., A.M. Davis, and M.J. Drake, Ion microprobe study of plagioclase-basalt partition experiments at natural concentration levels of trace elements, *Geochim. Cosmochim. Acta, 62*, 1175-1193, 1998.

Boynton, W.V., Geochemistry of the rare earth elements: meteorite studies, in *Rare earth element geochemistry*, edited by P. Henderson, pp. 63-114, Elsevier, 1984.

Caroff, M., H. Bellon, L. Chauris, J.P. Carron, S. Chevrier, A. Gardinier, J. Cotten, Y. Le Moan, and Y. Neidhart, Magmatisme fissural Triasico-Liasique dans l'ouest du massif armoricain (France) : pétrologie, géochimie, âge et modalité de mise en place, *Can. J. Earth Sci., 32*, 1921-1936, 1995.

Caroff, M., X. Le Gal, and J. Rolet, Magmatisme tholéiitique continental en contexte orogènique hercynien : l'exemple du volcanisme viséen de Kerroc'h, massif armoricain (France), *C. R. Acad. Sci. Paris, 322,* 269-275, 1996.

Davies, G.R., A. Gledhill, and C. Hawkesworth, Upper crustal recycling in southern Britain: evidence from Nd and Sr isotopes, *Earth Planet. Sci. Lett., 75,* 1-12, 1985.

Davies, G.R., M.J. Norry, D.C. Gerlach, and R.A. Cliff, A combined chemical and Pb-Sr-Nd isotope study of the Azores and Cape Verde hot spots: the geodynamic implication, in *Magmatism in the ocean basins,* edited by A.D. Saunders and M.J. Norry, pp. 231-255, Blackwell, Oxford, 1989.

Deckart, K., G. Féraud, and H. Bertrand, Age of Jurassic continental tholeiites of French Guyana, Surinam and Guinea: implications for the initial opening of the Central Atlantic Ocean, *Earth Planet. Sci. Lett., 150,* 205-220, 1997.

Demant, A., and D. Morata, Les dolérites tholéiitiques de Gaujacq et St-Pandelou (Landes, France). Pétrologic, géochimie et cadre géodynamique, *Bull. Soc. Geol. France, 167,* 321-333, 1996.

De Min, A., E. M. Piccirillo, A. Marzoli, G. Bellieni, P. R. Renne, M. Ernesto and L. S. Marques, The Central Atlantic Magmatic Province (CAMP) in Brazil: petrology, geochemistry, $^{40}Ar/^{39}Ar$ ages, paleomagnetism and geodynamic implications, Present Volume.

Downes, H., C. Dupuy, and A.F. Leyreloup, Crustal evolution of the Hercynian belt of Western Europe: Evidence for lower-crustal granulitic xenoliths (French Massif Central), *Chem. Geol., 83,* 209-231, 1990.

Downes, H., P.D. Kempton, D. Briot, R.S. Harmon, and A.F. Leyreloup, Pb and O isotope systematics in granulite facies xenoliths, French Massif Central : implication for crustal processes, *Earth Planet. Sci. Lett., 102,* 342 - 357, 1991.

Dupuy, C., J. March, J. Dostal, A. Michard, and S. Testa, Asthenospheric and lithospheric source for Mesozoic dolerites from Liberia (Africa): trace element and isotope evidence, *Earth Planet. Sci. Lett., 87,* 100-110, 1988.

Ellam, R.M., and K.G. Cox, An interpretation of Karoo picrite basalts in terms of interaction between asthenospheric magmas and the mantle lithosphere, *Earth Planet. Sci. Lett., 105,* 330-342, 1991.

Foland, K.A., F.G.F. Gibb, and C.M.B. Henderson, Patterns of Nd and Sr isotopic ratios produced by magmatic and postmagmatic processes in the Shiant Isles Main Sill, Scotland., *Contr. Min. Pet., 139,* 655-671, 2000.

Gerlach, D.C., J.C. Stormer, and P.A. Mueller, Isotopic geochemistry of Fernando de Noronha, *Earth Planet. Sci. Lett., 85,* 129-144, 1987.

Gerlach, D.C., R.A. Cliff, G.R. Davies, M. Norry, and N. Hodgson, Magma source of the Cape Verde archipelago: isotopes and trace element constrains, *Geochim. Cosmochim. Acta, 52,* 2979-2992, 1988.

Ghiorso, M.S., and R.O. Sack, Chemical mass transfer in magmatic processes IV. A revised and internally consistent thermodynamic model for the interpolation and extrapolation of liquid-solid equilibria in magmatic systems at elevated temperatures and pressures, *Cont. Min. Pet., 119,* 197-212, 1995.

Halliday, A.N., R.D. Gareth, S. Tommasini, C.R. Paslik, J.G. Fitton, and D.E. James, Lead isotope evidence for young trace element enrichment in the oceanic upper mantle, *Nature, 359,* 623-627, 1992.

Hames, W. E., C. Ruppel, P. R. Renne, New evidence for geologically instantaneous emplacement of earliest Jurassic Central Atlantic magmatic province basalts on the North American margin, *Geology, 28,* 859-862, 2000.

Hart, S.R., A large scale isotopic anomaly in the southern hemispheric mantle, *Nature, 309,* 753-757, 1984.

Hergt, J.M., D.W. Peate, and C.J. Hawkesworth, The petrogenesis of Mesozoic Gondwana low-Ti flood basalts, *Earth Planet. Sci. Lett., 105,* 134-148, 1991.

Hill, R.I., Starting plume and continental break-up, *Earth Planet. Sci. Lett.,* 398-416, 1991.

Hirose, K., and I. Kushiro, Partial melting of dry peridotites at high pressures: determination of compositions of melts segregated from peridotite using aggregates of diamond, *Earth Planet. Sci. Lett., 114,* 477-489, 1993.

Le Bas, M.J., R.W. Le Maitre, A. Streickeisen, and B. Zanettin, A chemical classification of volcanic rocks based on the total alkali silica diagram, *J. Pet., 27,* 745-750, 1986.

Leitch, A.M., G.F. Davies, and M. Wells, A plume head melting under a rifting margin, *Earth Planet. Sci. Lett., 161,* 161-177, 1998.

Lenoir, X., C. Garrido, J.L. Bodinier, and J.M. Dautria, Contrasting lithospheric mantle domains beneath the Massif Central (France) revealed by geochemistry of peridotite xenoliths, *Earth Planet. Sci. Lett., 181,* 359-375, 2000.

Lindsley, D.H., Pyroxene thermometry, *Am. Mineral., 68,* 477-483, 1983.

Luttinen, A.V., and H. Furnes, Flood basalts of Vestfjella: Jurassic magmatism across an Archean-Proterozoic lithospheric boundary in Dronning Maud Land, Antarctica, *J. Pet., 41,* 1271-1305, 2000.

Marzoli, A., P. Renne, E. Piccirillo, M. Ernesto, G. Bellieni, and A. De Min, Extensive 200-million-year-old continental flood basalts of the Central Atlantic Magmatic Province, *Science, 284,* 616-618, 1999.

Matte, P., Tectonics and plate tectonic model for the Variscan Belt of Europe, *Tectonophys., 126,* 329-374, 1986.

McHone, J.G., Non-plume magmatism and rifting during the opening of the Central North Atlantic Ocean, *Tectonophys., 316,* 287-296, 2000.

McKenzie, D., and R.K. O'Nions, Partial Melt Distribution from Inversion of Rare Earth Element Concentrations, *J. Pet., 32,* 1021-1091, 1991.

Moreau, M.G., J.Y. Berthou, and J.A. Malod, New paleomagnetic Mesozoic data from the Algarve (Portugal): fast rotation of Iberia between the Hautervian and the Aptian, *Earth Planet. Sci. Lett., 146*, 689-701, 1997.

Nimis, P., A clinopyroxene geobarometer for basaltic systems based on crystal-structure modeling, *Contrib. Min. Pet., 121*, 1115-125, 1995.

Onstott, T.C., M.L. Miller, R.C. Ewing, G.W. Arnold, and D.W. Alsh, Recoil refinements: implication for the ^{40}Ar/^{39}Ar dating technique, *Geochim. Cosmochim. Acta, 59*, 1821-1834, 1995.

Panjasawatwong, Y., L.V. Danyuushevsky, A.J. Crawford, and K.L. Harris, An experimental study of the effects of melt composition on plagioclase-melt equilibria at 5 and 10 kbar: implications for the origin of magmatic high-An plagioclase, *Contr. Min. Pet., 118*, 420-432, 1995.

Pegram, W.J., Development of continental lithospheric mantle as reflected in the chemistry of the Mesozoic Appalachian Tholeiites, U.S.A, *Earth Planet. Sci. Lett., 97*, 316-330, 1990.

Peng, Z.W., J. Mahoney, P. Hooper, C. Harris, and J. Beane, A role for lower continental crust in flood basalt genesis? Isotopic and incompatible element study of the lower six formations of the western Deccan Traps, *Geochim. Cosmochim. Acta, 58*, 267-288, 1994.

Reiners, P.W., B. Nelson, and M.S. Ghiorso, Assimilation of felsic crust by basaltic magma; thermal limits and extents of crustal contamination of mantle derived magmas, *Geology, 23*, 563-566, 1995.

Renne, P. R., C. C. Swisher, A. L. Deino, B. B. Karner, T. Owens and D. J. De Paolo, Intercalibration of Standards, Absolute ages and uncertainties in ^{40}Ar/^{39}Ar dating, *Chem. Geol., 145*, 117-152, 1998.

Rhodes, J.M., Geochemistry of Mauna Loa eruption: implication for magma storage and supply, *J. Geophys. Res., 93*, 4453-4466, 1988.

Sebai, A., G. Féraud, H. Bertrand, and J. Hanes, ^{40}Ar/^{39}Ar dating and geochemistry of tholeiitic magmatism related to the early opening of the Central Atlantic rift, *Earth Planet. Sci. Lett., 104*, 45-472, 1991.

Shaw, D.M., Trace element fractionation during anatexis, *Geochim. Cosmochim. Acta, 34*, 237-243, 1967.

Sichler, B., and M. Perrin, New early Jurassic paleopole from France and Jurassic apparent polar wander, *Earth Planet. Sci. Lett., 115*, 13-27, 1993.

Sinigoi, S., J.E. Quick, A. Mayer, and G. Demarchi, Density-controlled assimilation of underplated crust, Ivrea-Verbano zone, Italy., *Earth Planet. Sci. Lett., 129*, 183-192, 1995.

Stampfli, G.M., J. Mosar, A. De Bono, and I. Vavasis, Late Paleozoic Early Mesozoic plate tectonics of the Western Tethis, *Bull. Geol. Soc. Greece, 32*, 113-120, 1998.

Steiger, R.H., and E. Jäger, Subcommission on geochronology: convention on the use of decay constants in geo and cosmo-chronology, *Earth Planet. Sci. Lett., 36*, 359-362, 1977.

Sun, S.S., and W.F. McDonough, Chemical and isotopic systematics of oceanic basalts: implication for mantle composition and processes, in *Magmatism in the ocean basins*, edited by A.D. Saunders and M.J. Norry, pp. 313-345, Blackwell, Oxford, 1989.

Turpin, L., D. Velde, and G. Pinte, Geochemical comparison between minettes and kersantites from the Western European Hercynian Orogen: trace element and Pb-Sr-Nd isotope constraints on their origin, *Earth Planet. Sci. Lett., 87*, 73-86, 1988.

Wilson, M., Thermal evolution of the Central Atlantic passive margins: continental break-up above a Mesozoic superplume, *J. Geol. Soc. London, 154*, 491-495, 1997.

Zangana, N.A., H. Downes, M.F. Thirlwall, and E. Hegner, Relationship between deformation, equilibration temperatures, REE and radiogenic isotopes in mantle xenoliths (Ray Pic, Massif Central, France): an example of plume-lithosphere interaction?, *Contr. Min. Pet., 127*, 187-203, 1997.

Zindler, A., and S. Hart, Chemical geodynamics, *An. Rev. of Earth . Plan. Sci., 14*, 493-571, 1986.

Herve Bertrand, Ecole Normale Supérieure et UCBL, Laboratoire des Sciences de la Terre, UMR-CNRS 5570, Lyon, France; Herve.Bertrand@ens-lyon.fr

Mike Cosca, Institute of Mineralogy and Geochemistry, University of Lausanne, Switzerland; Mike.Cosca@imp.unil.ch

Denis Fontignie, Fred Jourdan, and Andrea Marzoli Section de Sciences de la Terre; Université de Genève, Switzerland; Jourdan6@etu.unige.ch; Andrea.Marzoli@terre.unige.ch; Denis.Fontignie@terre.unige.ch

Magma Flow Pattern in the North Mountain Basalts of the 200 Ma CAMP Event: Evidence From the Magnetic Fabric

Richard E. Ernst

Continental Geoscience Division, Geological Survey of Canada, Ottawa, Ontario, Canada

Jelle Zeilinga de Boer, Peter Ludwig, and Taras Gapotchenko

Department of Earth and Environmental Sciences, Wesleyan University, Middletown, Connecticut, USA

The relationship between flood basalts and feeder dikes is investigated through an anisotropy of magnetic susceptibility (AMS) study of the 200 Ma Central Atlantic Magmatic Province (CAMP) in the Fundy basin of southeastern Canada. Over a 200 km strike length the North Mountain basalts exhibit a consistent N-NE (or S-SW) emplacement direction flow pattern (based on the orientation of maximum AMS axes). Less definitively, the data suggest that the lower lavas (including the Grand Manaan basalts) have a NE-SW flow axis, while the middle and upper flow units have a more N-S flow axis. The Swallowtail dike of Grand Manaan Island exhibits sub-vertical flow based on AMS data. Its presumed extension along the axial zone of the Fundy basin is the possible feeder for at least the middle and upper North Mountain flows.

1. INTRODUCTION

The Central Atlantic Magmatic Province (CAMP) is the largest (in terms of area) preserved continental flood basalt event. Its basalts are exposed on four continents, in eastern North America (ENA), northern South America, west Africa, and southwestern Europe (Figure 1). The magmatism occurred as a rapid burst of activity around 200 Ma with a probable duration of only a few million years based on U-Pb and ^{40}Ar-^{39}Ar dating [e.g. *Dunning and Hodych,* 1990; *Sebai et al.* 1991; *Marzoli et al.* 1999; *Hames et al.* 2000]. This short duration of emplacement is supported by a single

The Central Atlantic Magmatic Province:
Insights from Fragments of Pangea
Geophysical Monograph 136
Published in 2003 by the American Geophysical Union
10.1029/136GM12

normal paleomagnetic polarity for most of the basalts in comparison with the geomagnetic time-scale at ca. 200 Ma [*Gradstein et al.* 1994]. Lacustrine cycles in sediments of the Newark basins of North America separating different flow units suggest that this period may have been no longer than about 600,000 years [*Olsen,* 1997]. This dominantly mafic province consists of a radiating dike swarm with a radius of up to 3000 km, along with flow remnants and sills in rift basins. This burst of magmatism followed a precursor rifting event at about 230 Ma and preceded rifting/opening of the Central Atlantic at about 175 Ma. The radiating swarm of dikes converges on a focal point near the present-day Blake Plateau.

ENA dikes and their extrusive equivalents in the rift valleys have been classified into two major geochemical groups [*Ragland et al.,* 1992]. In the southern Appalachians the dikes are predominantly olivine normative tholeiites,

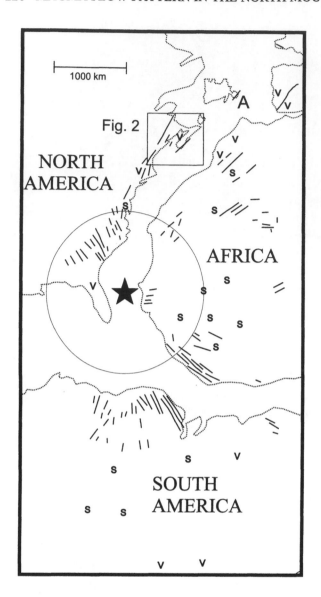

Figure 1. Distribution of dikes (lines), sills ('s') and volcanics ('v') of the 200 Ma Central Atlantic Magmatic Province (CAMP) in a pre-Atlantic Ocean reconstruction of Pangea. Overall center of dike convergence (near the Blake Plateau) marked by star. Study area indicated by box labelled Figure 2. "A" marks Avalon dike which is discussed in the text.

while those in the northern Appalachians are exclusively quartz normative tholeiites with two distinct affinities, titanium enriched and titanium depleted [*de Boer and Snider*, 1979; *de Boer et al.*, 1988; *Cummins et al.*, 1992; *Ragland et al.*, 1992; *Ernst and Buchan*, 2001]. Emplacement of the olivine normative magma is believed to have preceded that of the quartz normative sequence.

An origin of the CAMP province by a mantle plume has been favored on the basis of the large radiating dike pattern [*May*, 1971; *Oliveira et al.*, 1990; *Hill*, 1991; *Ernst and*

Buchan, 1997a], lateral dike flow [*Greenough and Hodych*, 1990; *de Boer et al.*, in press], very short duration of this huge event [e.g. *Marzoli et al.*, 1999; *Hames et al.*, 2000] and evidence for ~210 Ma regional uplift preserved only in the southern Newark basins and linked to CAMP plume arrival [*Hill*, 1991; *Rainbird and Ernst*, 2001]. In most plume models, the plume center is located near Florida below the Blake Plateau [e.g. *Greenough and Hodych*, 1990]. However, *de Boer* [1992], favors two hotspots, one off Florida (as in Figure 1) and the other in the Gulf of Maine, linked with the White Mountain Magma Series (Figure 2). Plume models typically require a central source from which material is dispersed laterally, either as partial melts generated at the plume center and subsequently distributed within the crust through laterally injected dikes [*Ernst and Buchan*, 1997a,b], or through outward mantle flow at the base of the lithosphere [*Ebinger and Sleep*, 1998] followed by partial melting and vertical emplacement within the crust. As an example of the latter, *Wilson* [1997] and *Oyarzun et al.* [1997] proposed northeast-directed sub-lithospheric channelling of plume flow away from the plume center along the incipient Central Atlantic rift

Non-plume origins of the CAMP have also been advocated for several reasons including the presence of additional dikes not fitting the radiating pattern and because the formation of the (pre-CAMP) Newark rift basins is not easily explained by the plume model [*McHone*, 2000]. Nonplume models require magma generation from multiple localized sources underlying the rifts [*Bertrand*, 1991; *Philpotts*, 1992]. *Cummins et al.* [1992] proposed the concept of a mantle "keel" which developed below the zone of maximal crustal extension and in which a low degree of partial melting generated the magmas which rose subvertically through the crust to the surface. *Hames et al.* [2000] also favored strong lithospheric control consistent with plume incubation or edge-driven convection models.

In a previous study [*de Boer et al.* in press] we investigated the flow pattern in feeder dikes in New England in order to differentiate between plume and non-plume models. Specifically, we used the anisotropy of magnetic susceptibility (AMS) technique to determine flow patterns in 25 sites along the Higganum-Holden and Christmas Cove dikes (Figure 2). Individually these dikes have been traced for 250 and 100 km, respectively, but their geochemical similarity suggests that they may represent a single dike nearly 500 km long. Although textural observations at the margin of an outcrop of the Higganum dike suggest a locally complex emplacement pattern [*Philpotts and Asher*, 1994], the along- and across-dike consistency of our AMS data suggests that the main flow pattern along both the Higganum-Holden and Christmas Cove segments was predominately lateral.

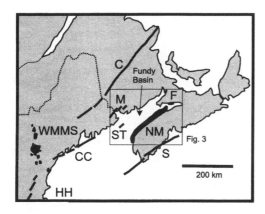

Figure 2. Regional map showing location of Shelburne (S), Caraquet (C), Minister Island (M), Swallowtail (ST), Christmas Cove (CC), and Higganum-Holden (HH) dikes. NM identifies North Mountain basalt which is found in the Fundy Basin. F locates Five Islands and other fault bounded outcropings of basalt on the north shore of the Bay of Fundy [*Greenough et al.*, 1989]. WMMS labels White Mountain Magma Series [*McHone and Butler*, 1984].

Based on AMS measurements, *Greenough and Hodych* [1990] determined subhorizontal flow in the Shelburne dike of Nova Scotia and the Minister Island dike (also referred to as the Passamaquoddy Bay dike) of New Brunswick and Maine (Figure 2). Furthermore, the Shelburne dike contains ramping structures tens of centimeters long that indicate flow towards the northeast [*Greenough and Hodych*, 1990].

These data from New England and Atlantic Canada indicating predominant lateral flow over a distance of at least 1000 km are consistent with a model of lateral injection within the crust along a level of neutral buoyancy for plume-related giant radiating dike swarms [e.g. *Ernst and Buchan*, 2001]. Further AMS studies are needed to test for lateral flow in other ENA dikes and to search for vertical flow regions such as might be expected for dikes in the vicinity of the plume center, and in the rift valleys where lithospheric extension was maximal.

The ENA dikes must have been the feeders for flows found in the Newark-type rift basins. However, a key question that remains is whether the dikes fed a single large flood basalt province originally distributed over the entire Circum-Atlantic region, but subsequently eroded ("broad terrane model advocated by *McHone*, 1996) or whether flows were only restricted to the rift basins. A partial answer is available from Atlantic Canada where *Pe-Piper and Piper* [1999] concluded that regional geochemical correlation of dikes and flows (including flows intersected during offshore drilling), indicates a widespread distribution of the earliest lavas (associated with Minister Island and Shelburne dykes), but an absence of flows associated with the younger ENA dikes (Caraquet, Anticosti and Avalon).

A contribution to this debate may result from studies that attempt to link individual basalt flows to specific feeder dikes. In this paper we investigate the emplacement pattern in the North Mountain basalts and nearby Grand Manaan Island basalts of southeastern Canada (Figure 3) using the AMS method in order to assist in the identification of feeder dike(s).

2.1 North Mountain Basalts

The North Mountain basalts occur in the Fundy Basin, the largest and deepest of the major early Mesozoic basins of the Newark Supergroup which extend from Florida to Nova Scotia. A reflection seismic survey of the Fundy Basin reveals the distribution of sediments and volcanic rocks within the basin [*Wade et al.*, 1996]. The isopach map for the North Mountain basalt (Figure 3) reveals that they extend underneath the entire basin, and gradually thicken to both the northwest and southwest. The basalt accumulation is abruptly terminated on the north and west sides of the Fundy basin along known major faults, the Headlands fault and the Grand Manaan fault (GMF), respectively. The cumulative basalt thickness adjacent to these faults reach up to 600 m. However, about 10 km to the east of the GMF fault, thicknesses locally reach up to 800 and 1000 meters.

The main exposure of North Mountain basalts occurs on the southern shore of the Bay of Fundy, where it crops out over a distance of about 200 km as a more or less continuous belt. The basalts are believed to be comprised of three flow units: a massive coarse-grained lower unit about 190 m thick; a middle unit consisting of at least 7 relatively thin and highly amygdaloidal flows (totaling about 50 m) and a thick upper, coarse-grained phenocryst-rich (pyroxene + plagioclase) flow of about 160 m. Unit thicknesses generally decrease northeastward. The lower flow thins to about 60 m in the Morden area (MO in Figure 3). The upper flow thins to about 30 m at Baxter's Harbor, 40 km northeast of Morden [*J.A. Colwell*, pers. comm.]. The number of thin flows from the middle unit more than double in the Morden area, where many are only a few meters thick. Cumulative thickness of the North Mountain basalt formation varies from about 400 m in the southwest to 275 m in the northeast [*Papezik et al.*, 1988]. Faulted masses of North Mountain basalt equivalents also occur along the north shore of the Bay of Fundy (Location F in Figure 2), and on Grand Manaan Island (Figure 3).

The North Mountain basalts and equivalents in coastal New Brunswick and on Grand Manaan Island belong to the high titanium quartz normative (HTQ) group of *Ragland and Whittington* [1983]. They are the chemical equivalents of the lowermost flow units, Talcott and Orange Mountain, in the Newark rift basins located further southwest. The

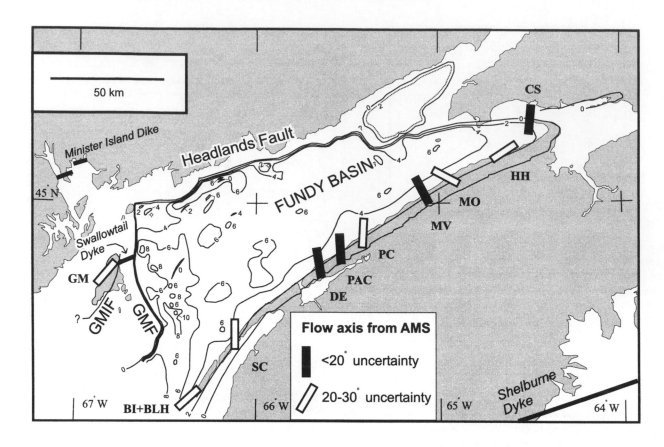

Figure 3. Flow directions determined from AMS maximum axes (this paper) compared with isopach distribution of North Mountain basalts throughout the Bay of Fundy [after *Wade et al.*, 1996]. Contours x 100 m. GMF is Grand Manaan Fault. GMIF is Grand Manaan Island Fault. Location shown in inset to Figure 2.

chemistry of the North Mountain basalt changes along strike. MgO content decreases northeastward by about 40 percent (9.8% to 6%) and TiO_2 content more than doubles in the northeastward direction (0.6 to 1.4%). The longitudinal chemical variations have been attributed to flowage differentiation during lateral northeastward flow of the magmas within the crust along feeder dikes [*Papezik et al.*, 1988]. The North Mountain basalt equivalents on Grand Manaan Island (GM in Figure 3) possess MgO contents varying from 8.2 to 10.6%, indicating that this region contains the least evolved flows [*Pe-Piper and Piper*, 1999]. The Grand Manaan flows are considered by *Pe-Piper and Piper* [1999] to be correlative to the lower unit of the North Mountain basalts.

1.2. Feeder Dikes

While it is widely inferred that the North Mountain and Grand Manaan basalts were fed from a dike (or dikes), no feeder has been identified underlying the North Mountain basalts and none of the nearby ENA dikes (Figure 2) has

been conclusively linked to the basalts. We should note that each of the three flow units in the North Mountain basalt may represent a different feeder, so it is possible that we are looking for three or more feeders. Below we review geochemical and stratigraphic evidence for the feeder zone location and provide additional data constraining magma flow direction.

Other than a direct physical link, the most definitive approach for linking a dike with a flow is compositional similarity. Of regional ENA dikes with observed or extrapolated continuity to the vicinity of the North Mountain basalts, the more distal Caraquet and Avalon dikes (Figures 1 and 2) can be ruled out on the basis of compositional dissimilarity (see Figure 3 in *Pe-Piper and Piper*, 1999]. However, the more proximal Minister Island dike of New Brunswick, the Shelburne dike of Nova Scotia, and the Christmas Cove dike along the coast of Maine are compositionally similar [*de Boer et al.* in press; *Pe-Piper and Piper*, 1999]. Similar longitudinal chemical variations seen in the North Mountain basalts are also observed in the Shelburne dike [*Pe-Piper and Piper*, 1999]. However, consideration of limited isotopic data eliminates the Shelburne dike as a feeder. Specifically, the initial Sr and Nd isotopic ratios of the North Mountain basalt are distinct from those of the Shelburne dike [*Pe-Piper and Piper,* 1999]. Likewise although geochemically similar the Sr isotopic composition of the Minister Island dike differs from that of the North Mountain basalts. Isotopic data are not yet available for the Christmas Cove dike. Although the composition of the Swallowtail dike on Grand Manaan Island is unknown, this dike is on line with the Christmas Cove dike about 100 km to the west (Figure 2) and its presumed southerly continuation as the Higganum-Holden dike [*de Boer et al.* in press] which has a similar geochemistry to the North Mountain Basalts. In summary, limited geochemical and isotopic data rule out the Caraquet, Avalon and Shelburne dikes, leaving the Minister Island, Christmas Cove and Swallowtail dikes as possible feeders for the North Mountain basalts.

Another constraint on locating the feeder dike(s) to the North Mountain basalts is based on the regional distribution of lava flow thickness with respect to major faults [e.g. *Wade et al.,* 1996, p. 216-218]. The greatest thickness of North Mountain basalts is located close to the northern and western faults of the Fundy Basin (Figure 3). This would suggest that the feeder for the basalts either lies along the ENE trending axial zone of the Fundy Basin (just south of the Headlands Fault) or just east of and parallel to the N-S trending Grand Manaan fault. A very speculative linear "feeder vent system" (dike) is located at the northeast end of the North Mountain Basalts (Five Islands location in Figure 2) [*Greenough et al.,* 1989]. *Wade et al.* [1996, p. 216-218] reviews the evidence for feeder dikes and favors a location along the axial zone of the Fundy basin (south of the Headlands fault) and furthermore concludes that thickening of the basalt adjacent to the N-S trending Grand Manaan Fault was the result of ponding due to basin geometry/paleotopography.

Good analogies are provided by the massive dike/sill complexes (Westrock and Barndoor Hill) east of the exposed scarp of the western border fault of the Connecticut rift basin (to the southwest). The magma rose along the eastward dipping listric fault, but diverged from it a few hundred meters below the surface, where the hanging wall was intersected by swarms of vertical extensional fissures. The magma rose into these fissures in the hanging wall and accumulated in linear dike/sill complexes parallel to the western border fault. By analogy, such complexes may exist just south of the Headlands fault in the Fundy basin, and would explain the irregular pattern in thickness of North Mountain basalts (Figure 3).

These arguments would favor the identification of the newly-recognized Swallowtail dike of Grand Manaan Island (and its presumed extension to the northeast along the axial zone of the Fundy Basin) as the feeder. The Swallowtail dike on Grand Manaan Island is subvertical, has a NE-trend, and a thickness of about 30 m. To the west it abuts the N-S Grand Manaan Island fault (GMIF in Figure 3). (The GMIF together with the Grand Manaan fault, GMF, bound an uplifted block which exposes basement rocks.) Significant dip-slip movement along the GMIF occurred subsequent to the ENA magmatic episode and therefore we presume that the Swallowtail dike continues west of the GMIF beneath the Grand Manaan volcanics. The continuation of the Swallowtail dike to the east-northeast is unknown; however its extrapolated trend would locate it within about 20 km of the northern boundary faults of the Fundy basin. Evidence for a second possible feeder dike on Grand Manaan Island is discussed below.

A third type of evidence relating North Mountain flows to feeder dike(s) derives from flow direction information. The northeastward thinning of the massive units, the significant increase in the number of thin flows in the middle unit and the longitudinal geochemical trend [*Papezik et al.,* 1988] suggest northeastward emplacement and therefore would imply a source dike to the southwest. However, this result is not consistent with the conclusions in the previous paragraph for a source on the northern side of the Fundy Basin.

In this paper we provide an additional type of flow direction indicator. AMS measurements are used to determine local flow axes in the North Mountain basalts, Grand Manaan basalts and dikes (including the Swallowtail dike). The pattern of flow in the basalts is evaluated in light of known feeder dikes (Figures 2 and 3) and in the context of end-member geometric models relating feeder dikes to flow patterns in volcanic rocks (Figure 4). These geometries include point source and linear source models. Point source models favor a localized source along a dike, while linear source models favor magma spilling from the sides of the dike. These models neglect the effect of topography, which may be a critical but difficult-to-constrain parameter controlling the flow pattern.

1.3. Anisotropy of Magnetic Susceptibility (AMS)

Magnetic susceptibility is a second rank tensor that relates the strength of an inducing magnetic field to the induced magnetization in rocks. AMS fabric is graphically represented as a triaxial ellipsoid and data are typically presented in terms of the azimuthal orientation and magnitude of the three principal axes of this ellipsoid [*Nye,* 1985; *Rochette et al.,* 1992; *Tarling and Hrouda,* 1993]. In unaltered basaltic dikes and flows, the AMS technique records the anisotropic distribution of the titano-magnetite grains.

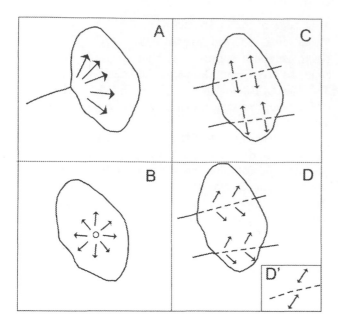

Figure 4. Simple models (neglecting topographic and drainage controls) for the feeding of a lava flow from a dike: A) point initiation from a dike intersecting a basin; B) point source underneath (possibly also reflecting a dike source, *Ernst and Buchan,* 1997b); C) fed from an underlying dike(s); and D) fed from an underlying dike(s) with a bias in the direction of transport. D' indicates situation in which the flow axis is independent of an underlying dike and therefore provides no evidence for a link with that dike.

Basaltic fabric is commonly dominated by lath- and tabular-shaped plagioclase crystals and the magnetite distribution mimics this fabric regardless of the timing of magnetite growth. Early-crystallizing magnetite grains will be redistributed by flow during plagioclase growth and late-crystallizing magnetite will infill between plagioclase grains [e.g. *Hargraves et al.,* 1991]. The result in both cases is that the distribution of the magnetite grains will mimic that of plagioclase and hence that of the overall rock texture. The anisotropy in the distribution of the isotropic magnetite crystals is the key control on the magnetic fabric [*Hargraves et al.,* 1991; *Stephenson,* 1994; *Cañón-Tapia,* 1996; *Cañón-Tapia et al.,* 1996, 1997]. With proper instrumentation the AMS technique can be sensitive to <1% anisotropy, making it useful for revealing very subtle fabrics in dikes and volcanic flows.

2. METHODOLOGY

Seventy four samples were collected from 10 sites in the North Mountain basalts distributed along the south shore of the Bay of Fundy (Figures 3 and 5) and 45 samples were collected from 7 sites on Grand Manaan Island (Figures 3 and 6). Five of the Grand Manaan sites are from lava flows, one is from the Swallowtail dike, and the last is from another possible dike (site DC). All samples were field-drilled and oriented using a magnetic compass held about 40 cm from the rock surface. Orientation was checked for each fifth core using a sun compass.

The low-field AMS fabric of one or two specimens from each core was measured using the Sapphire SI-2B Magnetic Susceptibility and Anisotropy Instrument. This instrument is designed to measure low-field magnetic susceptibility of a specimen in a number of different orientations with respect to an applied field with magnitude comparable to the Earth's field and with a frequency of 19,000 Hz. Mean results from each site were determined according to the tensor averaging procedure proposed by *Jelinek* [1978]. Calculations and plots were generated using the computer program discussed in *Lienert* [1991].

3. RESULTS

The data are displayed in Figures 5 and 6 and summarized in Table 1. The observation of both positive and negative values of ellipsoid shape T (and negative and positive values of B%) reflect the presence of both oblate and prolate fabrics. The North Mountain basalt sites show a consistent pattern of AMS in which the minimum axes are clustered vertically and the maximum and intermediate (not displayed) axes group in the horizontal plane. Such fabrics, in which the minimum axes are normal to the flow plane, are termed "normal" [*Rochette et al.,* 1992], and in such cases the maximum axes are presumed to align in the direction of magma transport.

The flow axis pattern for North Mountain basalts derived from the AMS data (Figures 3, 5 and 6) reveals a north-south subhorizontal flow in sites, SC, DE, PAC, PC, MV and CS, west-southwest (or east-northeast) flow in BLH and southwest-northeast flow in HH. Based on the best-determined sites (with angular uncertainties less than 20°) the emplacement pattern is dominantly N-S.

Grand Manaan volcanic sites exhibit slightly scattered data distributions although most sites favor a NE-SW axis of flow (Figure 6). The pooled data for the Grand Manaan basalts indicate a clear pattern: minimum axes perpendicular to the flow plane (i.e. subvertical) and a NE-SW flow axis (Table 1, Figure 3).

The Swallowtail dike (ST) also exhibits a 'normal' fabric in which the minimum susceptibility axes are perpendicular to the dike plane. In such cases, the fabric is considered to be due to flow and the maximum axes give the flow direction [*Knight and Walker,* 1989; *Ernst and Baragar,* 1992].

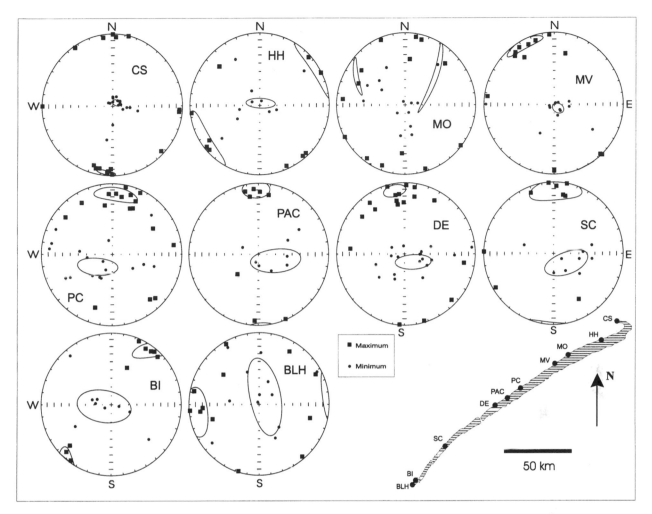

Figure 5. Equal area stereonet plots (lower hemisphere) of AMS site data from North Mounain basalts on the south shore of the Bay of Fundy, Nova Scotia, Canada. AMS maxima indicated by squares and AMS minima by circles. Intermediate data not shown because they are normal to the maximum and minimum axes. 95% confidence ellipses (thin lines) for data shown. Location of sample sites are shown in inset diagram (refer to Figures 2 and 3 for the regional setting). Sites are Cape Split (CS), Hall's Harbor (HH), Morden Harbor (MO), Margaretsville (MV), Phinney's Cove (PC), Delaps Cove (DE), Sandy Cove (SC), Brier Island (BI), and Brier Lighthouse (BLH).

The maximums in site ST are sub-vertical, from which we infer sub-vertical flow in the Swallowtail dike at site ST.

Site DC was originally believed to be a flow, but the presence of subhorizontal minimum susceptibilities suggests a dike-like pattern and a review of field notes indicates that site DC was distinctly more massive than adjacent units of Grand Manaan basalts. Therefore, this site has tentatively been reinterpreted as a dike (although a zone of preserved vertical convective flow within the basalt cannot be ruled out). The host rocks to the possible Deep Cove dike (at site DC) would be Grand Manaan basalts rather than the basement gneisses that host the Swallowtail dike (ST). From this it can be inferred, that if site DC is a dike, it fed lavas higher in the sequence than the preserved Grand Manaan basalts.

4. DISCUSSION

Overall the AMS data (Figures 5 and 6) from the North Mountain flows show a consistent N-NE (or S-SW) flow axis along 200 km along the south shore of the Bay of Fundy. Interestingly, this also suggests a consistent pattern of flow throughout the time of emplacement of the North Mountain basalt except possibly for the earliest basalts at the southwest end. There is an age progression in the sampling from southwest to northeast. The sites at the southwest end (on Brier Island) are from the lowermost massive unit. Sites further east (e.g. MO and MV) are from the middle amygdaloidal unit. Sites from the northeast end, such as Cape Split (CS), are from the upper massive/porphyritic unit. The overall pattern of a N-NE (S-SW) flow axis would

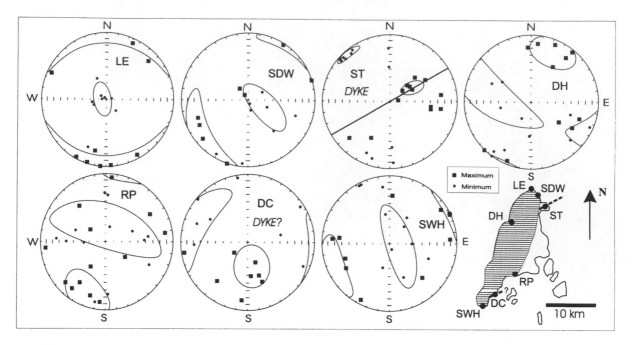

Figure 6. Equal area stereonet plots (lower hemisphere) of AMS site data from Grand Manaan flows (LE, SDW, DH, RP, DC and SWH) and dikes (ST and possibly DC). Symbols and lines have same meaning as in Figure 5. Inset diagram shows site locations on Grand Manaan Island. Sites are Long Eddy Point (LE), Seven Days Work Cove (SDW), Swallowtail (ST), Dark Harbor (DH), Red Point (RP), Deep Cove (DC), and Southwest Head (SWH).

be consistent with the Swallowtail dike and its presumed continuation to the northeast beneath the axial zone of the Fundy Basin as a feeder dike for North Mountain basalts. According to the schematic models in Figure 4, the inferred relationship would be most similar to model 4C. The Minister Island and Shelburne dikes would also be consistent with being feeders in a model 4C scenario. However, the Shelburne dike was earlier ruled out on the basis of isotopic differences. Furthermore, based on published magnetic fabric data, as discussed below, the Shelburne and Minister Island dikes are unlikely to have been feeders.

The relationship between the Grand Manaan lavas and the Swallowtail dike requires some additional discussion. The observed emplacement axes of flows on Grand Manaan Island do not suggest a simple relationship as depicted in Figure 4C and D. Data from both sides of the ST dike, sites LE and SDW to the north, DH, above, and RP and SWH to the south, show consistent NE flow axes. This flow pattern and its relationship to a proposed feeder (either the Swallowtail dike or possible Deep Cove dike) has a geometry similar to that depicted schematically in Figure 4D', a pattern that is not evidence for an underlying dike. In contrast, a flow pattern as depicted in Figure 4C or D, would be more consistent with evidence for an underlying dike. Therefore, either the Swallowtail dike was not the feeder for the Grand Manaan basalts or there was another control on the flow pattern. Perhaps the NE flow axis was caused

by deflection along a NE-SW topographic barrier. This barrier could have been the Grand Manaan Island fault, if this fault was active (west-side-down) during basalt emplacement.

Considering the Swallowtail dike as the feeder for the North Mountain basalt could explain its magnetic fabric indicating subvertical flow (Figure 6). Other regional ENA dikes, the Shelburne dike and the Minister Island dike, exhibit lateral flow based on limited measurements of anhysteretic magnetic susceptiblity for both dikes and, in the case of the Shelburne dike, also based on field structures [*Greenough and Hodych*, 1990]. Similarly, the flow direction in the Higganum-Holden and Christmas Cove dikes is dominantly lateral [*de Boer et al.*, in press]. Lateral flow would not be consistent with a dike spawning a lava flow. The subvertical flow observed only in the Swallowtail dike supports the idea of it as a local feeder for North Mountain basalts.

A more detailed examination of the AMS data suggests that it is also possible that the lower lavas have a slightly different flow pattern than the middle amygdaloidal and upper massive porphyritic flow units. Specifically, both the Brier Island (BI +BLH) and Grand Manaan lavas (Figure 6) which are both interpreted to belong to the lower unit of the North Mountain basalts show a scattered NE-SW average flow axis (Figure 3). In contrast, the most reliable data (uncertainty less than 20°) from the middle amygdaloidal unit

Table 1. Summary of AMS data for the North Mountain basalt and Grand Manaan basalts.

Site	N,n	D_{max}	I_{max}	A_{max}	D_{int}	I_{int}	A_{int}	D_{min}	I_{min}	A_{min}	K_{mean} ($\pm 1\sigma$)	K_{n-max}	K_{n-int}	K_{n-min}	P_J	T	A%	B%	H%
North Mountain Basalts																			
CS	8,17	185.6	4.3	9:3	275.9	3.9	9:5	48.1	84.2	5:3	3.00 (0.73)	1.012	1.002	0.985	1.028	0.280	1.82	-0.73	2.70
HH	5,10	237.8	2.3	28:14	147.7	1.6	28:10	22.6	87.2	17:6	3.20 (0.45)	1.004	1.001	0.996	1.008	0.207	0.55	-0.16	0.78
MO	9,14	299.3	19.4	22:3	190.5	42.5	48:6	47.2	41.1	46:3	2.23 (0.84)	1.011	0.999	0.990	1.021	-0.181	1.66	0.39	2.11
MV	6,12	332.7	6.0	18:7	242.4	2.1	18:5	133.1	83.6	7:5	3.43 (0.50)	1.006	1.001	0.993	1.013	0.180	0.87	-0.22	1.25
PC	7,17	3.7	15.9	21:10	98.2	15.5	25:18	230.3	67.5	23:10	2.24 (0.11)	1.003	1.000	0.998	1.005	-0.207	0.39	0.10	0.48
PAC	5,8	357.8	8.7	13:12	264.5	20.2	31:6	109.9	67.8	30:13	2.89 (9.29)	1.006	0.999	0.996	1.010	-0.384	0.85	0.39	1.02
DE	8,18	350.8	10.8	10:9	259.2	8.4	21:9	131.9	76.2	21:8	3.03 (0.21)	1.003	1.000	0.997	1.006	-0.099	0.47	0.06	0.60
SC	6,9	1.4	11.2	25:15	268.3	15.4	29:20	126.0	70.8	27:13	3.40 (0.25)	1.003	1.001	0.996	1.007	0.516	0.41	-0.34	0.66
BI	5,9	36.2	6.7	23:7	126.9	5.6	30:11	256.7	81.2	32:18	2.24 (0.13)	1.005	0.999	0.997	1.009	-0.462	0.71	0.38	0.82
BLH	6,12	262.9	8.4	22:18	171.7	8.2	46:22	38.0	78.3	46:18	1.53 (0.30)	1.003	1.000	0.997	1.006	-0.219	0.47	0.13	0.59
BI+	11,21	228.5	2.3	25:10	318.7	4.2	26:23	110.7	85.2	26:14	1.84 (0.43)	1.003	1.000	0.997	1.006	-0.137	0.46	0.08	0.58
BLH																			
MEAN	65,126	5.0	2.9	10:3	274.8	4.2	10:4	129.4	84.9	4:3	2.72 (0.73)	1.004	1.001	0.995	1.009	0.351	0.57	-0.30	0.86

Table 1. Continued.

Grand Manaan Island Basalts

Site	N, n	D_{max}	I_{max}		D_{int}	I_{int}		D_{min}	I_{min}		K_{mean}	K_{n-max}	K_{n-int}	K_{n-min}	P_J	T	A%	B%	H%
LE	5,9	181.2	0.3	68:17	91.2	5.0	68:11	274.2	85.0	21:11	1.39 (0.12)	1.004	1.003	0.993	1.012	0.731	0.63	-0.81	1.11
SDW	6,8	232.6	12.1	48:15	326.8	19.2	48:36	112.2	67.0	37:15	2.44 (1.40)	1.003	1.000	0.997	1.007	-0.043	0.51	0.03	0.67
DH	7,11	23.6	18.0	24:23	134.3	47.4	76:22	279.4	37.2	76:21	0.54 (0.21)	1.005	0.998	0.997	1.008	-0.786	0.71	0.59	0.75
RP	7,13	201.2	18.9	31:21	109.5	4.9	76:25	5.6	70.4	76:27	1.67 (0.59)	1.003	0.999	0.998	1.005	-0.714	0.45	0.35	0.49
SWH	6,11	246.7	10.5	31:17	338.0	7.2	54:31	101.8	77.2	54:16	1.17 (0.20)	1.008	0.998	0.994	1.014	-0.474	1.13	0.62	1.31
MEAN	31,52	221.3	3.3	23:11	311.6	4.9	23:19	97.5	84.1	19:11	1.40 (0.87)	1.004	1.000	0.997	1.007	-0.042	0.53	0.03	0.69

Grand Manaan Island Dykes

Site	N, n	D_{max}	I_{max}		D_{int}	I_{int}		D_{min}	I_{min}		K_{mean}	K_{n-max}	K_{n-int}	K_{n-min}	P_J	T	A%	B%	H%
ST	7,13	58.5	57.8	15:7	224.5	31.4	16:6	318.4	6.3	12:5	5.06 (2.35)	1.006	1.002	0.992	1.014	0.384	0.91	-0.53	1.40
DC	7,8	170.1	59.9	27:21	27.9	24.6	49:25	290.2	16.2	49:21	2.24 (0.43)	1.002	1.000	0.998	1.005	-0.216	0.37	0.10	0.46

Notes: N, n is the number of samples/specimens used for computation. Subscripts 'max', 'int' and 'min' denote the maximum, intermediate and minimum principal AMS axes, respectively. **D** and **I** are the direction and inclination of the axes (the selection of a direction vs. its antipode was arbitrary; our convention was to quote the axis direction for which the inclination was positive, i.e. downward). A consists of two numbers equalling the long and short radii of the 95% uncertainty ellipse. K_{mean} is the magnetic susceptibility in 10^{-2} SI units. K_{n-max}, K_{n-int}, K_{n-min} are the susceptibilities of the principal axes normalized such that their sum equals 3. **T** and P_J are the ellipsoid shape and degree of anisotropy (Jelinek, 1981). $T = (2LnK_{int}\text{-}LnK_{max}\text{-}LnK_{min})/(LnK_{max}\text{-}LnK_{min})$ which ranges from +1 (oblate) and −1 (prolate). (Ln is natural logarithm). $P_J = \exp$ [sqrt$\{2[(LnK_{max}\text{-}LnK_{mean})^2 + (LnK_{int} - LnK_{mean})^2 + (LnK_{min}\text{-}LnK_{mean})^2]\}$] where $LnK_{mean} = (LnK_{max} + LnK_{int} + LnK_{min})/3$. $P_J \geq 1$. The percentage of anisotropy, P_J, of sites varies from 1% to 9% (1.01 - 1.09). **A%** is degree of anisotropy $\{= 100[1\text{-}(K_{min}+K_{int})/2K_{max}]$, range 0, 100} and **B%** is ellipsoid shape, $\{= 100[1+(K_{min}\text{-}2K_{int})/K_{max}]$ which ranges from -100 (oblate) to +100 (prolate)} after Cañón-Tapia (1994). **H (%)** is $100*(K_{max}\text{-}K_{min})/K_{mean}$ (Owens 1974; Tarling & Hrouda 1993).

(e.g. sites MO and MV) and from the upper massive porphyritic unit (e.g. site CS) indicate a more N-S flow axis. Combining this observation with the possibility (discussed above) that the Swallowtail dike was not the feeder for the Grand Manaan basalts, we offer the following speculative model. We propose that the Swallowtail dike was the feeder for only the middle and/or upper North Mountain basalts and that the lower North Mountain unit (and correlated Grand Manaan flows) were either fed from a different dike (with a different trend) or the NE-SW flow pattern in this lower unit unit was influenced by preexisting topography.

5. CONCLUSIONS

Based on magnetic fabric (AMS) studies from 10 sites along the 200 km long NE-SW exposure of the North Mountain and 7 sites in the Grand Manaan Island basalts of Atlantic Canada, we infer dominant flow along a N-S axis and minor flow along a NE SW axis. The NE-SW pattern is only present in lower lavas (from the Grand Manaan sites and from sites at the southwest end of the North Mountain basalts). The feeder for the lower flows has not been identified. The longitudinal consistency of the N-S flow pattern from all but one of the sites in the middle and upper flows favors a linear feeder (a dike) rather than a point feeder source. Integration of thickness information in the basalts with the position of possible dike feeders suggests that the most likely feeder for the middle and upper flows is the Swallowtail dike of Grand Manaan Island that we assume continues underneath the northern side of the Fundy basin. The Swallowtail dike may also be the eastward continuation of the Christmas Cove - Higganum-Holden dike. Supporting the notion that the Swallowtail dike is a feeder for at least some of the North Mountain basalts, is its AMS pattern (Figure 5) suggestive of subvertical flow, which contrasts with lateral flow inferred for other nearby ENA dikes, the Christmas Cove (- Higganum-Holden) dike, and the Shelburne and Minister Island dikes.

Wider application of the AMS method to flows, dikes and sills of the CAMP event is critical to understanding the plumbing system(s) of this huge mafic magmatic episode.

Acknowledgments. We thank Brooks Ellwood and Ken Buchan for comments on the manuscript. This is Geological Survey of Canada Publication 2002022.

REFERENCES

Bertrand, H., The Mesozoic tholeiitic province of northwest Africa: a volcano-tectonic record of the early opening of central Atlantic. In *Magmatism in Extensional Structural Settings: the Phanerozoic African Plate* edited by A. B. Kampunzu and R. T. Lubala, pp. 148-188. Berlin: Springer Verlag, 1991.

Cañón-Tapia, E., Single-grain versus distribution anisotropy: a simple three-dimensional model. *Physics of the Earth and Planetary Interiors*, 94,149-158, 1996.

Cañón-Tapia, E., G. P. L. Walker, E. Herrero-Bervera, The internal structure of lava flows – insights from AMS measurements II: Hawaiian pahoehoe, toothpaste lava and 'a'ā. *Journal of volcanology and geothermal research*, 76, 19-46, 1997.

Cañón-Tapia, E., G. P. L. Walker, E. Herrero-Bervera, The internal structure of lava flows – insights from AMS measurements I: Near-vent a'a. *Journal of volcanology and geothermal research*, 70, 21-36, 1996.

Cañón-Tapia, E., AMS parameters: guidelines for their rational selection. *Pure Appl. Geophys.* 142, 365-382, 1994.

Cummins, L. E., J. D. Arthur, and P. C. Ragland, Classification and tectonic implications for early Mesozoic magma types of the Circum-Atlantic. In *Eastern North American Mesozoic Magmatism, Special Paper 268* , edited by J. H. Puffer and P. C. Ragland, pp. 119-135. Geological Society of America, 1992.

de Boer, J. Z. Stress configurations during and following emplacement of ENA basalts in the northern Appalachians. In *Eastern North American Mesozoic Magmatism. Special Paper 268*, edited by J. H. Puffer and P. C. Ragland, pp. 361-378, Geological Society of America, 1992.

de Boer, J., F. G. Snider, Magnetic and chemical variations of Mesozoic diabase dikes from eastern North America: evidence for a hotspot in the Carolinas? *Geological Society of America Bulletin*, 90, 185-198, 1979.

de Boer, J. Z., J. G. McHone, J. H. Puffer, P. C. Ragland, and D. Whittington. Mesozoic and Cenozoic magmatism. In *The Geology of North America: The Atlantic Continental margin.* Edited by R. E. Sheridan and J. A. Grow, pp. 217-241. Geological Society of America, 1988.

de Boer, J. Z, R. E. Ernst, A. G. Lindsey, Evidence for predominant lateral magma flow along major feeder dike segments of the Eastern North American swarm based on magnetic fabric. In *The Great Rift Valleys of Pangea in Eastern North America: Volume 1: Tectonics, Structure, and Volcanism of Supercontinent Breakup* edited by P.M. LeTourneau, and P.E. Olsen, Columbia University Press, in press.

Dunning, G. and J. P. Hodych, U/Pb zircon and baddeleyite ages for the Palisades and Gettysburg sills of the northeastern United States: Implications for the ages of the Triassic/Jurassic boundary, *Geology*, 18, 795-798, 1990.

Ebinger, C., and N. Sleep, Cenozoic magmatism in Africa: One plume goes a long way. *Nature*, 395, 788-791, 1998.

Ernst, R. E. and W. R. A. Baragar, Evidence from magnetic fabric for the flow pattern of magma in the Mackenzie giant radiating dyke swarm, *Nature*, 356, 511-513, 1992.

Ernst, R. E. and K. L. Buchan. Giant radiating dyke swarms: their use in identifying pre-Mesozoic large igneous provinces and mantle plumes. In *Large Igneous Provinces: Continental, Oceanic and Planetary Flood Volcanism, Geophysical Monograph 100*, edited by J. Mahoney and M. Coffin, pp. 297-333, American Geophysical Union, 1997a.

Ernst, R. E. and K. L. Buchan, Layered mafic intrusions: a model for their feeder systems and relationship with giant dyke swarms and mantle plume centres, *South African Journal of Geology*, 100, 319-334, 1997b.

Ernst, R. E. and K. L. Buchan, The use of dyke swarms in identifying and locating mantle plumes. In: *Mantle Plumes: Their Identification Through Time, Special Paper 352* edited by R. E. Ernst and K. L. Buchan K.L, pp. 247-265, Geological Society of America, 2001.

Gradstein, F. M., Agterberg, F. P., Ogg, J.G., Hardenbol, J., van Veen, P, Thierry, J., and Huang, Z-H., A Mesozoic time scale. Journal of Geophysical Research, 99, 24051-24074, 1994.

Greenough, J. D. and J. P. Hodych, Evidence for lateral injection in the early Mesozoic dykes of eastern North America. In *Mafic Dykes and Emplacement Mechanisms* edited by A. J. Parker, P. C. Rickwood and D. H. Tucker, pp. 35-46, Balkema, Rotterdam, 1990.

Greenough, J. D. L. M. Jones, and D. J. Mossman, Petrochemical and stratigraphic aspects of North Mountain Basalts from the north shore of the bay of Fundy, Nova Scotia, Canada, *Canadian Journal of Earth Sciences*, 26, 2710-2717, 1989.

Hames, W. E., P. R. Renne, C. Ruppel, New evidence for geologically instantaneous emplacement of earliest Jurassic Central Atlantic magmatic province basalts on the North American margin, *Geology*, 28, 859-862, 2000.

Hargraves, R. B., D. Johnson and C. Y. Chan. Distribution anisotropy: the cause of AMS in igneous rocks? *Geophysical Research Letters*, 18, 2193-2196, 1991.

Hill, R. I., Starting plumes and continental break-up, *Earth and Planetary Science Letters*, 104, 398-416, 1991.

Jelinek, V., Statistical processing of anisotropy of magnetic susceptibility measured on groups of sediments. *Studia Geophysica et Geodaetika*, 22, 50-62, 1978.

Jelinek, V., Characterization of the magnetic fabric of rocks. *Tectonophysics*, 79, 63-67, 1981.

Knight, M. D., and G. P. L. Walker, Magma flow directions in dikes of the Koolau Complex, Oahu, determined from magnetic fabric studies. *Journal of Geophysical Research*, 93, 4301-4319, 1988.

Lienert, B. R., Monte Carlo simulation of errors in the anisotropy of magnetic susceptibility: a second-rank symmetric tensor. *Journal of Geophysical Research*, 96, 19539-19544, 1991.

Marzoli, A., P. R. Renne, E. M. Piccirillo, M. Ernesto, G. Bellieni, A. De Min, Extensive 200-million-year-old continental flood basalts of the Central Atlantic Magmatic Province, *Science*, 284, 616-618, 1999.

May, P. R. Pattern of Triassic-Jurassic diabase dikes around the north Atlantic in context of predrift position of the continents. *Geological Society of America Bulletin*, 82, 1285-1292, 1971.

McHone, J. G., Broad-terrane Jurassic flood basalts across northeastern North America, *Geology*, 24, 319-322, 1996.

McHone, J. G., Non-plume magmatism and rifting during the opening of the central Atlantic Ocean. *Tectonophysics*, 316, 287-296, 2000.

McHone, J. G., and J. R. Butler. Mesozoic igneous provinces of New England and the opening of the North Atlantic Ocean, *Geological Society of America Bulletin*, 95, 757-765, 1984.

Nye, J. F. Physical Properties of Crystals, 2nd edition, Clarendon Press, Oxford, 329 p., 1985.

Oliveira, E. P., J. Tarney, and X. J. João. 1990. Geochemistry of the Mesozoic Amapá and Jari dyke swarms, northern Brazil: Plume-related magmatism during opening of the central Atlantic. In *Mafic Dykes and Emplacement Mechanisms* edited by A. J. Parker, P. C. Rickwood, D. H. Tucker, pp. 173-183, Balkema, Rotterdam, 1990.

Olsen, P. E. Stratigraphic record of the early Mesozoic breakup of Pangea in the Laurasia-Gondwana rift system. *Annual Review of Earth and Planetary Science Letters*, 25, 337-401, 1997.

Owens, W. H. Mathematical model studies on factors affecting the magnetic anisotropy of deformed rocks. *Tectonophysics*, 24, 115-131, 1974.

Oyarzun, R., M. Doblas, J. L. Lopez-Ruiz, and J. M. Cebria, Opening of the central Atlantic and asymmetric mantle upwelling phenomena: implications for long-lived magmatism in western North Africa and Europe. *Geology*, 8, 727-730, 1997.

Papezik, V. S., J. D. Greenough, J. A. Colwell, and T. J. Mallinson. North Mountain basalt from Digby, Nova Scotia: models for a fissure eruption from stratigraphy and petrochemistry, *Canadian Journal of Earth Sciences*, 25, 74-83, 1988.

Pe-Piper, G. and D. J. W. Piper, Were Jurassic tholeiitic lavas originally widespread in southeastern Canada?: a test of the broad terrane hypothesis. *Canadian Journal of Earth Sciences*, 36, 1509-1516, 1999.

Philpotts, A. R. 1992. A model for emplacement of magma in the Mesozoic Hartford Basin. In *Eastern North American Mesozoic Magmatism. Special Paper 268*, edited by J. H. Puffer, and P. C. Ragland, pp. 137-148, Geological Society of America, 1992.

Philpotts, A. R., and P. M. Asher. Magmatic flow-direction indicators in a giant diabase feeder dike, Connecticut. *Geology*, 22, 363-366, 1994.

Rainbird, R, and R. E. Ernst. The sedimentary record of mantle-plume uplift, In *Mantle Plumes: Their Identification Through Time. Special Paper 352* edited by R.E. Ernst, and K.L. Buchan, pp. 227-245, Geological Society of America 2001.

Ragland, P. C. and D. Whittington, Early Mesozoic diabase dikes of eastern North America: magma types. Geological Society of America, Abstracts with programs, 15, 666, 1983.

Ragland, P. C., L. E. Cummins, and J. D. Arthur. Compositional patterns for early Mesozoic diabases from South Carolina to central Virginia. In *Eastern North American Mesozoic Magmatism, Special Paper 268*, edited by J. H. Puffer, and P. C. Ragland, pp. 309-332, Geological Society of America, 1992.

Rochette, P., M. Jackson, and C. Aubourg, Rock magnetism and the interpretation of anisotropy of magnetic susceptibility, *Reviews of Geophysics*, 30, 209-226, 1992.

Sebai, A., G. Feraud, H. Bertrand, and J. Hanes, $^{40}Ar/^{39}Ar$ dating and geochemistry of tholeiitic magmatism related to the early opening of the Central Atlantic rift. *Earth and Planetary Science Letters*, 104, 455-472, 1991.

Stephenson, A. Distribution anisotropy: two simple models for magnetic lineation and foliation. *Physics of the Earth and Planetary Interiors*, 82, 49-53, 1994.

Tarling, D. H., and F. Hrouda, *The Magnetic Anisotropy of Rocks*. Chapman and Hall, 217 pp., 1993.

Wade, J. A., D. E. Brown, A. Traverse, and R. A. Fensome, The Triassic-Jurassic Fundy Basin, eastern Canada: regional setting, stratigraphy and hydrocarbon potential. *Atlantic Geology*, 32, 189-231, 1996.

Wilson, M. Thermal evolution of the Central Atlantic passive margins: continental break-up above a Mesozoic super-plume, *Journal of the Geological Society of London*, 154, 491-495, 1997.

Richard E. Ernst, Geological Survey of Canada, 601 Booth Street, Ottawa, Ontario Canada K1A OE8, (rernst@nrcan.gc.ca).

Jelle Zeilinga de Boer, Peter Ludwig, and Taras Gapotchenko, Department of Earth and Environmental Sciences, Wesleyan University, Middletown, CT 06459-0139, USA (jdeboer@wesleyan.edu).

Volatile Emissions From Central Atlantic Magmatic Province Basalts: Mass Assumptions and Environmental Consequences

J. Gregory McHone

Department of Geology and Geophysics, University of Connecticut, Storrs, Connecticut

Mesozoic basins that contain extrusive basalts of the 200 Ma Central Atlantic Magmatic Province (CAMP) presently total about 320,000 km^2. However, CAMP dikes and sills similar to those that fed the basin basalts are also spread widely across an area greater than 10 million km^2 within four continents. In addition, basalts of the east coast margin igneous province (ECMIP) of North America, which cause the east coast magnetic anomaly, covered about 110,000 km^2 with 1.3 million km^3 of extrusive lavas. If only half of the continental CAMP area was originally covered by 200 m of surface flows, the total volume of CAMP and ECMIP lavas exceeded 2.3 million km^3. Weighted averages for the volatile contents of 686 CAMP tholeiitic dikes and sills, in weight %, are: $CO_2 = 0.117$; $S = 0.052$; $F = 0.035$; and $Cl = 0.050$. Atmospheric emissions of volatiles from flood basalts are conservatively estimated as 50 % to 70 % of the volatile content of the sub-volcanic magmas, mainly exsolved into gaseous plumes from lava curtains at the erupting fissures. Total volcanic emissions of these gases therefore ranged between 1.11×10^{12} and 5.19×10^{12} metric tons, enough for major worldwide environmental problems. Radiometric and stratigraphic dates indicate that most CAMP volcanic activity was brief, widespread, and close to the Tr-J boundary, which is marked by a profound mass extinction. More precise information about the timing, duration, and chemical emissions of volcanic episodes is needed to support a model for CAMP in the extinction event.

INTRODUCTION

The initial breakup of Pangaea in Early Jurassic time provided a legacy of basaltic dikes, sills, and lavas over a vast area around the present North Atlantic Ocean (Fig. 1). Although some connections among these basalts had long been recognized, Rampino and Stothers [1988] were possibly the first to rank them among major flood basalt provinces as a group. Marzoli et al. [1999] showed that basaltic sills of similar age (near 200 Ma, or earliest Juras-

The Central Atlantic Magmatic Province:
Insights from Fragments of Pangea
Geophysical Monograph 136
Copyright 2003 by the American Geophysical Union
10.1029/136GM013

sic) and composition (intermediate-Ti quartz tholeiite) also occur across the vast Amazon River basin of Brazil, and they proposed a group acronym of CAMP (Central Atlantic Magmatic Province). The province has been described by McHone [2000] as extending within Pangaea from modern central Brazil northeastward about 5000 km across western Africa, Iberia, and northwestern France, and from Africa westward for 2500 km through eastern and southern North America as far as Texas and the Gulf of Mexico (Fig. 1). If perhaps not the largest by volume, the CAMP certainly encompasses the greatest area known – possibly 10 x 10^6 km^2 -- of any continental large igneous province.

Nearly all CAMP rocks are tholeiitic in composition, with widely separated areas where basalt flows are preserved, and many large groups of diabase (dolerite) sills or sheets, small lopoliths, and dikes throughout the province. CAMP volcanism occurred in the middle of rifting activity of Pangaea during the lower Mesozoic, and the enormous province size, varieties of basalt, and brief time span of CAMP magmatism invite speculation about mantle processes that could produce such a magmatic event as well as break up a supercontinent [Wilson, 1997; McHone, 2000].

Throughout the Phanerozoic, the greatest mass extinctions have virtually coincided with the greatest eruptions of continental flood basalts [Stothers, 1993; Courtillot, 1994]. As the precision of radiometric dates for these events has improved in recent years, their correlation with extinctions has generally improved as well [Courtillot et al., 1996; Olsen, 1999], to the point where some events are timed to within a few hundred thousand years [Pálfy et al., 2000; Courtillot et al., 2000]. This correlation includes a possible link between the CAMP and widespread faunal extinctions at the Triassic-Jurassic boundary, as suggested by Stothers [1993] and Courtillot [1996], and in new studies by McElwain et al. [1999], Pálfy et al. [2000] and Wignall [2001].

The exact mechanism by which a flood basalt could cause a mass extinction remains speculative, but the most likely scenario involves the injection into the upper atmosphere of large amounts of SO_2 and/or CO_2, causing drastic, if temporary, climatic cooling and/or heating [Rampino et al., 1988; Palais and Sigurdsson, 1989; McElwain, 1999]. The heights of lava fountains at very large fissure eruptions may be 1 to 2 km, and the super-hot volatiles escaping from the vented magma can easily rise to 11 km or more [Woods, 1993] and enter into global circulation. The climatic cooling mechanism of sulfur aerosols has been calibrated by Rampino et al. [1988] and Palais and Sigurdsson [1989], using records of historic eruptions. Thordarson et al. [1996] applied this calibration to the Laki, Iceland flood basalt and climatic cooling event of 1783-1784, and another application for the larger Roza basalt flow of the Miocene-age Columbia River basalt province has been made by Thordarson and Self [1996]. Although they described volcanic CO_2 as a possible cause of global warming, Caldeira and Rampino [1990] concluded that CO_2 emissions of the voluminous Deccan basalts (K-T boundary age) were probably spread over too much time to produce a catastrophic increase in climatic temperature. A pertinent discussion about the timing of CAMP magmatism is presented by Baksi in this volume.

This paper is an exploration of the potential for basalts of the Central Atlantic Magmatic Province to have produced a catastrophic climate change, even if no conclusion about is yet possible. Eroded remnants of CAMP basalt that flowed onto the surface, along with their sub-fissure dike sources, are preserved in basins around the central North Atlantic Ocean, and the total extrusive mass is estimated by extrapolating a similar proportion of lava across the entire central Pangaean province. The volatile-element contents of sub-volcanic sills and dikes are reasonably well known for the Mesozoic basins in eastern North America, and through analogy with other flood basalts, become the basis for an estimate of the amounts of CAMP volatiles injected into the atmosphere. The goal is to outline such estimates, and to discuss problems that remain before the CAMP can be determined to have caused the Tr-J boundary extinction event.

CAMP BASALTS

Ages

The age precision for the main pulse of magmatism throughout the province has improved significantly from studies by Sutter [1988], Dunning and Hodych [1990], Sebai et al. [1991], Hodych and Dunning [1992], Fiechtner et al. [1992], Deckart et al. (1997], Marzoli et al. [1999], and Hames et al. [2000]. Most of the modern dates fit between 196 and 202 Ma. Olsen [1997] described stratigraphic evidence that major lavas of the northern CAMP basins were all produced within a span of 580,000 years. Similar basalts can be correlated throughout the province, and so it appears that much or most of the CAMP volcanism occurred in less than a million years around 200 to 201 Ma (beginning Jurassic). A few radiometric dates near 196 Ma in eastern North America [Sutter, 1988; Hames et al., 2000] and western Africa [Deckart et al., 1997] may indicate a smaller, later event with magmas that remain similar to the major tholeiite types. Marzoli et al. [1999] note that several dates for dikes in the central CAMP, along the future coastal region between Africa and South America (Fig. 1), are close to 192 Ma, which could be a younger age for some high-Ti tholeiites within CAMP.

Ironically, the age of the Tr-J mass extinction event is partly determined by its proximity to the oldest basalt flows in the Mesozoic basins of the northern CAMP. Olsen [1997] and his colleagues have analyzed Milankovitch climatic cycles in detail from drill cores in several basins, which show that the earliest basalt lavas of each basin are in essentially identical stratigraphic positions, and thus are identical in age. This finding agrees with the excellent petrological correlation of basin basalts [Puffer, 1992], which are therefore co-magmatic across the northern CAMP. The extinction horizon lies from a few meters to tens of meters beneath the basalt in several widely separated basin locations [Olsen, 1997; Mossman et al., 1998],

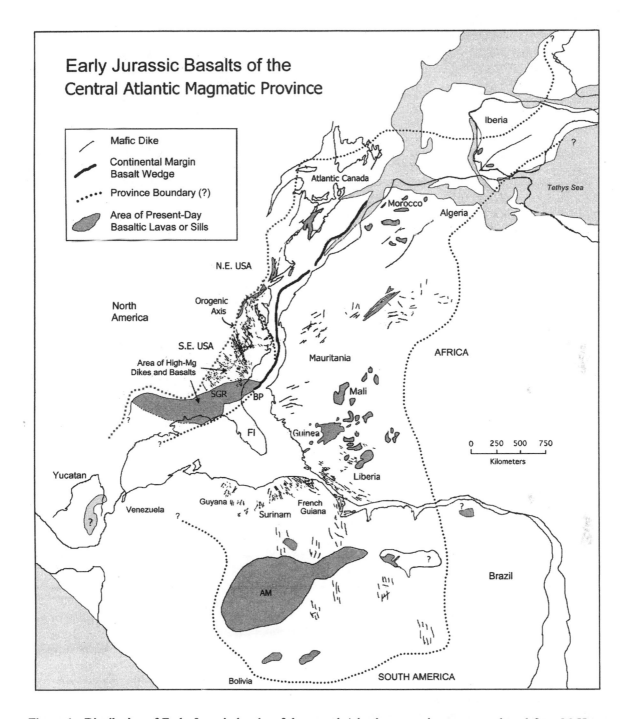

Figure 1. Distribution of Early Jurassic basalts of the central Atlantic magmatic province, adapted from McHone (2000). The predrift configuration of Pangaea is by Klitgord and Schouten (1986). The location of the North American margin volcanic wedge is from Holbrook and Kelemen (1993). Other features are from Deckart et al. [1997], Olsen [1997], and Marzoli et al. [1999], and other studies referenced in the text. Abbreviations: BP = Blake Plateau; FL = Florida; SGR = South Georgia Rift; S.E = Southeast; N.E. = Northeast.

which may correspond to a few thousands to tens of thousands of years at typical rates of sedimentation. Thus, the age precision of the Tr-J boundary and extinction falls within the precision range of radiometric dates for CAMP, except that we know the boundary slightly precedes the terrestrial lavas of the northern CAMP basins [*Pálfy et al.*, 2000].

Geographic Extent

Basaltic lavas of the CAMP are best preserved within Mesozoic rift basins of eastern North America and northwestern Africa [*Manspeizer*, 1988]. Some CAMP lavas are also known over older basement terrains in Africa and South America [*Bertrand*, 1991; *Olsen*, 1997; *Marzoli et al.*, 1999], demonstrating that CAMP dikes reached the surface in places outside of the rift basins. The CAMP lavas are therefore remnants of much larger flood basalts that have been mostly removed by erosion [*Rampino and Stothers*, 1988; *McHone*, 1996; *McHone and Puffer*, 2002]. Sources for the former and present surface basalts are represented by the huge swarms of diabase dikes along eastern North America, northern South America, western Africa, and western Europe, around but also well beyond the central Atlantic rift zone (Fig. 1). Numerous large sills of CAMP basalt (such as the famous Palisades sill) occur within the rift basins, but even greater sills with volumes of 10^4 to 10^5 km^3 or more are known, with areas exceeding 10^5 km^2 in basement terrains of western Africa [*Deckart et al.*, 1997] and northern South America [*Marzoli et al.*, 1999].

Studies now suggest that dikes or flows of the CAMP occur as far as northwestern France [*Caroff et al.*, 1995; *Jourdan et al.*, this volume], the Mississippi Embayment (*Sundeen*, 1989; *Baksi*, 1997), the Gulf of Mexico (*Schlager et al.*, 1984) and west-central Brazil [*Montes-Lauar et al.*, 1994; *Marzoli et al.*, 1999]. During its formation, large sections of the CAMP extended into both the northern and southern hemispheres. The entire igneous province may stretch beyond 5000 km in length by 2500 km in width, but its actual limits are not yet known.

The total lengths of some dikes or single dike sets that fed CAMP basalts exceed 700 km (Fig. 1), and remarkably, many maintain essentially the same distinct composition along their great lengths. Others, such as the 250-km long Shelburne dike of Atlantic Canada, show variations consistent with liquid-crystal fractionation similar to surface flows [*Pe-Piper and Piper*, 1999]. In the pre-rift Pangaean region of Morocco, northeastern USA, and Atlantic Canada, at least three distinct quartz tholeiite subtypes comprise basin basalts that apparently flowed

within a 580,000 year period throughout the region [*Philpotts and Martello*, 1986; *Olsen*, 1997]. Each of the three northern basalt subtypes also occurs in individual dikes that are from 250 to 700 km long in northeastern North America, adjacent northwestern Africa, and western Europe, and these dikes were certainly co-magmatic fissure sources for the basalts [*Philpotts and Martello*, 1986; *Bertrand*, 1991; *McHone*, 1996]. There may be additional sub-types, as observed in dikes that apparently have no basalts in exposed basins [*McHone*, 1996; *Pe-Piper and Piper*, 1999]. Surface basalts from most other sections of the CAMP in Figure 1 are not preserved, are poorly preserved, or are buried beneath later sediment, but based on the areas near basins, the major dike swarms are everywhere likely to have reached the surface as fissure eruptions.

The East Coast Margin Igneous Province

The widespread groups of dikes fed fissure eruptions and flood basalts that apparently preceded the initial formation of Atlantic ocean crust, which started during the Early to Middle Jurassic along sections of the central Atlantic rift [*Withjack et al.*, 1998; *Benson*, this volume]. There is also evidence for a close link between continental CAMP magmatism and a basaltic border province (volcanic rift margin) adjacent to new ocean crust along the eastern margin of North America. Austin et al. [1990] and Holbrook and Kelemen [1993] determined that sub-aerial volcanic flows comprise at least the upper section of seaward-dipping seismic reflectors, or a basalt wedge, along most of the central Atlantic continental margin. In the southeastern USA, strong sub-horizontal seismic reflectors of continental flood basalts of the South Georgia Mesozoic basin intersect or overlap the seaward-dipping reflectors [*Oh et al.*, 1995]. The SDR represents a very thick (to 25 km) basalt and plutonic cumulate wedge that is uniformly large (about 55 km wide) along roughly 2000 km of the eastern North American margin (Fig. 1), as shown by the East Coast Magnetic Anomaly and seismic reflections, and it has been referred to as the "east coast margin igneous province," or ECMIP, by Holbrook and Kelemen [1993].

Volume Estimates

Lava flows of CAMP are generally thin in comparison with other large flood basalts. The northern basins contain 1 to 3 lava units of about 50 to 200 m each, with the thickest flows (North Mountain basalt) approaching 1000 m in the center of the Fundy basin of Atlantic Canada [*Wade et al.*, 1996]. However, as demonstrated by their chemistry and other co-magmatic correlations, widely-separated northern basin basalts are derived from individual dike systems that extend between basins for 700 km or more, which strongly indicates that their lava products were also once continuous between the basins (*McHone*, 1996]. In addition, the horizon of CAMP basalt in the subsurface South Georgia rift basin is around 200 m thick, and it covers at least 100,000 km^2 of the southern U.S.A. [*Austin et al.*, 1990; *Oh et al.*, 1995]. McHone [2000] suggested that this basalt also extends westward under the southern U.S. coastal plain for at least another 100,000 km^2.

Lava flows are also preserved in other areas of the CAMP, where they flowed across older rocks outside of Mesozoic basins [*Montes-Lauar et al.*, 1994; *Baksi and Archibald*, 1997; *Olsen*, 1997; *Marzoli et al.*, 1999]. In addition, very large sills in South America [*Marzoli et al.*, 1999] and western Africa [*Deckart et al.*, 1997] comprise great volumes of basalt that were likely shallow sources for fissure volcanoes. It is apparent that great swarms of dikes and sills remain where surface flows once existed before their removal by 200 Ma of tectonic uplift and erosion. However, as pointed out by Gudmundsson et al. [1999], many smaller dikes that radiate from shallow sub-volcanic magma chambers never reach the surface and must be discounted as sources for lavas. In addition, topographic highs probably existed in some sections of the Pangaean rift zone before and during CAMP magmatism, which would have precluded lava accumulations in those regions. A conservative estimate, therefore, could be that about half of the present CAMP area of 10 million km^2 was originally covered by tholeiite lava flows averaging 200 m in total thickness.

Table 2 summarizes the measurements, estimates, and calculations of CAMP magmas that were compiled into spreadsheet format for this study. Basalt types and volatiles are discussed below. Although estimates of basin areas and volumes of sills and dikes are included in Table 2, only the surface lava calculations are pertinent to the volatile emission calculations. Because very conservative assumptions were used, it is likely that greater rather than smaller amounts of surface lavas and volatiles were actually produced by the CAMP.

Geochemical Groups

Quartz tholeiites in eastern North America were initially subdivided by Weigand and Ragland [1970] into three groups based on relative compositions: 1) HTQ = "high-Ti" quartz-normative tholeiites, 2) LTQ = "low-Ti" quartz-normative tholeiites, and 3) HFQ = "high-Fe" quartz-normative tholeiites. This classification system was also followed by Grossman et al. [1991] for descriptions of their data set of 960 analyses from basins of the eastern U.S., and used for this paper. The HFQ tholeiites are now believed to represent melts derived through fractional crystallization from either of the other two quartz groups [*Ragland et al.*, 1992; *Puffer*, 1992], and for this study, the appropriate HFQ analyses were included with those of the other groups. Olivine-normative tholeiites (OLT), found mainly in the southeastern USA, form the other major division in many earlier studies, and this group has also been subdivided [see *Warner et al.*, 1985, among others]. However, as discussed elsewhere [*Ragland et al.*, this volume], the OLT and LTQ types appear to be gradational and closely related, and so they are best considered as variants of the same low TiO$_2$ (0.4 to 0.8 wt. %) basalt group.

The HTQ tholeiites have also been referred to as initial Pangaean Rift (IPR) tholeiites [*Puffer*, 1994], because in the Mesozoic basins of northeastern North America and northwestern Africa, the HTQ/IPR flows are the oldest in the stratigraphy. However, this group (with its derivative HFQ) is now known to be the most widespread across the CAMP, and much of it could be of a different age outside of the northeastern USA and Morocco. In fact, relative to other flood basalt provinces [*Peate and Hawkesworth*, 1996], the Ti contents of HTQ magmas (typically 0.9 to 1.5 wt. %) are not high. Other CAMP basalts with quite high TiO$_2$ contents (2 to 5 wt. %) are now recognized in a central region of the CAMP, particularly in Liberia, Guiana, Surinam, and possibly Brazil [*Choudhuri*, 1978; *DuPuy et al.*, 1988; *Mauche et al.*, 1989; *Oliveira et al.*, 1990; *Bellieni et al.*, 1992]. As a result, a new three-fold classification can be used for all CAMP tholeiites: *LTi* for low-Ti olivine and quartz tholeiites (old OLT + LTQ + some HFQ); *ITi* for intermediate-Ti quartz tholeiites (old HTQ + most HFQ); and *HTi* for high-Ti quartz and olivine tholeiites (Table 1).

The LTi and ITi basalt analyses that include volatiles are plotted on diagrams for this study. Figure 2 illustrates their classification on an SiO$_2$-total alkalies diagram [*Le-Bas et al.*, 1986], in which most LTi tholeiites plot as "Basalt" with a few samples in "Picro-basalt" (i.e. olivine rich), and ITi basalts fall along the boundary of "Basalt and "Basaltic Andesite," with some samples trending into "Basaltic Trachy-andesite" and also into higher-SiO$_2$ types. Selected analyses of HTi dikes (outlined by a dotted line on Fig. 2) fall within the "Basalt" field.

As a generality, LTi olivine tholeiite dikes are abundant from South Carolina to central Virginia in the southeastern USA (west-central Pangaean CAMP), LTi quartz tholeiites are common in most of the eastern USA, and the LTi type is more scattered elsewhere in the CAMP. HTi tholeiites

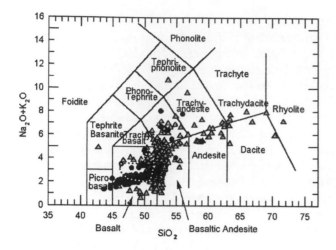

Figure 2. CAMP tholeiite groups plotted on an SiO_2-Total Alkalies classification diagram (LeBas et al., 1986). Symbols: gray triangles = ITi dikes and sills; solid circles = LTi dikes; dotted black and white line in the basalt field encloses HTi dike analyses

are concentrated along the western Africa-northern South America margins (east-central Pangaean CAMP) and possibly into the subsurface of central Florida; and ITi tholeiite dikes, sills, and basalt flows predominate everywhere else in the CAMP. This is a semi-concentric configuration, as shown schematically in Figure 1. Some tholeiites that are intermediate between LTi and ITi are found in the transition zone between the core of LTi and outer zones of ITi within the CAMP; i.e., in central Georgia through eastern Alabama and northern Virginia through Connecticut [*Puffer*, 1992; *Ragland et al.*, 1992]. The central position of HTi rocks in Guyana, Liberia, and Surinam also invite special interest in their relationship to the other major magma types, and in their own implications for mantle geodynamics.

CAMP VOLATILES

Publications that include volatile-element analyses of CAMP basalts are uncommon. By far, the largest group of such analyses is presented by Grossman et al. [1991], which is a summary of geochemical studies by the U. S. Geological Survey from 1984 to 1990 of dikes and sills associated with Mesozoic basins in the eastern United States. Except for platinum-group elements, many of the 960 analyses listed by Grossman et al. [1991] are incomplete, but 686 examples list some combination of CO_2, H_2O+, S, Cl, and F, each in weight percent.

Field and chemical methods used for samples in the USGS data set are described by Gottfried et al. [1991]. Most major elements were analyzed via standard "rapid

rock" XRF techniques [*Shapiro*, 1975]. H_2O+ (water bound in minerals) was measured by differential heated samples and filter weights as described by Shapiro [1975], and CO_2 was analyzed via a colorimetric technique of Engleman et al. [1985]. S was determined from SO_2 by combustion in a sulfur analyzer [*Kirschenbaum*, 1983]. Cl was determined as chloride by the selective ion method of Aruscavage and Campbell [1983]. Fluorine was analyzed as fluoride by the selective ion method of Kirschenbaum [1988]. Estimates of analytical precision and error are not described in detail by Gottfried at al. [1991], but essentially the oxides and volatile elements (averaged in Table 1) were rounded to two significant decimal places, in weight %. For volatiles that commonly occur only as trace amounts, such as F and S, this round off resulted in many reported analyses of 0.00, 0.01, and other values in intervals of a few hundreds percent. This commonality of values is not significant for averages of many analyses (Table 1), but it becomes more apparent in element plots of individual samples.

Using the group classification outlined by Weigand and Ragland [1970], analyses with one or more volatiles include 530 samples labeled as HTQ (high-Ti quartz-normative tholeiite); 14 samples labeled as HFQ (high-Fe quartz-normative tholeiite); 36 samples labeled as LTQ (low-Ti quartz-normative tholeiite; and 106 samples labeled as OLT (olivine-normative tholeiite). The large proportion of HTQ samples is due to an emphasis on sills and other low-angle sheet intrusions, which for an unknown reason were derived mainly from HTQ dikes in eastern North American basins. As discussed previously, the HFQ and HTQ samples are combined as the ITi (intermediate titanium) tholeiite super-group, while the OLT and LTQ samples are combined as the LTi (low titanium) super-group. Averages of these two sample groups are presented in Table 1.

To examine sample volatile distributions and possible magmatic differentiation trends, the LTi and ITi samples were plotted against Mg#'s ($100 \times Mg/Mg+Fe^*$) of their analyses, in which the Mg# varies mainly from olivine (in LTi) and pyroxene (in ITi) fractionation [*Puffer*, 1992; *Ragland et al.*, 1992]. The range of H_2O+ is similar in both the groups (Fig. 3A), but water contents in the ITi increase toward lower Mg#'s. CO_2 appears to have the same range of values in a total sample plot (Fig. 3B), but with a stretched Y scale, it can be seen that the LTi group has a higher proportion of higher CO_2 values than ITi group samples (Fig. 3C).

S, F, and Cl are also not evenly distributed. In Figure 4, it is apparent that S is more abundant in LTi samples in general, while Cl and to a lesser extent F are more abun-

Table 1. Compositions of CAMP Basalts and Comparison Basalts

| | LTi | | | ITi | | | HTi | | | Laki glass | Roza dikes |
	Mean	s.d.	n	Mean	s.d.	n	Mean	s.d.	n		
SiO$_2$	48.84	2.27	142	52.61	2.54	574	51.87	2.04	60	49.68	51.45
TiO$_2$	0.62	0.29	142	1.26	0.62	574	3.21	0.48	60	2.96	3.40
Al$_2$O$_3$	16.06	1.67	142	14.06	1.62	574	14.32	1.27	60	13.05	12.80
FeO*	9.92	1.08	142	10.73	1.93	627	12.14	1.65	60	13.78	14.46
MnO	0.16	0.05	142	0.18	0.03	574	0.19	0.02	60	0.22	0.25
MgO	9.46	2.44	130	6.72	3.13	574	4.11	1.16	60	5.78	4.07
CaO	10.92	1.30	142	9.92	2.11	604	7.64	1.09	60	10.45	8.32
Na$_2$O	2.07	0.46	142	2.44	0.90	627	2.87	0.38	60	2.84	2.73
K$_2$O	0.46	0.62	142	0.83	0.64	574	1.65	0.56	60	0.42	1.36
P$_2$O$_5$	0.10	0.08	142	0.17	0.11	574	0.58	0.19	60	0.28	0.75
H$_2$O+	0.981	0.779	130	0.850	0.570	535	(0.19)			0.19	
CO$_2$	0.091	0.153	133	0.124	0.671	535	(0.148)			0.148	
S	0.067	0.041	135	0.034	0.032	421	(0.111)			0.168	0.111
F	0.023	0.055	91	0.030	0.022	411	(0.102)			0.066	0.102
Cl	0.030	0.037	37	0.064	0.086	429	(0.024)			0.031	0.024
Mg#	61.98	7.61		50.19	15.43		37.65			42.77	33.38
Density	2.683	0.037		2.647	0.045		2.652			2.713	2.695

LTi and ITi analyses are from Grossman et al. [1991], as described in the text. HTi analyses are from Choudhuri [1978], DuPuy et al. [1988], Mauche et al. [1989], and Oliveira et al. [1990], with volatiles in parentheses assumed from the Laki and Roza values. Laki (Iceland) glass inclusion data are from Thordarson et al. [1996]; Roza dike selvage data (Columbia River basalt group) are from Thordarson and Self [1996]. Components are in weight percent. FeO* is total iron as FeO. Mg# is 100xMg/(Mg+Fe*). Density is calculated on a normalized basis, using the method of Bottinga and Weill [1970].

dant in the ITi samples. This is also clear in a plot of those three volatiles against Mg#'s (Fig. 5), which indicates an increase of both Cl and F with lower Mg#'s. No such trend is apparent for S, which is generally higher in the LTi type. Averages of these elements in Table 1 support these plotted distributions.

The HTi (high titanium) tholeiites of the central Pangaean rift zone lack published analyses of volatile elements. However, their bulk compositions are similar to flood basalts of the Miocene-age Roza flow of the Columbia River basalt province [*Thordarson and Self,* 1996] and also the historic Laki flow of Iceland [*Thordarson et al.,* 1996], and for the purposes of this review, similar volatile contents have been assumed for the CAMP HTi basalts (Table 1).

In addition to measurements of aerosols at active volcanoes and in ice cores [*Pyle et al.,* 1996], several studies compare volatile contents of source dikes and magmatic glass inclusions with volatiles in the lavas from those dikes. Martin [1996] described sulfur in several dikes and flows of the Wanapum basalt of the Columbia River basalt group, with original dike S contents of about 300 to 2800 ppm in source dikes dropping to 70 to 590 ppm in corresponding flows. The average of 60 to 70 % loss was

primarily at the fissure vents, with up to 2×10^9 metric tons of SO$_2$ emitted per eruptive event. A study of Columbia River basalt group flows by Thordarson and Self [1996] suggests that vent emissions represented about 90 % of the original magmatic sulfur (as SO$_2$), 37 % of the chlorine (as HCl), and 30 % of the fluorine (as HF). Similarly, Thordarson and others [1996] found that emissions at the Laki, Iceland eruption included over 85 % of the original magmatic S, 50 % of the original Cl and F, and 80 % of the original CO$_2$.

Because there are no studies comparing volatiles of CAMP source dikes with volatiles in their resultant lava flows, a conservative estimate of 70 % release of S, CO$_2$, and H$_2$O, and 50 % release of F and Cl from basalt type averages in Table 1 is used for calculations of CAMP volcanic emissions in Table 2. Surface lava proportions of 70 % ITi, 20 % LTi, and 10 % HTi were assumed for the purpose of calculating weighted total emissions. Magmatic densities for the three types were calculated in Table 1, and a weighted average of 2.655 tons/m^3 was used in converting volumes into basalt mass units. The total calculated volatile emissions by basalts of the CAMP range from 1.11 $\times 10^{12}$ metric tons for F, to 5.19×10^{12} tons for CO$_2$ (Table 2). As discussed below, this total is for all volcanic events

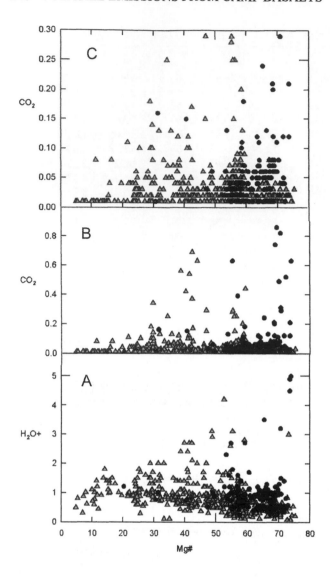

Figure 3. Mg# (100xMg/Mg+Fe*) vs.H_2O+ (A) as water bound in mineral structures, and CO_2 (B and C, scale change) as mainly in carbonate minerals, for LTi and HTi CAMP basalts of the eastern USA. Symbols are explained in Fig. 2.

spread over the entire age range of the CAMP, much of which postdates the Tr-J mass extinction.

DISCUSSION

Sulfur Emissions

The enormous scale for sulfur ejections from the CAMP encourages an assumption that they caused at least several drastic cooling events that lasted a few seasons or years each, as caused by higher optical densities of the upper atmosphere in proportion to historic eruptions (*Stothers et*

al., 1986). Cooling events would have been repeated for each episode of CAMP volcanism, with gaps between the episodes of perhaps centuries to many thousands of years. Sulfur-based aerosols diminish rapidly within a year to several years [*Pyle et al.*, 1996]. A fissure eruption on the scale of CAMP dikes would be active for months or years [*Thordarson and Self*, 1996], and could overlap in time with other eruptions across the province to cause longer periods of sulfur injections into the atmosphere. The actual amount of cooling can only be quantified by determining the precise duration and volume of each CAMP magmatic event. Alternately, evidence for or against cooling may become available through new studies of organic fossils, sediment isotope chemistry, or Milankovitch cyclo-stratigraphy.

Although the total CAMP sulfur emission of 2.31×10^{12} tons (Table 2) might imply cooling of 20 °C or more (Fig. 6), individual CAMP lava flows probably were closer in size to the Roza flow of the Columbia River basalt group [*Thordarson and Self*, 1996], which indicate releases of about 10^{10} tons of S into the atmosphere for each major fissure eruption. Such volumes, if rapidly injected during brief eruptions, could cause global cooling of between 2 and 8 °C for a few months to years if the projected extrapolation from historic eruptions is accurate in Figure 6 (essentially with the surface darkness of a moon-lit night) [*Rampino et al.*, 1988]. In addition, it is not clear whether very large volumes of sulfide aerosols are likely to cause

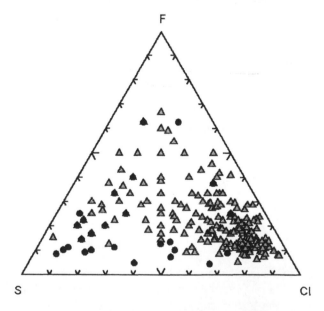

Figure 4. Ternary plot of sulfur, fluorine, and chlorine in LTi and ITi CAMP basalts of the eastern USA. Symbols are explained in Fig. 2.

the extreme temperature declines of Figure 6. The largest CAMP eruptive event may be represented by IPR lavas of northeastern North America, which covered an area about 20 times larger than the Roza flow, but only about 3 times its volume [*McHone*, 1996]. Thus, temperature declines between 2 °C and 10 °C are reasonable expectations.

Carbon Dioxide Emissions

Long-term Holocene climatic temperature changes have been closely correlated with atmospheric CO_2, especially from studies of Antarctic ice cores [EPA, 1990]. During the past 160,000 years, local Antarctic temperatures varied from -8 °C to +2 °C relative to modern temperatures as CO_2 varied from 190 to 305 ppmv (parts per million-volume). Such observations have led to suggestions that extreme global heating could have resulted from massive CO_2 emissions by flood basalts, with an example of Deccan Traps (India) volcanism causing or contributing to the K-T mass extinction [ex. *Officer and Drake*, 1985, among others]. Even a few degrees of rapid increase in climate temperatures could disrupt reproduction in reptilian animals, as well as cause other major life-cycle stress in terrestrial and marine fauna and flora.

Caldeira and Rampino [1990] estimated a total emission of CO_2 from the Deccan basalts of 2.6 to 8.8 x 10^{12} metric tons, which is comparable to CAMP CO_2 calculations in Table 2. They note that the solubility of CO_2 in basaltic magma at surface pressure is only about 0.03 weight %, so that the estimate of original Deccan basalt CO_2 of 0.2 % [*Leavitt*, 1982] requires the loss of up to 80% of that volatile during volcanic activity. However, when calculations include buffering effects from the oceans, plants, and inorganic weathering, volcanic emission time spans of 10 kyr to 500 kyr result in climatic temperature increases of only 0.7 to 0.1 °C from Deccan CO_2 [*Caldeira and Rampino*, 1990. Fig. 1].

More recently, McElwain and others [1999] studied changes in fossil plant stomata from samples around the Tr-J extinction boundary, which they interpret to indicate a 4-fold increase in atmospheric CO_2 at the boundary. Such a change might correspond to a greenhouse warming of 3 to 4 °C, which would be widely lethal to many plants and animals, and McElwain and others [1999] suggest a relationship to CAMP volcanic CO_2 emissions. The uncertainty of exact timing for particular volumes of CAMP volcanism remains a problem for modeling cause and effect, but assumptions for buffering effects as used in the calculations of Caldeira and Rampino [1990] are also imprecise. Rapid eruptions might temporarily overwhelm

some buffers, or there may be triggers of reinforcing events such as releases of marine methane.

Halogen Emissions

Fluorine poisoning of livestock and acid damage to crops led to severe famine as a direct result of the Laki, Iceland eruption of 1783-1784 [*Grattan and Charman*, 1994]. Similar fluorine emission problems continue to be a major concern during modern eruptions in Iceland.

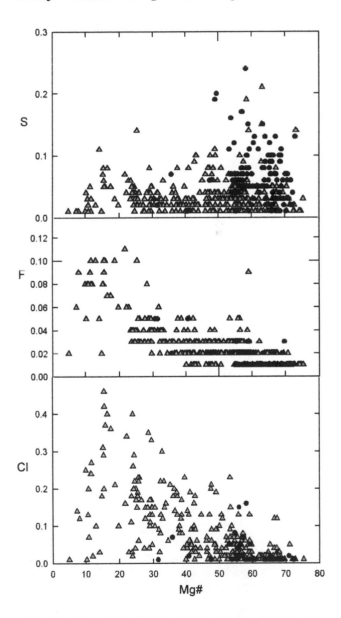

Figure 5. Mg# (100xMg/Mg+Fe*) vs. chlorine, fluorine, and sulfur in LTi and ITi basalts of the eastern USA. Symbols are explained in Fig. 2.

Table 2. Sizes, Volumes, and Emissions Estimated for CAMP

BASALT LAVAS	Basin	Basalt Type	Area km2	Avg. T km	Vol. km3
	Fundy	ITi	22500	0.4	9000
	Hartford	ITi/LTi	4500	0.3	1350
	Newark	ITi/LTi	5600	0.3	1680
	Gettysburg	ITi/LTi	2400	0.1	240
	Culpeper	ITi/LTi	22500	0.2	4500
	SoGeorgia	LTi	100000	0.2	20000
	Argana	ITi	70000	0.2	14000
	Offshore	LTi	100000	0.1	10000
	non-basin	ITi/LTi	20000	0.1	2000
LAVA TOTALS			347500		62770

BASALT SILLS	Region	Basalt Type	Area km2	Avg. T km	Vol. km3
	NE USA	ITi	17500	0.2	3500
	SE USA	LTi	2000	0.2	400
	Africa	ITi	150000	0.3	45000
	SoAmerica	ITi	1000000	0.5	500000
SILL TOTALS			1169500		548900

BASALT DIKES	Number	Length km	Total Length	Depth km	Width km	Vol. km3
Very long dikes	10	500	5000	50	0.05	12500
Long dikes	20	200	4000	50	0.04	8000
Medium dikes	100	50	5000	50	0.02	5000
Short dikes	300	20	6000	50	0.01	3000
DIKE TOTALS	430		20000			28500

EAST COAST MARGIN I.P.	Length km	Width km	Thick. km	Area km2	Vol. km3	Aerial Vol.
	2000	55	25	110000	2750000	1375000

EMISSIONS	No. Amer.	ECMIP	Europe	Africa	So Amer.	Total CAMP
CAMP area km2	1400000	1100000	700000	4500000	3500000	11210000
Lava vol km3 (1)	140000	1375000	70000	450000	350000	2385000
Lava mass tons (2)	3.72E+14	3.65E+15	1.86E+14	1.20E+15	9.30E+14	6.34E+15
avg CO2, wt.% (3)	0.117	0.117	0.117	0.117	0.117	0.117
Total CO2 emission (4)	3.05E+11	2.99E+12	1.52E+11	9.80E+11	7.62E+11	5.19E+12
avg S, wt.% (3)	0.052	0.052	0.052	0.052	0.052	0.052
Total S emission (4)	1.35E+11	1.33E+12	6.77E+10	4.35E+11	3.39E+11	2.31E+12
avg F, wt.% (3)	0.035	0.035	0.035	0.035	0.035	0.035
Total F emission (4)	6.51E+10	6.40E+11	3.26E+10	2.09E+11	1.63E+11	1.11E+12
avg Cl, wt.% (3)	0.050	0.050	0.050	0.050	0.050	0.050
Total Cl emission (4)	9.30E+10	9.14E+11	4.65E+10	2.99E+11	2.33E+11	1.58E+12
avg H2O+, wt.% (3)	0.823	0.823	0.823	0.823	0.823	0.823
Total H2O emission (4)	2.14E+12	2.11E+13	1.07E+12	6.89E+12	5.36E+12	3.65E+13

Notes: (1) assuming 1/2 CAMP area is covered with 0.2 km of lava (except ECMIP = 12.5 km); (2) weighted average density of 2.658 metric tons/m3; (3) weighted averages (see text); (4) in metric tons, with proportions as in the text.

Large ejections of halogens may also affect atmospheric chemistry, leading to ozone depletion [*Sigurdsson*, 1990]. Gaseous halogens will be converted in the atmosphere to halides such as HCl, and rainfall made acidic by Laki HCl and HF (in addition to H_2SO_4) may also have caused severe crop and tree damage in Great Britain [*Grattan and Charman*, 1994]. However, these effects must be temporary because halide aerosols are rapidly dissipated by

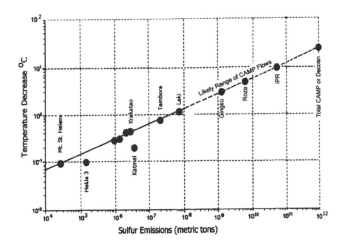

Figure 6. Volcanic sulfur vs. climatic temperature decrease, adapted from Palais and Sigurdsson [1989]. The dashed line is extrapolated from the labeled historic eruptions, and it may exaggerate the hypothetical temperature decrease. The Gingko and Roza flow data are from the Columbia River basalt group [*Martin*, 1996; *Thordarson and Self*, 1996], and the IPR basalt of northeastern North America is calculated after McHone [1996].

rainfall, and serious effects may be confined to local regions down-wind from the fissures.

Summary

Table 3 summarizes the environmental effects of volcanism. Note that the time scales involved (a few months to a few millennia) remain far smaller than age uncertainties for CAMP events (hundreds of millennia). Although the amounts of injected materials can only be estimated, and the eruptive time scales are speculative, the volcanic

plume mechanisms and environmental effects are well documented in historic eruptions and studies of ice cores.

Many fossils of plants and animals, including a limited variety of tracks from the earliest Jurassic dinosaurs, are found in strata deposited between major CAMP lava flows in basins of the northern CAMP [*Olsen*, 1997]. Thus, within a few hundred thousand years after the Tr-J boundary extinction, new populations of some species had already spread across Pangaea, and they survived several subsequent volcanic events of the CAMP that are represented in the northern basins.

CONCLUSION

The enormous scale for atmospheric injections of aerosols from CAMP volcanic fissure eruptions makes it likely, although unproven, that they caused both short-term cooling events for several years or decades each, and longer-term heating events for hundreds to thousands of years. Destruction of habitat by lavas and ash falls, changes in rainfall patterns, and poisoning of plants and animals from halide precipitation were major environmental problems on a regional scale. There is also some likelihood for extensive wildfires across Pangaea from the long fissures, similar to the proposed effect from bolide impacts.

The most extreme problems that actually caused the Tr-J mass extinction must predate a large portion of the CAMP volcanic activity, as shown by basin stratigraphy in the northern CAMP areas. Unless new stratigraphic evidence in the southern regions or in buried sections of the CAMP can be found to correlate the extinction more exactly with volcanism, a cause-and-effect link must rely on better precision in radiometric dates.

Table 3. Summary of Environmental Effects of CAMP Volcanism

Feature	Mechanism	Timescale	Geography	Evidence	Effect
Halogens (mainly Cl and F)	Exsolution from magma at vent	Months to years	Local to regional	Paleontology	Poison fauna and flora
Ash and dust	Injection into atmosphere	Months to years	Regional	Stratigraphy	Light block and cooling
Water	Exsolution from magma	Months to years	Regional to global	Sedimentology, Paleontology	Wetter climate
Sulfur Dioxide	Exsolution from magma	Years to decades	Hemispheric to global	Acidic Leaching	Light block and cooling
Carbon Dioxide	Exsolution from magma	Centuries	Global	Plant stomata	Climatic warming
Lava flows	Fissure eruption	Millennia	Local to regional	Stratigraphy	Habitat destruction

Acknowledgments. Correlations and groupings of CAMP basalts are based on many years of interaction with Paul Ragland, the instrumental authority in this field. My enthusiasm for attributing the Tr-J mass extinction to CAMP volcanism has been tempered by the important work of Paul Olsen and his colleagues, whose input I greatly appreciate. Our knowledge of CAMP volatiles is due to the foresight of David Gottfried, Al Froelich, and others at the U. S. Geological Survey. An anonymous reviewer suggested important corrections to the manuscript.

REFERENCES

Aruscavage, P. J. and E. Y. Campbell, An ion selective electrode method for the determination of chlorine in geological material, *Talanta, 30*, 745-749, 1983.

Austin, J.A., Jr., P. L. Stoffa, J. D. Phillips, J. Oh, D. S. Sawyer, G. M. Purdy, E. Reiter, and J Makris, Crustal structure of the southeast Georgia embayment-Carolina trough: Preliminary results of a composite seismic image of a continental suture (?) and a volcanic passive margin, *Geology, 18,* 1023-1027, 1990.

Baksi, A. K., and D. A. Archibald, Mesozoic igneous activity in the Maranhão province, northern Brazil: $^{40}Ar/^{39}Ar$ evidence for separate episodes of basaltic magmatism. *Earth Planet. Sci. Lett,. 151,* 139-153, 1997.

Bellieni, G., M.H.F. Macedo, R. Petrini, E. M. Picrillo, G. Cavazzini, P. Comin-Chiaramonti, M. Ernesto, J. W. P. Macedo, G. Martins, A. J. Melfi, I. G. Pacca, and A. De Min, Evidence of magmatic activity related to Middle Jurassic and Lower Cretaceous rifting from northeastern Brazil (Ceara-Mirim): K/Ar age, paleomagnetism, petrology, and Sr-Nd isotope characteristics, *Chem. Geol, 9,* 9-32, 1992.

Bertrand, H., The Mesozoic tholeiitic province of northwest Africa: A volcano-tectonic record of the early opening of the central Atlantic, in *The Phanerozoic African Plate*, edited by A. B. Kampunzo and R. T. Lubala, pp. 147-191, Springer-Verlag, New York, 1991.

Bottinga, Y., and D. F. Weill, Densities of liquid silicate systems calculated from partial molar volumes of oxide components, *Am. J. Sci., 269,* 169-182, 1970.

Caldeira, K., and M. R. Rampino, Carbon dioxide emissions from Deccan volcanism and a K/T boundary greenhouse effect, *Geophys. Res. Lett., 17,* 1299-1302, 1990.

Caroff, M., H. Bellon, L Chauris, J.-P. Carron, S. Chevrier, A. Gardinier, J. Cotten, Y. Le Moan, and Y. Neidhart, Magmatisme fissural Triasico-Liasique dans l'ouest du Massif Armoricain (France): Petrologie, age, et modalities des la mise en place, *Can. J. Earth Sci. 32,* 1921-1936, 1995.

Choudhuri, A., Geochemical trends in tholeiite dikes of different ages from Guyana, *Chem. Geol., 22,* 79-85, 1978.

Courtillot, V. E., Mass extinctions in the last 300 million years: One impact and seven flood basalts? *Isr. J. Earth Sci., 43,* 255-266, 1994.

Courtillot, V. E., J. Jaeger, Z. Yang, G. Féraud, and C. Hofman, The influence of continental flood basalts on mass extinctions: Where do we stand? in *The Cretaceous-Tertiary Event and Other Catastrophes in Earth History*, edited by G. Ryder and others, pp. 513-525, *Geol. Soc. Am. Special Paper, 307,* 1996.

Courtillot, V. E., Y. Gallet, R. Rocchia, G Féraud, E. Robin, C Hofmann, N. Bhandari, and Z. G. Ghevariya, Z.G., Cosmic markers, $^{40}Ar/^{39}Ar$ dating and paleomagnetism of the K/T sections in the Anjar Area of the Deccan large igneous province, *Earth Planet. Sci. Lett., 182,* 137-156, 2000.

Deckart, K., G. Féraud, and H. Bertrand, Age of Jurassic continental tholeiites of French Guyana/Surinam and Guinea: Implications to the initial opening of the central Atlantic Ocean, *Earth Planet. Sci. Lett., 150,* 205-220, 1997.

Dunning, G. R., and J. P. Hodych, U/Pb zircon and baddeleyite ages for the Palisades and Gettysburg sills of the northeastern United States: Implications for the age of the Triassic/Jurassic boundary, *Geology, 18,* 795-798, 1990.

DuPuy, C., J. Marsh, J. Dostal, A. Michard, and S. Testa, Aesthenospheric and lithospheric sources for Mesozoic dolerites from Liberia: Trace element and isotopic evidence, *Earth Planet. Sci. Lett., 87,* 100-110, 1988.

Engleman, E. E., L. L. Jackson, and D. R. Norton, Determination of carbonate carbon in geological materials by colourmetric titration, *Chem. Geol., 53,* 125-128,1985.

Fiechtner, L., H. Friedrichsen, and K. Hammerschmidt, K., Geochemistry and geochronology of Early Mesozoic tholeiites from Central Morocco, *Geol. Rundschau, 81,* 45-62, 1992.

Gottfried, D., A. J. Froelich, and J. N. Grossman, Geochemical data for Jurassic diabase asociated with Early Mesozoic basins in the eastern United States: Geologic setting, overview, and chemical methods used, *U. S. Geol. Survey, Open-File Report, 91-322-A,* 1991.

Grattan, J. P. and D. J. Charman, Non-climatic factors and the environmental impact of volcanic volatiles: implications of the Laki fissure eruption of AD 1783, *The Holocene, 4(1),* 101-106, 1994.

Grossman, J. N., D. Gottfried, and A. J. Froelich, Geochemical data for Jurassic diabase associated with Early Mesozoic basins in the eastern United States, *U. S. Geol. Survey, Open-File Report, 91-322-K,* 1991.

Gudmundsson, A., L. B. Marinoni, and J. Marti, Injection and arrest of dykes: implications for volcanic hazards, *J. Volc. Geotherm. Res., 88,* 1-13, 1999.

Hames, W. E., P. R. Renne, and C. Ruppel, New evidence for geologically instantaneous emplacement of earliest Jurassic Central Atlantic magmatic province basalts on the North American margin, *Geology, 28,* 859-862, 2000.

Hodych, J. P., and G. R. Dunning, Did the Manicouagan impact trigger end-of-Triassic mass extinction? *Geology, 20,* 51-54, 1992.

Holbrook, W. S., and P. B. Kelemen, Large igneous province on the U.S. Atlantic margin and implications for magmatism during continental breakup, *Nature, 364*, 433-436, 1993.

Kirschenbaum, H., The classical chemical analysis of rocks – the old and the new, *U. S. Geol. Survey Bull., 1547*, 55 p., 1983.

Kirschenbaum, H., The determination of fluoride in silicate rocks by ion selective electrode: an update, *U. S. Geol. Survey Bull., 88-588*, 5 p., 1988.

Klitgord, K.D., and H. Schouten, Plate kinematics of the central Atlantic, in *The Geology of North America*, vol. M, *The Western North Atlantic Region*, edited by P. R. Vogt and B. E. Tucholke,.pp. 351-378, Geol. Soc. Am., Boulder, Colo., 1986.

LeBas, M. J., R. W. LeMaitre, A. Streckeisen, and B. Zanettin, A chemical classification of volcanic rocks based on the total alkali silica diagram, *J. Petrol., 27*, 745-750, 1986.

Leavitt, S. W., Annual volcanic carbon dioxide emission: an estimate from eruption chronologies, *Environ. Geol., 4*, 15-21, 1982.

Manspeizer, W., Triassic-Jurassic rifting and opening of the Atlantic: An overview, in *Triassic-Jurassic Rifting*, edited by W. Manspeizer, pp. 41-79, New York, Elsevier, 1988.

Martin, B. S., Sulfur in flows of the Wanapum basalt formation, Columbia River basalt group: Implications for volatile emissions accompanying the emplacement of large igneous provinces (abstract), *Geol. Soc. Am. Abs. with Prog., 28 (7)*, A419, 1996.

Marzoli, A., P. R. Renne, E. M. Piccirillo, M Ernesto, G. Bellieni, and A. De Min, Extensive 200 million-year-old continental flood basalts of the central Atlantic magmatic province, *Science, 284*, 616-618, 1999.

Mauche, R., G. Faure, L. M. Jones, and J. Hoefs, 1989, Anomalous isotopic compositions of Sr, Ar, and O in the Mesozoic diabase dikes of Liberia, West Africa: *Contrib. Min. Petrol.., 101*, 12-18, 1989.

McElwain, J. C., D. J. Beerling, and F. I. Woodward, Fossil plants and global warming at the Triassic-Jurassic boundary, *Science, 285*, 1386-1390, 1999.

McHone, J. G., Broad-terrane Jurassic flood basalts across northeastern North America, *Geology, 24*, 319-322, 1996.

McHone J. G., Non-plume magmatism and rifting during the opening of the Central Atlantic Ocean, *Tectonophysics, 316*, 287-296, 2000.

McHone, J. G., and J. H. Puffer, Flood basalt provinces of the Pangaean Atlantic rift: Regional extent and environmental significance, in *The Great Rift Valleys of Pangea in Eastern North America*, edited by P. M. LeTourneau and P. E. Olsen, Columbia University Press, New York, in press, 2002.

Montes-Lauar, C. R., I. G. Pacca, A. J. Melfi, E. M. Piccirillo, G. Bellieni, R. Petrini, and R. Rizzieri, The Anari and Tapirapua Jurassic formations, western Brazil: paleomagnetism, geochemistry, and geochronology, *Earth Planet. Sci. Lett., 128*, 357-371, 1994.

Mossman, D.J., R. G. Grantham, and F. Langenhorst, A search for shocked quartz at the Triassic-Jurassic boundary in the Fundy and Newark basins of the Newark Supergroup, *Can. J. Earth Sci., 35*, 101-109, 1998.

Officer, C. B., and C. L. Drake, Terminal Cretaceous environmental events, *Science, 227*, 1161-1167, 1985.

Oh, Jinyong, J. A. Austin, Jr., J. D. Phillips, M. E. Coffin, and P. L. Stoffa, Seaward-dipping reflectors offshore the southeastern United States: Seismic evidence for extensive volcanism accompanying sequential formation of the Carolina trough and Blake Plateau basin. *Geology, 23*, 9-12, 1995.

Oliveira, E. P., J. Tarney, and X. J. Joao, Geochemistry of the Mesozoic Amapa and Jari dyke swarms, northern Brazil: Plume-related magmatism during the opening of the central Atlantic, in *Mafic dikes and emplacement mechanisms*, edited by A. J. Parker, P. C. Rickwood, and D. H. Tucker, pp. 173-183, Balkemia, Rotterdam, Netherlands, 1990.

Olsen, P. E., Stratigraphic record of the early Mesozoic breakup of Pangea in the Laurasia-Gondwana rift system, *Ann. Rev. Earth Planet. Sci., 25*, 337-401, 1997.

Olsen, P. E., Giant lava flows, mass extinctions, and mantle plumes, *Science, 284*, 604-605, 1999.

Palais, J. and H. Sigurdsson, Petrologic evidence of volatile emissions from major historic and pre-historic volcanic eruptions, *Am. Geophys. Union Monograph, 52*, 31-56, 1986.

Pálfy, J., J. K. Mortensen, E. S. Carter, P. L. Smith, R. M. Friedman, and H. W. Tipper, Timing the end-Triassic mass extinction: First on land, then in the sea? *Geology, 28*, 39-42, 2000.

Peate, D. W., and C. J. Hawkesworth, C.J., Lithospheric to asthenospheric transition in low-Ti flood basalts from southern Paraná, Brazil, *Chem. Geol., 127*, 1-24, 1996.

Pe-Piper, G., and D. J. W. Piper, Were Jurassic tholeiitic lavas originally widespread in southeastern Canada?: a test of the broad terrane hypothesis, *Can. J. Earth Sci., 36*, 1509-1516, 1999.

Philpotts, A. R. and A. Martello, Diabase feeder dikes for the Mesozoic basalts in southern New England, *Am. Jour. Sci., 286*, 105-126, 1986.

Puffer, J. H., Initial and secondary Pangaean basalts, in *Pangaea: Global Environments and Resources*, Can. Soc. Petrol. Geol. Mem., *17*, 85-95, 1994.

Puffer, J. H., Eastern North American flood basalts in the context of the incipient breakup of Pangaea, in *Eastern North American Mesozoic Magmatism*, edited by J. H. Puffer and P. C. Ragland, pp. 95-118, *Geol. Soc. Am. Sp. Paper, 268*, 1992.

Pyle, D. M., P. D. Beattie, and G. J. S. Bluth, Sulphur emissions to the stratosphere from explosive volcanic eruptions, *Bull. Volcanol. 57*, 663-671, 1996.

Ragland, P. C., L. E. Cummins, and J. D. Arthur, Compositional patterns for Early Mesozoic diabases from South Carolina to central Virginia, in *Eastern North American Mesozoic Mag-*

matism, edited by J. H. Puffer and P. C. Ragland, pp. 309-332, *Geol. Soc. Am. Sp. Paper, 268*, 1992.

Rampino, M. R., S. Self, and R. B. Stothers, Volcanic winters, *Ann. Rev. Earth Planet. Sci., 16*, 73-99, 1988.

Rampino, M. R., and R. B. Stothers, R.B., Flood basalt volcanism during the past 250 million years, *Science, 241*, 663-668, 1988.

Schlager, W., R. T. Buffler, and 15 others, Deep Sea Drilling Project, Leg 77, southeastern Gulf of Mexico, *Geol. Soc. Am. Bull., 95*, 226-236, 1984.

Sebai, A., G. Féraud, H. Bertrand, and J. Hanes, ^{40}Ar/^{39}Ar dating and geochemistry of tholeiitic magmatism related to the early opening of the central Atlantic rift, *Earth Planet. Sci. Lett., 104*, 455-472, 1991.

Shapiro, L., Rapid analysis of silicate, carbonate, and phosphate rocks – revised edition, *U. S. Geol. Survey Bull., 1401*, 76 p., 1975.

Sigurdsson, H., Assessment of the atmospheric impact of volcanic eruptions, in *Global Catastrophes in Earth History*, edited by V. L. Sharpton and P. D. Ward, *Geol. Soc Am. Sp. Paper, 247*, 99-110, 1990.

Stothers, R. B., Flood basalts and extinction events, *Geophys. Res. Lett. 20*, 1399-1402, 1993.

Stothers, R. B., J. A. Wolff, S. Self, and M. R. Rampino, Basaltic fissure eruptions, plume heights, and atmospheric aerosols, *Geophys. Res. Lett., 13*, 725-728, 1986.

Sundeen, D. A., Note concerning the petrography and K-Ar age of Cr-spinel-bearing olivine tholeiite in the subsurface of Choctaw County, north-central Mississippi, *Southeast. Geol., 30*, 137-146, 1989.

Sutter, J. F., Innovative approaches to the dating of igneous events in the early Mesozoic basins of the eastern United States, in *Studies of the Early Mesozoic basins of the eastern United States*, edited by A. J. Froelich and G. R. Robinson, Jr., pp. 194-199, *U. S. Geol. Survey Bull., 1776*, 1988.

Thordarson, Th., and S. Self, Sulfur, chlorine, and fluorine degassing and atmospheric loading by the Roza eruption, Columbia River Basalt Group, Washington, USA, *J. Volcanol. Geotherm. Res., 74*, 49-73, 1996.

Thordarson, Th., S. Self, N. Óskarsson, and T. Hulsebosch, Sulfur, chlorine, and fluorine degassing and atmospheric loading by the 1783-1784 AD Laki (Skaftár fires) eruption in Iceland, *Bull. Volcano., 58*, 205-225, 1996.

Wade, J. A., D. E. Brown, A. Traverse, and R. A. Fensome, R.A., The Triassic-Jurassic Fundy Basin, eastern Canada: regional setting, stratigraphy, and hydrocarbon potential, *Atlantic Geol., 32*, 189-231, 1996.

Warner, R. D., D. S. Snipes, S. S. Hughes, J. C. Steiner, M. W. Davis, P. R. Manoogian, and R. A. Schmitt, Olivine-normative dolerite dikes from western South Carolina: Mineralogy, chemical composition and petrogenesis, *Contr. Min. Petrol., 90*, 386-400, 1985.

Weigand, P. W., and P. C. Ragland, Geochemistry of Mesozoic dolerite dikes from eastern North America, *Contr. Min. Petrol., 29*, 195-214, 1970.

Wignall, P. B., Large igneous provinces and mass extinctions, *Earth-Sci. Rev., 53*, 1-33, 2001.

Withjack, M. O., R. W. Schlische, and P. E. Olsen, Diachronous rifting, drifting, and inversion on the passive margin of central eastern North America: An analog for other passive margins, *Am. Assoc. Petrol. Geol. Bull., 82*, 817-835, 1998.

Wilson, M., Thermal evolution of the Central Atlantic passive margins: Continental break-up above a Mesozoic super-plume, *J. Geol. Soc. London, 154*, 491-495, 1997.

Woods, A. W., A model of the plumes above basaltic fissure eruptions, *Geophys. Res. Lett., 20*, 1115-1118, 1993.

J. Gregory McHone, Earth Science Education and Research, PO Box 647, Moodus, CT 06469-0647.

Volcanism of the Central Atlantic Magmatic Province as a Potential Driving Force in the End-Triassic Mass Extinction

József Pálfy

Hungarian Natural History Museum, Department of Geology and Paleontology, Budapest, Hungary

Radiometric dating suggests that eruptions in the Central Atlantic magmatic province (CAMP) are synchronous with the ~200 Ma end-Triassic mass extinction. Although stratigraphic evidence for major flows prior to the extinction horizon is still lacking, the vast extent of the province allows the assumption of cause-and-effect relationship between volcanism and extinction, mediated by drastic environmental change. A recently recognized negative carbon isotope anomaly at the Triassic–Jurassic boundary is interpreted to reflect combined effects of volcanically derived CO_2 input, methane release through dissociation of gas hydrates in a global warming episode, and a possible marine productivity crisis. Maximum duration of the Rhaetian stage is estimated as only 2 m.y., and the isotope event appears short, lasting for less than 100 k.y. A variety of marine and terrestrial fossil groups (e.g., radiolarians, corals, bivalves, and plants) experienced correlated and sudden extinction at the end of Triassic, although some groups (e.g., ammonoids and conodonts) underwent a prolonged period of declining diversity. Post-extinction faunas and floras are cosmopolitan. Biotic recovery was delayed and the earliest Hettangian is a lag phase characterized by low diversity, possibly due to sustained environmental stress. The hypothesis of CAMP as the principal driving force in the end-Triassic extinction appears more consistent with paleontological and isotopic observations than alternative models. The temporally adjacent large igneous provinces, the Siberian Traps at the Permian–Triassic boundary and the Early Jurassic Karoo–Ferrar province, are also linked to extinction events, albeit of differing magnitude.

INTRODUCTION

The Triassic–Jurassic (Tr–J) transition was a remarkable time in earth history: the end-Triassic extinction was one of the five most severe mass extinctions in the Phanerozoic, and broadly coeval volcanism led to the formation of the Central Atlantic magmatic province (CAMP), one of the most extensive large igneous provinces (LIP), which also

The Central Atlantic Magmatic Province:
Insights from Fragments of Pangea
Geophysical Monograph 136
Copyright 2003 by the American Geophysical Union
10.1029/136GM014

heralded the breakup of Pangea. Although the correlation of several extinction events and flood basalt provinces was noted some time ago [*Rampino and Stothers*, 1988; *Courtillot*, 1994], it was only recently that improved radiometric dating of several LIPs better constrained their timing and synchrony with extinction events [e.g., *Renne et al.*, 1995; *Pálfy and Smith*, 2000; *Wignall*, 2001]. Similarly, a large set of new isotopic dates from the CAMP [*Marzoli et al.*, 1999] agrees within error with the age of the marine extinction at the Tr–J boundary [*Pálfy et al.*, 2000a]. The recognition of the vast areal extent and volume of the CAMP added strength to the proposition of causal link between this volcanic episode and the extinction [e.g.,

Olsen, 1999], which is nevertheless far from being unanimously accepted. Alternative explanations, such as anoxia [Hallam, 1995], sea level change [Hallam and Wignall, 1999], and bolide impact [Ward et al., 2001; Olsen et al., 2002a] continue to be favoured by some workers.

This contribution attempts to summarize the available evidence for the case of volcanically induced environmental change in the terminal Triassic, which could have triggered a mass extinction in both the marine and terrestrial ecosystems. As a necessary starting point for discussion, a short review of radiometric dating of the Tr–J boundary and the CAMP volcanics is given in support of the case for their synchrony. Proxy indicators of environmental change, such as stable isotopic and strontium isotopic records across the Tr–J boundary, are also reviewed. The duration of the terminal Triassic stage, the Rhaetian, is assessed, because a temporal framework is necessary to distinguish between competing hypotheses. Predictions of the volcanic forcing model are tested against the observed extinction patterns across the Tr–J boundary. Finally, the alternative models for the cause of end-Triassic extinction are briefly discussed, and this event is compared to others for which a link between LIP formation and biotic extinction has been proposed or established.

RADIOMETRIC DATING EVIDENCE FOR SYNCHRONY OF EXTINCTION AND VOLCANISM

Dating the Triassic–Jurassic Boundary

Until recently, the numeric age of the Tr–J boundary was poorly constrained. The lack of available radiometric ages with reasonable stratigraphic control necessitated the estimation of boundary age through interpolation between distant Late Triassic and Early Jurassic tie-points. Successive estimates ranged from 213 Ma [Harland et al., 1982] through 208 Ma [Harland et al., 1990] to 205.4 Ma [Gradstein et al., 1994]. A concerted effort to recalibrate the Jurassic time scale yielded several biochronologically constrained U–Pb ages from the topmost Triassic Rhaetian as well as the basal Jurassic Hettangian stage [Pálfy et al., 2000b]. Notably, a tuff layer immediately below the sharp change in radiolarian faunas marking the Tr–J boundary in a marine section in the Queen Charlotte Islands yielded a U–Pb age of 199.6±0.4 Ma, which effectively provided a direct estimate of the system boundary age in the marine realm [Pálfy et al., 2000a]. This age is supported by a set of three other U–Pb ages from the Rhaetian and five from the Hettangian, which, taken together, also suggest that the boundary lies near 200 Ma [Pálfy et al., 2000a, b].

Dating the CAMP

Major advances have been made recently in radiometric dating of CAMP. In fact it was the dating effort that led to the recognition of the vast extent of CAMP, allowing the inclusion of substantial tracts of volcanics from South America [Deckart et al., 1997; Marzoli et al., 1999]. Demonstrating a short, single magmatic episode permitted to establish the CAMP as a LIP ranking among the largest ones in the Phanerozoic [Marzoli et al., 1999; Olsen, 1999]. A set of 41 dates (mostly ^{40}Ar/^{39}Ar ages) cluster around 199.0±2.4 Ma indicating the age of peak volcanic activity [Marzoli et al., 1999].

The northern segment of CAMP received more attention from geochronologists, as extrusive and subvolcanic rocks within the Newark Supergroup have been dated from numerous localities. Key U–Pb dates were obtained from the lowermost flow unit, the North Mountain Basalt (201.7+1.4/–1.1 Ma, Hodych and Dunning [1992]) and an equivalent feeder dike system (Palisades sill: 200.9±1.0 Ma; Gettysburg sill: 201.3±1.0 Ma, Dunning and Hodych [1990]). Hames et al. [2000] obtained very similar ^{40}Ar/^{39}Ar ages of 201.1±2.1 and 198.8±2.0 Ma from the Watchung flows. They also report the first set of ^{40}Ar/^{39}Ar ages from dikes in South Carolina, which yield a weighted average age of 199.5±2.0 Ma. The increased precision of radiometric ages provides mounting evidence that the bulk of CAMP may represent a brief volcanic episode, perhaps not more than 1 m.y. in duration.

A more detailed review of CAMP radiometric dates is given by Baksi [2002]. An important conclusion drawn from all relevant recent studies is that the age of CAMP volcanism and the age of the Tr–J boundary are statistically indistinguishable.

STRATIGRAPHIC RELATIONSHIP OF EXTINCTION AND VOLCANISM

The relative age of volcanic rocks may be determined from their relationship to fossiliferous sedimentary rocks. Because CAMP eruptions occurred in continental environment, direct linkage of volcanics with sediments cannot be expected to be pervasive. Nevertheless, CAMP flows are often intercalated with lacustrine strata of the Newark Supergroup in eastern North America. The lake record is best studied in the Newark Basin, where cyclostratigraphy suggests that the oldest extrusive rocks are some 20 k.y. younger than the palynologically defined Tr–J boundary [Olsen et al., 1996]. Olsen et al. [2002b] cite evidence that such relationship also holds in other basins. Clearly, if the extrusive event is demonstrably younger than the Tr–J boundary throughout the entire province, then the presently recognised basaltic volcanism could not trigger the extinction. However, a significant portion of the CAMP has either not yet been studied in detail, or has no preserved relationship with dateable sediments. The very short time lag (20 k.y.) between the Tr–J boundary and local initiation of volcanic activity in the northern segment of the province does not exclude the possibility that significant eruptions elsewhere preceded the extinction peak. Indeed, considering that the spread of radiometric ages indicate that most flood basalt provinces were active for up to several million years, this remains a strong possibility.

STABLE ISOTOPIC RECORD OF THE TRIASSIC–JURASSIC TRANSITION

Stable isotope stratigraphies, especially that of carbon, record changes in global geochemical cycles. Geological boundaries marked by mass extinction events commonly correspond to major isotopic events [*Holser et al.*, 1996], thus linking underlying environmental change to perturbations in the carbon cycle and to mass extinctions [*Hallam and Wignall*, 1997]. In the context of the CAMP as a potential trigger of the end-Triassic extinction, it is therefore crucial to assess the carbon isotopic evolution across the Tr–J boundary. Previously, the stable isotope stratigraphy of the Tr–J transition was poorly understood, with only scant evidence for a negative $\delta^{13}C$ excursion [e.g., *McRoberts et al.*, 1997; *McElwain et al.*, 1999]. A series of recent studies confirmed the existence of a significant and short-lived negative carbon isotope anomaly. Similar results obtained from the Queen Charlotte Islands in western Canada [*Ward et al.*, 2001], Hungary [*Pálfy et al.*, 2001], England, and Greenland [*Hesselbo et al.*, 2002] attest to the global nature of the isotopic event (Figure 1). Combined evidence from marine carbonate, bulk organic matter, and fossil wood suggests that both the shallow marine and atmospheric carbon reservoirs were affected. Thus the carbon anomaly is of use for global stratigraphic correlation, also between marine and terrestrial strata [*Hesselbo et al.*, 2002]. Moreover, it is possible to address the question whether the carbon isotope anomaly is directly or indirectly related to CAMP volcanism. Ward et al. [2001] argued that the negative excursion is best explained by marine primary productivity collapse, perhaps brought about by an extraterrestrial impact, similar to the better known Cretaceous–Tertiary boundary scenario. The lack of preserved 200 Ma oceanic crust precludes a crucial test for the "Strangelove ocean" with depressed productivity, i.e., the disappearance of a carbon isotopic gradient between the surface and deep ocean. However, the similar, correlative

signal in terrestrial fossil wood [*Hesselbo et al.*, 2002] suggests that the isotopic event was not restricted to the oceanic reservoir, hinting at a more far-reaching mechanism than a marine productivity crisis alone.

Large volumes of CO_2 emitted through CAMP volcanism is expected to lower the $^{12}C/^{13}C$ ratio because mantle-derived carbon is relatively light (–5‰). Modeling and mass balance calculations suggest, however, that this is not sufficient to account for the observed shift of –1.5 to –4‰ [*Kump and McArthur*, 1999; *Beerling and Berner*, 2002]. An additional mechanism capable of drastically lowering the $^{12}C/^{13}C$ ratio of both marine and atmospheric reservoirs is the release of methane through dissociation of gas hydrates. Methane, stored in hydrate form in marine sediments under specific pressure–temperature conditions, has a carbon isotopic ratio of –60‰. Warming of bottom waters, a likely effect of volcanically-induced gradual global warming, may trigger the dissociation of gas hydrates, which in turn may induce runaway greenhouse conditions. Such scenario was first developed to explain the Late Paleocene Thermal Maximum [*Dickens et al.*, 1995], and similar mechanism has been suggested to operate during other events of large and abrupt negative carbon isotope anomalies (e.g., Permo–Triassic: Krull and Retallack [2000]; Toarcian: Hesselbo et al. [2000]). The available data permit that a similar hypothesis be proposed for the Tr–J boundary, whereby CAMP volcanism may have a direct effect in providing light carbon but also act as a trigger for methane release via pushing oceanic bottom water temperatures beyond a threshold for gas hydrate stability [*Pálfy et al.*, 2001]. The resulting catastrophic climate warming is recorded in stomatal density of plants [*McElwain et al.*, 1999], which shows a pronounced decrease corresponding to a calculated fourfold increase in $p CO_2$ at the Tr–J boundary. Further support for this scenario is provided by a carbon cycle model developed by Beerling and Berner [2002], who found that the magnitude of the observed negative carbon anomaly cannot be accounted for by volcanic CO_2 emission from CAMP, nor can it be fully explained by an additional marine productivity crisis. Only the introduction of substantial amounts of isotopically light methane (–60‰) derived from dissociation of gas hydrates could reconcile the model results with the observed record.

DURATION OF THE CARBON ISOTOPE ANOMALY AND THE RHAETIAN STAGE

Duration of the isotope anomaly and extinction rates within the Rhaetian are two key issues, among others, that require we know the length of the terminal Triassic stage.

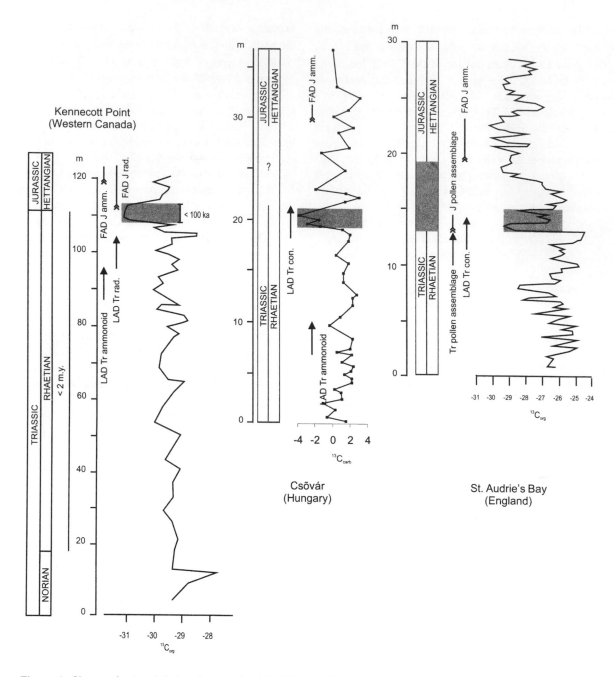

Figure 1. Changes in the global carbon cycle at the Triassic–Jurassic boundary as recorded in the carbon isotopic composition of marine carbonate and marine and terrestrial organic matter (data from *Pálfy et al.* [2001], *Ward et al.* [2001], and *Hesselbo et al.* [2002]).

Disparate estimates in recent time scales derive from interpolation within a sparse isotopic age dataset and range from 1.5 m.y. [*Harland et al.*, 1990] to 3.9 m.y. [*Gradstein et al.*, 1994]. Instead, the duration of the Rhaetian is determined here using the astronomical time scale from the Newark Basin [*Olsen and Kent*, 1999] and a definition of the Norian–Rhaetian boundary through magnetostratigraphic

correlation with fossiliferous, although discontinuous, marine Tethyan sections. Such correlation was previously believed to be insufficient [*Kent et al.*, 1995; *Muttoni et al.*, 2001] but, despite the uncertainties, this method is useful in providing an independent estimate for the duration of the Rhaetian. The magnetostratigraphy of key Tethyan sections which include the uppermost Norian (Sevatian) were studied at Scheiblkogel in the Austrian Alps [*Gallet et al.*, 1996] and Kavur Tepe in Turkey [*Gallet et al.*, 1993]. The latter one is located in a tectonically complex area, hence the possibility of 180° rotation or origin in the southern hemisphere and subsequent northerly displacement introduced ambiguity for polarity determination. A revision of the original interpretation through comparison of a recently studied companion section [*Gallet et al.*, 2000] now confirms a northern hemisphere origin (which was also tentatively suggested by Muttoni et al. [2001]). Herein a correlation is proposed between the Tethyan Norian composite magnetostratigraphy of Gallet et al. [2000] and the Newark Basin reference scale [*Kent and Olsen*, 1999] (Figure 2). Biostratigraphically the Norian–Rhaetian boundary is marked by an ammonite-rich layer in the Scheiblkogel section [*Gallet et al.*, 1996]. Thus defined, the top of Norian corresponds to a normal magnetochron which is correlated here to E22n in the Newark Basin reference scale if the composite Tethyan Norian section [*Gallet et al.*, 2000] is anchored to the Carnian–Norian boundary. This practice follows Muttoni et al. [2001] but also considers a slightly different conodont-based boundary definition [*Orchard et al.*, 2001]. Muttoni et al. [2001] tentatively arrived at a somewhat similar correlation, where the Newark magnetochron E22n was compared to penultimate normal chron in the Tethyan Upper Norian. The solution proposed here implies that the Norian–Rhaetian boundary is younger than suggested by Kent and Olsen [1999]. Existing different definitions of the Norian–Rhaetian boundary [*Dagys and Dagys*, 1994] and difficulties in palynological correlation of the Newark Basin with the Tethyan sections may explain the discrepancy. The Rhaetian so defined in the Newark Basin comprises less than five 400 k.y. long eccentricity cycles, i.e., the duration of the Rhaetian age is no more than 2 m.y.

Among the Tr–J sections with available carbon isotope data, only Kennecott Point in the Queen Charlotte Islands [*Ward et al.*, 2001] comprises the entire Rhaetian and is lithologically sufficiently monotonous to allow assumption of constant sedimention rate. Converting thickness of strata to time and taking the maximum duration of Rhaetian as 2 m.y., it is likely that the isotope anomaly lasted less than 100 k.y. (Figure 1). Although no more precise estimate is possible yet, the isotopic event (and, as discussed later, also the extinction) appears to be sudden affair.

STRONTIUM AND OSMIUM ISOTOPIC RECORD OF THE TRIASSIC–JURASSIC TRANSITION

The evolution of the strontium isotopic composition of seawater, as recorded by well-preserved marine organisms,

may also be used to track past environmental change. The ratio of marine $^{87}Sr/^{86}Sr$ is controlled by the varying influx radiogenic ^{87}Sr, a weathering product of exposed crustal rocks delivered by rivers, and the juvenile, non-radiogenic ^{86}Sr mainly derived from submarine volcanism at mid-ocean ridges [e.g., *Jones et al.*, 1994]. The first component is primarily sensitive to changes in climatic parameters, such as temperature and precipitation at low latitudes, whereas the second is largely controlled by rates of plate tectonic processes. Additionally, the Sr isotopic ratio of riverine influx is governed by the areal distribution of exposed rock types and their respective ages. Thus the sudden emplacement of continental basalts that contain no radiogenic Sr yet cover vast areas at low paleolatitude, such as

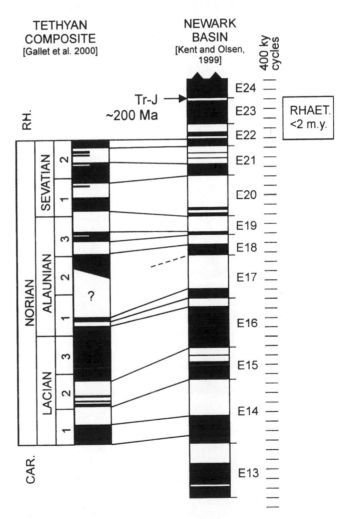

Figure 2. Duration and boundary ages of the terminal Triassic Rhaetian stage determined using integrated Late Triassic cyclostratigraphy [*Olsen and Kent*, 1999] and magnetostratigraphy [*Kent and Olsen*, 1999; *Gallet et al.*, 2000], and radiometric dating [*Pálfy et al.*, 2000a] of the Triassic–Jurassic boundary.

the forming of CAMP, is expected to significantly lower the riverine $^{87}Sr/^{86}Sr$ ratio [*Taylor and Lasaga*, 1999]. Increased weathering rates that result from volcanically-induced global warming amplify its effect on the oceanic Sr isotopic composition. Hence short and intense episodes of flood basalt volcanism are likely to be reflected in the Sr isotope record. The Late Triassic and Early Jurassic Sr isotope stratigraphy was studied in detail by Korte [1999] and Jones et al. [1994], respectively. Indeed, there appears to be a significant inflection of the curve at the Tr–J boundary. The Late Norian and Rhaetian are characterized by elevated $^{87}Sr/^{86}Sr$ ratios, although some fluctuation or noise is present in the available data. A rather monotonous, long-term decline, that characterizes the Sr curve throughout much of the Early Jurassic, starts near the system boundary. Therefore Sr isotope stratigraphy provides independent, although not unambiguous, evidence that the onset of large-scale CAMP volcanism triggered major environmental change and it coincided with the Tr–J boundary.

The osmium isotopic composition of seawater also changes through geologic time. In addition to inputs from igneous activity and continental weathering, the Os budget is also affected by meteorite flux. A further difference compared to the Sr system is the shorter oceanic residence time of Os, which results in a more rapid response to perturbations. In mudrock sequences in SW England, Cohen and Coe [2002] documented a marked decrease in the $^{187}Os/^{188}Os$ ratio coincident with the Tr–J boundary. The shift towards non-radiogenic values is accompanied by parallel increase in both the Os and Re abundances. Comparable to the Sr isotopic signature, these geochemical data are also interpreted to record the igneous activity of CAMP and the weathering of large volumes of basaltic rocks, furnishing additional evidence for CAMP-driven environmental change and its synchrony with the Tr–J boundary.

OBSERVED EXTINCTION PATTERNS OF KEY FOSSIL GROUPS

The end-Triassic extinction ranks fourth in magnitude among the major mass extinction events, with a 22% loss of marine families and 53% loss of genera, which is projected to represent the elimination of approx. 80% of marine species [*Sepkoski*, 1996]. Noteworthy caveats regarding Sepkoski's compilation are its relatively coarse time resolution and the dependence of some extinction metrics on the then accepted, but poorly constrained length of stratigraphic units. In Sepkoski's database the youngest Triassic stage is the Norian, there taken to include the Rhaetian, although the latter has subsequently been reinstated as the terminal Triassic stage [*Dagys and Dagys*, 1994]. The combined Norian–Rhaetian is undoubtedly of longer duration than most other stages. Consequently, the end-Triassic percent extinction may have been overestimated, while the extinction rate may have been underestimated, if there was indeed a short terminal Triassic event. The mass extinction status of the Tr–J event is strongly supported by its global nature and the simultaneous crisis of terrestrial organisms [*Benton*, 1995].

A clear need exists to characterize the end-Triassic biotic events at a higher resolution, at least separating Rhaetian extinctions from true Norian ones, but preferably assessing faunal and floral change at the zonal level. The taxonomic fabric of the end-Triassic crisis is summarized by Hallam [1996] and Hallam and Wignall [1997]. However, more recent research warrants the reappraisal of their conclusions in several cases. Here we limit the discussion to those groups which possess the best record, and allow at least a qualitative assessment of extinction processes at a higher temporal resolution. The key parameter is the extinction rate, with the aim of a possible distinction between sudden, dramatic loss of taxa and gradual decline.

The different tempo of biotic response to extinction forcing permit distinction between "press events" characterized by sustained stress and shorter, more abrupt "pulse events" [*Erwin*, 1998]. An interesting feature of the terminal Triassic event is that the available fossil record suggests both press and pulse event characteristics. There are clades which exhibit abrupt extinction while others apparently underwent an extended period of gradual decline. Here I discuss radiolarians, calcareous nannofossils, reef organisms, bivalves, and terrestrial plants that tipify the first group, while conodonts and ammonoids represent the latter.

Perhaps the best documented sudden turnover from highly diverse Late Triassic faunas to impoverished Early Jurassic ones occurs among the radiolarians [*Carter*, 1994; *Carter et al.*, 1998]. In one section in the Queen Charlotte Islands, western Canada, the radiolarian turnover is precisely correlated with the negative carbon isotopic anomaly [*Ward et al.*, 2001]. Radiolarian extinction is clearly global. Apart from the eastern Pacific, the radiolarian record is well known from the western Pacific [*Hori*, 1992] and the Tethys [*Tekin*, 1999]. Significantly, in all sections where high resolution data are available, the extinction is demonstrably abrupt and severe.

Calcareous nannofossils, that first appeared in the Late Triassic, were almost annihilated in the terminal Triassic event [*Bown*, 1996]. Of the eight known Rhaetian species only two survived the Tr–J boundary. Such high percent extinction among this phytoplankton group indicates that the base of the marine food web was severely affected.

Reefs are known to suffer preferentially at most mass extinctions and the end-Triassic event is no exception. Major buildups formed at equatorial and low latitude Tethyan shelves: Kiessling et al. [1999] lists 117 reefs of Norian and Rhaetian age, many of them were concentrated in the western Tethys. Their abrupt demise at the Tr–J boundary is one of the most spectacular aspects of the biotic crisis. In its aftermath, the geographic distribution of reefs remained severely constricted for more than 20 m.y. [Kiessling, 2001]. The crisis is equally apparent among the constituent taxa of reef ecosystems. Scleractinian corals, which attained remarkably high diversity following their Late Triassic evolutionary innovation of zooxanthellate symbiosis, suffered a sharp reduction in number of genera, and their diversity did not rebound before the Toarcian [Stanley, 1988].

Triassic bivalves have a good fossil record, and their end-Triassic extinction has long been noted [Hallam, 1981]. McRoberts' [2001] compilation of revised taxon ranges reveal an extinction in excess of 30% among the Rhaetian genera, but the coarse temporal resolution does not easily allow an assessment of the tempo of extinction. Diversity decreased from a Triassic peak in the Carnian, but the decline in origination rate is more pronounced than the increase in extinction rates during the Norian and Rhaetian. In a regional study in the Southern Alps, it was possible to deconvolute this long-term effect from a sudden terminal Triassic extinction episode, which preferentially affected the infaunal species [McRoberts et al., 1995].

High turnover and extinction rates at the Tr–J boundary are also evident in terrestrial ecosystems. The detailed palynological record of lacustrine strata in the Newark Basin reveals a regional extinction of 60% of the taxa at the system boundary [Fowell et al., 1994]. European data also indicate a major change in palynoflora [Visscher and Brugman, 1981]. A dramatic turnover in macroflora has long been recognized in Greenland [Harris, 1937], despite later claims that deny any significant change [Ash, 1986]. Vertebrate faunas have also been subject to controversial views regarding the nature and magnitude of their change across the Tr–J boundary. Olsen et al. [1987] report evidence for a sudden change in tetrapods while Padian [1994] suggests a more gradual change. Benton [1994] confirms a pronounced turnover at the Tr–J boundary, but argues that an even more important tetrapod extinction event took place in the Carnian. Apart from focusing at the taxonomic loss alone, Shubin and Sues [1991] studied the biogeographic patterns across the Tr–J boundary. They noted that the Late Triassic tetrapod assemblages, which exhibited a high degree of endemism, were replaced by largely cosmopolitan Early Jurassic assemblages.

Patterns of gradual decline always need to be thoroughly tested for the Signor–Lipps effect, as the sudden disappearance of a taxon may be masked by the chance factors of fossil preservation and collection so that it appears gradual [Signor and Lipps, 1982]. Among the marine invertebrates, ammonoids are known to have been severely decimated at the Tr–J boundary. It was suggested that perhaps only a single lineage of deep-water phylloceratids survived into the Hettangian and became the rootstock of a spectacular Early Jurassic radiation [Tozer, 1971]. Undoubtedly, the end-Triassic represents one of the worst crises in the group's history. Early Jurassic post-extinction faunas are markedly different from Late Triassic ones. A key issue again is the rate of species loss. At the species level, Rhaetian faunas appear impoverished relative to Norian and earlier ones, hence a gradual decline of ammonoids towards the terminal Triassic is often assumed. Although at some localities heteromorph choristoceratids are the sole representative of this otherwise diverse group, several families persist through the end of Rhaetian, signaling that perhaps the ammonoid extinction was more abrupt than a literal reading of the fossil record may suggest. A quantitative analysis of a comprehensive database with high temporal resolution is required to settle this issue.

Conodonts are the real victims of the end-Triassic event, as their long and prosperous history spanning some 300 m.y. was terminated. Their demise appears to be characterized by tapering diversity and diminishing abundance [Clark, 1987]. By Rhaetian time both diversity and population sizes were at very low levels. In their final extinction the extremely low Late Triassic origination rate appears to weigh more importantly than the extinction rate that remained relatively constant [Clark, 1987]. De Renzi et al. [1996] speculate that intrinsic biological factors may be more important that environmental ones, whereby the conodont animal may have proved competitively inferior to groups of newly emerged, modern organisms. This theory is supported by the observation that the disappearance of the final conodont may postdate the terminal Triassic environmental crisis. The last conodont was recorded slightly above the negative carbon isotope anomaly in both Hungary [Pálfy et al., 2001] and England [Hesselbo et al., 2002; Swift, 1989].

PREDICTIONS OF A VOLCANIC FORCING HYPOTHESIS CONFRONTED WITH OBSERVATIONS

Continental flood basalt volcanism is a geological process that cannot be observed in operation today. Study of the rock record of numerous LIPs formed in geological history suggests that outpouring of large volumes of magma (up to several million km^3) occurred within short time intervals (100 k.y. to a few m.y. at most). Extinction is hypothesized as the biotic response to various environmental changes triggered by intense volcanic activity. The most commonly cited kill mechanism is global climate change caused by volcanic outgassing of CO_2 and SO_2. Carbon dioxide is a greenhouse gas that contributes to climate warming at the 10^1–10^5 yr time scale, whereas SO_2 induces short-term (10^{-1}–10^1 yr) cooling through formation of sulphate aerosols. There is no geological evidence for the cooling event, albeit it may have contributed to ecologic instability. Halogen emissions have too short atmospheric residence time to cause significant environmental damage, i.e., acid rain, elsewhere than in the immediate region of the eruptions. The volatile emissions of CAMP are discussed in detail by McHone [2002], who provided estimates for CO_2 and SO_2 degassing.

Although the uncertainties of radiometric dating still render CAMP duration estimates poorly constrained, it is not unreasonable to speculate that peak volcanic intensity may have been sustained between 100 k.y. and 1 m.y. The best paleobiological evidence for global warming comes from a study of fossil plants. McElwain et al. [1999] measured a reduction of stomatal density in leaves and suggested that it responded to a fourfold increase in atmospheric CO_2. A coherent model to explain this "super greenhouse" episode together with the sharp negative $\delta^{13}C$ anomaly postulates that CO_2 emission from CAMP led to global warming that destabilized gas hydrate reservoirs, and the resultant release of methane triggered a positive feedback for runaway greenhouse conditions. Beerling and Berner [2002] provide modeling evidence for the viability of this scenario. Sudden extinction registered in several fossil groups may also be explained by crossing a climatic threshold that led to ecosystem collapse.

An additional, potentially significant effect of volcanism on the marine ecosystem is exerted by the increased flux of biolimiting elements, such as iron, to the seawater. Primary productivity in the open ocean is normally limited by lack of these nutrients, and is known to increase dramatically after experimental "iron fertilization" of the ocean [Coale et al., 1996]. Enhanced nutrient delivery as fallout following volcanic eruptions and from weathering products of the areally extensive volcanic province may lead to eutrophication. Phyto- and zooplankton were heavily affected at the Tr–J boundary event. A productivity crash followed by a bloom of opportunist taxa is consistent with the fossil data, even though some of the primary producers belonged to non-skeletonized groups with poor fossil record. Eutrophication may have played a role in the demise of reefs, similarly to what was suggested for the Late Devonian reef crisis [Caplan et al., 1996]. Organic-rich facies, a predic-

table outcome of enhanced productivity, exist in the earliest Jurassic [Hallam, 1995], although they are not as widespread as following the Late Devonian or end-Permian events. The possible role of large-scale eutrophication requires further research, as it would obviously also effect models of changes in the carbon cycle across the Tr–J boundary. Figure 3 summarizes a possible model of the Tr–J extinction event, assuming that CAMP volcanism was its principal driving force.

The biotic recovery following the terminal Triassic extinction may also be assessed for compatibility with the CAMP forcing. Assuming that volcanism was sustained for up to a million (or even more) years, environmental conditions likely remained stressful for such extended period of time. Indeed, for various fossil groups the early Hettangian is best described as survival or lag phase with depressed diversity levels. Diversity did not start to increase in earnest until the middle or late Hettangian. Ammonoid and radiolarian faunas follow this pattern with impoverished assemblages in the earliest Jurassic. Radiometric dates broadly constrain this interval to c. 2 m.y. [Pálfy et al., 2000b], whereas Weedon et al. [1999] obtained a cyclostratigraphic minimum estimate of 1.3 m.y. for the entire Hettangian. An even longer gap exists in the reef record, as framework reefs are not known to reappear until the Late Sinemurian, some 7–8 m.y. later [Stanley, 1988].

Apart from the loss of taxa and drop in diversity, the end-Triassic extinction is also manifest in the loss of bioprovincialism. The replacement of endemic latest Triassic faunas by cosmopolitan earliest Jurassic ones is a noted feature both among terrestrial vertebrates [Shubin and Sues, 1991] and, albeit less clearly documented, among marine radiolarians and ammonoids. Such observations call for truly global causal agents that affect both the marine and terrestrial realms, and thus are compatible with the CAMP-induced extinction hypothesis.

COMPARISON WITH OTHER EXTINCTIONS LINKED TO LARGE IGNEOUS PROVINCES

Feasibility of the role of CAMP in triggering the end-Triassic extinction is supported by the synoptic observation that many Phanerozoic LIPs are temporally linked to extinctions [Rampino and Stothers, 1988; Courtillot, 1994]. In a recent review, Wignall [2001] found that advances in radiometric dating of the volcanic and biotic events better constrained and confirmed the synchrony in most cases, yet the volume of extrusive rocks, proportion of pyroclastics, and duration of event do not appear to be proportional to extinction intensity. The link between extinctions and episodes of flood basalt volcanism is therefore a strong but imperfect one. A notable trend is that efficiency of LIPS as extinction agents is diminished from the Late Jurassic. Accepting that the most deleterious effect of large-scale volcanism is massive global warming via CO_2 outgassing, amplified by subsequent CH_4 release, a possible explanation is that the evolutionary rise of calcareous nannoplankton and plankton foraminifera during the Late Jurassic introduced a feedback loop for drawing down atmospheric CO_2 levels by opportunistic increase of planktonic carbonate secretion [*Wignall*, 2001]. For this reason, I limit the brief discussion here to the two LIP–extinction event pairs nearest in time to the CAMP and predating the Late Jurassic: the end-Permian and the Early Jurassic (Toarcian) events.

The temporal link between the end-Permian extinction and Siberian Trap volcanism [*Renne et al.*, 1995] and the Toarcian extinction and the Karoo–Ferrar province [*Pálfy and Smith*, 2000] is well established. Considerable uncertainties remain in the estimates of original volumes of these provinces: the Siberian Trap may have reached 1 to 4 x 10^6 km^3 [*Wignall*, 2001], whereas the Karoo–Ferrar rocks are thought to exceed 2.5 x 10^6 km^3 in volume [*Encarnación et al.*, 1996]. Thus volumetrically both LIPs are comparable to the CAMP, but the associated extinction events are disparate in magnitude. The end-Permian is the largest Phanerozoic mass extinction, whereas the Toarcian one is only a minor event. A review of the affected groups [e.g., *Hallam and Wignall*, 1997] also reveals differences as well as similarities. For example, ammonoids and reef-dwelling organisms suffered heavily at both the Permian–Triassic and Tr–J boundary but were much less affected by the Toarcian crisis. Similarly, there are widespread losses in terrestrial faunas and floras in the earlier two events and only minor changes in the Early Jurassic.

Proxy indicators of environmental change point to several common features. In all three cases, a negative carbon isotopic excursion and an inflection in the Sr curve are registered. However, in the Toarcian, there is also a positive carbon isotopic anomaly associated with an oceanic anoxic event and anoxia is known to be extremely widespread at the Permian–Triassic boundary. Changes in the oceanic Sr isotopic ratio are in the opposite sense at the Tr–J boundary compared to the other two events. Different paleolatitudinal position of the LIPs involved — low-latitude CAMP versus relatively high-latitude Siberian Traps and Karoo–Ferrar province — may partially explain the dissimilarities.

ALTERNATIVE MODELS FOR EXTINCTION CAUSATION

Apart from linking it to CAMP volcanism, several other models have been proposed to explain the end-Triassic extinction. The most commonly evoked alternative causes are sea level changes, marine anoxia, and extraterrestrial impact.

Sea level change, in form of a rapid regression–transgression couplet near the Tr–J boundary, may be genetically linked to lithospheric bulging and collapse in relation with CAMP eruptions [*Hallam and Wignall*, 1999]. Anoxic conditions, although evidence from around the Tr–J boundary is limited, may be related to the earliest Jurassic transgressive phase [*Hallam*, 1995] thus likely postdate the extinction. CAMP could have played a role as global warming promotes sluggish ocean circulation. Neither mechanism provides an adequate explanation for the Tr–J extinction, as both fail to explain the terrestrial biotic crisis.

Bolide impact as a possible cause of the end-Triassic extinction has been suggested repeatedly [see review in *Hallam and Wignall*, 1997]. New evidence, a modest Ir anomaly from the Newark Basin, is reported by Olsen et al. [2002a]. The negative carbon isotope anomaly was also interpreted as a consequence of impact-induced productivity collapse [*Ward et al.*, 2001] but, as discussed above, alternative scenarios involving CAMP-derived CO_2 input, global warming, and methane release may better explain the observed pattern [*Pálfy et al.*, 2001]. In fact many signatures of environmental change, extinction patterns, and the isotopic record may be perplexingly similar. At present, unambiguous evidence for accepted impact signatures (Ir anomaly, shocked quartz, microspherules) is lacking at the Tr–J boundary. Changes in seawater Os isotopic composition and Os and Re abundance at the system boundary and in the earliest Jurassic are more consistent with sustained supply of mantle-derived Os and Re from CAMP eruptions and weathering of juvenile basalt than with a single meteoritic source at the boundary from a putative impact event [*Cohen and Coe*, 2002].

CONCLUSIONS

Recent research produced several lines of evidence that make the causal link between the CAMP volcanism and the terminal Triassic extinction an attractive hypothesis.

(1) The synchrony between the major CAMP eruptive phase and the Tr–J boundary is indicated through precise ^{40}Ar/^{39}Ar and U–Pb dating.

(2) The stable isotopic record across the Tr–J boundary reveals a major perturbation in the global carbon cycle. The large negative anomaly is modeled to reflect volcanically-derived CO_2 input, amplified by methane release through dissociation of gas hydrates in a global warming episode, perhaps supplemented by a productivity crisis.

(3) There is paleobotanical support from leaf stomatal density studies for highly elevated atmospheric CO_2 concentration and runaway greenhouse climate. This may be

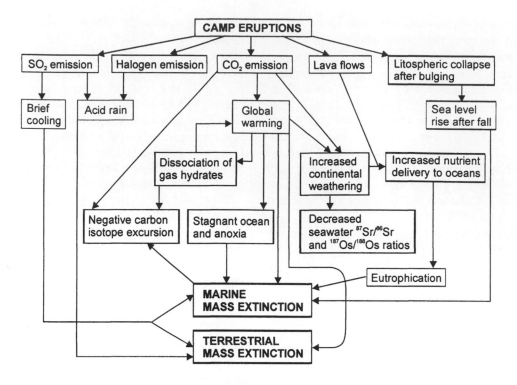

Figure 3. Schematic overview of CAMP and related environmental, biotic, and isotopic events around the Triassic–Jurassic boundary. Adapted from a model of Wignall [2001].

the single most important environmental consequence of CAMP volcanism, triggering the biotic crisis.

(4) A re-assessment of the terminal Triassic time scale suggests a short (<2 m.y.) duration for the Rhaetian stage. The isotopic event lasted for less than 100 k.y. Various marine and terrestrial fossil groups exhibit sudden extinction, compatible with a "pulse event".

(5) The subsequent biotic recovery was delayed and the early Hettangian is characterized by low-diversity, cosmopolitan post-extinction assemblages, likely reflecting prolonged environmental stress.

(6) Large igneous provinces formed before the CAMP (the Siberian Traps at the Permian–Triassic boundary) and after it (the Early Jurassic Karoo–Ferrar province) are also synchronous with extinction events, although of differing magnitude.

The hypothesis of CAMP as the principal driving force in the end-Triassic extinction explains the growing body of paleontological and isotopic observations better than alternative models. Some of the goals for future research are (1) to substantiate that the first major eruptive phase preceded the extinction peak; (2) to better resolve the diversity histories, extinction and recovery trajectories of various fossil groups across the Tr–J boundary; and (3) to assess the effects on volcanically derived supply of

biolimiting nutrients, such as iron, on the marine food web and ecosystem.

Acknowledgements. I thank G. McHone and P. Renne for inviting this contribution. Discussions with D. Beerling, A. Cohen, S. Hesselbo, and P. Olsen were helpful. Reviews by C. McRoberts, P. Olsen, D. Kent, and A. Hallam, and scientific editing by W. Hames improved the manuscript. This research was carried out during the tenure of an Alexander von Humboldt Fellowship at the Museum für Naturkunde, Berlin and a Bolyai Research Fellowship from the Hungarian Academy of Sciences. This is a contribution to IGCP 458 project.

REFERENCES

Ash, S., Fossil plants and the Triassic–Jurassic boundary, in *The beginning of the age of the dinosaurs, faunal change across the Triassic–Jurassic boundary*, edited by K. Padian, pp. 21–30, Cambridge University Press, New York, 1986.

Baksi, A., Evaluation of radiometric ages for the central Atlantic magmatic province: Timing, duration, and possible migration of magmatic centers, in *The Central Atlantic Magmatic Province*, edited by W.E. Hames, J.G. McHone, P.R. Renne and C. Ruppel, AGU, Washington, 2002, in press.

Beerling, D.J., and R.A. Berner, Biogeochemical constraints on the Triassic–Jurassic boundary carbon cycle event, *Global Biogeochem. Cycles*, 2002, in press.

Benton, M.J., Late Triassic to Middle Jurassic extinctions among continental tetrapods: testing the pattern, in *In the shadow of the dinosaurs*, edited by N.C. Fraser and H.-D. Sues, pp. 366-397, Cambridge University Press, Cambridge, 1994.

Benton, M.J., Diversification and extinction in the history of life, *Science, 268*, 52–58, 1995.

Bown, P.R., Recent advances in Jurassic calcareous nannofossil research, in *Advances in Jurassic Research*, edited by A.C. Riccardi, pp. 55–66, Transtec Publications, Zürich, 1996.

Caplan, M.L., R.M. Bustin, and K.A. Grimm, Demise of a Devonian–Carboniferous carbonate ramp by eutrophication, *Geology, 24*, 715–718, 1996.

Carter, E.S., Evolutionary trends in latest Norian through Hettangian radiolarians from the Queen Charlotte Islands, British Columbia, *Géobios, Mém. Spéc., 17*, 111–119, 1994.

Carter, E.S., P.A. Whalen, and J. Guex, Biochronology and paleontology of Lower Jurassic (Hettangian and Sinemurian) radiolarians, Queen Charlotte Islands, British Columbia, *Geol. Surv. Can. Bull., 496*, 1–162, 1998.

Clark, D.L., Conodonts: the final fifty million years, in *Paleobiology of conodonts*, edited by R.J. Aldridge, pp. 165–174, Ellis Horwood, Chichester, 1987.

Coale, K.H., K.S. Johnson, S.E. Fitzwater, and 16 others, A massive phytoplankton bloom induced by an ecosystem-scale iron fertilization experiment in the equatorial Pacific Ocean, *Nature, 383*, 495–501, 1996.

Cohen, A.S., and A.L. Coe, New geochemical evidence for the onset of volcanism in the Central Atlantic magmatic province and environmental change at the Triassic–Jurassic boundary, *Geology, 30*, 2002, in press.

Courtillot, V., Mass extinctions in the last 300 million years: One impact and seven flood basalts?, *Isr. J. Earth Sci., 43*, 255–266, 1994.

Dagys, A.S., and A.A. Dagys, Global correlation of the terminal Triassic, in *Recent Developments on Triassic Stratigraphy*, edited by J. Guex and A. Baud, *Mém. Géol. (Lausanne), 22*, 25–34, 1994.

De Renzi, M., K. Budurov, and M. Sudar, The extinction of conodonts – in terms of discrete elements – at the Triassic–Jurassic boundary, *Cuad. Geol. Ibérica, 20*, 347–364, 1996.

Deckart, K., G. Féraud, and H. Bertrand, Age of Jurassic continental tholeiites of French Guyana, Surinam and Guinea: Implications for the initial opening of the Central Atlantic Ocean, *Earth Planet. Sci. Lett., 150*, 205–220, 1997.

Dickens, R.G., J.R. O'Neil, D.K. Rea, and R.M. Owen, Dissociation of oceanic methane hydrate as a cause of the carbon isotope excursion at the end of the Paleocene, *Paleoceanography, 10*, 965–971, 1995.

Dunning, G.R., and J.P. Hodych, U/Pb zircon and baddeleyite ages for the Palisades and Gettysburg sills of the northeastern United States: Implications for the age of the Triassic/Jurassic boundary, *Geology, 18*, 795–798, 1990.

Encarnación, J., T.H. Fleming, D.H. Elliot, and H. Eales, Synchronous emplacement of Ferrar and Karoo dolerites and the early breakup of Gondwana, *Geology, 24*, 535–538, 1996.

Erwin, D.H., The end and the beginning: recoveries from mass extinctions, *Trends Ecol. Evol., 13*, 344–349, 1998.

Fowell, S.J., B. Cornet, and P.E. Olsen, Geologically rapid Late Triassic extinctions: Palynological evidence from the Newark Supergroup, in *Pangea: Paleoclimate, Tectonics, and Sedimentation During Accretion, Zenith, and Breakup of a Supercontinent*, edited by G.D. Klein, *Geol. Soc. Am. Spec. Pap., 288*, 197–206, 1994.

Gallet, Y., J. Besse, L. Krystyn, H. Theveniaut, and J. Marcoux, Magnetostratigraphy of the Kavur Tepe Section (Southwestern Turkey) – A Magnetic Polarity Time-Scale for the Norian, *Earth Planet. Sci. Lett., 117, pp,* 443–456, 1993.

Gallet, Y., J. Besse, L. Krystyn, and J. Marcoux, Norian Magnetostratigraphy from the Scheiblkogel Section, Austria – Constraint on the Origin of the Antalya Nappes, Turkey, *Earth Planet. Sci. Lett., 140*, 113–122, 1996.

Gallet, Y., J. Besse, L. Krystyn, J. Marcoux, J. Guex, and H. Theveniaut, Magnetostratigraphy of the Kavaalani Section (Southwestern Turkey) – Consequence for the Origin of the Antalya Calcareous Nappes (Turkey) and for the Norian (Late Triassic) Magnetic Polarity Timescale, *Geophys. Res. Lett., 27*, 2033–2036, 2000.

Gradstein, F.M., F.P. Agterberg, J.G. Ogg, J. Hardenbol, P. van Veen, J. Thierry, and Z. Huang, A Mesozoic time scale, *J. Geophys. Res., B, 99*, 24,051–24,074, 1994.

Hallam, A., The end-Triassic bivalve extinction event, *Palaeogeogr., Palaeoclimatol., Palaeoecol., 35*, 1–44, 1981.

Hallam, A., Oxygen-restricted facies of the basal Jurassic of northwest Europe, *Hist. Biol., 10*, 247–57, 1995.

Hallam, A., Major bio-events in the Triassic and Jurassic, in *Global Events and Event Stratigraphy in the Phanerozoic*, edited by O.H. Walliser, pp. 265–283, Springer, Berlin, 1996.

Hallam, A., and P.B. Wignall, *Mass Extinctions and Their Aftermath*, 320 pp., Oxford University Press, Oxford, 1997.

Hallam, A., and P.B. Wignall, Mass extinctions and sea-level changes, *Earth-Sci. Rev., 48*, 217–250, 1999.

Hames, W.E., P.R. Renne, and C. Ruppel, New evidence for geologically instantaneous emplacement of earliest Jurassic Central Atlantic magmatic province basalts on the North American margin, *Geology, 28*, 859–862, 2000.

Harland, W.B., A.V. Cox, P.G. Llewellyn, C.A.G. Pickton, A.G. Smith, and R. Walters, *A Geologic Time Scale*, 131 pp., Cambridge University Press, Cambridge, 1982.

Harland, W.B., R.L. Armstrong, A.V. Cox, L.E. Craig, A.G. Smith, and D.G. Smith, *A Geologic Time Scale 1989*, 263 pp., Cambridge University Press, Cambridge, 1990.

Harris, T.M., The fossil flora of Scoresby Sound East Greenland. Part 5: Stratigraphic relations of the plant beds, *Medd. Grønl., 112*, 1-114, 1937.

Hesselbo, S.P., D.R. Gröcke, H.C. Jenkyns, C.J. Bjerrum, P. Farrimond, H.S. Morgans Bell, and O.R. Green, Massive dissociation of gas hydrate during a Jurassic oceanic anoxic event, *Nature, 406*, 392–395, 2000.

Hesselbo, S.P., Robinson, S.A., Surlyk, F., and Piasecki, S, Terrestrial and marine mass extinction at the Triassic–Jurassic boundary synchronized with initiation of massive volcanism, *Geology, 30*, 2002, in press.

Hodych, J.P., and G.R. Dunning, Did the Manicouagan impact trigger end-of-Triassic mass extinction?, *Geology, 20*, 51–54, 1992.

Holser, W.T., M. Magaritz, and R.L. Ripperdan, Global isotopic events, in *Global events and event stratigraphy in the Phanerozoic*, edited by O.H. Walliser, pp. 63–88, Springer, Berlin, 1996.

Hori, R., Radiolarian biostratigryphy at the Triassic/Jurassic period boundary in bedded cherts from the Inuyama area, central Japan, *J. Geosci., Osaka City Univ., 35*, 53–65, 1992.

Jones, C.E., H.C. Jenkyns, and S.P. Hesselbo, Strontium isotopes in Early Jurassic seawater, *Geochim. Cosmochim. Acta, 58*, 1285–1301, 1994.

Kent, D.V., and P.E. Olsen, Astronomically tuned geomagnetic polarity timescale for the Late Triassic, *J. Geophys. Res., B, 104*, 12831–12841, 1999.

Kent, D.V., P.E. Olsen, and W.K. Witte, Late Triassic–earliest Jurassic geomagnetic polarity sequence and paleolatitudes from drill cores in the Newark rift basin, eastern North America, *J. Geophys. Res., B, 100*, 14,965–14,998, 1995.

Kiessling, W., E. Flügel, and J. Golonka, Paleoreef maps: Evaluation of a comprehensive database on Phanerozoic reefs, *AAPG Bull., 83*, 1552–1587, 1999.

Kiessling, W., Paleoclimatic significance of Phanerozoic reefs, *Geology, 29*, 751–754, 2001.

Korte, C., $^{87}Sr/^{86}Sr$-, $\delta^{18}O$- und $\delta^{13}C$-evolution des triassischen Meerwassers: geochemische und stratigraphische Untersuchungen an Conodonten und Brachiopoden, *Bochum. Geol. Geotechn. Arb., 52*, 1–171, 1999.

Krull, E.S., and G.J. Retallack, $\delta^{13}C$ depth profiles from paleosols across the Permian–Triassic boundary: Evidence for methane release, *Geol. Soc. Am. Bull., 112*, 1459–1472, 2000.

Kump, L.R., and M.A. Arthur, Interpreting carbon-isotope excursions: carbonates and organic matter, *Chem. Geol., 161*, 181–198, 1999.

Marzoli, A., P.R. Renne, E.M. Piccirillo, M. Ernesto, G. Bellieni, and A. De Min, Extensive 200-million-year-old continental flood basalts of the Central Atlantic Magmatic Province, *Science, 284*, 616–618, 1999.

McElwain, J.C., D.J. Beerling, and F.I. Woodward, Fossil plants and global warming at the Triassic–Jurassic Boundary, *Science, 285*, 1386–1390, 1999.

McHone, J.G., Volatile emissions of Central Atlantic Magmatic Province basalts: Mass assumptions and and environmental consequences, in *The Central Atlantic Magmatic Province*, edited by W.E. Hames, J.G. McHone, P.R. Renne and C. Ruppel, AGU, Washington, 2002, in press.

McRoberts, C.A., C.R. Newton, and A. Alissanaz, End-Triassic bivalve extinction: Lombardian Alps, Italy, *Hist. Biol., 9*, 297–317, 1995.

McRoberts, C.A., Triassic bivalves and the initial marine Meso-zoic revolution: A role for predators?, *Geology, 29*, 359–362, 2001.

McRoberts, C.A., H. Furrer, and D.S. Jones, Palaeoenvironmental interpretation of a Triassic–Jurassic boundary section from Western Austria based on palaeoecological and geochemical data, *Palaeogeogr., Palaeoclimatol., Palaeoecol., 136*, 79–95, 1997.

Muttoni, G., D.V. Kent, P. Di Stefano, M. Gullo, A. Nicora, J. Tait, and W. Lowrie, Magnetostratigraphy and biostratigraphy of the Carnian/Norian boundary interval from the Pizzo Mondello section (Sicani Mountains, Sicily), *Palaeogeogr., Palaeoclimatol., Palaeoecol., 166*, 383–399, 2001.

Olsen, P.E., Giant lava flows, mass extinctions, and mantle plumes, *Science, 284*, 604–605, 1999.

Olsen, P.E., and D.V. Kent, Long-period Milankovitch cycles from the Late Triassic and Early Jurassic of eastern North America and their implications for the calibration of the early Mesozoic time scale and the long-term behavior of the planets, *Phil. Trans. Roy. Soc. London, A, 357*, 1761–1786, 1999.

Olsen, P.E., N.H. Shubin, and M.H. Anders, New Early Jurassic tetrapod assemblages constrain Triassic–Jurassic tetrapod extinction event, *Science, 237*, 1025–1029, 1987.

Olsen, P.E., R.W. Schlische, and M.S. Fedosh, 580 ky duration of the Early Jurassic flood basalt event in eastern North America estimated using Milankovitch cyclostratigraphy, in *The Continental Jurassic*, edited by M. Morales, *Mus. North. Ariz. Bull., 60*, 11–22, 1996.

Olsen, P.E., C. Koeberl, H. Huber, A. Montanari, S.J. Fowell, M. Et-Touhami, and D.V. Kent, The continental Triassic–Jurassic boundary in central Pangea: recent progress and preliminary report of an Ir anomaly, in *Catastrophic Events and Mass Extinctions: Impacts and Beyond*, edited by C. Koeberl and K.G. MacLeod, *Geol. Soc. Amer. Spec. Pap., 356*, 2002a, in press.

Olsen, P.E., D.V. Kent, M. Et-Touhami, and J. Puffer, Cyclo-, magneto-, and biostratigraphic constraints on the duration of the CAMP event and its relationship to the Triassic–Jurassic boundary, in *The Central Atlantic Magmatic Province*, edited by W.E. Hames, J.G. McHone, P.R. Renne and C. Ruppel, AGU, Washington, 2002b, in press.

Orchard, M.J., J.P. Zonneveld, M.J. Johns, C.A. McRoberts, M.R. Sandy, E.T. Tozer, and G.G. Carrelli, Fossil succession and sequence stratigraphy of the Upper Triassic of Black Bear Ridge, northeast British Columbia, a GSSP prospect for the Carnian–Norian boundary, *Albertiana, 25*, 10–22, 2001.

Padian, K., What were the tempo and mode of evolutionary change in the Late Triassic to Middle Jurassic?, in *In the shadow of the dinosaurs: Early Mesozoic tetrapods*, edited by N.C. Fraser and H.-D. Sues, pp. 401-407, Cambridge University Press, Cambridge, 1994.

Pálfy, J., and P.L. Smith, Synchrony between Early Jurassic extinction, oceanic anoxic event, and the Karoo–Ferrar flood basalt volcanism, *Geology, 28*, 747–750, 2000.

Pálfy, J., J.K. Mortensen, E.S. Carter, P.L. Smith, R.M. Friedman, and H.W. Tipper, Timing the end-Triassic mass extinction: First on land, then in the sea?, *Geology, 28*, 39–42, 2000a.

Pálfy, J., P.L. Smith, and J.K. Mortensen, A U–Pb and ^{40}Ar/^{39}Ar time scale for the Jurassic, *Can. J. Earth Sci., 37*, 923–944, 2000b.

Pálfy, J., A. Demény, J. Haas, M. Hetényi, M. Orchard, and I. Vetö, Carbon isotope anomaly and other geochemical changes at the Triassic–Jurassic boundary from a marine section in Hungary, *Geology, 29*, 1047–1050, 2001.

Rampino, M.R., and R.B. Stothers, Flood basalt volcanism during the last 250 million years, *Science, 241*, 663–667, 1988.

Renne, P.R., Z. Zichao, M.A. Richards, M.T. Black, and A.R. Basu, Synchrony and causal relations between Permian–Triassic boundary crises and Siberian flood volcanism, *Science, 269*, 1413–1416, 1995.

Sepkoski, J.J., Jr., Patterns of Phanerozoic extinction: a perspective from global data bases, in *Global Events and Event Stratigraphy in the Phanerozoic*, edited by O.H. Walliser, pp. 35–51, Springer, Berlin, 1996.

Shubin, N.H., and H.-D. Sues, Biogeography of early Mesozoic continental tetrapods, *Paleobiology, 17*, 214–230, 1991.

Signor, P.W., and J.H. Lipps, Sampling bias, gradual extinction patterns and catastrophes in the fossil record, in *Geological Implications of Impacts of Large Asteroids and Comets on the Earth*, edited by L.T. Silver and P.H. Schultz, *Geol. Soc. Am. Spec. Pap., 190*, 291–296, 1982.

Stanley, G.D., The history of early Mesozoic reef communities: a three step process, *Palaios, 3*, 170–183, 1988.

Swift, A., First record of conodonts from the late Triassic of Britain, *Palaeontology, 32*, 325–334, 1989.

Taylor, A.S., and A.C. Lasaga, The role of basalt weathering in the Sr isotope budget of the oceans, *Chem. Geol., 161*, 199–214, 1999.

Tekin, U.K., Biostratigraphy and systematics of late Middle to Late Triassic radiolarians from the Taurus Mountains and Ankara region, Turkey, *Geol.–Paläontol. Mitt. Innsbruck, Sonderband, 5*, 1–296, 1999.

Tozer, E.T., One, two or three connecting links between Triassic and Jurassic ammonoids?, *Nature, 232*, 565–566, 1971.

Visscher, H., and W.A. Brugman, Ranges of selected palynomorphs in the Alpine Triassic of Europe, *Rev. Palaeobot. Palynol., 34*, 115–128, 1981.

Ward, P.D., J.W. Haggart, E.S. Carter, D. Wilbur, H.W. Tipper, and T. Evans, Sudden productivity collapse associated with the Triassic–Jurassic boundary mass extinction, *Science, 292*, 1148–1151, 2001.

Weedon, G.P., H.C. Jenkyns, A.L. Coe, and S.P. Hesselbo, Astronomical calibration of the Jurassic time-scale from cyclostratigraphy in British mudrock formations, *Phil. Trans. Roy. Soc. London, A, 357*, 1787–1813, 1999.

Wignall, P.B., Large igneous provinces and mass extinctions, *Earth-Sci. Rev., 53*, 1–33, 2001.

J. Palfy, POB 137, Budapest, H-1431 Hungary. E-mail: palfy@paleo.nhmus.hu

Printed and bound by CPI Group (UK) Ltd, Croydon, CR0 4YY